Ecology and Ethics

Volume 2

Ecology and Ethics

This series is devoted to continuing research at the interfaces of ecology and ethics (embedded in the multiple fields of philosophy and ecology) to broaden our conceptual and practical frameworks in this transdisciplinary field. Confronted with global environmental change, the academic community still labors under a tradition of strong disciplinary dissociation that hinders the integration of ecological understanding and ethical values to comprehensively address the complexities of current socio-ecological problems. During the 1990s and 2000s, a transdisciplinary integration of ecology with social disciplines, especially economics, has been institutionalized via interdisciplinary societies, research programs, and mainstream journals. Work at this interface has produced novel techniques and protocols for assessing monetary values of biodiversity and ecosystem services, as illustrated by the Millennium Ecosystem Assessment. At the beginning of the 2010s, however, an equivalent integration between ecology and philosophy still remains elusive. This series undertakes the task to develop crucial theoretical and practical linkages between ecology and ethics through interdisciplinary, international, collaborative teamwork. It aims to establish a new forum and research platform to work on this vital, but until now insufficiently researched intersection between the descriptive and normative domains. The scope of this series is to facilitate the exploration of sustainable and just ways of co-inhabitation among diverse humans, and among humans and other-than-human co-inhabitants with whom we share our heterogeneous planet. It will address topics integrating the multiple fields of philosophy and ecology such as biocultural homogenization, Planetary or Earth Stewardship.

More information about this series at http://www.springer.com/series/11941

Ricardo Rozzi • F. Stuart Chapin III
J. Baird Callicott • S.T.A. Pickett
Mary E. Power • Juan J. Armesto
Roy H. May Jr.
Editors

Earth Stewardship

Linking Ecology and Ethics in Theory and Practice

Springer

Editors
Ricardo Rozzi
Department of Philosophy
and Religion Studies
University of North Texas
Denton, TX, USA

Institute of Ecology and Biodiversity
Santiago, Chile

Universidad de Magallanes
Punta Arenas, Chile

J. Baird Callicott
Department of Philosophy
and Religion Studies
University of North Texas
Denton, TX, USA

Mary E. Power
Department of Integrative Biology
University of California
Berkley, CA, USA

Roy H. May Jr.
Departamento Ecuménico de Investigaciones
San José, Costa Rica

F. Stuart Chapin III
Institute of Arctic Biology
University of Alaska Fairbanks
Fairbanks, AK, USA

S.T.A. Pickett
Cary Institute of Ecosystem Studies
Millbrook, NY, USA

Juan J. Armesto
Departamento de Ecología
Pontificia Universidad Católica
Santiago, Chile

Institute of Ecology and Biodiversity
Santiago, Chile

ISSN 2198-9729 ISSN 2198-9737 (electronic)
Ecology and Ethics
ISBN 978-3-319-12132-1 ISBN 978-3-319-12133-8 (eBook)
DOI 10.1007/978-3-319-12133-8

Library of Congress Control Number: 2015935508

Springer Cham Heidelberg Dordrecht London New York

Printed on acid-free paper

Springer International Publishing AG Switzerland is part of Springer Science+Business Media (www.springer.com)

Foreword

Ethics is embedded in a system of beliefs and practices that orient our behaviors with respect to every external factor in our lives, or what we would like those behaviors to be. In order to examine properly where global and local societies are now in the development of ethics, we must first consider the situation of the human species here on Earth.

In the 4.54-billion-year history of Earth, primates appeared about 85 million years ago and members of our genus, *Homo*, 2.3 million years ago, in Africa. *Homo erectus* and *H. ergaster* migrated out of Africa 1.3–1.8 million years ago, but the evolutionary developments that led to the appearance of modern humans continued in Africa. What are recognized as members of our species, *H. sapiens*, lived in Africa from 400,000 to 200,000 years ago, when we estimate that modern *Homo sapiens* appeared for the first time. Migrating out of Africa from 100,000 to 50,000 years ago, they colonized the whole world, replacing earlier members of the genus *Homo* everywhere they went and, ultimately, becoming the only surviving representative of the human line.

Before our ancestors developed the skills of domesticating animals and growing crops, evidently starting with the dog, they lived as small bands of mostly 20–40 individuals, rarely coming into contact with other bands and in general mostly without social interactions between them. At the time agriculture was begun, it is estimated that the entire human population of Earth amounted to only 3–4 million people, scattered widely over the six habitable continents. These people would have found patterns of behavior and ethics appropriate for their circumstances, but it is likely that many of these have survived and that they are much less appropriate in the modern world than they were originally. Following the development of crop agriculture about 12,000 years ago, some of them would have become inappropriate or even destructive as the conditions in which people lived changed rapidly. At present, though, very different styles of living are characteristic of different groups of people in mountainous, coastal, and other regions, as the chapters of this Earth Stewardship book make clear.

Over the approximately 12,000 years since the domestication of plants and animals made the formation of settled villages, towns, and cities possible by providing a dependable supply of food, the various elements of what we now consider civilization developed gradually. When large numbers of people lived together in a single place, they could specialize in their activities and thus produce benefits for the population as a whole as well as for themselves. Poets, storytellers, religious and civic leaders, farmers, and builders appeared, and began to produce the trappings of cities and nations that characterize the world in which we live now. Ultimately, about 5,000 years ago, written languages were developed more or less at the same time in Mesopotamia and along the Nile, with cuneiform script forming the basis of most modern writing and hieroglyphics remaining a more local language along the Nile.

The invention of written language began defining for people that part of their history that was well known and understood, as the feats of kings and generals, conquest and defeat began to be recorded as they occurred. Whatever happened more than 5,000 years ago was either remembered or imagined, coming down to the people of later generations in stories and myths. Through these tales and myths, they tried to understand the meaning of life and to develop plausible stories about what had happened on Earth before the means existed to record them in a permanent form. These events pretty clearly define the erroneous belief that the world was created about 5,000 years ago that is so strangely held by fairly large numbers of well-intentioned people.

In the Bible, some of which seems to have been written at the time of David and Solomon about 4,000 years ago, two different versions of human's role on Earth are presented in the first two chapters of Genesis. Presumably the views of two different authors dating from different periods, one (the first) celebrates human's domination of Earth, to be subdued for their purposes, and the other (the second) counsels us to save and care for the Earth. This second interpretation aligns with Earth stewardship.

It is likely that after some 8,000 years of building ever-larger fields and running herds of animals over the semi-arid hills of the Near East ecological damage was obvious. In the face of these developments, it is not strange that people would have begun to recognize the need for sustainable practices locally as their numbers grew. When people existed only as widely-scattered bands foraging in natural communities, individuals and groups would have gained benefit by gathering and hoarding whatever supplies of food or other valuable commodities that they could find. Similar behaviors in the very different modern world have become highly destructive and are widely recognized as inappropriate. However, no one seems to have developed a suitable theory of what might be done about it – in some ways essentially the subject of this Ecology and Ethics book series.

In this book focusing on Earth stewardship, an effort has been made to represent a range of different land ethics and procedures practiced in different parts of the world and to use them as the basis for considering what we could learn from one another, and what we could do together. What I consider a particularly useful discussion of this aspect, and one that perhaps assists in understanding the conditions for developing general modes of globally-suitable behavior, is that of May. He points out that in Latin America, sectors of the dominant Roman Catholic Church, which traditionally have defended social justice, have in recent years integrated concern for the natural environment into their social justice agendas. Indeed, as many Evangelical Christians emphasize, neglecting the environment is clearly at odds with the traditional admonition to care for the poor.

The science of ecology, less than a century old in its predictive form, is a necessary ingredient for the evolution of any generally effective land ethic. As Covich brings out so well in his fine review of Frank Golley's lifelong contribution, and as the various chapters on long-term ecological research in this book illustrate, we must continue making important scientific advances in ecology throughout the world. It is the knowledge we gain of these principles that put us in a position to respond to the challenges we face.

Despite this knowledge, it is by no means a simple matter to reconcile the principles of ecology with those of practical politics (as documented by Kingsland). In this context, the strong efforts of Aguirre to integrate environmental knowledge with ethics through hermeneutics and the novel methodology of field environment philosophy seems very useful to me. Although there is clearly much about living systems that we do not know (Li et al.) – for example, I estimate that we have named only about 2 million of the estimated 12 million species of eukaryotic organisms – there is a great deal of available knowledge that we can apply to enhancing the sustainability of these systems. This knowledge can be applied to building a sustainable Earth (Callicott). Such an Earth, however, must also feature social justice and the encouragement of individual talents for children, women, and men everywhere if it is to succeed. We evolved into a complex biological world that not only supports us but determines our features, and we must use practices like those proposed by the Earth Stewardship Initiative of the Ecological Society of America (Chapin and collaborators) to provide a stable basis for civilization in the future. As Rozzi has put it, echoing Leopold, we need to take all of nature into consideration from an ethical perspective, honoring and preserving it for our own sakes. His biocultural ethic emphasizes that we are co-inhabitants in the natural world, no matter how complex our inventions may become. We should not neglect the understanding that that realization brings, in order to avoid being at our very great peril. In the face of these relationships, we are so dominant that we must manage the Earth's living systems actively and sustainably.

How can we work together to modify our collective behavior, driven by competitive and essentially greedy nations and individuals into what many see as an unstable nightmare? Several chapters open avenues for answering these questions by documenting pathways that are being forged by socio-ecological research networks (Hideaki, Maass and Equihua, Redman and Miller, Orenstein and Groner, Barbosa and Villagra, Goralnik et al.), religious alliances (Kerber, Tucker), policy actions (Viola and Basso), environmental citizenship and participation (Hargrove, Taylor), and new forms of conservation (Enkerlin et al., Berchez et al., Valenti and da Rocha), based on both traditional and contemporary ecological knowledge and values (Gao, Mamani, Sarmiento). However, no situation like the one we confront today has ever existed in the past, so that our future, with that of everything we hold as important, is at stake.

In much of the world, family planning is still regarded as wrong or unaffordable in the face of individual strategies for survival. In view of this, how do we reach a stable population, when we are already using more than 1.5 times what the world can produce on an ongoing basis (http://www.footprintnetwork.org), unevenly distributed in different countries and regions, and adding a net of 200,000 people per day to our current population of approximately 7.2 billion people? We don't even know that the world can indefinitely support its present human population, much less the even more appalling population numbers, an estimated 9 billion people 36 years from now in 2050. As for limiting consumption, what politician could run successfully on the basis of limiting individual consumption? Perhaps each subconsciously envisions himself in a hunter-gatherer world, so "Follow me over the next hill, and we'll all find plenty of food for everyone." In any case, limiting our consumption, although the time to do so has long since passed for many of us, is absolutely necessary but for the world as a whole seems largely unattainable. As for the development of necessary new technologies, perhaps the current shifting of the world view toward dealing in a meaningful way with global climate change offers hope for the future. In any case, I view the concepts of Ogden et al. as necessary, in understanding properly global differences in degrees and kinds of consumption, but also perhaps visionary, in their implicit assumption that people given the proper array of sound ecological knowledge will behave in increasingly appropriate ways.

In view of these factors, I believe that only a major, ultimately worldwide shift in our ethics and morals will bring about change. At the first Earth Day in the United States, April 1970, some 20 million people turned out for an individual activity somewhere, a tenth of the nation's population at that time, and politicians were quick to take notice and pass strong environmental legislation. The philosopher-biologist E.O. Wilson in his book *Social Conquest of the Earth* (2012) offers the diversity of populations that occur in some major cities as part of the hope for the future. In principle, such situations offer the possibility of overcoming prejudice and working together to achieve necessary common goals, as those proposed by Earth Stewardship. Many people remain unconcerned even with the poor and needy in their own areas, much less worldwide, but despite this we are all tied together in operating what Adlai Stevenson aptly termed "Spaceship Earth," and we must ultimately all succeed if any of us is to do so.

In view of these relationships, what I am calling for is nothing less than a worldwide moral revolution, one to which the impressive contributions of this volume linking ecology and ethics, in theory and practice, have advanced importantly. Given the structure of the society that we have evolved over the years, nothing less is likely to insure success and the continuation into the indefinite future of what we value so deeply and appropriately in our civilization.

Missouri Botanical Garden Peter H. Raven
St. Louis, MO, USA

Acknowledgments

This volume conveys the work of many researchers, students, and people involved with and committed to the care of the Earth. We are grateful to the Cary Institute of Ecosystem Studies, Millbrook, New York, that in May 2011 hosted the 14th Cary Conference, Linking Ethics and Ecology for a Changing World. The conference was organized by the Institute of Ecology and Biodiversity (IEB Chile) and the University of North Texas (UNT), and provided the intellectual stimulus and collaborative atmosphere that forged the initial steps of this book. At IEB and the University of Magallanes (UMAG) in Chile, we are especially thankful to Francisca Massardo (UMAG editorial office of the Sub-Antarctic Biocultural Conservation Program, SBCP), who provided invaluable editorial assistance during the final revision of chapters and production of this volume. At SBCP-UNT office, Kelli Moses, managing editor, and Rachel Weaver, project manager for this book, coordinated and provided knowledgeable editorial assistance during the preparation of manuscripts. At UNT, the SBC Program and the Center for Environmental Philosophy (CEP) provided financial support and advice; we especially thank its director Eugene Hargrove. At Springer, we appreciate the encouragement and editorial guidance offered by Valeria Rinaudo and Ineke Ravesloot for the present volume and for the *Ecology and Ethics* book series. The preparation of this book has also benefited from multiple collaborations. The Ecological Society of America provided a vital impetus for discussing the integration of ecology and ethics into Earth Stewardship. The Chilean Long-Term Socio-Ecological Research (LTSER) network partly supported the international course on ecology and ethics associated with the VI Latin- (Inter) American Environmental Philosophy Conference held at the Universidad of Magallanes in 2013, which enhanced the north–south discussion of concepts regarding Earth Stewardship. The International Long-Term Ecological Research (ILTER) network further provided a cosmopolitan context, and support for the discussion of practices to implement Earth Stewardship. In 2014, the Universidad Biblica Latinoamericana (San José, Costa Rica) hosted a research visit and editorial collaboration on religious foundations and alliances for Earth Stewardship. At UC Berkeley, the Eel River Critical Zone Observatory (National Science Foundation CZP EAR-1331940) provided support for our consideration of river social-ecological systems, and we thank the Eel River Recovery Project (www.eelriverrecovery.org), a volunteer citizen science group working to inform stewardship throughout the watershed. We acknowledge the financial support of the Chilean Millennium Science Initiative (ICM) grant P05-002, and the National Commission for Scientific and Technological Research (CONICYT, Chile) grant PFB-23 awarded to IEB-Chile. This book is also a contribution to the research program of the Chilean LTSER Network (http://www.ltser-chile.cl/) and the international Sub-Antarctic Biocultural Conservation Program, coordinated by IEB, UMAG, and UNT (http://chile.unt.edu/). We acknowledge as well the support of the Cary Conference initiating this book by US National Science Foundation, Division of Environmental Biology (DEB-1057538), and Science Technology Program (SES-10581630). Any opinions, findings, and conclusions or recommendations expressed in this material are those of the author(s) and do not necessarily reflect the views of the National Science Foundation.

Contents

Contributors

Jorge F. Aguirre Sala Department of Humanities, University of Monterrey, Garza García, NL, Mexico

Diana Allen Healthy Parks Healthy People US, US National Parks Service, Washington, DC, USA

Angela Andrade Conservation International, Bogotá, Colombia

Juan J. Armesto Departamento de Ecología, Pontificia Universidad Católica, Santiago, Chile

Institute of Ecology and Biodiversity, Santiago, Chile

Tim Badman World Heritage Programme, International Union for Conservation of Nature (IUCN), Gland, Switzerland

Olga Barbosa Instituto de Ciencias Ambientales y Evolutivas, Universidad Austral de Chile, Valdivia, Chile

Institute of Ecology and Biodiversity, Santiago, Chile

Centro de Desarrollo Urbano Sustentable (CEDEUS), Valdivia, Chile

Jill S. Baron Fort Collins Science Center, U.S. Geological Survey, Fort Collins, CO, USA

Larissa Basso Climate Change and International Relations Research Programme, Institute of International Relations, University of Brasília, Brasília, DF, Brazil

Flavio Berchez Botany Department, Instituto de Biociências, Universidade de São Paulo, São Paulo, SP, Brazil

Paula Bueno Parques Nacionales Naturales de Colombia, Bogotá, Colombia

J. Baird Callicott Department of Philosophy and Religion Studies, University of North Texas, Denton, TX, USA

Kathryn Campbell Healthy Parks Healthy People, Parks Victoria, Melbourne, Australia

F. Stuart Chapin III Institute of Arctic Biology, University of Alaska Fairbanks, Fairbanks, AK, USA

Scott L. Collins Department of Biology, University of New Mexico, Albuquerque, NM, USA

Alan P. Covich Odum School of Ecology, Institute of Ecology, University of Georgia, Athens, GA, USA

Fernando José Rodrigues da Rocha Department of Philosophy, Universidade Federal do Rio Grande do Sul, Porto Alegre, Brazil

Nigel Dudley World Commission on Protected Areas (WCPA), International Union for Conservation of Nature (IUCN), Natural Solutions, Bristol, UK

Ernesto C. Enkerlin-Hoeflich World Commission on Protected Areas, IUCN (International Union for Conservation of Nature), Monterrey, México

Miguel Equihua Instituto de Ecología, A.C. (INECOL), Xalapa, Veracruz, Mexico

Jamison Ervin The Nature Conservancy, Burlington, VT, USA

Shan Gao Department of Philosophy, School of Politics and Public Management, Soochow University, Suzhou, China

Natália P. Ghilardi-Lopes Centro de Ciências Naturais e Humanas, Universidade Federal do ABC, São Bernardo do Campo, SP, Brazil

Lissy Goralnik Department of Forest Ecosystems and Society, Oregon State University, Corvallis, OR, USA

Hannah Gosnell Geography, Environmental Sciences, and Marine Resource Management, College of Earth, Ocean and Atmospheric Sciences, Oregon State University, Corvallis, OR, USA

Elli Groner Ramon, The Dead-Sea and Arava Science Center, Mitzpe Ramon, Israel

Eugene C. Hargrove Center for Environmental Philosophy, University of North Texas, Denton, TX, USA

Terence Hay-Edie United Nations Development Programme (UNDP), New York, NY, USA

Nik Heynen Department of Geography, University of Georgia, Athens, GA, USA

Marc Hockings World Commission on Protected Areas (WCPA), International Union for Conservation of Nature (IUCN), University of Queensland, Brisbane St Lucia, QLD, Australia

David W. Inouye Department of Biology, University of Maryland, College Park, MD, USA

Stig Johansson Agricultural and Environmental Services Department, World Bank, Washington, DC, USA

Karim-Aly Kassam Department of Natural Resources, College of Agriculture and LifeSciences, Cornell University, Ithaca, NY, USA

Karen Keenleyside Parks Canada's National Parks Directorate, Gatineau, QC, Canada

Guillermo Kerber Programme on Ecological Justice, World Council of Churches, Geneva, Switzerland

Sharon E. Kingsland History of Science and Technology Department, Johns Hopkins University, Baltimore, MD, USA

Dan Laffoley World Commission on Protected Areas (WCPA), International Union for Conservation of Nature (IUCN), Peterborough, UK

Penny Langhammer School of Life Sciences, Arizona State University, Tempe, AZ, USA

Kelen Leite Estação Ecológica Tupinambás, ICMBio, São Sebastião, SP, Brazil

Ben Li Department of Information Processing Science, University of Oulu, Oulu, Finland

Manuel Maass Centre for Ecosystem Research (CIEco), Universidad Nacional Autónoma de México (UNAM), Morelia, Michoacan, Mexico

Kathy MacKinnon World Commission on Protected Areas (WCPA), International Union for Conservation of Nature (IUCN), Cambridge, UK

Vicenta Mamani-Bernabé Instituto Superior Ecuménico Andino de Teología (ISEAT), La Paz, Bolivia

Andrés Mansilla Laboratorio de macroalgas Antarticas y Subantarticas, Universidad de Magallanes, Punta Arenas, Chile

Institute of Ecology and Biodiversity, Santiago, Chile

Francisca Massardo Universidad de Magallanes, Puerto Williams, Chile

Institute of Ecology and Biodiversity, Santiago, Chile

Roy H. May Jr. Departamento Ecuménico de Investigaciones (DEI), San José, Costa Rica

Thaddeus R. Miller Nohad A. Toulan, School of Urban Studies and Planning, College of Urban and Public Affairs, Portland State University, Portland, OR, USA

Eduard Mueller World Commission on Protected Areas (WCPA), International Union for Conservation of Nature (IUCN), San José, Costa Rica

Michael Paul Nelson Department of Forest Ecosystems and Society, Oregon State University, Corvallis, OR, USA

Laura Ogden Department of Anthropology, Dartmouth College, Hanover, NH, USA

Daniel E. Orenstein Faculty of Architecture and Town Planning, Technion – Israel Institute of Technology, Haifa, Israel

Ulrich Oslender Department of Global and Sociocultural Studies, Florida International University, Miami, FL, USA

Terry Parr Lancaster Environment Centre, Centre for Ecology and Hydrology, Lancaster, UK

S.T.A. Pickett Cary Institute of Ecosystem Studies, Millbrook, NY, USA

Fábio Valenti Possamai Department of Philosophy and Religion Studies, University of North Texas, Denton, TX, USA

Mary E. Power Department of Integrative Biology, University of California, Berkeley, CA, USA

Charles L. Redman School of Sustainability, Arizona State University, Tempe, AZ, USA

Paul Robbins Nelson Institute for Environmental Studies, University of Wisconsin-Madison, Madison, WI, USA

Ricardo Rozzi Department of Philosophy and Religion Studies, University of North Texas, Denton, TX, USA

Institute of Ecology and Biodiversity, Santiago, Chile

Universidad de Magallanes, Punta Arenas, Chile

Leslie Ryan Department of Forest Ecosystems and Society, Oregon State University, Corvallis, OR, USA

Trevor Sandwith Global Protected Areas Program, International Union for Conservation of Nature (IUCN), Gland, Switzerland

Fausto O. Sarmiento Neotropical Montology Collaboratory, University of Georgia, Athens, GA, USA

Evangelina Schwindt Grupo de Ecología em Ambientes Costeros, Centro Nacional Patagónico (CENPAT-CONICET), Puerto Madryn, Argentina

Hideaki Shibata Field Science Center for Northern Biosphere, Hokkaido University, Sapporo, Japan

Peter J. Taylor Program in Critical and Creative Thinking, Program on Science, Technology and Values, College of Education and Human Development, University of Massachusetts, Boston, MA, USA

Mary Evelyn Tucker Yale School of Forestry and Environmental Studies, Yale University, New Haven, CT, USA

Monica G. Turner Department of Zoology, University of Wisconsin, Madison, WI, USA

Marjo Vierros International Organizations Center, United Nations University Institute of Advanced Studies (UNU-IAS), Yokohama, Japan

Paula Villagra Instituto de Ciencias Ambientales y Evolutivas, Universidad Austral de Chile, Valdivia, Chile

Eduardo Viola Climate Change and International Relations Research Programme, Institute of International Relations, University of Brasília, Brasília, DF, Brazil

Leigh Welling Climate Change Response, U.S. National Parks Service, Fort Collins, CO, USA

Paige West Department of Anthropology, Barnard College and Columbia University, New York, NY, USA

Stephen Woodley World Commission on Protected Areas (WCPA), International Union for Conservation of Nature (IUCN), Ottawa, Canada

Chapter 1
Introduction: Linking Ecology and Ethics for an Interregional and Intercultural Earth Stewardship

Ricardo Rozzi, F. Stuart Chapin III, J. Baird Callicott, S.T.A. Pickett, Mary E. Power, Juan J. Armesto, and Roy H. May Jr.

Abstract Earth Stewardship implies a paradigm shift in linking facts and values, multiple forms of ecological knowledge and practices, and broadening the mission of the ecological sciences. However, two core limitations need to be addressed: (i) geographical gaps in the coverage of long-term ecological and socio-ecological research (LTER, LTSER, and other long-term environmental research networks) across the planet; (ii) philosophical gaps in the epistemological, political, and ethical dimensions of LTSER. If the rates of anthropogenic damage to the biosphere are to be reduced, both research and its application on a planetary scale requires transdisciplinary as well as inter-hemispheric, and intercultural inputs. Also both scientific and traditional ecological knowledge are dynamic. The integration of biocultural diversity is not an integration of a collection of biological, physical, or cultural objects; it is the incorporation of dynamic, often conflictive, processes of intercultural dialogue, negotiation, and poetic creativity. These intercultural, interdisciplinary, inter-institutional, and international processes generate forms of ecosystem co-management, which constitute Earth stewardship. Three areas of discussion contribute to finding the way forward: (1) embracing the multiple forms of

R. Rozzi (✉)
Department of Philosophy and Religion Studies, University of North Texas, Denton, TX, USA

Institute of Ecology and Biodiversity, Santiago, Chile

Universidad de Magallanes, Punta Arenas, Chile
e-mail: rozzi@unt.edu

F.S. Chapin III
Institute of Arctic Biology, University of Alaska Fairbanks, Fairbanks, AK, USA
e-mail: terry.chapin@alaska.edu

J.B. Callicott
Department of Philosophy and Religion Studies, University of North Texas, Denton, TX, USA
e-mail: callicott@unt.edu

R. Rozzi et al. (eds.), *Earth Stewardship*, Ecology and Ethics 2,
DOI 10.1007/978-3-319-12133-8_1

understanding and co-inhabiting the biosphere; (2) undertaking the transdisciplinary work of long-term socio-ecological research networks; and (3) integrating ethics and ecological sciences through environmental citizenship. Bringing these broad areas together will contribute to overcoming the geographical and philosophical gaps that limit effective Earth Stewardship.

Keywords Biocultural ethics • Ecological economics • Environmental justice • Intercultural • Long-term socio-ecological research (LTSER)

Earth Stewardship implies a paradigm shift that links facts and values, multiple forms of ecological knowledge and practices, and broadens the mission of the ecological sciences. To confront global environmental change it is necessary, but not sufficient, to conduct long-term socio-ecological research. It is also necessary to act. Earth stewardship calls ecologists to engage not only in the production of knowledge, but also in public discourse, as well as in decision making, education, and governance. As a means of engaging science and society in rapidly reducing the rates of anthropogenic damage to the biosphere, the Ecological Society of America launched the Earth Stewardship Initiative in 2009 (Power and Chapin 2009; Chapin et al. 2011a, b).[1] Since then, this call for action has been appealing not only to ecologists, but also to anthropologists, sociologists, engineers, economists, religion scholars, philosophers,

[1] Note that the ESA defines Earth Stewardship as a science. Chapin et al. (2011a, p. 89) define it as "science that facilitates the active shaping of trajectories of social-ecological change to enhance ecosystem resilience and human well-being." The concept has since evolved to be "a strategy to shape the trajectories of change…;" i.e., the application of sustainability science to problem solving (Chapin et al. in this volume [Chap. 12]). In this book we focus on it as a transdisciplinary science, embedded in social and cultural action. Within the ESA, Earth Stewardship has as antecedents the notions of ecosystem stewardship (Chapin et al. 2009) and planetary stewardship (Power and Chapin 2009), and beyond the ESA it is paralleled by the Planetary Stewardship Initiative developed internationally as part of the scientific planning for Future Earth (Steffen et al. 2011). See chapters by Callicott and by Chapin et al. in this volume [Chaps. 11, 12].

S.T.A. Pickett
Cary Institute of Ecosystem Studies, Millbrook, NY, USA
e-mail: picketts@caryinstitute.org

M.E. Power
Department of Integrative Biology, University of California, Berkeley, CA, USA
e-mail: mepower@berkeley.edu

J.J. Armesto
Departamento de Ecología, Pontificia Universidad Católica, Santiago, Chile

Institute of Ecology and Biodiversity, Santiago, Chile
e-mail: jarmesto@bio.puc.cl

R.H. May Jr.
Departamento Ecuménico de Investigaciones (DEI), San José, Costa Rica
e-mail: royhmay@gmail.com

conservation biologists, and other professionals, decision makers, and citizens interested in the combination of environmental, economic, and social sustainability.

This book contributes to advancing the Stewardship Initiative toward a planetary scale. What is happening today in the Amazon depends partly on environmental policies in North America, Asia, and other continents. What happens to the climate in North America, Asia, and other continents depends in part on the conservation of forests in the Amazon. Therefore, today, inter-hemispheric, intercultural, and transdisciplinary collaborations for Earth Stewardship are an imperative. The call for socio-environmental stewardship at a planetary scale faces, however, two core limitations that need to be addressed:

(i) geographical gaps in the coverage of long-term ecological and socio-ecological research (LTER and LTSER) across the planet;
(ii) philosophical gaps in the coverage of epistemological, political and ethical dimensions in LTSER (Rozzi et al. 2012).

Geographical gaps exist because more than 90 % of LTER or LTSER sites are located in the Northern Hemisphere. As Li et al. (Chap. 13) discuss in this volume, the International Long-Term Ecological Research network (ILTER) offers an ideal research, information, and infrastructural platform for the Earth Stewardship initiative; however, it presents a marked Northern Hemisphere bias, with more than 90 % of the ILTER publications generated by researchers from the Northern Hemisphere. Furthermore, within this hemisphere 89 % of ILTER publications are generated by researchers associated with LTER networks in temperate regions, and only 1 % are in equatorial regions. Consequently, the distribution of ILTER sites is more associated with political and economic resources than with the geographic distribution of biodiversity.

Regarding philosophical gaps, until now the social component considered in socio-ecological studies worldwide has been primarily economic (Rozzi et al. 2012).[2] Furthermore, as documented by Li et al. (Chap. 13), social research is still incipient in long-term socio-ecological research programs. For example, <0.5 % of ILTER publications are included in social sciences databases. Noticeably, however,

[2] ESA's Earth Stewardship call gives special "consideration to both ecological and socioeconomic" (Chapin et al. 2011a). Similarly, the European LTSER platform was designed "as a research infrastructure to support integrated socioeconomic and ecological research and monitoring of the long-term development of society–nature interaction within the context of global environmental change" (Haberl et al. 2009, p. 1798). These quotes show that socio-ecological is subsumed by "socio-economic" in foundational documents of Earth Stewardship and LTSER (see also Parr et al. 2002; Redman et al. 2004; Lui et al. 2007; Ohl et al. 2007). It is also striking that in socio-ecological research, the fields of philosophy, including ethics, are most often absent. For example, in a recent comprehensive review of the state of the art in long-term socio-ecological research in the US and Europe by Singh et al. (2013), philosophy is not included, and the word ethics is not used. The integration of socioeconomic research into the LTSER framework during the last decades represents a significant step forward for the inclusion of the human component in LTER (See Redman and Miller in this volume [Chap. 17]). Our book complements these approaches by incorporating philosophy and ethics as disciplines into the theory and practice of LTSER and Earth Stewardship.

>99 % of all ILTER publications in the arts and the humanities are generated by researchers working in the Southern Hemisphere. This volume calls attention to the opportunities for stronger partnership and complementarity in long-term socio-ecological research and stewardship initiatives across the planet. The southern regions can demonstrably add to the integration of social, ethical, and artistic dimensions to transdisciplinary socio-ecological research at ILTER and other networks, providing a broader intercultural and participatory foundation for Earth Stewardship.

This publication has its origin in the 14th Cary Conference held at the Cary Institute of Ecosystem Studies, Millbrook, New York, in 2011.[3] During the conference we acknowledged utmost the importance of global scale and interregional dialogue integrating ecology and ethics. As a follow up, we created the *Ecology and Ethics* book series with the publishing house Springer. This volume is the second in the series. It is conceived as a companion to the first one, *Linking Ecology and Ethics for a Changing World* (Rozzi et al. 2013), which placed greater emphasis on core concepts of ecological sciences and environmental philosophy. It was organized using conceptual frameworks provided by the notion of worldview and by a biocultural approach to environmental ethics.[4] This second volume places stronger emphasis on the practice of ecology and ethics. It was stimulated by the challenges and opportunities raised by the Earth Stewardship Initiative of the Ecological Society of America (ESA). Indeed, this book elaborates a conceptual framework at the planetary scale for continuing to build Earth Stewardship as part of the centennial celebration of the ESA.

More fully understanding and respecting biocultural diversity, with the multiple forms of land stewardship it implies, will allow us more effectively and justly to confront local and global socio-environmental challenges. Through dialogical processes and partnerships it will be possible to achieve novel forms of stewardship. Both scientific and traditional ecological knowledge are dynamic. The integration of biocultural diversity is not an integration of a collection of biological, physical, or cultural objects. Rather, it is the incorporation of dynamic, often conflicting, processes of intercultural dialogue, negotiation, and poetic creativity. These

---

[3]The 14th Cary Conference was jointly organized by three institutions: the Cary Institute of Ecosystem Studies (New York), the Institute of Ecology and Biodiversity (IEB-Chile), and the University of North Texas (UNT). The Cary Institute has a tradition of frontier research on ecosystem science and coupled human-nature systems. IEB is a leading Latin American research center that coordinates and supports the Long-Term Socio-Ecological Research network (LTSER-Chile) in southwestern South America. The UNT Department of Philosophy and Religion Studies and its Center for Environmental Philosophy represent a world-leading center for environmental ethics. With the joint coordination of the Sub-Antarctic Biocultural Conservation Program (www.chile.unt.edu), these three institutions are supporting this *Ecology and Ethics* book series (see Rozzi et al. 2013).

[4]The formal proposal of the biocultural ethic interrelates the habits and habitats with the identities and wellbeing of the co-inhabitants, human and other-than-human beings. Consequently, the conservation of habitats and access to them by communities of co-inhabitants becomes an ethical imperative. The biocultural ethic's proposal demands to incorporate this imperative into development policies as a matter of socio-environmental justice (see Rozzi 2013).

intercultural and interdisciplinary processes generate forms of co-management of ecosystems, which contribute to planetary stewardship.

Our ultimate goal is to contribute to dynamic, intercultural, and interregional approaches to planetary stewardship initiatives. We have organized the book into three parts. Part I presents contrasting forms of understanding and co-inhabiting the biosphere, forms that often remain outside of academia and prevailing government discourses. Part II examines the Earth Stewardship Initiative, relating it to transdisciplinary work conducted at ILTER sites and networks around the globe. Part III introduces environmental citizenship and participatory approaches, policy and conservation actions, religious belief systems and alliances, and exemplary lives of people who have made, and are making, a difference for practicing Earth stewardship. These approaches and initiatives place the value of life, human and other-than-human, above the value of capital, and have the capacity to implement Earth stewardship practices driven by that reoriented value hierarchy.

1.1 Part I: A Biocultural Approach to Earth Stewardship

Earth stewardship is a biocultural practice because it operates at the interface of biophysical and cultural domains. Different forms of stewardship have evolved from ancient, collective practices in Global Western, Southern, and Eastern societies. Ricardo Rozzi indicates that preserving the diversity of ways of understanding the natural world and of co-inhabiting with it is an essential aspect of the stewardship of both local places and the entire Earth. Part I examines multiple current forms of ecological knowledge and practices in various regions of the world—such as crab- and oyster-harvesting communities living on the Chesapeake Bay, the ancient agricultural tradition of *satoyama* that today molds the life of remnant rural communities in Japan, and lifeways of the Aymara and Quechua people in the high Andean Plateau that relate to the Earth as a living being and regard themselves as integrally connected to the forces of nature. In these living ecological worldviews and practices we can find vital elements to enrich our understanding of Earth stewardship today.

Focusing on local ecological knowledge in North America, Sharon Kingsland calls attention to the complex history of integrating ecological sciences and vernacular conservation practices. Based on a case study in Chesapeake Bay (eastern United States), Kingsland criticizes the split between two cultures: that of scientists and that of "watermen" whose livelihoods rely on harvesting of shellfish. The first culture is based on faith in theoretical models and logical arguments, while the second on knowledge grounded in everyday experience. The historical analysis of this case illustrates how this split was overcome through collaborative work that led to the establishment of co-management practices involving watermen, scientists, and policy makers. Kingsland remarks that scientists are now being challenged to overcome disciplinary constraints in order to be able to produce innovative responses to address the environmental, economic, and social challenges of the twenty-first century.

Scientists must interact with local communities in more respectful and open-minded ways in order to better assist and participate in Earth stewardship.

Hideaki Shibata presents an elegant example of how scientific and traditional ecological knowledge can complement each other. His overview of Japanese ecosystems and cultures introduces the experience of his country's Long-term Ecological Research network (JaLTER), which explores social-ecological interactions along with the more usual focus of LTER programs on biophysical patterns and processes. Shibata shows that traditional ecological knowledge continues to be important to the biogeochemistry of landscapes, and that environmental ethics and belief systems that respect nature can be guiding references for plans to develop a sustainable future. The example of JaLTER's incorporating traditional ecological knowledge in its core research mandate is a powerful one.

In Chinese philosophical traditions, as well as in everyday life, the aesthetic appreciation of nature is central. Shan Gao examines how aesthetic appreciation of nature is also aesthetic appreciation of *ch'i*, a core concept in Chinese philosophy that has no physical form, is invisible, and is always in an unceasing process of movement that produces and reproduces life. Both Shibata and Gao examine ways of understanding nature that include visible and invisible realities (the *kami* among Japan's Ainu population), and how such understanding shapes social-ecological relationships. Shibata affirms that "from ancient times, there has been an established traditional religion that fosters respect for diverse natural objects, including both visible and invisible entities, through a belief in nature deities that reside in various natural places such as mountains, forests, lakes, and oceans."

Visible and invisible realms of reality also play an essential role in Andean worldviews in South America. Based on his research on sacred sites, Fausto Sarmiento introduces the dynamic integration of the physical, the psychological, and the spiritual realms in the Quechua worldview. The triad of body, mind, and spirit is not exclusive to humans. As a member of an Aymara community and a feminist in Bolivia, Vicenta Mamani presents another Andean worldview. She shows how Aymara life is framed in rituals regarding the *Pachamama* or Mother (*mama*) Earth (*pacha*), based on a close and communicative relationship with nature. Humans and their social organizations, nature, and the spirits of humans, nature, and the divine, constitute an indissoluble unit. The Aymara worldview is based on male-female pairing because reciprocity, duality, and complementarity are fundamental concepts. Even personhood is not granted to the individual but to the couple in Aymara societies. Collective complementarity is the basis of labor relations and forms of collective labor continue to be an important dimension of community life and for respecting the *Pachamama* as a living being. Hard work, honesty and truthfulness, generosity and hospitality are Aymara values that reinforce collective labor practices. These values could acquire a broader cultural significance within the Earth Stewardship Initiative.

Also starting from the Aymara worldview, Roy May discusses the concept of Earth Stewardship. He highlights how this concept arises from peasant societies where the earth or land is fundamental to their well-being. Many ancient peasant traditions (including that of Adam and Eve), emphasize the earth as the substance of human genesis and as the means that make life possible. Humanity is seen as being part of a network of interrelationships binding together the earth and the human, in such a way that a good and just life is facilitated, as conveyed by the Andean idea of *Pachamama*. In Aymara and other Andean societies, May highlights the multiple and important functions of llamas. Even the dung they produce is worthy of respect and care because it contributes to the cycle of life through enriched soil for cultivation. It is this integral respect for the animal that makes the llama a fitting metaphor for stewardship. Reading stewardship from the praxis of peasant societies such as the Aymara, provides a perspective that emphasizes mutuality, care and protection, and advocacy for the wellbeing of the Earth and its many forms of life.

The closing chapters of Part I address a core question. If there are so many forms of traditional ecological knowledge with associated traditions of ecologically sustainable practices, why do we face environmental crises in Asia, Latin America, and around the globe? The chapters by Ricardo Rozzi and by Laura Ogden and collaborators provide complementary answers. Rozzi combines the conceptual frameworks of the biocultural ethic and of liberation philosophy to argue that the core problem is axiological, that is, a matter of values. Today, the value of capital is ranked above the value of life. As Argentinean-Mexican philosopher Dussel (2003) has demonstrated, this scale of values is in disagreement with the theological and philosophical roots of Western civilization. Therefore, Rozzi argues that it is necessary to re-establish the right hierarchy of values; that is, to rank the value of life above the value of capital. This conclusion coincides with the perspective of influential US environmental philosophers, such as Holmes Rolston (1985) or Mark Sagoff (2008). As concisely stated by Poole et al. (2013, p. 356) in the closing chapter of the first book of this *Ecology and Ethics* series, "inverting the value hierarchy—i.e., treating economic value as the primary value as we usually do—is as incorrect as planting a tree with its roots in the air."

From the perspective of political ecology, Laura Ogden et al. argue that social-ecological changes associated with global assemblages—that is, globally extensive and multiform governance arrangements—disproportionately impact poorer nations and communities along the development continuum, or the "Global South," as well as others who depend largely on natural resources for subsistence. Complementarily, they show how transnational networks of grassroots organizations resist the negative social and environmental impacts of global assemblages, thus fostering social-ecological resilience. Thus, new community-based global assemblages have emerged as alternative governance mechanisms to counteract the hegemony of corporate, economic versions of the global order.

In summary, the biocultural approach undertaken in Part I suggests that to build a solid Earth Stewardship initiative, we need to identify more precisely the main agents responsible for socio-environmental problems at all scales, from local to global. They are not humanity in general, but specific agents—unequal power relationships, exclusionary institutional arrangements, inequitable and unjust economic strategies. Rozzi concludes that omitting this specification in the diagnosis of global environmental change would be a mistake as serious as a physician blaming microorganisms in general for a disease, rather than identifying the specific organisms that are actually responsible for an infection. As Aldo Leopold (1949, p. 258) stated, "health is the capacity of the land for self-renewal. Conservation is our effort to understand and preserve this capacity." A biocultural approach to Earth stewardship helps to achieve a better diagnosis of specific threats and opportunities for conserving the health of the land and people.

1.2 Part II: Integrating Stewardship Across Disciplines and Scales

The chapters in Part I lay out a broad range of topics that form the threads of a stewardship tapestry. These threads are diverse, both conceptually and culturally, suggesting that the formulation of effective approaches to Earth Stewardship will vary with time, place, scale, and audience. It is unlikely that a single formula or strategy of stewardship will be universally effective, but rather that different conceptual threads of stewardship will vary in their importance depending on context. The chapters in Part II explore stewardship across scales, disciplines including the humanities and ecological sciences, and the timely relationship between stewardship and the Long-Term Socio-Ecological Research (LTER) networks.

Paradigm shifts, such as that implied by Earth Stewardship, often require examining the past in order to transform the present and project into the future. J. Baird Callicott traces the history of tension between ecological science and advocacy in the Ecological Society of America (ESA) from its birth nearly a century ago to the present. Callicott examines the work of the first president of the ESA, Victor Shelford. Today, we can learn from Shelford by understanding how he combined theory and practice in his proposal to create the Committee for the Preservation of Natural Conditions for Ecological Study in 1917. In today's terminology, Shelford developed a pioneer transdisciplinary approach by working closely with federal and state governmental agencies to implement "nature sanctuaries" as "research reserves" that were protected from impacts by people. However, as Callicott points out, in contrast to Shelford's early aim to preserve natural reserves free of human influence, stewardship efforts now recognize the importance of integrating humans as essential components of ecosystems.

Chapin et al. describe how renewed concern about human impacts on the biosphere led to the Earth Stewardship Initiative of the Ecological Society of America (ESA). This chapter, coauthored by current and past presidents of the ESA, discusses multiple approaches that were used to develop a platform for stewardship action, as illustrated in four case studies. Approaches included clarification of the stewardship concept through articles and a website, open discussion and elaboration of the stewardship concept at ESA's annual meetings, engagement of ESA members in activities organized by ESA sections, and outreach beyond ecology through collaborations and demonstration projects with academics and practitioners from other disciplines as well as with other groups in civil society.

The following chapters describe the application of diverse stewardship approaches in contrasting cultural contexts, drawing primarily on experiences from the

International Long-Term Ecological Research (ILTER) network. Maass and Equihua discuss the conceptual framework that has guided the ILTER in its stewardship efforts. They undertake a transdisciplinary research approach to understanding socio-ecosystems, representing an important epistemological shift from earlier LTER paradigms that focused on ecology, with people viewed as external influences rather than integral components of the system.

An initiative at a global scale presupposes information about the different regions of the planet. Ben Li and collaborators examine the cumulative publication output of the ILTER network—some 30,000 publications—to document striking gaps in terms of regions of the world where knowledge is produced and published, and the type of information that is included in ILTER research. This chapter provides a quantitative diagnosis of critical geographical and conceptual gaps in ILTER that an Earth Stewardship initiative should aim to fill. Ways to integrate ecological sciences and ethics should be found in order to solve intercultural and interdisciplinary conceptual gaps. To address these gaps, Jorge Aguirre describes field environmental philosophy (FEP), a methodological approach developed in Latin America that underscores the value of the integration of poetic work with scientific and philosophical research into education and conservation. Aguirre enriches the FEP methodology by incorporating hermeneutics—i.e., theory of text interpretation—with a dual purpose:

(i) to deconstruct a narrow economic-utilitarian rationality, which is not inherent to human nature but emerged in a particular historical and cultural context, and
(ii) to develop innovative practices of biocultural conservation that are informed scientifically and ethically, illustrated with examples from Mexico and Chile.

The arts and humanities also contribute to interdisciplinary research at sites of the US LTER network. Based on an extensive series of questionnaires, Lissy Goralnick et al. describe novel collaborations among ecologists, artists, writers, and philosophers to frame the stewardship discussion in a very different context, using multiple media to explore distinct ways to communicate concerns about Earth's future. They focus on empathy as a key element because empathy touches those scientists and students who have developed a commitment and sense of responsibility to stewardship. This focus coincides with one of the methodological elements highlighted by Aguirre regarding FEP, in resonance with the essay "Thinking like a Mountain," written by Aldo Leopold—another ESA president who articulated a stewardship ethic, as Callicott explains.

To achieve interdisciplinary work, Charles Redman and Thaddeus Miller emphasize the methodological importance of understanding the specific meanings of concepts used with contrasting connotations by different disciplines. They note that infrastructure has both technological and cultural implications. Within a new interdisciplinary framework, they propose that infrastructure should be considered in the context of three equally important domains: social, ecological, and technological/infrastructural systems (SETS).

Part II closes with two chapters that present case studies of interdisciplinary work in remote arid, rural Mediterranean, and urban ecosystems. Daniel Orenstein and Elli Groner describe an LTER site on the border between Jordan and Israel that provides a venue for developing trust and collaborations in a politically contentious region of the world. They describe discussions about what kind of knowledge is important to local stakeholders. In some cases, experts can clarify which management actions should receive highest priority and which require either more research to fill knowledge gaps or greater dialogue to overcome gaps in values between locals and scientists. For example, the aesthetic value of landscapes often is more relevant to citizens and decision makers than to scientists. Orenstein and Groner propose a social-based research approach to ecosystem services within the LTSER platform that provides a framework for integrating the values and opinions of local communities into the local research and policy agenda. This social-based approach to ecosystem services assessment—which has proven to be a catalyst for constructive, community-level engagement—could be further applied within the Earth Stewardship initiative. Olga Barbosa and Paula Villagra highlight the relevance of combining bottom-up with top-down approaches. Capacity-building in local communities is as important as building relationships with regional and national government institutions and private business organizations.

In summary, Part II explores the integration of historical and cultural analyses, philosophical methodologies, and long-term socio-ecological research platforms, with practices that are essential for creating a stronger stewardship commitment that is conceptually grounded in diverse realities, and is relevant to addressing the practical issues faced by today's global and local societies. Effective approaches depend deeply on cultural context, requiring interdisciplinary exploration, study, partnership, and infrastructural implementation throughout the world.

1.3 Part III: Integrating Ecology and Ethics as a Foundation for Earth Stewardship Action

This final part introduces concepts, ongoing initiatives, and future perspectives for stewardship actions. Earth stewardship, as much citizenship, entails rights and responsibilities. Eugene Hargrove introduces the concept of Earth citizenship as a metaphor for an ecological governance of the planet as its capacity to support human life is pushed to the limits. Following Mark Sagoff, he contrasts the notions of *citizen* and *consumer.* Furthermore, Hargrove argues that stewardship has a religious connotation, whereas citizenship is religiously neutral, and therefore can be widely accepted across the many cultures of the world. Peter Taylor emphasizes the importance of engagement and participation of people, cultivating collaborators, transversality, and fostering curiosity for dynamic environmental planning and management. Engagement and participation of people also is central to the approach to protected areas taken by the International Union for Conservation of Nature (ICUN). Ernesto Enkerlin and collaborators introduce the "Promise of Sydney," the focus of the 2014 World Parks Congress in Australia, to emphasize that protected areas can be an effective strategy to put Earth stewardship into action. They argue that protected areas not only are necessary for conservation aims, but they also contribute critically to human well-being and social justice in the Anthropocene.

Flavio Berchez and collaborators support ICUN's argument with experiences in South American Marine Protected Areas (MPA), where MPAs are essential for protecting biodiversity, informing policy making, managing coastal fisheries, and supporting ecological education and scientific tourism programs.[5] Berchez et al. indicate that to achieve these goals in MPAs it is imperative to include not only scientists, but also policy makers, teachers, and importantly, graduate students, a point that coincides with the perspective that Chapin et al. have for the ESA Earth Stewardship Initiative. However, socio-ecological problems are complex. Based on their experience in political sciences and global climate change in Brazil, Eduardo Viola and Larissa Basso underscore this complexity, and identify plutocratic regimes as a central problem: "economic sectors have excessive influence over governmental decision-making and the political system." A governance and ethical shift toward a low carbon economy and consciousness is required.

To achieve a change in global consciousness, religions are playing a major role. Uruguayan theologian Guillermo Kerber explains the Climate Initiative of the World Council of Churches, which brings together most of the Christian communities in more than 110 countries. Additionally, the Climate Initiative is an inter-faith effort including Buddhism, Hinduism, Judaism, and Islam. Kerber explains that the concept of Earth stewardship is at the core of religious messages.[6] These affirm that humans are not owners of the Earth, but rather care-takers of the Earth. This idea is shared by groups of Christians, Jews, and Muslims, who have come together to address the challenges of Global Climate Change. The role of religions in Earth Stewardship is further introduced by US theologian Mary Evelyn Tucker. She identifies six core values that are widely shared by religious traditions: *reverence*, *respect*, *restraint*, *redistribution*, *responsibility*, and *renewal*. These values were adopted by the Earth Charter initiative of the United Nations, matching them with six corresponding components for human-Earth flourishing: *cosmological context*, *ecological integrity*, *social equity*, *economic justice*, *democracy*, and *non-violence and peace*.

Dorothy Stang (1931–2005), a US Roman Catholic religious who in 1966 went to the Amazon, is portrayed by Roy May within a tradition of the Latin American

[5] The South American Marine Protected Areas framework is similar to the one developed in the US National Marine Protected Areas (MPAs) Center of the National Oceanic and Atmospheric Administration (NOOA) created in 2001. NOOA's MPA Center defines marine stewardship as "careful and responsible management to ensure goals and objectives are being achieved for the benefit of current and future generations." The MPA Center focuses its objectives on enhancing MPA stewardship by strengthening capacity for planning, management and evaluation. (http://marineprotectedareas.noaa.gov/sciencestewardship/)

[6] Since the recognition of the environmental crisis in the 1960s, Earth Stewardship has been a primary metaphor among Christian churches that have been concerned with sustainability and the wellbeing of life in the planet (see Ball et al. 1992; Northcott 1996; Hessel and Ruether 2000). A majority of Christian theologians, as well as lay thinkers have been supportive of a stewardship environmental ethic (see Berry 1981; Attfield 1983; Callicott 1994; Rasmussen 1996; Petrie 2000). However, some criticize the concept of stewardship as being anthropocentric and hierarchic (see Palmer 1992).

Church of non-violent and peaceful defense of social justice, and in recent years also environmental justice. This Latin American tradition began as early as the sixteenth century, with Bartolome de las Casas—a Dominican who undertook the role of "Protector of the Indians" in the Maya territories of southern Mexico and Central America. Today, this tradition of defending the culture and wellbeing of indigenous and other local communities has acquired a relevant role in Latin America and worldwide through liberation theology. Theological texts associated with Earth Stewardship concepts are grounded in peasant communities and indigenous cultures. Advocacy for the human rights of Brazilian peasants and indigenous people, and the fight to preserve the Amazon rainforest, were undertaken by one of the most important environmental leaders in South American history, Chico Mendes (1944–1985). Fábio Valenti Possamai and Fernando da Rocha portray the life and work Francisco "Chico" Alves Mendes Filho—a rubber tapper who became a grassroots union organizer—in counterpoint to Jose Lutzenberger (1926–2002)—a German-Brazilian agronomist, politician, and environmentalist who was appointed Minister for the Environment in Brazil in the early 1990s. Both made endurable contributions to the concept and implementation of sustainable extractive reserves in Amazonia.

The biographies of Stang, Mendes, and Lutzenberger are important for understanding the difficulties of implementing an Earth-stewardship environmental ethic. The first two were murdered for defending the poor and the Earth, while the latter was criticized harshly and finally marginalized by the Brazilian political establishment. Their lives teach us much about stewardship and what it may cost to practice it. They also teach us about the importance of transdisciplinary and international alliances. Frank Golley (1930–2006), an ecologist who served as president of the International Association of Ecology, the International Society of Tropical Ecology, and the Ecological Society of America (ESA), pioneered academic international networking, and recognized the great value of learning from other cultures and of involving different kinds of people in ecological research. Alan Covich, also a former president of the ESA, describes how Golley reached out far beyond the confines of his Georgia-based university. His academic interest integrated ecological sciences and environmental ethics; as a scientist his stewardship praxis was broad and deep. If we want to understand what stewardship means, we should review the lives of these and other people. They show us that Earth Stewardship is not only what we think and write about the Earth, but, foremost, what we do, individually and collectively, on behalf the Earth's creatures, its biocultural diversity, and its climate. They also reveal the essential role played by international collaborations and exchanges, by building institutional platforms, and by complementing disciplines and life experiences.

Earth Stewardship requires personal commitment and involvement. It is to do science and philosophy committed personally to the well-being of all the Earth's co-inhabitants—human and other-than-human—and to the biogeochemical processes that make life, as we know it and cherish it, possible on what Holmes Rolston calls "the home planet."[7] The lives of Golley, Stang, Mendes, and Lutzenberger

[7] See Rolston (2013).

make this evident. Personal commitment takes us beyond theory and puts us into practice, into actively practicing Earth Stewardship. It is a stimulating perspective because the viewpoint from personal involvement, commitment, and experience will shape not only the science, but the very philosophy of Earth Stewardship. Praxis integrates theory and practice, and in so doing transforms both. And both are needed. In this sense Earth Stewardship is praxis.

1.4 Concluding Remarks

Latent throughout this book is the importance of praxis, that is, the way people actively relate in and to the natural world. Research and ideas are necessary but not sufficient. Earth stewardship is not only an idea, but a way of co-inhabiting in the world.

Core ideas for Earth stewardship emerge from praxis, a praxis that is ancient and contemporary, collective and individual.

Ethical values, citizenship traditions, metaphors and poetic creativity, contemporary and traditional vernacular ecological knowledge, political ecology, institutional networks, local communities, and exemplary lives complement the scientific perspectives of the Earth Stewardship initiative.

The hybridization of disciplines and traditions will stimulate and strengthen a paradigm shift that fosters dynamic, intercultural, and interregional approaches to Earth Stewardship. The integration of ecology and ethics into Earth Stewardship inaugurates a new transdisciplinary stage of long-term socio-ecological research at a global scale, and a biocultural approach that includes all beings with whom humans co-inhabit the biosphere.

References

Attfield R (1983) The ethics of environmental concern. Columbia University Press, New York

Ball I, Goodal M, Palmer C et al (eds) (1992) The Earth beneath. A critical guide to green theology. SPCK, London

Berry W (1981) The gift of good land. Counterpoint Press, Berkeley

Callicott JB (1994) Earth's insights: a survey of ecological ethics from the Mediterranean basin to the Australian outback. University of California Press, Berkeley

Chapin FS III, Kofinas GP, Folke C (eds) (2009) Principles of ecosystem stewardship: resilience-based natural resource management in a changing world. Springer, New York

Chapin FS, Power ME, Pickett STA et al (2011a) Earth stewardship: science for action to sustain the human-earth system. Ecosphere 2:89

Chapin FS, Pickett STA, Power ME et al (2011b) Earth stewardship: a strategy for social-ecological transformation to reverse planetary degradation. J Environ Stud Sci 1:44–53

Dussel E (2003) Algunos Principios para una Ética Ecológica Material de Liberación (Relaciones entre la Vida en la Tierra y la Humanidad). In: Pixley J (ed) Por un mundo otro. Alternativas al mercado global. Quito, Ecuador, Comunidad Cristiana Mesoamericana y Consejo Latinoamericano de Iglesias, pp 29–44

Haberl H, Graube V, Díaz-Delgado R et al (2009) Towards an integrated model of socioeconomic biodiversity drivers, pressures and impacts: a feasibility study based on three European long-term socio-ecological research platforms. Ecol Econ 68:1797–1812
Hessel DT, Ruether RR (eds) (2000) Christianity and ecology: seeking the well-being of Earth and humans. Harvard University Press, Cambridge
Leopold A (1949) A sand county almanac and sketches here and there. Oxford University Press, New York
Lui J, Dietz T, Carpenter SR et al (2007) Complexity of coupled human and natural systems. Science 317:1513–1516
Northcott M (1996) The environment and Christian ethics. Cambridge University Press, Cambridge
Ohl C, Krauze K, Grünbühl C (2007) Towards an understanding of longterm ecosystem dynamics by merging socio-economic and environmental research: criteria for long-term socio-ecological research sites selection. Ecol Econ 63:383–391
Palmer C (1992) Stewardship: a case study in environmental ethics. In: Ball I, Goodal M, Palmer C et al (eds) The Earth beneath. A critical guide to green theology. SPCK, London
Parr TW, Ferretti M, Simpson IC et al (2002) Towards a long-term integrated monitoring programme in Europe: network design in theory and practice. Environ Monit Assess 78:253–290
Petrie A (2000) Releasing heaven on Earth. God's principles for restoring the land. Baker Book House, Grand Rapids
Poole AK, Hargrove E, Day P et al (2013) A call for ethics literacy in environmental education. In: Rozzi R, Pickett STA, Palmer C et al (eds) Linking ecology and ethics for a changing world: values, philosophy, and action, ecology and ethics. Springer, Dordrecht, pp 349–372
Power ME, Chapin FS III (2009) Planetary stewardship. Front Ecol Environ 7:399
Rasmussen L (1996) Earth community, Earth ethics. WCC, Geneva
Redman CL, Grove JM, Kuby LH (2004) Integrating social science into the long-term ecological research (LTER) network: social dimensions of ecological change and ecological dimensions of social change. Ecosystems 7:161–171
Rolston H (1985) Valuing wildlands. Environ Ethics 7:23–58
Rolston H (2013) Foreword. In: Rozzi R, Pickett STA, Palmer C et al (eds) Linking ecology and ethics for a changing world: values, philosophy, and action, vol 1. Springer, Dordrecht, pp vii–xi
Rozzi R (2013) Biocultural ethics: from biocultural homogenization toward biocultural conservation. In: Rozzi R, Pickett STA, Palmer C et al (eds) Linking ecology and ethics for a changing world: values, philosophy, and action, vol 1. Springer, Dordrecht, pp 9–32
Rozzi R, Armesto JJ, Gutiérrez J et al (2012) Integrating ecology and environmental ethics: Earth stewardship in the southern end of the Americas. BioScience 62(3):226–236
Rozzi R, Pickett STA, Palmer C et al (eds) (2013) Linking ecology and ethics for a changing world: values, philosophy, and action, vol 1. Springer, Dordrecht
Sagoff M (2008) The economy of the Earth: philosophy, law, and the environment. Cambridge University Press, New York
Singh SJ, Haberl H, Chertow M et al (eds) (2013) Long term socio-ecological research. Studies in society-nature interactions across spatial and temporal scales. Springer, Dordrecht
Steffen W, Jansson Å, Deutsch L et al (2011) The Anthropocene: from global change to planetary stewardship. Ambio 40:739–761

Part I
A Biocultural Approach to Earth Stewardship

Chapter 2
Ecological Science and Practice: Dialogues Across Cultures and Disciplines

Sharon E. Kingsland

Abstract Promoting earth stewardship entails re-examining economic arguments, such as the "tragedy of the commons" logic, which are coercive, out of step with cultural values, and often lack empirical support. A counter-example is the effort by Chesapeake Bay watermen to resist privatization of the commons, while adopting an alternative strategy more in keeping with their cultural values. Creating trust between scientists and watermen has been difficult, however. Research from the social sciences, notably by the late Elinor Ostrom and colleagues, and William Burch Jr., suggests that human ecology can be developed in a way that is more attuned to human values. Citizens have important roles in fostering good stewardship when they can mobilize support, as illustrated in Jane Jacobs's writing about urban communities, and by citizen-led creation of a nature reserve in Toronto, Canada. Two challenges in promoting earth stewardship are to create trust between scientific experts and resource users, and to create an academic culture that values interaction between scholarly disciplines.

Keywords Biocultural conservation • Common-pool resources • Ecological economics • Tragedy of the commons • Urban ecology

Promoting earth stewardship[1] involves enhancing public understanding of the ecosystems of our world and how they support us. Too often our dependence on these systems comes to our attention only in a crisis, as happened during the American Dust Bowl of the 1930s, a stunning example of the collapse and loss of ecosystem services that prompted the U.S. Department of Agriculture to pay more attention to soil conservation and curbing the bad habits of over-plowing, over-grazing, and over-cutting of timber. Two generations later we express our debt to nature and duty to future generations when we assert the importance of protecting ecosystems so as

---

[1] Throughout the book *Earth stewardship* refers to stewardship at a planetary scale, while *earth stewardship* refers to a local community stewardship at the scale of ecosystems or landscapes.

S.E. Kingsland (✉)
History of Science and Technology Department, Johns Hopkins University, Baltimore, MD, USA
e-mail: sharon@jhu.edu

R. Rozzi et al. (eds.), *Earth Stewardship*, Ecology and Ethics 2,
DOI 10.1007/978-3-319-12133-8_2

to maintain ecosystem services. Earth stewardship involves making this awareness part of the fabric of our society.

Historical and contemporary examples from modern ecology and environmentalism suggest that a multicultural perspective that combines values and perspectives both from the culture of science and the culture of ordinary citizens can be more productive than an approach that seeks to apply scientific expertise without regard to local culture and custom. Research in social science also underscores the importance of studying human behavior, institutional structures, and the conditions that lead to effective stewardship, rather than relying on over-simplified logical arguments, such as the "tragedy of the commons" argument popularized by Garrett Hardin (1968). In this chapter I consider one example that illustrates the problem of trying to impose the "tragedy of the commons" logic on resistant resource users, before highlighting some of the interdisciplinary scholarship that has provided an alternative approach to understanding problems of governing the commons. I end by considering a couple of instances of citizen-led stewardship that has countered modern forces of development in order to foster an ecological viewpoint that conforms to human values and needs.

2.1 Two Cultures: Scientists and Watermen

Today, ecology affirms that humans are intrinsic parts of ecosystems, and their values and cultural beliefs, which motivate their actions, are aspects of these systems (McDonnell and Pickett 1993). As Ricardo Rozzi (2013) has pointed out, preserving human cultures is an important aspect of earth stewardship. This endeavor can be seen as a problem of ethics that invites collaboration between ecologists and philosophers. He cautions however that environmental ethics must depart from the philosophical traditions that have separated humans from other animals, in order to understand what connects humans, the environment, and the species that co-inhabit the environment. With such understanding one can appreciate how the conservation of cultural diversity – the diverse cultures that have evolved in close interaction with other species and environments – must be part of the broader goal of working toward sustainable practices. Rozzi refers to this viewpoint as "biocultural ethics" which serves the goal of "biocultural conservation." He views this approach as also capable of addressing problems of social and environmental justice that are inextricably linked to broader problems of environmental stewardship. The scientifically trained ecologist who enters local communities with this goal in mind has to be capable of discovering, internalizing, and perhaps even recovering the environmental perspectives of local communities. That such discovery is a means of enlightenment for ecologists has been emphasized in several essays in the volume *Linking Ecology and Ethics for a Changing World*, for example Stuart Chapin's reflections on his interactions with native communities in Alaska (Chapin et al. 2013).

It has proved more difficult to accord non-indigenous local communities, even those that have lived off the land for generations, the same attention and respect.

One reason has been the dominance of the logical argument known as the "tragedy of the commons," one modern version of which was articulated forcefully by Garrett Hardin (1968). According to this logic, resources that are held in common will inevitably be over-exploited by the resource users, and therefore a solution can come only from the outside, either in the form of privatization or some type of coercion or regulation of human action. This argument long predates Hardin's influential article and has created a stumbling block to bringing scientific and lay communities together for productive dialogue. A good example is the history of acrimonious debate about oyster conservation in the Chesapeake Bay on the east coast of the United States. The still-unsolved problem is how to preserve this once-productive estuary and the once-abundant oysters that are so important for maintaining water quality. Environmental historian Christine Keiner (2001, 2009) has analyzed the "oyster question" in this region and argues persuasively that one persistent blind spot in trying to solve environmental problems has been failure to recognize the crucial link between conserving oysters and valuing and protecting the culture of the local watermen or commercial fishermen. One reason has been that scientists and policy makers have been in thrall to the "tragedy of the commons" logic.

As Keiner points out, the need to conserve the Chesapeake oyster population was recognized well over a century ago. William Keith Brooks, a zoologist at the Johns Hopkins University, published a book on oysters in 1891 that was an early interdisciplinary work (Brooks 1996). He drew on biology and political economy to argue for the importance of sustaining the Chesapeake oysters. Brooks's studies of oyster reproduction revealed that the Chesapeake oyster could be cultivated, and he concluded that the best conservation strategy was aquaculture, which meant privatizing the commons. For many years he advocated privatizing oyster beds because he thought it would bring prosperity to the impoverished watermen living on the eastern shore of the Chesapeake Bay. Other scientists picked up the refrain: the solution was to enclose the commons and develop oyster culture.

But Maryland's watermen, the people he was trying to help, were vehemently against the idea of privatization because it was expensive and threatened many aspects of their worldview. Quite simply, privatization and aquaculture required capital, for underwater farming was expensive, and the watermen did not have capital. Those with the ability to afford the high costs of aquaculture were the packers and canners, the capitalists of the oyster economy, who if allowed to farm oysters would gain control of the oyster beds. If aquaculture were instituted, the watermen would become the equivalent of farm workers, employees working for the capitalists. Watermen fiercely valued their independence, as they still do today, and being corporate employees was much against their self-image and their culture. As it happened, the watermen had considerable political clout in the state legislature, because Maryland's system of representation favored the rural counties, where the watermen lived, over the city of Baltimore, where the scientists lived. While the scientists in Baltimore continued to define the "oyster question" as a question of privatization, or oyster culture, the watermen continued to resist a "solution" that threatened their core values and their culture.

The scientists had failed to grasp that conserving the oyster also meant conserving the waterman and his culture: the two were part of one system. A solution that tried to conserve the oyster at the expense of the waterman's culture was no solution, as long as the watermen held political power. This historical episode is typically seen as a classic case of the "tragedy of the commons", whereby greedy resource users (the watermen) exploited the commons, the Chesapeake oyster beds, and eventually destroyed the resource on which they depended (e.g., Wennersten 2001). Keiner suggests that the reality is more complicated. Indeed, scientists were forced to acknowledge the watermen's arguments, but she perceived this cross-cultural dialogue as a positive step. Solutions were developed based on replenishing and reseeding the oyster beds, and watermen themselves took part in these conservation efforts, while not sacrificing their sense of independence and their cultural values. Scientists, watermen, and policymakers ended up working together to create a unique system, which turned into a well-managed commons. This system, although not perfect, was a reasonable compromise that worked for several decades.

While not romanticizing the watermen as model conservationists, Keiner seeks to correct the perception that they were enemies of nature. Watermen valued conservation and contributed an understanding of the bay that may not have been couched in the language of science but was nonetheless grounded in their experience on the water. As Keiner suggests, "The ways in which oystermen and their legislative allies crafted a viable alternative to private cultivation can be seen as a case of co-management," and she argues that we need more historical analysis of natural resource management regimes in which "local knowledge played a greater role than elite scientific expertise" (Keiner 2009, p. 10). She further cautions that we should avoid looking back on this long debate and concluding that because the watermen resisted the scientific advice, therefore science was compromised and conservation efforts failed. Moreover, we cannot be sure that the solution advocated by scientists until the mid-twentieth century – private cultivation and enclosure of the commons – would have solved the problem. Her case study was the first to give full weight to the Chesapeake watermen's perspective.

Ultimately the oyster population did crash, but late-twentieth century population declines must be attributed to many changes in the region. High population densities were producing significant levels of stress on the ecosystem and transforming the watershed. One consequence was that during a period of prolonged drought, possibly linked to human-caused climate change in the 1980s, conditions favored the spread of lethal parasites that devastated the oyster populations. Understanding the many causes contributing to extreme weather conditions requires a broad understanding of modern industrial society, including what is occurring globally, not just in the immediate Chesapeake region. Likewise casting environmental problems in terms of the remorseless logic of arguments like the "tragedy of the commons" oversimplifies and distorts the reality. Keiner also insists that the oyster question cannot be removed from its political context or be seen as a purely scientific problem.

Arriving at a method of co-management that combines the perspectives of scientists and resource users requires a degree of trust between these different stakeholders. Today scientists and watermen work together, although distrust between the two groups has been hard to overcome and has not completely evaporated. Anthropologist Michael Paolisso (Greer 2003) has worked with communities of Chesapeake watermen and has tried to identify the core beliefs of both watermen and scientists that are important for conservation practice. In this case the conservation problem centered on the Chesapeake blue crab, whose populations can fluctuate dramatically from year to year. Paolisso noted some similarities in core beliefs but also striking differences between the two groups, most notably in the watermen's belief that "God and nature" were the best "managers" of natural resources. While it can seem nonsensical to a scientist to make such a statement, the watermen were trying to express the idea that population fluctuations of the blue crab were not predictable, nor were they capable of being controlled by humans. In addition, the watermen had a faith-based view of natural cycles which came from their daily experiences on the water. To the watermen, the scientists' faith in their models was perplexing. While the two sides maintained their points of difference, these dialogues about core values and perceptions of nature nonetheless helped them to find some common ground. The kind of work Paolisso does to facilitate dialogues across the cultures of science and watermen helps to build respect and trust on both sides.

2.2 Challenging the Tragedy of the Commons Logic

The importance not just of understanding other cultures, but of building respect and trust between different communities, is emphasized in a growing literature that has been challenging the tragedy of the commons logic since the 1980s. This critique forms the basis for a powerful analysis and synthesis by social scientists such as the late Elinor Ostrom and her colleagues and collaborators. Ostrom shared the Nobel Prize in Economic Sciences in 2009 for her analysis of economic governance, especially the governance of common-pool resources (such as local fisheries, pastures, irrigation systems, and forests). *Common-pool resources* refers to cases where one person's consumption subtracts from the availability of consumable benefits to others, but where it is difficult to exclude people from access to the resource.

Ostrom (1990) recognized that the tragedy of the commons argument, in tandem with other economic arguments in the same vein, had become dominant without being properly tested by empirical studies. Essentially these models were being used metaphorically to invoke an image of looming disaster, and when such images were used as the basis of policy, this made the models dangerous in her view. Empirical studies that would provide tests of these dire predictions had in fact been accumulating, but they were dispersed across different disciplines, were not coordinated, and on their own no single case study offered definitive conclusions. This situation started to change in the 1980s, as a result of a National Research Council (NRC) Panel on Common Property Resource Management, which published its report in 1986 (National Research Council 1986). The panel's steering committee first met in 1983 and quickly found that there were a large number of existing case studies relevant to their task (Poteete et al. 2010). Scholars from different disciplines joined to assess the results of these worldwide studies. This panel also promoted a framework that had been developed in the 1970s by several social scientists, including Ostrom, called the Institutional Analysis and Development Framework. The framework was meant to provide a conceptual map that would help to organize thinking about how individuals or groups dealt with collective-action problems. Participation in the NRC workshops led Ostrom to try to synthesize the findings from the case studies, and this effort in turn stimulated an extensive research program on common-pool resources.

The work of Ostrom and her collaborators involved consolidating data from empirical studies ranging across many disciplines, carefully designed laboratory experiments to test the assumptions of economic theory, extensive field studies, and theory development. In her ground-breaking book, *Governing the Commons*, she argued that any theory of human organization should be "based on realistic assessment of human capabilities and limitations in dealing with a variety of situations that initially share some or all aspects of a tragedy of the commons" (Ostrom 1990, pp. 23–24). Her approach was interdisciplinary. It used the strategy then being developed by social scientists under the rubric of "the new institutionalism," which called attention to the importance of understanding the details of institutional structures. But she also adopted the strategy biologists used when they linked empirical work to a theoretical understanding of the biological world. Her inspiration in biology came not from ecological studies but rather from the idea of selecting a simple organism in which a process to be studied occurs in a clear or exaggerated way. Her equivalent "representative organism" was instead a human situation, namely a small-scale common-pool resource situation having certain characteristics. Among the many lessons of her decades of work, as summarized in her Nobel Lecture (Ostrom 2010), was that humans have much greater capacity to solve dilemmas than early economic theories suggested. Moreover she concluded that "designing institutions to force (or nudge) self-interested individuals to achieve better outcomes" may be far less effective than facilitating "the development of institutions that bring out the best in humans" (Ostrom 2010, pp. 435–436).

Ostrom's work and that of her colleagues focused on human behavior and institutions and recognized the importance of a multi-level attack that includes the socio-ecological context. This broad and ambitious program complements the literature of ecological science and suggests that dialogue between these disciplines would be highly productive. The kinds of problems that Ostrom studied, and the issues she confronted in the course of that study, such as how to deal with complexity and contingency, are closely parallel to the problems and methodological challenges that ecologists face. An excellent summation is the book *Working Together: Collective Action, the Commons, and Multiple Methods in Practice* (Poteete et al. 2010). This book offers much food for thought for ecologists who are grappling with problems of stewardship and should suggest many ways of engaging with economists.

Like Ostrom, sociologists have also warned against over-simplification of environmental problems and the need to engage with communities in a productive way. William R. Burch, Jr., a sociologist at Yale University, was an early critic of the way environmental debates had split into either extremely pessimistic or overly optimistic viewpoints which both adopted a simplified view of things. Reflecting on the environmental crisis emerging in the 1960s, he became interested in the interpenetration of myth, social systems, and ecosystems, developing these ideas in a book, *Daydreams and Nightmares: A Sociological Essay on the American Environment* (Burch 1971). Burch warned academics to be wary of falling into the trap of blaming environmental problems on a flawed human "nature," on single causes like overpopulation, or on various villains and conspirators. Over-simplifying the cause of the problem would not help to solve it, and pinning one's hope on technological fixes was not likely to work either.

Burch's arguments were also relevant to the simplified logic of the tragedy of the commons. Hardin's discussion in 1968 had not been restricted to problems of resource use. Most provocatively he extended his logic to the problem of over-population and concluded that solving that problem also entailed some form of coercion. While much debate at that time focused on the population "explosion," as though humans were cancers on the earth, Burch countered that human reproduction was not a strictly biological phenomenon, over-population was not to be blamed on irresponsible behavior among the underclass, and the solution would not come from handing out the latest birth control technology. Always there were social, cultural, economic, and political dimensions to these problems that had to be understood. Environmental problems had broad ramifications that resisted reductionist thinking. Burch's ideas have been adapted by ecologists and applied to the development of a Human Ecosystem Model, or a framework for studying human-dominated ecosystems, such as those in urban environments (Pickett et al. 1997). The Baltimore Ecosystem Study, part of the Long-Term Ecological Research program, uses this framework. Its objective is to analyze how humans, including their institutions and cultures, operate as parts of ecosystems, but without judging that activity in the stark negative terms that were intrinsic to Hardin's logic.

2.3 Power to the People

This rich literature in social science, in combination with ecological and environmental discussions, opens the possibility for an approach to environmental literacy that would try to put more agency in the hands of the citizenry, or encourage people to be self-educators through their interactions with their environments. It is easy to see the problem of environmental literacy as conveying knowledge from experts to people who are ignorant, in order to get ignorant people to alter their behavior. That kind of knowledge flows in one direction, and the approach would be something like this: teach more about environmental science (and related subjects like natural history), at an earlier age, teach it better, and keep driving home the message throughout people's lives. There is nothing wrong with such teaching, but another goal would be to make people realize that they are capable of making correct decisions even without a lot of expertise, if they can critically analyze what is around them and link what they observe to their values, their culture, and what they think is worth preserving. That is, ecological knowledge is not just about conveying information; it involves a way of thinking. Sometimes it becomes necessary to challenge the experts, and as Rozzi (2013) argues, to actively reverse the trends toward biocultural homogenization. Such actions may require an active push from the level of ordinary citizens.

This was the message of Jane Jacobs' landmark book of 1961, *The Death and Life of Great American Cities*. Jacobs fought against the idea that the planning expert always knew best, and that tearing apart urban communities with such things as expressways or any structure that alienated people from their environment was a progressive step. She was not a scientist, but a keen observer of the urban scene, and she perceived that what might appear as clutter could also appeal to urban dwellers, providing aesthetic pleasures and sources of interest as they navigated their environments. A city's downtown core should welcome people in, not drive people away. Her idea was that one did not have to be an expert to evaluate what worked and did not work in an urban environment. The key skill was to be able to observe how people used spaces: where did they cluster, shop or play, and what did they avoid? What made some areas seem dangerous and others inviting? She had faith that if people were attuned to their environments and how they worked, they would make good decisions about the future of those communities. Again the emphasis was to preserve what was culturally valuable, rather than to tear things down for the sake of a modern look that is devoid of unique local characteristics.

A final example illustrates the way an educated and ecologically sensitive public can steer decisions toward ends that promote earth stewardship by asserting the value of biocultural conservation. In Toronto, Canada, an urban wilderness called the Leslie Street Spit was created at first by accident and then with the support of a group of citizens called Friends of the Spit, who formed in 1977 (Carley 1998; Courval 1990). The spit is a human-made peninsula jutting into Lake Ontario, which started as a breakwater for harbor expansion in the 1950s but then became a construction landfill site when the harbor plans were abandoned. In time, vegetation started to grow and the process of ecological succession got underway. The peninsula attracted various wildlife species and became a bird watcher's paradise. Although it continued to be used for construction landfill, limited public access on bus tours was allowed starting in 1973, followed by cyclists and hikers the next year. In 1977 Friends of the Spit formed and began to lobby for greater public access, but otherwise they hoped to keep the land in an undeveloped state, allowing it to mature as an urban wilderness park. They had to fend off efforts to develop the area for recreational use, for instance plans to build a multi-purpose aquatic park, and held firmly to the principle that the best possible thing was to leave it alone and let nature take its course. This struggle was by no means easy and required determined efforts by citizens over many years. Today, after decades of lobbying, the 5-km peninsula is exactly what these citizens envisioned, a unique urban wilderness that draws nature-lovers to what is in effect an ecological experiment, ever evolving and maturing.

This is a different example of what Rozzi has called biocultural conservation, a case where citizens have chosen to adopt a culture of earth stewardship because they recognized the value of biological diversity. It should remind us that cities are very good environments in which to promote ecological awareness, and that there is nothing quite so interesting as an ecological experiment in progress, especially when it emerges as a result of people's intrinsic love of nature.

2.4 Conclusion

Historical and contemporary examples, ranging from late-nineteenth-century debate about oyster conservation in the Chesapeake Bay to the creation of a wilderness reserve in modern Toronto, illustrate the importance of involving resource users and ordinary citizens in decisions about earth stewardship. Expert authority should be challenged when it depends on uncritical application of simplified logical arguments such as the "tragedy of the commons" argument. One powerful lesson emerging from four decades of research is the need for careful evaluation and testing of such logical arguments, which can be long-lived even when unsupported by evidence. As the history of the Chesapeake watermen illustrates, crude applications of such logical arguments can destroy trust between experts and resource users. Such trust is a necessary step toward the goal of biocultural conservation. The work of Ostrom and her colleagues demonstrates that the predictions of conventional theory may be quite wrong and that people are capable of adopting good cooperative solutions. Their work supports Jane Jacobs's inspired idea that one must observe how people actually behave, rather than assume how they might behave, when thinking about how cities function.

In order to understand people and their environments, another type of cross-cultural dialogue must be fostered between disciplines within the academic and professional spheres. As Poteete et al. (2010) recognize, creating opportunities for interdisciplinary interaction is difficult, given our expectations of how academic careers are made. Specialization is often valued over interdisciplinary collaboration. Yet solving the problems of earth stewardship, which involves understanding human behavior and human potential, must involve interactions at the frontiers where the ecological and social science disciplines meet. One of the biggest challenges within the culture of academe is to recognize the value of pioneering efforts at these interdisciplinary and intercultural frontiers. Just as we face the challenge of communication between the cultures of science and the lay public, we also face challenges within academe to find mechanisms or institutional structures that can help to build trust between scholars in different disciplines.

References

Brooks WK (1996) The oyster: a popular summary of a scientific study. Johns Hopkins University Press, Baltimore\London

Burch WR Jr (1971) Daydreams and nightmares: a sociological essay on the American environment. Harper and Row, New York

Carley J (1998) The Leslie street spit. Friends of the spit website, page About FOS. http://www.friendsofthespit.ca/fos_about.htm. Accessed 15 May 2014

Chapin FS III, Cochran P, Huntington OH et al (2013) Traditional knowledge and wisdom: a guide for understanding and shaping Alaskan social-ecological change. In: Rozzi R, Pickett STA, Palmer C et al (eds) Linking ecology and ethics for a changing world: values, philosophy and action. Springer, Dordrecht, pp 49–62

Courval J (1990) Friends of the spit: the greening of a community-based environmental group. In: Gordon D (ed) Green cities: ecologically sound approaches to urban space. Black Rose Books, Montreal, pp 243–254

Greer J (2003) A life among the watermen. Chesapeake Quart 2(3):3–14

Hardin G (1968) The tragedy of the commons. Science 162:1243–1248

Jacobs J (1961) The death and life of great American cities. Random House, New York

Keiner C (2001) Chesapeake. In: Miller C, Cioc M, Showers K (eds) History in dispute. Water and the environment since 1945: global perspectives, vol 7. St. James Press, Detroit, pp 40–50

Keiner C (2009) The oyster question: scientists, watermen, and the Maryland Chesapeake Bay since 1880. University of Georgia Press, Athens\London

McDonnell MJ, Pickett STA (eds) (1993) Humans as components of ecosystems: the ecology of subtle human effects and populated areas. Springer, New York

National Research Council (1986) Proceedings of the conference on common property resource management, April 21–26, 1985. National Academy Press, Washington, DC

Ostrom E (1990) Governing the commons: the evolution of institutions for collective action. Cambridge University Press, Cambridge

Ostrom E (2010) Beyond markets and states: polycentric governance of complex economic systems. In: Grandin K (ed) Les prix Nobel: the Nobel prizes 2009. Nobel Foundation, Stockholm, pp 408–444

Pickett STA, Burch WR Jr, Dalton SE et al (1997) A conceptual framework for the study of human ecosystems in urban areas. Urban Ecosyst 1:185–199

Poteete A, Janssen MA, Ostrom E (2010) Working together: collective action, the commons, and multiple methods in practice. Princeton University Press, Princeton\Oxford

Rozzi R (2013) Biocultural ethics: from biocultural homogenization toward biocultural conservation. In: Rozzi R, Pickett STA, Palmer C et al (eds) Linking ecology and ethics for a changing world: values, philosophy, and action. Springer, Dordrecht\Heidelberg\New York\London, pp 9–32

Wennersten JR (2001) The Chesapeake: an environmental biography. Maryland Historical Society, Baltimore

Chapter 3
Biogeochemistry and Traditional Ecological Knowledge and Practices in Japan

Hideaki Shibata

Abstract Analysis of biogeochemical dynamics between biotic and abiotic processes through Long-Term Ecological Research (LTER) is essential for understanding relations between humans and nature. Traditional ecological knowledge (TEK) and practices also relate to nutrient and material flows across human and ecological systems. Here, I present current developments in biogeochemical research programs with special attention to recent findings in Japan on human-ecological interactions. I also provide examples of Japanese TEK and practices, and discuss their current problems and future directions. Japanese LTER encompasses a number of ecosystems: 20 core and 36 associate-sites, and uses multiple monitoring and experimental techniques to assess long-term and large-scale dynamics of ecosystem structures, functioning, and biodiversity. In comparison, the scope of TEK in assessing socio-ecological interactions is still limited. However, two examples are introduced in this chapter. The traditional Japanese agricultural landscape, or *Satoyama*, is representative of the application of traditional ecological knowledge and practices for both food production and providing a critical habitat for diverse wildlife through the sustainable use of natural resources and nutrients. The practices of the *Ainu,* who are indigenous to northern Japan, also exemplify an environmental ethics and belief system that respects nature and can be a guiding reference to develop a sustainable future. Evolving more interdisciplinary approaches and recognizing regional and local differences in traditions and cultures will be key challenges for our stewardship of sustainable environments, locally, regionally, and globally.

Keywords Biogeochemical processes • Coupled human and environmental systems • Nitrogen cycling • *Satoyama*

H. Shibata (✉)
Field Science Center for Northern Biosphere, Hokkaido University, Sapporo, Japan
e-mail: shiba@fsc.hokudai.ac.jp

R. Rozzi et al. (eds.), *Earth Stewardship*, Ecology and Ethics 2,
DOI 10.1007/978-3-319-12133-8_3

3.1 Introduction

Biogeochemistry is a scientific discipline concerned with material flows and cycles across ecosystems in relation to various human activities such as food and energy production and consumption, changing land use and land cover, and exploitation of natural resources and products. The activities of human societies have undoubtedly disturbed natural biogeochemical cycles at different spatial and temporal scales, and are continuing to do so now and in the future. The intensity of these disturbances is mostly driven by environmental factors such as geographical patterns of climate, geology, and vegetation, as well as by human factors such as consumer demand, economy, and environment awareness. Human ethics concerning nature also vary across time and space. Moreover, traditional knowledge systems pertaining to nature may also differ from the prevailing system. Therefore, an understanding of the regional specificity of natural ecosystem processes, and of different knowledge systems, is very important. In the next section of this chapter, I review some of the salient literature on current findings of Japanese biogeochemical studies and of traditional ecological knowledge and practices.

3.2 Japanese Long-Term Ecological Research Network

Long-term ecological research (LTER) is a site-based, sound research methodology for understanding ecosystem structures and functioning. LTER is particularly applicable to contexts of significant spatial and temporal variation driven by diverse natural environments and anthropogenic activities. The Japan Long-Term Ecological Research Network (JaLTER) was established in 2006 to facilitate site-based long-term ecological studies in Japan. It was registered as a formal member of the International LTER Network (ILTER) in 2007. There is in fact a long tradition of research on site-based ecosystem ecology in Japan. The International Biological Program in Japan (JIBP), implemented during the 1960s and 1970s, produced many outstanding findings on net primary production and nutrient cycles in a variety of Japanese ecosystems (e.g., Kira et al. 1978; Enoki et al. 2014). However, JIBP was followed mainly by short-term ecological studies conducted by individuals and small research groups that did not sustain a continuous network of long-term field stations by applying JIBP resources and outcomes. A subsequent increase in awareness of various environment issues such as acid rain, global warming, and biodiversity degradation prompted action to establish the LTER network in Japan through collaborations among groups of researchers within universities, governmental research institutes, and other research organizations. As a result, JaLTER was officially endorsed by the Biodiversity Center of Japan within the Ministry of the Environment, while financial resources for its operation, including competitive funds, were supplied by participating members at each JaLTER site. This funding situation is quite different from that of the US-LTER, for which the major funding source is the National Science Foundation.

JaLTER currently has 56 sites divided into 20 core sites and 36 associate sites (Fig. 3.1). A core site is a fully-operational LTER site, whereas an associate site is a LTER site operated for limited focus or under developing process to a core site. Ecosystem types include diverse kinds of forest, paddy fields, grasslands, lakes, freshwater, estuaries, and marine environments. Management organizations also are varied, and include universities' experimental forests and marine laboratories, government research institutes such as the National Institute for Environment Studies and the Forestry and Forest Products Research Institute, and other research organizations. The goals of JaLTER are:

(i) To produce scientific knowledge based on multidisciplinary long-term and large-scale research.
(ii) To create a well-designed database to share and exchange original data to support scientific communities, the general public, and policy makers.
(iii) To find better solutions for critical ecological and environmental problems.
(iv) To promote education regarding long-term and large-scale changes of ecosystems and environments.
(v) To facilitate collaboration and coordination among scientists engaged in long-term ecological research.

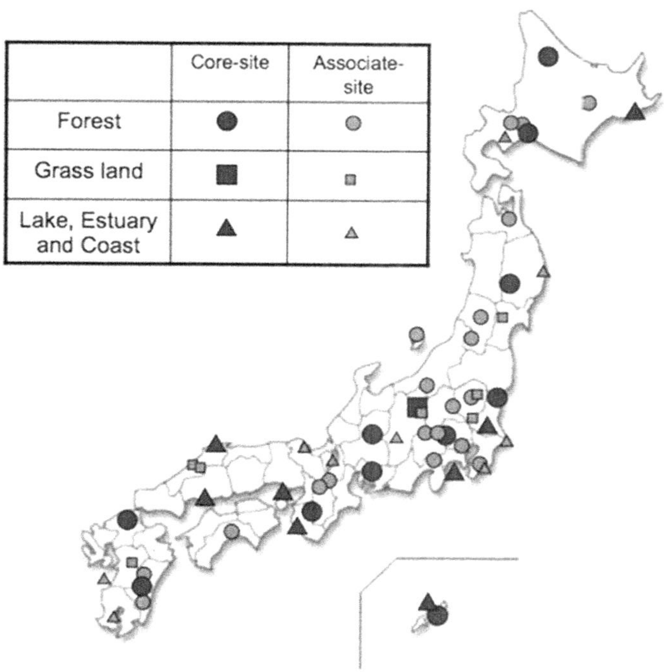

Fig. 3.1 Site distribution of the Japan Long-Term Ecological Research Network (March 2014)

The main research themes addressed within JaLTER include:

1. Responses and feedback on biodiversity and ecosystem functioning under conditions of climate change.
   - Evaluation of regional ecosystem changes caused by warming.
   - Developing an understanding of driving factors of temporal and spatial biodiversity changes in diverse ecosystems.
2. Hydro-biogeochemical processes and interactions within terrestrial and aquatic ecosystems.
   - Clarification of the mechanisms of material cycling changes in ecosystems.
   - Development of a model for conserving and rehabilitating cascading dynamics in watershed ecosystems.
3. Development and establishment of an ecosystem monitoring network and techniques that address multiple scales and dimensions.
   - Validation and parameterization of a photosynthesis model in terrestrial ecosystems using satellite remote sensing.
   - Development of a common environmental monitoring system and an integrated database across various ecosystems.

As listed above, the current JaLTER program primarily focuses on biodiversity, community dynamics, ecosystem ecology, ecophysiology, water/carbon/energy fluxes, biogeophysical dynamics, and other natural science-based topics. Moreover, JaLTER's activities are linked to other research programs and observation networks. These include CO_2 flux networks (AsiaFlux and Japan Flux), biodiversity monitoring networks (the Monitoring Sites 1000 Project funded by Japan's Ministry of the Environment), and collaborative initiatives with satellite remote sensing programs, for example, GCOMC-RA funded by the Japan Aerospace Exploration Agency. However, some research gaps are apparent in the current program, especially regarding geographical and topical site distribution.

Most research sites were established in areas with ecosystems such as native forests, plantations, or marine environments to elucidate natural ecosystem behaviors and processes. Agricultural, urban, and transient (seminatural or disturbed) sites are still quite limited despite the importance of understanding dynamic features in coupled human and environmental systems subjected to natural and anthropogenic disturbances. Typically, long-term measurements of CO_2 fluxes, atmospheric deposition, water discharge, or water quality tend to be mostly concentrated in relatively homogeneous and undisturbed ecosystems rather than in areas with mosaic ecosystems, spatially disturbed landscapes, or temporally transient conditions. More research themes aimed at developing an integrated understanding of ecosystem sustainability, resilience, and vulnerability of coupled social and ecological systems would be needed. Further, the current JaLTER program does not include studies of traditional ecosystem knowledge and processes. Although funding for such network-level projects is currently limited, the importance of conducting studies aimed at developing sustainable stewardship of local, regional, and global ecosystems that are coupled with human societies is steadily growing.

3.3 Biogeochemical Processes Coupled with Anthropogenic Disturbances

Biogeochemistry offers a powerful scientific approach for understanding ecosystem structures, functioning, and their human impacts through the analysis of material dynamics and their driving mechanisms within different environments. In particular, for most biota, nitrogen (N) is a major limiting nutrient within biogeochemical cycles relating to various biotic and abiotic processes and their interactions in and across ecosystems. Moreover, N behavior is closely linked to a number of human activities. For example, excess production and consumption of fossil fuels and food are known to affect ecosystem structures and functioning through several processes and mechanisms (Schlesinger 1997; Chapin et al. 2002; Shibata et al. 2011a). Shibata et al. (2001a) have shown that nitrate concentration in forest streams has increased near the Kanto area in Japan. This finding suggests that these forests have mostly reached N saturation (Ohte et al. 2001) as a result of the atmospheric N pollution produced in the Tokyo metropolitan region. Recently, air pollution (e.g., $PM_{2.5}$) has increased to dangerous levels for human health in China (Zhao et al. 2013), thus raising the risk of ecosystems and environments in the Japanese archipelago being further impacted by long-range transport of N pollutants from the Asian continent.

Japan consists of an archipelago of five main islands (Hokkaido, Honshu, Sikoku, Kishu, and Okinawa) and over 6,800 small islands that varies widely from north to south (20–45°N). This archipelago includes many active and inactive volcanos forming part of the circum-Pacific orogenic zone. Forests cover about 66 % of the total land area of Japan and mostly located in mountainous regions. There are diverse topographies from steep alpine and subalpine mountainous areas to middle-downstream flat plains. The climate belongs mostly to the temperate zone, but has a range from sub-tropical to sub-boreal zones, all of which are affected by the Asian monsoon climate. These natural environmental characteristics strongly influence the diverse structures and functioning of Japanese ecosystems. Regarding forest ecosystem responses to anthropogenic air pollution, Shibata et al. (2001b) conducted a comprehensive analysis of proton (H^+ ion) budgets in Japanese forest ecosystems. Their findings in relation to ambient levels of anthropogenic acidic deposition indicated that relatively young bedrock derived primarily from volcanic rock contributed to a higher acid neutralization capacity through base cation exchanges and chemical weathering in soil.

Timber production is another major form of human disturbance to biogeochemical processes and stream water quality in forested catchments. In northern Hokkaido, which is at Japan's northernmost point, *Sasa* dwarf bamboo species (*Sasa kurilensis* and *S. senanensis*) are predominant forest floor plants that densely cover the ground in most forest ecosystems. They often facilitate slow forest recovery after disturbances caused by timber harvesting by providing shade during tree seed germination (Yoshida et al. 2005). Fukuzawa et al. (2006) found that the presence of *Sasa* dwarf bamboo in this northern region prevented nitrate leaching after the clear-cutting of trees in the whole catchment. This occurred through increased N uptake by *Sasa* dwarf bamboo after the timber harvest, suggesting that certain characteristics of the local vegetation structure provide unique biogeochemical responses to disturbances caused by forestry practices.

The Asian monsoon climate provides unique seasonality of precipitation and temperature. This leads to different temporal responses of Japanese forest catchments to anthropogenic N pollution compared with those noted in the US and European regions (Ohte et al. 2001; Park et al. 2010; Mitchell 2011). In the US, increased nitrate leaching during the growing season is regarded as an early warning indicator of N saturation in forested watersheds (see, e.g., Stoddard 1994). On the contrary, researchers have observed that higher temperatures with higher precipitation during the summers than other seasons produce more nitrate leaching from the soil to streams during growing season in Japanese forested regions. This is especially evident along the eastern side of Japan, even in less polluted environments (Ohte et al. 2001).

Japan's diverse climate pattern also induces different N biogeochemical responses in the soil to winter climate change (Shibata et al. 2013; Makoto et al. 2014; Urakawa et al. 2014). A decrease in snowpack amplifies soil freeze-thaw cycles because of the corresponding decrease in the heat insulation function of snowpack. This leads to an alteration of soil microbial N processes (N mineralization, nitrification, or denitrification) during the dormant season (Shibata et al. 2013; Makoto et al. 2014). Responses of soil microbial N processes to winter climate changes vary in soils found in different locations in Japan, as recently highlighted by Urakawa et al. (2014).

3.4 Traditional Ecological Knowledge and Practices

The characteristic diversity of the Japanese environment in terms of climate, geology, topography, and vegetation is strongly associated with a historical and continuing traditional culture, environmental ethic, and perspectives on nature and religion among the Japanese people. From ancient times, there has been an established traditional religion that fosters respect for diverse natural objects, including both visible and invisible entities, through a belief in nature deities that reside in various natural places such as mountains, forests, lakes, and oceans. This belief of the Japanese influences their fundamental attitudes and ethic toward nature, and forms the basis for many traditional festivals celebrated locally by farmers, fisher folk, and other community groups. Therefore, many nature deities are respected for the benefits from nature that they provide for human well-being. These include use of natural products, exploitation of natural resources, harvests, hunting, averting of natural disasters, or granting of wishes for sound health. Frequent natural hazards that occur in Japan such as volcano eruptions, earthquakes, floods, or tsunamis also form the basis of human awareness and respect for these nature deities. Many traditional Japanese festivals are also centered on the wish to avoid natural hazards, including extreme climatic phenomena such as drought. Therefore, the development of traditional ecological knowledge and practices in Japan has been largely founded on an environmental ethic based on respect for nature.

The *Ainu*, who are indigenous to northern Japan, are an example of a people who have developed traditional ecological knowledge and practices. They originally depended on hunting, fishing, and gathering of natural products for their livelihoods, and evolved traditional ways of enabling these activities to be sustainable (Kojima 2011). However, mainly since the Meiji period in the late nineteenth century, the traditional activities and rights of the Ainu were severely eroded by the Japanese government through mass colonization of their living spaces by populations from southern Japan (Yamada 2011). The *Ainu* revere various wildlife species, such as the brown bear (*Ursus arctos*), Blakiston's fish owl (*Ketupa blakistoni*), dog salmon (*Oncorhynchus keta*), and other species as nature deities or *Kamui*. During their hunting and fishing activities and subsequent treatment of brown bears and salmon, the *Ainu* people handle them respectfully as *Kamui* based on their traditional customs and knowledge (Kojima 2011). These activities also have contributed to the formation of a unique environmental ethic, and to diverse cultural traditions among them. Watanabe (1972) noted the existence of two interrelated systems working at different levels between the *Ainu* and their habitat. The first was an *ecological system* entailing a structure of relationships that humans developed with their habitat through technological activities. The second was a *system of social solidarity* forged between humans and nature through relationships between them based on human beliefs and rituals centering on their habitat. The importance and uniqueness of the *Ainu*, their activities, and knowledge are still topics under discussion because they have been extremely restricted as a result of the vast reclamation (not for *Ainu*) of their habitat by the Japanese government in modern times. Abundant natural resources and a smaller human population size do provide an important background context for the achievement of a sustainable lifestyle among the *Ainu* in the past. However, the *Ainu*'s perspective on nature and their methods of using natural resources could also provide a reference and guide for developing sustainable ecosystem management and stewardship under changing social and environmental conditions in local, regional, and even global contexts (Chapin et al. 2009).

To provide a second example of traditional ecological knowledge and practices, *Satoyama* landscapes are defined as traditional Japanese rural landscapes composed of a mosaic of coupled social and ecological features (Fig. 3.2). These specifically include agricultural land (rice paddy, and crop fields), irrigation ponds, secondary forests, and human settlements (Fukamachi et al. 2001; Ichikawa et al. 2006; Duraiappah et al. 2012). The above is an accepted definition of *Satoyama*, although various other definitions have also been proposed (Kobori and Primack 2003; Katoh et al. 2009; Saito and Shibata 2012). Historically, *Satoyama* landscapes were prevalent in Japanese rural areas, and delivered a bundle of ecosystem functions and services. This mosaic landscape with its combination of land cover categories also contributes foundationally to the Japanese perspective on nature

Traditional irrigation systems within a *Satoyama* landscape, obtained their water supply from the upper forested catchment, which was diverted through irrigation

Fig. 3.2 Example of *Satoyama* landscape representing a mosaic of multiple land-uses such as paddy field, farmland, irrigation, human residences, secondary forest and others (September 2005, Hirai village, Wakayama prefecture, Japan)

canals and reused to cultivate rice in paddy fields. During the cascading flow of this water use, some nutrients (typically N and phosphorus) were removed from the water via plant nutrient uptake, soil microbial immobilization or denitrification in anaerobic soil conditions within paddy fields. This helped to maintain better water quality downstream (see, e.g., Natuhara 2013). Traditional ecological practices in *Satoyama* landscapes entailed the collection of shrubs, brushwood, branches, litter, and undergrowth plants from the surrounding forests (typically secondary deciduous forests) near agricultural land and human settlements for use as domestic fuel and fertilizer for paddy and crop fields. This meant that the flow of nutrients and organic matter from nearby forests to human habitation and agriculture land contributed to sustainable food and energy production in the *Satoyama* landscape. At the same time, the harvesting of these materials from forests resulted in more sunlight availability for germination of tree seeds, thereby preserving biodiversity and avoiding forest coverage by a single dominant species (Fukamachi et al. 1999). The irrigation system and ponds also provided a good habitat for various species of wildlife, insects, and other life forms (Kobori and Primack 2003; Duraiappah et al. 2012).

Historical utilization of natural resources from the surrounding forest for fuel and fertilizer in *Satoyama* landscape also affected the current N cycle and their response to the environmental changes (Shibata et al. 2011b; Urakawa et al. 2014). Past long-term intensive utilization of forest resource in secondary forests around

Kyoto city (previous capital of Japan), western Japan resulted in tight N cycle among plant–soil–microbe system and decreased the inorganic N fertility in surface soil of the current ecosystem (Shibata et al. 2011b). In these ecosystems, the responses of soil N cycle to the environmental change such as winter climate changes were lower than that of other ecosystems (Shibata et al. 2011b; Urakawa et al. 2014). Understandings these path dependencies (legacy effect) of the current ecosystem structure and functioning by the past human impact would help to predict the future directions of the coupled social and ecological system on long-term aspects.

Kadoya and Washitani (2011) have developed what they refer to as the *Satoyama* index (SI) using 1 km by 1 km gridded land cover data on a global scale. They observed that the SI was closely associated with traditional agricultural systems with high conservation values, not only in Japan but also in other countries. This suggests that the landscape form and characteristics of the *Satoyama* system may even be universal for traditionally coupled socio-ecological systems in rural landscapes.

The anthropogenic flow of nutrients and their use across systems occurred not only between the *Satoyama* and their surrounding production systems, but also between urban areas and surrounding agricultural spaces during the Edo period (1603–1867) of early modern Japan. During this period, human waste produced within urban society (Edo city, currently Tokyo) was purchased by farmers, transported to the surrounding agriculture lands, and used as fertilizer to produce food that was then consumed by urban residents (Tajima 2007) (Fig. 3.3). Tajima (2007) has shown how the marketing of human waste as fertilizer became more sophisticated

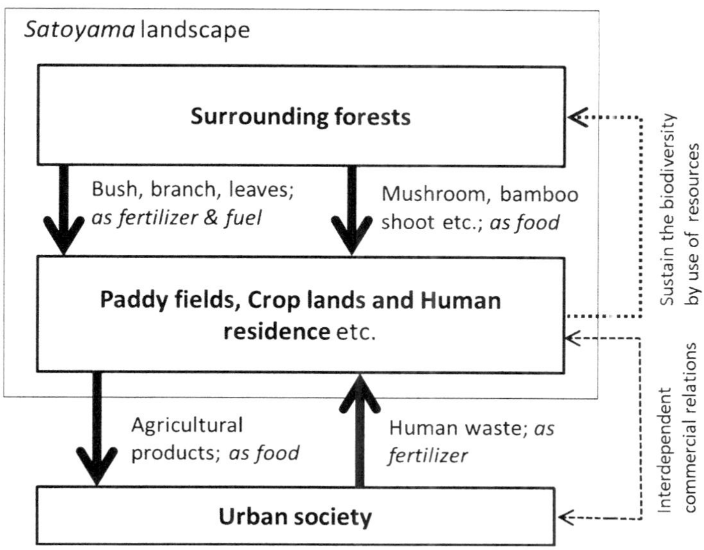

Fig. 3.3 Typical examples of nutrient flows as fertilizer, fuel, and food, and their technological and economic relations across *Satoyama* and urban systems during Japan's early modern period (The early seventeenth century to the late nineteenth century)

with the development of agricultural and transportation technologies during the late nineteenth century. These technologies contributed not only to an autonomous and sustainable flow of nutrients/food through commercial transactions between urban and rural areas, but also helped to prevent pandemic disease outbreaks by removing human excrement from the city (Tajima 2007). Currently, human wastes are not used as fertilizer in Japanese farm land. Readily available alternative fertilizers produced industrially though the Harber–Bosch process and other techniques have succeeded in increasing food production to sustain a rapid growth in the human population. However, the extensive use of chemical fertilizers in agricultural land has also resulted in excess nutrients leaching from the soil to the groundwater, streams, lakes, and estuaries, with a higher associated risk of water eutrophication (Shindo et al. 2009).

More recently, Japanese traditional systems such as *Satoyama* landscapes have become degraded and declined, especially in the aftermath of the fuel and fertilizer revolution during the mid-twentieth century (Duraiappah et al. 2012). Increased supplies of fossil fuels and chemical fertilizer have resulted in lower levels of use of natural resources from surrounding rural environments. Moreover, Japan's rapid industrialization following World War II contributed to extensive migration to urban areas and resulted in serious depopulation and an aging population in rural areas. Consequently, many *Satoyama* systems have become unsustainable or have disappeared (Itoga and Yazawa 1984; Ichikawa et al. 2006; Duraiappah et al. 2012).

3.5 Toward Sustainable Ecosystem Management

Despite the decline of traditional production systems in Japanese rural areas, some promising new perspectives and activities have also recently emerged. Kobori and Primack (2003) have discussed developing a citizen participatory program to conserve traditional *Satoyama* landscapes as critical wildlife habitats and for providing a bundle of ecosystem services. This could be implemented through the ample donations made by urban residents, as well as through active participation in recreating paddy fields with traditional irrigation ponds in rural areas.

The recognition and awareness of citizens of the need to support such initiatives through public funding are growing as a result of a range of public education programs, media information, and other participatory activities (Kobori and Primack 2003; Duraiappah et al. 2012). Key challenges for the scientific community will be to disseminate precise information to the public, not just through scientific papers, but also using different methods of communication. Most natural sciences tend to advance through a narrow and in-depth focus and separation from other disciplines. However, to develop sustainable stewardship and management of nature with public consent and participation, an interdisciplinary approach is necessary. This should entail reference to traditional ecological ethics, knowledge, and practices, and be developed through collaborations among multiple stakeholders, including people and scientists with diverse backgrounds. Recognition of regional and cultural differences in traditional and contemporary ecological knowledge would also help us to co-design and co-produce strategies involving a broad range of communities for embarking on global and regional ecosystem stewardship.

Acknowledgments I gratefully acknowledge financial support obtained from the Environmental Research and Technology Development Fund (S-9-3) of the Ministry of the Environment, Japan. I thank technical staff of Wakayama Experimental Forest, Hokkaido University for providing the photo in Fig. 4.2. This chapter contributes to the synthesis of the Global Land Project (IGBP/IHDP) and the Nitrogen Initiative of International Long-Term Ecological Research Network (ILTER).

References

Chapin FS III, Matson P, Mooney HA (eds) (2002) Principles of terrestrial ecosystem ecology. Springer, New York

Chapin FS III, Kofinas GP, Folke C (eds) (2009) Principles of ecosystem stewardship. Springer, New York

Duraiappah AK, Nakamura K, Takeuchi K et al (eds) (2012) Satoyama-Satoumi Ecosystems and Human Well-Being: socio-ecological production landscapes of Japan. United Nations University Press, Tokyo

Enoki T, Nakashizuka T, Nakano S et al (2014) Progress in the 21st century: a roadmap for the Ecological Society of Japan. Ecol Res 29:357–368

Fukamachi K, Oku H, Shimomura A et al (1999) Relations between forest management methods and ecological features of SATOYAMA beech forests in the Kamiseya and Ikaga districts, Kyoto. J Jpn Inst Lands Archit 62:687–692 (in Japanese with English summary)

Fukamachi K, Oku H, Nakashizuka T (2001) The change of a Satoyama landscape and its causality in Kamiseya, Kyoto Prefecture, Japan between 1970 and 1995. Landsc Ecol 16:703–717

Fukuzawa K, Shibata H, Takagi K et al (2006) Effects of clear-cutting on nitrogen leaching and fine root dynamics in a cool-temperate forested watershed in northern Japan. For Ecol Manag 225:257–261

Ichikawa K, Okubo N, Okubo S et al (2006) Transition of the Satoyama landscape in the urban fringe of the Tokyo metropolitan area from 1880 to 2001. Landsc Urban Plan 78:398–410

Itoga R, Yazawa Y (1984) A study on the reevaluation of traditional environmental conservation system in Japanese rural village. Zoen Zasshi 47:231–236 (in Japanese)

Kadoya T, Washitani I (2011) The Satoyama index: a biodiversity indicator for agricultural landscapes. Agric Ecosyst Environ 140:20–26

Katoh K, Sakai S, Takahashi T (2009) Factors maintaining species diversity in Satoyama, a traditional agricultural landscape of Japan. Biol Conserv 142:1930–1936

Kira T, Ono Y, Hosokawa T (eds) (1978) Biological production in a warm-temperate evergreen oak forest of Japan, vol 18, JIBP synthesis. University of Tokyo Press, Tokyo

Kobori H, Primack RB (2003) Participatory conservation approaches for *Satoyama*, the traditional forest and agricultural landscape of Japan. AMBIO 32:307–311

Kojima K (2011) Natural resource use by Ainu. In: Yumoto T (ed) What's environment history? Bunichi Sogo Shuppan, Tokyo, pp 177–185 (in Japanese)

Makoto K, Kajimoto T, Koyama L et al (2014) Winter climate change in plant–soil systems: summary of recent findings and future perspectives. Ecol Res 29:593–606

Mitchell MJ (2011) Nitrate dynamics of forested watersheds: spatial and temporal patterns in North America, Europe and Japan. J For Res 16:333–340

Natuhara Y (2013) Ecosystem services by paddy fields as substitutes of natural wetlands in Japan. Ecol Eng 56:97–106

Ohte N, Mitchell MJ, Shibata H et al (2001) Comparative evaluation on nitrogen saturation of forest catchments in Japan and North America. Water Air Soil Pollut 130:649–654

Park JH, Duan L, Kim B et al (2010) Potential effects of climate change and variability on watershed biogeochemical processes and water quality in northeast Asia. Environ Int 36:212–225

Saito O, Shibata H (2012) Satoyama and Satoumi, and ecosystem services: a conceptual framework. In: Duraiappah AK, Nakamura K, Takeuchi K et al (eds) Satoyama-Satoumi Ecosystems and Human Well-Being: socio-ecological production landscapes of Japan. United Nations University Press, Tokyo, pp 17–60

Schlesinger WH (1997) Biogeochemistry: an analysis of global change, 2nd edn. Academic, San Diego

Shibata H, Kuraji K, Toda H et al (2001a) Regional comparison of nitrogen export to Japanese forest streams. Sci World 1:572–580

Shibata H, Satoh F, Sasa K et al (2001b) Importance of internal proton production for the proton budget in Japanese forested ecosystems. Water Air Soil Pollut 130:685–690

Shibata H, Toda H, Inagaki Y et al (2011a) Soil nitrogen dynamics of forest ecosystems under environmental changes. J For Res 16:331–332

Shibata H, Urakawa R, Toda H et al (2011b) Changes in nitrogen transformation in forest soil representing the climate gradient of the Japanese archipelago. J For Res 16:374–385

Shibata H, Hasegawa Y, Watanabe T et al (2013) Impact of snowpack decrease on net nitrogen mineralization and nitrification in forest soil of northern Japan. Biogeochemistry 116:69–82

Shindo J, Okamoto K, Kawashima H et al (2009) Nitrogen flow associated with food production and consumption and its effect on water quality in Japan from 1961 to 2005. Soil Sci Plant Nutr 55:532–545

Stoddard JL (1994) Long-term changes in watershed retention of nitrogen: its causes and aquatic consequences. In: Baker LA (ed) Environmental chemistry of lakes and reservoirs. American Chemical Society, Washington, DC, pp 223–284

Tajima K (2007) The marketing of urban human waste in the early modern Edo/Tokyo metropolitan area. Environ Urbain/Urban Environ 1:13–30

Urakawa R, Shibata H, Kuroiwa M et al (2014) Effects of freeze-thaw cycles resulting from winter climate change on soil nitrogen cycling in ten temperate forest ecosystems throughout the Japanese archipelago. Soil Biol Biochem 74:82–94

Watanabe H (1972) The Ainu ecosystem. University of Tokyo Press, Tokyo

Yamada S (2011) Modern Hokkaido and Ainu people. Hokkaido University Press, Sapporo, p 469 (in Japanese)

Yoshida T, Iga Y, Ozawa M et al (2005) Factors influencing early vegetation establishment following soil scarification in a mixed forest in northern Japan. Can J For Res 35:175–188

Zhao PS, Dong F, He D et al (2013) Characteristics of concentrations and chemical compositions for $PM_{2.5}$ in the region of Beijing, Tianjin, and Hebei, China. Atmos Chem Phys 13:4631–4644

Chapter 4
Aesthetic and Moral Appreciation of Nature in Philosophical Traditions of China

Shan Gao

Abstract In Chinese philosophy, nature is viewed as an organic system that is always in a self-generating process of production and reproduction of life. This view of nature is best expressed by the Chinese philosophers as *sheng-sheng-bu-xi.* The metaphysical foundation for this perspective of nature is *ch'i,* a core concept in Chinese philosophy as well as in Chinese everyday culture and worldview. The Chinese aesthetic appreciation of nature is also aesthetic appreciation of *ch'i. Ch'i* has no physical form and is invisible and it is always in an unceasing process of movement which produces and reproduces life. In Chinese philosophical traditions, especially Confucianism and Daoism, these two characteristic of ch'i are aesthetically expressed and appreciated as emptiness and creativity. The Chinese aesthetic appreciation of emptiness and creativity of *ch'i* can be best illustrated in traditional Chinese landscape painting. Nature has certain features or structures which trigger certain emotions in the subject; and the subject whose mind has structures similar to the structures or features of nature projects their cherished values onto nature.

Keywords Biocultural ethics • *ch'i* • Creativity • Earth stewardship • Emptiness • Intercultural • Self-realization

Confronted with global environmental change, inaugurating the second decade of our twenty-first century, the Ecological Society of America (ESA) launched the "Earth Stewardship" initiative (Chapin et al. 2011). However, Earth stewardship requires examination of the terms themselves and the interrelationship between them. These terms and relationships always are culturally embedded. Currently most of the discussion of the Earth stewardship initiative proposed by the ESA is being conducted within the Western scientific paradigm. However, in order to be truly planetary and to broaden the spectrum of worldviews, it is necessary to step outside this paradigm. This demands the incorporation of a biocultural ethic (Rozzi 2012), which better represents and values the biocultural heterogeneity of the many

S. Gao (✉)
Department of Philosophy, School of Politics and Public Management, Soochow University, Suzhou, China
e-mail: sgao@suda.edu.cn

R. Rozzi et al. (eds.), *Earth Stewardship*, Ecology and Ethics 2,
DOI 10.1007/978-3-319-12133-8_4

biomes and peoples of the planet (Rozzi et al. 2012). To better incorporate the diversity of ecological worldviews and cultures into a partnership-based Earth Stewardship initiative, in this chapter I present an aesthetic and a morally embedded appreciation of nature, based on the Chinese philosophical traditions of Confucianism and Daoism.

4.1 The Concept of *ch'i* in Chinese Philosophy and Aesthetic Appreciation

Chinese philosophers often use the word *ch'i* to form other words such as *zhi ch'i* (ambition), *sheng ch'i* (animation), and *kong ch'i* (air). From such usages, we can see that the concept of *ch'i* contains both material and spiritual meanings. *Ch'i* has been translated into English as "vital force," "material force," "material energy," and "spirit." However, like many fundamental Chinese philosophical terms, these translations are still not adequate to capture the rich meanings of *ch'i*. To explore its features will be very helpful for us in trying to understand the meaning of *ch'i*.. *Ch'i* in Chinese philosophy is characterized by emptiness and creativity. The two features are interconnected.

First, *Ch'i* has no physical form and it is invisible. Chinese philosophers often use *xu* to express the concept of *ch'i*. *Xu* is often translated as "void," "emptiness," and "vacuity." For the sake of simplicity, I use the word emptiness to express the idea of *xu* in this chapter. All these translations are misleading because they are easily misunderstood as non-existence. However, *xu* often contrasts with *shi* (fullness). *Shi* has the meaning of being solid, manifest, visible, tangible, and fully realized while *xu* suggests subtle, hollow, invisible, intangible, and unmanifested. Therefore, *xu* doesn't refer to non-being or nothingness. Chang Tsai (1020–1077) writes, "If we realize that the Great Vacuity is identical with material force, we know that there is no such thing as non-being" (Chan 1963, p. 503). The empty feature of *ch'i* makes the interconnection, interpenetration, and resonance among things and space possible. As T'ang Chun-I points out, "Whenever a thing is in intercourse with another, it is always that the thing by means of its void contains the other and apprehends it" (T'ang and Chang 1956, pp. 113–136). For T'ang Chun-I, only through emptiness within the things and space can one thing absorb the other so that the interpenetration can become possible. Since *ch'i* is empty, all concrete things which are made up of *ch'i* have emptiness within them. Therefore, all concrete things are interconnected with the space that surrounds the object. Chang Tsai addressed this interconnection among things and spaces in the following passage:

> If it is argued that all phenomena are but things perceived in the Great Vacuity, then since things and the Vacuity would not be mutually conditioned, since the physical form and the nature of things would be self-contained, and since these, as well as Heaven and man, would not be interdependent, such an argument would fall into the doctrine of the Buddha who taught that mountains, rivers, and the total stretch of land are all subjective illusions. (Chan 1963, p. 502)

According to Chang Tsai, the vacuity and things are interdependent rather than mutually exclusive. Things are not self-contained but can be influenced by the vacuity because things have emptiness within them.

Second, by creativity, I mean that *ch'i* is always in an unceasing process of movement that produces and reproduces life. The creative power in *ch'i* is due to the *yin-yang* principle. This feature of *ch'i* can be seen in the following passages:

> As the Great Vacuity, material force is extensive and vague. Yet it ascends and descends and moves in all ways without every ceasing. (Chan 1963, p. 503)
>
> Material force moves and flows in all directions and in all manners. Its two elements unite and give rise to the concrete. Thus the multiplicity of things and human beings is produced. In their ceaseless successions the two elements of yin and yang constitute the great principle of the universe. (Chan 1963, p. 505)

Because *ch'i* has inexhaustible power within it, it is always in an unceasing process of movement. The moving power of *ch'i* is due to the *yin-yang* principle intrinsic to *ch'i* rather than caused by external intelligence. *Yin* and *yang* are neither material things nor the principles of our mind, but are the opposing modes that are intrinsic to *ch'i*. The combination of *yin-yang* explains why *ch'i* is creative. By *yin* alone or by *yang* alone, creative power in *ch'i* ceases to exist so that things will not be produced and reproduced. The combination of *yin* and *yang* takes many forms that are beyond human comprehension.

In Chinese philosophy, the self-generating movement and transformation of *ch'i* manifests Tao, which is the combination of truth, goodness, and beauty. Confucianism and Daoism have different interpretations of Tao. Chan discussed the difference in one of his comments on *The Doctrine of the Mean*:

> In no other Confucian work is the Way (Tao) given such a central position. This self-directing Way seems to be the same Tao in Taoism. But the difference is great. As Ch'ien Mu has pointed out, when the Taoists talk about Tao as being natural, it means that Tao is void and empty, whereas when Confucianists talk about Tao as being natural, they describe it as sincerity. (Chan 1963, p. 109)

Tao, as the manifestation of the movement and transformation of *ch'I*, is the same in Confucianism and Daoism. The question is: why does Taoism describe Tao as emptiness while Confucianism characterizes Tao as sincerity? My answer is that Confucianism focuses on the creative feature in *ch'i*, while Taoism stresses the emptiness feature in *ch'i*. Sincerity is the translation for the Chinese character *cheng*. The word *cheng* frequently appears in *The Doctrine of the Mean*. It is regarded as the Way (Tao) of heaven (nature). *Cheng* is often described as "ceaseless." The Neo-Confucian scholar Chang Tsai explained *cheng* as "the way according to which heaven can last forever and is unceasing." Obviously, *Cheng* is seen as process. What kind of process is it? The answer can be found in this passage:

> The Way of Heaven and Earth may be completely described in one sentence: They are without any doubleness and so they produce things in an unfathomable way. (Chan 1963, p. 109)

The Chinese characters for "without any doubleness" in this passage are *bu er* which means "sincerity." From here, we can find the answer that *cheng* is a ceaseless process of production and reproduction of things. For this reason, some scholars translate *cheng* as creativity. Tu Wei-Ming explains *cheng* in terms of creativity:

> ...we can conceive of it as a form of creativity...it is that which brings about the transforming and nourishing processes of heaven and earth. As creativity, Ch'eng is "ceaseless" (*pu-hsi* [*buxi*]). Because of its ceaselessness it does not create in a single act beyond the spatiotemporal sequence. Rather, it creates in a continuous and unending process in time and space... it is simultaneously a self-subsistence and self-fulfilling process of creation that produces life unceasingly. (Ames and Hall 2001, p. 35)

From this passage, we can see that *cheng* is the expression of the feature of creativity in *ch'i*. Compared with Confucianism, Daoism emphasizes the emptiness feature of *ch'i*. Lao Tzu and Chuang Tzu often used the word *emptiness* to describe Tao:

> Tao is empty (like a bowl), It may be used but its capacity is never exhausted. It is bottomless, perhaps the ancestor of all things. (Ames and Hall 2001, p. 141)
> How Heaven and Earth are like a bellows! While vacuous, it is never exhausted. When active, it produces even more. (Ames and Hall 2001, p. 141)
> Emptiness, stillness, limpidity, silence, inaction-these are the level of Heaven and earth, the substance of the Way and its virtue. (Chuang 1968, p. 142)

In the above passages, Lao Tzu compares the Tao to a bowl and a bellows to convey the idea that *ch'i* has the inexhaustible and infinite power of producing life. However this power is hidden because of its emptiness.

The Chinese aesthetic appreciation of emptiness and creativity of *ch'i* can be best illustrated in traditional Chinese landscape painting. The important principle of Chinese painting which is recorded in the Ku Hua P'in by Hsieh Ho (479–502) is Spirit Resonance. Spirit is another English translation for the Chinese word *ch'i*. According to this principle, the excellent painting should embody the features of *ch'i*. Chinese painters use some methods to capture those features of *ch'i*. First, as I noted earlier, the invisible *ch'i* is always in an unceasing process of movement which nourishes and sustains all kinds of life. The important method adopted by Chinese painters to capture this feature of *ch'i* is called *liu bai* which means to leave the empty space in the painting. Painters often use fog, clouds and water to indicate empty space, which provides a rhythm and a breath to the overall painting. Doing so conveys the moving and creative power of *ch'i*. Second, based on Chinese philosophy, everything is in flux and change since every concrete thing in nature is made up of *ch'i*. Some Chinese painters draw different heights for mountains and trees to convey the idea of flux and change. The Ming Dynasty painter Chang Tung Chi declared:

> The rising and falling of mountains in the distance conveys a sensation of power. The varying height of the trees in a forest is expressive of feeling. (Tsung 1995, p. 36)

Chinese painters used this method to convey the dynamism and change in things. Other painters such as Ni Zan (1301–1374) painted the trees in a way that makes them sparse and bare so that they have a kind of dynamism to them. When we look at the trees, they appear to be stretching out toward the mountains.

4.2 Intellectual Intuition and the Knowledge of *ch'i*

In the first part, I examined the meaning of *ch'i* from the metaphysical and aesthetic dimensions. In this part, I explore the meaning of *ch'i* from the epistemological dimension by answering the question: how do the Chinese people know *ch'i*? According to the Chinese philosophers, ultimate reality can be known through intellectual intuition. By intellectual intuition, I mean our innate capability to directly grasp the ultimate reality. Compared with the dominant rational way of knowing in Western philosophy, Chinese philosophers depend on intellectual intuition to know things outside of us. Some of them believe that the knowledge we obtain through intellectual intuition is equal to and even superior to what we gain through the senses and reflective mind. Intellectual intuition can be acquired in different ways based on Confucianism and Daoism.

For Confucianists, we can have intellectual intuition if we are persistently cultivating humanity. The Chinese character for humanity is *jen* which is "not only the innermost sensitivity but also an all-pervading care" (Tu 1985, p. 163) Tu Wei-Ming considers the "unfolding of humanity" as the self-disclosure of ultimate reality. The Confucian way of knowing the ultimate reality is best illustrated in the following passage by Chang Tsai.

> By enlarging one's mind, one can enter into all the things in the world [to examine and understand their principle]. As long as anything is not yet entered into, there is still something outside of the mind. The mind of ordinary people is limited to the narrowness of what is seen and what is heard. The sage, however, fully develops his nature and does not allow what is seen or heard to fetter his mind. He regards everything in the world to be his own self. This is why Mencius said that if one exerts his mind to the utmost, he can know nature and Heaven. Heaven is so vast that there is nothing outside of it. Therefore the mind that leaves something outside is not capable of uniting itself with the mind of Heaven. Knowledge obtained through one's moral nature does not originate from seeing or hearing. (Chan 1963, p. 515)

To enlarge one's mind is to fully develop our moral nature, which is characterized by the innate capability to be sensitive to other forms of life. If we fully develop our ability, we will be able to embrace all things in the world in a caring way. In other words, we can identify ourselves with everything in the world. In terms of our fully developed moral nature, we can have intellectual intuition, which enables us to understand the ultimate reality of the world.

In contrast to Confucianism, Daoism stresses emptiness, which is the ideal structure of the mind, which is characterized by no-structure. In other words, emptiness is both a state of mind that is devoid of all the knowledge we get through sense experience and rational argument. With emptiness in the mind and the body, we can directly understand the ultimate reality. Chuang Tzu calls this ability of direct understanding *shen*, which is often translated into English as "spirit." In terms of spirit, T'ang Chun-I (1967, p. 273) gives an excellent definition that it is a function of mind which "meets the things in their changing process with intuitive and sympathetic understanding but without attachments." The belief that emptiness of the mind can enable us to have the intuitive knowledge of the ultimate reality is best demonstrated by Chuang Tzu in the following passage:

> Make your will one! Don't listen with your ears, listen with your mind. No, don't listen with your mind, but listen with your spirit. Listening stops with the ears, the mind stops with recognition, but spirit is empty and waits on all things. The Way gathers in emptiness along. Emptiness is the fasting of the mind. (T'ang 1967, pp. 57–58)

"Make your will one" refers to intense concentration. Ears represent the knowledge we obtain through sense organs while mind refers to the knowledge we get through reason. Spirit is the translation for the Chinese word *ch'i*. The Way (Tao) means the ultimate reality.

Emptiness in this passage not only refers to the psychological state of mind but also the state of body. Body and mind in Chinese philosophy are interconnected. The tranquil state of mind can generate the excellent flow of *ch'i* in our body. Therefore, when our mind does not have self-conscious thought from sense perceptions and reason, our body will be full of *ch'i*. When our mind is empty and full of *ch'i*, we remove our "cognitive flaws" such as "our tendencies to be 'full' of ourselves or 'stuck on' our values and ideals." As the result of emptiness, *shen* is followed which enables our mind to see the ultimate reality of the world. For the Daoist philosopher, the knowledge we get through *shen* is equal to or even superior to knowledge we get through sense perception and rational argumentation. Chuang Tzu (369–286 BCE) states:

> What you can look at and see are forms and colors; what you can listen to and hear are names and sounds. What a pity!-that the men of the world should suppose that form and color, name and sound are sufficient to convey the truth of a thing. It is because in the end they are not sufficient to convey truth that "those who know do not speak, those who speak do not know." But how can the world understand this! (Chuang 1968, p. 152)

For Chuang Tzu, Tao as the ultimate truth cannot be known by words. Words are the result of our sensory perception and understanding. The ultimate truth can only be intuited instead of being known by words. However, Tao can still be known by intellectual intuition. Many Western scholars tend to label this way of knowing as the mystical knowledge that has no truth value. To me, that is a bias and prejudice. We cannot say that everybody can arrive to achieve this state. It requires years of mental training. The founder of Buddhism after 6 years of meditation suddenly got enlightenment one day and he realized that everything is one and in a constant process of change and movement.

4.3 Empathy and the Aesthetic Appreciation of Nature

As I mentioned earlier, the Tao is the value term that represents the unity of truth, goodness, and beauty. Beauty is based on truth and goodness. Confucianism uses sincerity to describe the Tao, while Taoism uses emptiness to do it. Sincerity and emptiness represent the highest beauty. A question may arise here: Why are sincerity and emptiness regarded as the objects of aesthetic appreciation? I argue that it is through empathy. Empathy in the Chinese aesthetic appreciation of nature is characterized by the following two features. The first is that nature has certain features or structures which trigger certain emotions in the subject. Second, the subject whose mind has the similar structures or features with nature projects its cherished values onto nature. Sincerity and emptiness are both cherished values in human life in Confucianism and Daoism, respectively. As the result of empathy, the two objective features of *ch'i* become the objects of aesthetic appreciation. In the following I will examine why sincerity and emptiness are the cherished values in Confucianism and Daoism. I argue that emptiness and sincerity an help people to achieve self-realization.

In terms of emptiness, Chuang Tzu develops the concept. He thinks that emptiness is both a sublime state of mind and a state of body. As a sublime state of mind, emptiness is characterized by the following two features. (1) The subject is unconcerned with fame, reputation, profit, wealth, recognition, and cultural values. (2) The subject has no structure in his or her mind. That is, the subject is empty of all the knowledge we obtain through senses and rationality. As a result of emptiness, the subject has attained a tranquil state of mind, which has no self-conscious thought, deliberation, and application of rigid standards.

For Chuang Tzu, emptiness can help us achieve self-realization in the following two aspects. First, emptiness can help us realize natural endowments so that we can perform daily activities with excellence. This idea is best illustrated by Chuang Tzu's story of a wood carver named Ch'ing. Based on this story, woodworker Ch'ing carved a piece of wood and made a bell stand. Everyone was amazed by his work. When asked whether he has any secret for the creation of such an excellent work, he replied that the "secret" to making an excellent bell stand lies in the preservation of *ch'i* in his body. The way to do so is to cultivate a tranquil state of mind through emptying the mind of all elements of distractions, including praise, blame, congratulations, rewards, and even the body. When his mind is completely empty of all those elements of distractions, the *ch'i* can be well preserved which brings out the excellent work. From here, we can see that *ch'i* is both a state of mind and also a state of body.

Second, emptiness can help us realize the biological powers that can relate us to the vitality of nature. By the biological power, I refer to the excellent flow of *ch'i* within one's body, which renews and vitalizes one's body. Chuang Tzu states:

> If the gentleman can in truth keep from rending apart his five vital organs, from tearing out his eyesight and hearing, then he will command corps-like stillness and dragon vision, the silence of deep pools and the voice of thunder. His spirit will move in the train of Heaven, gentle and easy in inaction, and the ten thousand things will be dust on the wind. (Chuang 1968, p. 116)

The five vital organs are a concept of traditional Chinese medicine. They are believed to be the main dwelling house of *ch'i*. If our mind is not distracted by external things, we become very tranquil from the outside. However, *ch'i* has an excellent circulation in our body, which is accompanied by *shen* (spirit).

Chuang Tzu best illustrates the relationship between emptiness of the mind and its realization of biological power by using the opposite words *silence* and *thunder*.

Like emptiness, sincerity is also the cherished value in Confucianism. It can help people achieve self-realization. For the Chinese philosophers, sincerity is a process of transformation and nourishment. This process can help Confucians achieve self-realization through fully developing emotions. The defining nature in Confucianism is emotions. In terms of emotion, I will examine three questions. (1) What is emotion? (2) Why is emotion considered as a valued element of mind? (3) What is the highest development of emotion?

First, emotion in Confucianism mainly refers to our innate feelings to show sympathetic bonding with other human beings and other creatures. The natural affections toward our parents and brothers and sisters are the defining nature of a human being. It gives human beings dignity. Confucianism has a high respect for this affection. Second, for Confucianism, our sympathetic bonding toward parents and others needs to be developed so that it can be extended to the parents and children in other people's families. However, it is still not the fulfillment of emotion. For Mencius, we also need to extend these affections to animals, trees, and everything in nature. He articulates his ideal in the following passage:

> In regard to inferior creatures, the superior man is kind to them, but not loving. In regard to people generally, he is loving to them, but not affectionate. He is affectionate to his parents. And loving disposed to people generally. He is loving disposed to people generally, and kind to creatures. (Chan 1963, pp. 80–81)

This passage shows Mencius' understanding of love is graded instead of being impartial. For him, we can be affectionate toward our parents, but it would be wrong if we show much more affection to strangers and other creatures. However, love and kindness are still based on the natural affections we have toward our parents instead of reason. It is the extension of familial love. The neo-Confucians Chuang Tsai expresses a similar view in the famous Confucian ideal, "Forming one body with nature."

> Heaven is my father and Earth is my mother, and even such a small creature as I finds an intimate place in their midst. Therefore that which fills the universe I regard as my body and that which directs the universe I consider as my nature. All people are my brothers and sisters, and all things are my companions. (Chan 1963, p. 497)

In this passage, Chang Tsai identifies himself with the entire universe. His identification is based on affectionate belongings that can be seen from his analogy of parents, brothers, and companions. The emotions of love toward all people and creatures in the entire universe are based on the metaphysical foundation of nature, which I have discussed above. Based on the concept of *ch'i*, we know that everything in nature is full of vitality and life. Therefore, the emotions of love for all things in nature are also the emotions of love of life. Our capability to emotionally identify ourselves with all things in nature represents the highest development of emotion, which is also the state of self-realization, which is called *Jen* in Confucianism. For Confucianism, achieving the highest development of emotion is a long transformation process, which is expressed in Confucianism as *cheng*.

4.4 Chinese Nature Aesthetics and Its Contribution to Biocultural Ethic

Ricardo Rozzi (2013, p. 10) has proposed a biocultural ethic that highlights that "many cultures – including, Buddhist, and some Western philosophical traditions – have ecological worldviews that recognize humans, plants, waters, and other beings as co-inhabitants." Furthermore, he stresses that:

> A biocultural ethic demands an inter-cultural dialogue. The global environmental change we face today is caused by particular agents (social groups, corporations, individuals), not by the species, *Homo sapiens*, in general. Unsustainable practices that are detrimental to the life of other human and other-than-human beings need to be sanctioned and/or remedied. Complementarily, in the context of global socio-environmental change, the worldviews, forms of knowledge, values, and ecological practices of cultures that are sustainable should be respected, and eventually adopted through intercultural exchanges. (Rozzi 2013, p. 10)

My brief analysis of the aesthetic appreciation of nature based on the philosophical traditions of Confucianism and Daoism aims to contribute to the two points highlighted by biocultural ethics: an inter-cultural dialogue, and the consideration of worldviews, forms of knowledge, values, and ecological practices of cultures that cultivate sustainable ways of life. Regarding the first point, my analysis helps to better understand the difficulties of inter-cultural dialogue and of translation among philosophical traditions; even among Chinese philosophies. The understanding about intercultural translation raises a note of caution regarding statements that have been used by the ESA's Earth Stewardship Initiative to affirm that "the concept of stewardship is familiar to the general public and has essentially the same meaning in lay terms as we intend in its scientific usage" (Chapin et al. 2011, p. 45). Stewardship can acquire different meanings in different Western and non-Western cultures (see May Jr, Chap. 7 in this volume). In my analysis the complementary Daoist and Confucianist focuses on sincerity nd emptiness regarding the notion of *ch'i* show the value of diversity of meanings, at the same time that the requisite of precision in the use of terms, for a genuine inter-cultural dialogue and an effective implementation of Earth Stewardship initiative.

Regarding the second point the consideration of the diversity of worldviews, values, and sustainable ecological practices of heterogeneous cultures, my analysis of the Chinese philosophical traditions of Confucianism and Daoism concurs with analyses conducted in South America that highlight the fundamental need to incorporate non-economic values, including spiritual and ethical values, into an Earth Stewardship initiative (Rozzi et al. 2012, p. 233). As mentioned earlier in this chapter, the concept of *ch'i* involves aesthetic, spiritual, and ethical values. To implement an intercultural Earth Stewardship initiative, Chinese philosophical traditions offer words such *shen*, notions of emptiness and empathy, and practices such as intellectual intuition that contribute to make steps toward a planetary partnership in this valuable initiative for the sustainable life in our shared planet.

References

Ames RT, Hall DL (2001) Focusing the familiar: a translation and philosophical interpretation of the Zhongyong. University of Hawaii Press, Honolulu

Chan W-T (1963) A source book in Chinese philosophy. Princeton University Press, Princeton

Chapin FS, Pickett STA, Power ME et al (2011) Earth stewardship: a strategy for social-ecological transformation to reverse planetary degradation. J Environ Stud Sci 1:44–53

Chuang T (1968) The complete works of Chuang Tzu (trans: Watson B). Columbia University Press, New York

May Jr RH (2015) Andean Llamas and earth stewardship. In: Rozzi R, Chapin FS III, Callicott JB et al (eds) Earth stewardship: linking ecology and ethics in theory and practice. Springer, Dordrecht, pp 77–86

Rozzi R (2012) Biocultural ethics: the vital links between the inhabitants, their habits and regional habitats. Environ Ethics 34:27–50

Rozzi R (2013) Biocultural ethics: from biocultural homogenization toward biocultural conservation. In: Rozzi R, Pickett STA, Palmer C, Armesto JJ, Callicott JB (eds) Linking ecology and ethics for a changing world: values, philosophy, and action, vol 1. Springer, Dordrecht, pp 9–32

Rozzi R, Armesto JJ, Gutiérrez J et al (2012) Integrating ecology and environmental ethics: Earth stewardship in the southern end of the Americas. BioScience 62(3):226–236

T'ang C-i (1967) The individual and the world in Chinese methodology. In: Moore CA (ed) The Chinese mind: essentials of Chinese philosophy and culture. University Press of Hawaii, Honolulu, pp 57–58

T'ang C-i, Chang T (1956) Theory of mind and its metaphysical basis. Philos East West 6(2):113–136

Tsung P (1995) Space consciousness in Chinese art. In: Liyuan Zhu, Blocker Gene (eds) Contemporary Chinese aesthetics. Peter Lang Publishing, New York, p 36

Tu W-M (1985) Confucian thought: selfhood as creative transformation. State University of New York Press, Albany

Chapter 5
The Antlers of a Trilemma: Rediscovering Andean Sacred Sites

Fausto O. Sarmiento

Abstract The zoomorphic metaphor of deer anatomy explains Andean identity as a coupled environmental system. This is a result of mystic realism or magic pragmatism, which often obscures participation of the local cultures of the Andes cordillera, particularly in (re)defining their Andean self with strong biocultural anchors. Just like the antlers, the trilemma of Andean identity exemplifies the need for a deeper understanding of the stewardship of ecological processes that has been molded to fit geographical and cultural demands of ancestral societies. Quechua traditional ecological knowledge serves as guiding principles to define and implement sacred sites in the region that cherishes its heritage landscapes. (*Kichwa* is the phonetic writing of 'Quechua' (in Peru) or 'Quichua' (in Ecuador, Bolivia, Chile, and Argentina), the language of the Andean people (*runa simi*). I avoid the hegemony of Castilianized words, as I support the recovery of local identity and the invigoration of vernacular culture, including the use of the non-written language of the *Inka*. In this text, I use *italics* to highlight the phonetic *Kichwa* alphabet, while Spanish terms appear inside single quotation marks for emphasis. Scientific names are also italicized.) The relationship of the triangular representation of cultural identity, associated with the binary concept of opposite values or *Yanantin*, and the driver that accentuates spiritual dimension or *Masintin*, the development of ritualized practices observing natural phenomena creates the wholeness among Andeanity, Andeaness, and Andeanitude. By explaining the syncretism observed in contemporary societies of the Andes Mountains, the creation, (re)creation and (pro)creation of harmonious implications between people and the environment are realized. Finally, to assess actual and potential contributions of the discourse in the sacred narrative of biodiversity conservation for Earth Stewardship, several sacred sites exemplify the application of the new trend for biocultural heritage as the driver for cultural landscape management and sustainability scenarios in the Andes.

Keywords Andes • Heritage landscapes • Identity • Sarmiento's trilemma • Sacred site conservation • Traditional ecological knowledge

F.O. Sarmiento (✉)
Neotropical Montology Collaboratory, University of Georgia, Athens, GA, USA
e-mail: fsarmien@uga.edu

R. Rozzi et al. (eds.), *Earth Stewardship*, Ecology and Ethics 2,
DOI 10.1007/978-3-319-12133-8_5

> "*mallki, wak'a, apukuna, pakarina, machay, ushnu, tumi, waman, waylaka, yanantin, tinku, takanakuy, masintin, warachikuy, raymi, chakana, wiphala, tupak.*
>
> Too much to say, to few words to use. I rather vow to silence and meditate…"
>
> Don Segundo, *Utawalu runa yachak* (2013)

5.1 Introduction

Much has been said about Andean identity linked to the mixture of Native and non-Native members inhabiting one of the longest stretch of mountainous landscapes in the world, as if the appropriation of space warrants a homogenization of their many cultural manifestations through a plethora of linguistic, artistic, social, economic, ecological, and belief systems (Bordsdorf and Stadel 2014). There is, of course, primacy in linking environmental ethics with biodiversity conservation in indigenous societies as a way to reverse planetary degradation (Chapin et al. 2011; Rozzi et al. 2012). Indeed, many of the original people of the Americas have been portrayed as custodians of one of the richest hotspots of life in the planet. This requires a long-term commitment to Earth Stewardship grafted onto the mountain landscapes of South America (Rozzi 2012). I posit that Andean identities should be constructed, represented, claimed, and contested with care, using different lenses. One way to aid in building identity discourses is offered by the incorporation of traditional ecological knowledge (TEK) in the understanding of landscape dynamics, focusing on the intersection of nature, science, politics, and religion (Gudynas 2011). This approach makes explicit the sacred geographies of Andean peoples, and both will inform and contest argumentation towards the application of a bodily metaphor (Bastien 1978). This metaphor uses a shamanic symbol, the 'venado' or whitetail deer of the Andes (*Odocoyleus virginianus tropicalis*), one of forty subspecies listed for the Virginia deer, that is often found in mountain landscapes of Colombian, Ecuadorian, and Peruvian tropandean ecoregions. Just like the 'tarugo' (*Hippocamelus antisensis*) of the Central Andes or the 'huemul' (*Hippocamelus bisulcus*) of the Southern Andes, whose bucks or stags exhibit pointed horns that, on a yearly basis, bifurcate or trifurcate according to age, the 'venado' provides the analogy of its antlers as representing dendritic patterns borne of rhizomic processes of yearly renewal and constancy. The metaphor of deer antlers helps to understand how one unit of entity—such as the Andes—could encapsulate the notion of three variant components—such as Andeanity, Andeaness and Andeanitude, in the renewal cycle of the trilogy that explain 'the Andean' based on body, mind, and spirit (Sarmiento 2013a). The so-called Sarmiento's trilemma fusses three streams of identity in one (Fig. 5.1). It is used to accentuate the incorporation of the sacred into the realization of knowledge and practices of biocultural heritage worth protecting for sustainability scenarios (Sarmiento 2012). In this chapter I seek to engage coupled dynamics of socio-ecological systems (SES) methods with reified variants of fluid ethnic identities (*sensu* Cánepa 2008) that collectively represent the sacred geographies of the Andes, with special emphasis on the revival of Amerindian

Fig. 5.1 Sarmiento's Trilemma is the representation of the bodily metaphor of Body, Mind, and Spirit. It is applied to understanding Andean identity by the dynamic interaction of Yanantin, the complementary dualism of the physical and the psychological, with the influence of Masantin, the enthusiastic synergy of the spiritual. These interactions determine that Andean identity be negotiated between *Andeanity* for physical, *Andeaness* for psychological, and *Andeanitude* for spiritual realms. The background image depicts the Andean lapwing (*Vanellus respiendes*), an endemic Andean species that carries an intricate biocultural heritage (Photo: Fausto Sarmiento)

traditions and the incorporation of sacred sites conservation concerned with heritage cultural landscapes (Cachiguango et al. 2001; Sarmiento 2013b).

Cultural landscape narratives reflect increased participation of local and indigenous communities associated with a protected area of national or international importance, particularly considered emblematic as world heritage sites for listing within the Heritage Convention (Rössler 2006). Landscape itself is a term increasingly used to describe a manageable unit of the SES that integrates art, science, economy, religion, and philosophy into a concrete, tangible element of regional planning and sustainable development. The incidence of ecology on society and vice-versa (Odum and Sarmiento 1998) has prompted a (re)formulation of conservation territories and a (re)conceptualization of political ecology as the explanatory of farmscape transformation in the larger cordillera of the Andes. This about-face of nature/culture divide has, in reality, helped to include landscape and inscape in the arsenal of deep ecologists at the end of the twentieth century (Naess 1995) and landscape ecologists at the beginning of the twenty-first century (Naveh et al. 2002). Moreover, it is part of the lingo of today's environmental ethicists (Rozzi 2013).

5.2 Detaching Deer Discourses

> … the deer travels around the range looking for food…The snow is deep but he goes from place to place singing. They sing all things in the song; name every place he stops, everything that he eats…. (Isabel Kelly 1965, p. 67)

The deer continues to be an animal of multiple uses in the Andes, not only for its venison for food and hides for clothing provided as material goods, but also for the mythology associated with the characteristic behaviors of the animal, with ethograms that motivate ethnoecological insight. In Mesoamerican cultures, a deer day had been included in the calendar, and throughout the Andes, a deer dance is typically associated with the fertility and vivacity of young lovers (Harrison 1989). Bingham and Roberts (2010) point to the swiftness of the deer, as well as to its shy and aloof nature, to connect the animal with visualization of hallucinogenic revelations in shamanistic practices of cleansing. Often when in concentration or trance, respectively, most erudite 'maestros' (*amautakuna*) or teachers of *Pisak,* near *Kutsku* in Peru, and wise 'curanderos' or medicine men and women (*yachaqkuna*) of the *Iluman* region of the *Imbabura* watershed in Ecuador, would use the bodily metaphor of bucks while explaining the renewal cycle of life (*pachakutic*) or would mimic the witty nature of stags while performing ceremonial cleansing. The condition of being quick and somewhat apart from the scene when needed is attributed to the essence of a buck's behavior revered in ritualized bathing with wild plants and flowers. Some *Yachak* still retain the traditional antler sign hanging by the doorway, but most keep the seeds of the deer's eye climbing shrub (*Mucuna pruriens*), along with obsidian, candles, liquor, and cigarettes, readily available on their tool tables for conjuring the future.

Behavior mimicry for the layperson is only the visible manifestation of the common symbolism of the deer. The fact that mature antlers from large stags are cherished for the deeper symbolism of the triad that belongs to the same entity, that keeps renewing itself year after year, is important to the initiated. It represents a thought closely related to the Catholic view of henotheism, whereby the holy trinity associates three different gods into one (the bodily lamb, Christ; the idea, Father; and the spirit, the Holy Ghost). This is interpreted as an approximation to much older monolatry philosophies from Egyptian and Sumerian faiths that linked this tripartite entity to the pyramidal shape with a resplendent eye to illuminate not only the mental map, but also the physical landscape and the intimate corners of the soul.

Andeanists (Urton 1988; Webb 2012) believe that traditional Andean societies possibly have been influenced by bilatry, represented by the *Inka* mythology of *Yanantin*: the belief of complementary dualism, now conceived as binaries of simple thought, such as light/darkness, male/female, inner/outer, in/out or good/bad, as the expression of Andean harmony (Platt 1986). Furthermore, in seeking to harmonize opposites, the notion of *Masintin* is key to understanding that the process of finding harmony is important in making the complementary dualism work. Imperfect relations lack *Masintin*; despite *Yanantin* being clearly present, the difficulty to create, (re)create or (pro)create harmonically, makes it necessary to have *Masintin,* so the pairs become coupled in a perfect fit or *Ayni* (Webb 2012).

Geographers and other Andeanist scholars have approached religion as causality of Andean identity and also have pointed to the important contribution of miscegenation and syncretic rites (Cachiguango et al. 2001; Borsdorf, In Press). Anthropologists of different disciplinary lineages (Murra et al. 1986) have explored the notion that the ability to pair opposites requires admitting the presence of in-betweens. A few references exist to this Andean trait from the material world. The *Waylaka* is half man, half woman. *Waylaka* represents more than the ability of a transvestite to dress like a *Chinchiru* woman who behaves like a strong young male from the *Kutsku* area, because *Waylaka* stumbles upon the plaza waving a white flag on a pole and places it on top of a *mujun*, a pile of dirt setting the boundaries of the in-betweens. Another important example comes from the *mallki*, the mummified body of ancestors that are kept in people's dwellings. They are in-between the dead and the alive, forming a triad that explains the Andean philosophy of space and time (*pacha*), having the past in front (where it can be seen) and the future behind (where it cannot be seen). The mummy (*mallki*) is the personification of animistic guidance given from the other world to followers in this one. For example, former emperors (*Sapa Inka*) were always mummified and kept in the main temple of the capital city (*kurikancha*) or in the most important places (*wak'a*) located on mountaintops, on top of pyramidal structures (*ushnu*), or on the main corner of fortress buildings (*machay*), often in trapezoidal niches in the wall, emulating the shape of *Inka* windows. On ceremonial mountain festival days, these mummies were paraded in the streets for prayers and other celebrations around Andean townships. When foreigners arrived, they were taken to the 'room of knowledge' or *rixsiwasi*, with the goal of consulting the *mallki* who became an oracle, foretelling the reaction of the community towards the newcomers. Often decorated with jewelry, fine clothing, pottery full of fermented corn drink (*chicha*), and food grains (*tarwi, kinwa*), the decorated *mallki* was placed in the most important room of the house (*pacarina*). This is a place of adoration and reverence, where heirlooms and cherished llama figurines, feathers, and other elements considered of value for the afterlife, such as coca leaves (*kuka*) and lime, are guarded. Having the in-betweens, therefore, becomes a fact of life, just like zombie categories that are taken from time to time as needed to fit current paradigms of development (Gudynas 2011). Thus, the ability to break the dualism by having a spiritual in-between, helped to translate the Catholic tradition of the sacred Trinity to understanding the essence of being Andean, reflecting a negotiation between the body and the mind as *Yanantin*, negotiated by the spirit as *Masintin* (Webb 2013). For instance, in colonial times, religious processions by Catholic priests and non-native pious parishioners, gained popularity and fervor amongst the natives; their tradition of parading mummies, made religious processions an important bridge of syncretism. Non-natives and natives alike walked the streets chanting to Roman Catholic gods, saints, and colonial statues of the virgin Mary, while the deeper connection was observed by the indigenes in the longing for their Andean oracles *mallki, pachamama,* and *intitayta* with substitute representations of *Masintin* expressed in statues or paintings of the 'Escuela Quiteña' taken from the altars onto the streets (Vásquez 1998).

Effective pairing of material elements requires the acceptance of immaterial world forces that explain the ultimate pairing process. I posit here that syncretism has pushed the idea of the Trinity further into the triad created by the poles of *Yanantin* with the catalysis of *Masintin,* and that the effective interweaving of Catholic and pagan imageries of saints and semi-gods reflect syncretism in objects of adoration. I also argue that this attribute of a spiritual dimension bringing coalescence to dualism also explains the easiness of imbuing Catholic myth into the Andean cosmological vision of *Masintin* and the facility with which thousands of indigenous souls were easily converted into the Catholic faith. Thus far, it has been problematic to link the exotic white dove (*Streptopelia risoria)* or the white lamb (*Ovis aries*) as symbols of the Holy Spirit to the Andean worldview. The substitution of the lamb by the white alpaca (*vicugna pacos*) conveyed a strong sense of classism, because white alpacas were solely owned by the royal class or 'orejones' of the *Inka* elite and were used exclusively by the higher courtesans; nevertheless, it was easy to relate the Holy Spirit to the deer. Despite the fact that animistic religions have tended to confer human qualities to animal behaviors, the deconstruction of deer discourses makes it possible to understand the symbolism of the antlers of 'venado' (*Odocoyleus virginianus tropicalis*), the feet of the mule deer or *suchi* (*Masama rufina*), the cries of the dwarf deer or *pudu* (*Pudu mephistopheles*), the white puff of the flagging deer's tail, the seeds of the deer's eye shrub, and the fetuses of *llamas* (*Lama glama*), as evocative of *Masintin*; hence, the spiritual dimension. Thus, this trifecta allows the *Yanantin* dilemma of opposites between mind and body to be interpreted not only dialectically, but also with the triangular Sarmiento's trilemma that fits the trialectic of Andean identity making.

The solving of the trilemma requires that either antagonistic dualism of war (*takanakuy*) or complementary dualism of dance (*tinkuy*) be experienced (Webb 2013); but these would not be successful without the *Masintin* that prompts the trifecta. When the appropriate balance is found, peaceful coexistence or harmony between opposites is often called *Ayni*, or the 'Andean Way.' It is very important to begin work or other important tasks with a 'despacho' ceremony or *Yachay*, a kind of blessing of offerings (*pagapu*) when starting a job, to bring the female (*pachamama*) and the male (*intitayta*) into harmony. If the *ayni* is achieved in personal life, the realization of a balanced existence of the entire family (*kawsay*) is plausible. However if the entire *Ayllu,* or even the whole nation, reaches *Ayni* in all important facets of life, then the greater good life (*Sumak Kawsay*) is achieved. Current narratives of sustainable development include the ideography of *sumak kawsay* as the target of progress, the way to achieve the 'Buen Vivir' into a sustainable future (Gudynas and Acosta 2012).

5.3 Mountain Myths

> All mountain landscapes hold stories: the ones we read, the ones we dream, and the ones we create. (Michael Kennedy 2010, p. 1)

There are several elements that form part of the Andean worldview. Most of them are built from the ancestral chaos that was ordered by *Wairakucha*, the most important figure of the Andean pantheon. It was 'Viracocha' who created the sun (*inti*) venerated by the *Inka* as the primordial source of energy, the moon (*killa*) and the stars (*chaska*). In another vein, the *Wairakucha*, translated as the 'murmur of the water', or the 'sacred breath' of the sea, lake or spring, reaches the mountain air that mixes elements of the environment in their adiabatic forcing, in an orderly sequence to end chaos between fire, earth, water, and air (Kusch 1962). Northern Chilean folk often associate the images of clouds shrouding the denudated slopes of the Atacama (*Kamancha'ka*) as the perpetual dialogue of the elements, as if the murmur of the water keeps the 'Lomas' vegetation of the dry coastal ecosystems working in harmony. The upward flow of thermal winds, for instance, is often a good omen that will allow the coming of iconic birds, such as *kundur (Vultur gryphus), warru (Geranoaetus melanoleucus),* and *karak'ara (Phalcoboenus carunculatus),* three of the most important mystic avian epic semigods. This is contrary to the bad omen of katabatic, downslope winds that bring shrouding clouds rushing into the brooks, ravines, or even valleys during the night, allowing evil creatures (*tutasupay, sacharuna* or *macha'chig*) to wander around the mountains.

Another component of the Andean cosmological vision relates to the effect of incoming rainfall and the accompanying lightning strikes from the cumulous clouds of the tropical circulation. I have argued elsewhere (Sarmiento 2002) that the lack of scientific reference to natural fire in the Andes is striking, and possibly explained by the fact that the majority of electrical storms are cloud-to-cloud. However, the presence of fulgurite on some mountain tops evidence that cloud-to-ground episodes are indeed possible. Yet, no reported natural fire from a lightning bolt (*illap'u*) hitting the forest exists, hence my argument for the anthropogenic driver of landscape change in the highlands remains (Sarmiento 2000). Newcomers to a mountain site are often exposed to unexpected rain, landslide, flood, or equipment failure if they have not been harmonized with *pagapu,* the mountain god. To avoid upsetting this telluric force of the *Apu*, it is advisable to maintain a respectful acceptance of this bad luck as a mystic chastisement (*Yahshish'ka*) for not having observed the ritual.

The concept of gold ore (*kuri*) as the 'sweat of the sun' reflects the hard work required for getting the mineral from placers and other mining sites. In the same vein, the concept of silver ore (*kulk'i*) as the 'tears of the moon' found in deep caves and other mining sites of the mountains (*kuya*), has been utilized as a metaphor for Andean identity, due to the fact that mineral richness was the motivating factor of European colonization of the region that caused much effort and pain (sweat and tears) on the part of the indigenes (Pizzey 1988).

In the Andes, the idea of the existence of sacred sites has been articulated best by the presence of elements associated with origins, mainly mountains (*urku*), water (*kucha*), or caves (*machay*). The spirit of the mountain (*apu*) requires payments or offerings to calm risks or to pacify conflicts: this occurs regularly, such as by the suicidal phenomenon of the Andean lapwings (*Vanellus resplendens*) diving into some lakes, or the rare nesting of the Andean hillstar hummingbirds (*Oreotrochilus estella*) inside some caves; these are considered to be offerings to *Katekil* (lake god) or *Micakil* (cave god). However, in extreme cases extemporaneous offerings (*kayay*) are required; these include positioning household figurines into a room of the house that ritually faces the dawn (*pakari*). These also include building structures associated with the solstice/equinox cycles. Their purpose is to capture the influence of the sun (*intiwatu*) on the apex of the ceremonial buildings (*inkapirka*), or on the secluded plaza within the stone forts (*pukara*) that protected isolated villages or resting settlements (*tambu*) along the network of trails (*chakiñan*), wider roads (*chaupiñan*), and more formal paved ways to market places (*katuñan*), or the long imperial *Inka* mountain highway (*kapakñan*), reportedly crossing from southern Colombia to Central Western Argentina (Sarmiento 2003; Sarmiento and Hidalgo 1999).

High impact natural events, such as earthquakes, volcanic eruptions, or episodic drought resulting in heavy damage or imminent mortal danger, required extraordinary offerings (*wakayay*) to placate divine wrath, including animal sacrifices, or even human payments of female child priestesses who were trained from an early age for this very purpose (*kapakucha*). Infanticide by child sacrifice is a widespread practice dating from antiquity, not only in the Andes, but also in Mesoamerican and Caribbean cultures. Child sacrifice, as illustrated by 'Juanita'—the ice maiden found on top of Mt. Ampato, who was killed with a blow to the head and left on the summit, was restricted to being a major offering to the mountain spirits (Ceruti, in press). Other ice maidens now have been discovered, along with remnants of traditional offerings of *inka* mummies and sacred figurines, textiles, feathers, and pottery (*kapakuchina*), on top of sacred mountains along the cordillera.

5.4 Telluric Tutelage

> The souls of the ancestors come to dwell on their summits, and people fear the volcanic eruptions, storms, and droughts with which the mountain spirits punish ritual omissions and transgressions. (Constanza Ceruti In press)

As a representation of the spirit of the mountain that allows elemental harmony, the use of mountain spirits permits the notion of *Apu* as the telluric presence that guards the fate of the people living in their vicinity. A mountain could become an *Apu* exerting esoteric influence on each of the different sites, such as the *Utawalu* and the *Kayampi,* both of which are *Kichwa* ethnic groups living on different slopes of Mt. Imbabura, and settled on opposite shores of *Imbakucha* lake in Ecuador (Sarmiento et al. 2008). *Apu* is the god of safety, the protector of the kinship (*ayllu*), or the guardian of the good fortune of the valley's inhabitants. Each mountain edifice (*urku*) is represented by specific textiles woven as belts (*watu*), with pictograms of the main characteristic of the area, i.e., figures of animals, plants, or landscape features, a vivid tradition still kept by the women of *Chinchiru* and around the different villages of *Pisak*, the sacred valley of the *Inka*. The members of the *Ayllu* are compelled to use the same design in clothing and garb, such as the hat worn by young walkers (*chullu*) on mountain hikes, or the hat worn by senior officials or married women (*k'umpu*) to the market place, plaza festivities, and for bartering (*tiangix*), or the hat worn by single young women (*iñaka*) as a flat cover on their heads signaling their prompt initiation into adulthood. With the influence of the modern world, even fedora hats are now used as cultural markers for some groups that show a much more urbanized flair and networking abilities, such as the market-oriented (*mindala*), globe-troting 'Otavalos' (*Utawalukuna*) of northern Ecuador. However, a more elaborated physical ritual is the initiation rite of young males (*warachicuy*) that includes an exuberant display of red and golden fabrics for shorts (*wara*) and colorful coats (*kushma*). Several guardian mountains in the region (*apukuna*) can be taken as members of the pantheon where a sacral feature, or more importantly, sacred mountains, are located, such as Mt. Ausangate in Peru, surrounded by other lesser mountains in the range (Sallnow 1987). Ritual processions to the mountain with offerings to placate maladies or to stimulate fortune are observed yearly during the *Kyllur Rit'i* festival in the *Sinakara* valley of Peru, coinciding with the Catholic observance of 'Corpus Christi' every June (Randall 1987). The pilgrims, dressed as masked parishioners, perform either as trickster clowns or policemen (*ukuku* or *chapa*), with other pilgrims performing as old settlers or hill keepers (*machula* or *ñaupa machu*), form long lines ascending to the glaciers. There they pray, sing, dance, drink, and play. In the past, they used to fight; however, the Catholic Church banned this fighting practice decades ago. Many pilgrims also break chunks of ice to bring the 'holy water' to their families in the valley floor (Ceruti and Reinhard 2005).

The mountain air (*wairaurku*), as opposed to the murmur of the water (*wairakucha*), could become temperamental and influence the people of Andean valleys in a different fashion. Sometimes the good spirit (*tin*) could blow warm spells that make people sick with hot disease cured by the *yachak* with cold drinks and showers. Other times, the bad spirit (*xiagra*) could blow cold spells that make people sick with chills that are cured with hot infusions or direct blows of fire from the mouth of the *yachak*, along with insults and beatings with stinging nettle, herbs and, shrubs that will return warmth to the patient. Here, the antlers are used ritually to pinch the patient's back with their tips or knocking the patient's legs with the antler's trunk, in order to counter the effects of either *tin* or *xiagra*. Sometimes, unwanted pregnancies are blamed on the diabolic mountain spirit (*supay*), thought to be witty and playful like a fawn or a small doe, or some immaterial presence, such as the rainbow (*kuychi*) or the groundwater seepage (*macha'chig*) in cloud forest areas (*sacha*) where the deer hides before going to the highland grasslands (*jallka*) and cold, windy meadows (*sallka*).

The antlers also were used to represent the shifting of day and night along an axis of stars that readily can be seen along the milky-way in the clear skies above the drier plains of the *Puna* region. This is the opposite of the cloudy skies of the shrouded, fog-ridden slopes of the *Paramo* region. This constant rain (*para*) and drizzle (*garwa*) explain the *Paramuna* of moist mountains on the northern ranges. Again, crepuscular periods were considered as in-betweens for light and darkness *Yanantin,* hence, important rituals took place in ceremonial centers at dusk or dawn, as it is exemplified by the *Utawalu runakuna's* initiation rites in the *Piguchi* waterfall of Ecuador, or the mythical replenishment of cold mountain water into the sacred vase (*puyñu*), and more mundane domestic cups (*pilchi*), at the ceremonial baths of *Kundurmachay*, near the *Saksawaman* ruins.

5.5 Recent Ruins Rediscovery

> The time has come for the people from outside to understand, without wrongful translations, what is our view of the territory and why the violations to our Law of Origin are so serious. (Amado Villafaña 2009, p. 1)

Throughout the Andean realm, archaeologists increasingly are finding sites deemed to be mountain ceremonial centers. Hidden underneath a cloth of montane forests or covered by layers of pumice or volcanic ash, these sites were considered simply ruins from the perspective of the Western scientists who could not fathom the existence of major monumental architecture in areas so deep into the montane tropical cloud forest ecosystem. Perhaps because of the Western nature-culture dichotomy, the shift from an ecocentric preservation paradigm towards a new bio-cultural conservation practice, requires a (re)reading of the landscape of ruins.

For many indigenous groups in the Andes, from the northern tropics to the southern (sub)antarctic, reifying abandoned 'rock piles' as manifestations of ceremonial places, serves to (re)invigorate the identity of indigenes by (re)creating the notion of the "sacredness" of these sites. In the northern Andes, in the cloud forest belt of the Tairona National Park in northern Colombia, the ruins of the lost city have been (re) stored under the tutelage of the Kággaba -*Kogui* elders, who along with three other indigenous nations (Wiwa -*Arzario*, W^ntuka -*Arwaco* and Kaku' chucwa -*Kankuamo*), referred to from colonial times as 'Tayrona'. These groups have inherited the Sierra Nevada de Santa Marta territory, now officially recognized as a national park and indigenous reserve or 'resguardo'. The discovery of the 'Ciudad Perdida' ruins have (re)vived the identity of the *Kogui* to the point that they have become the guardians of the 'sacred city' where the older brothers and the spirit of the mountain endures (OGT 2009).

Ecuadorian sacred sites are also experiencing (re)newed protagonism, such as the ruins on the summit of 'Catequilla' hill in the middle of the world, protected by the forts of *Rumikuchu* and of *Nibli,* along the *Wayllabamba* river gorge. Despite ancient rituals to the shade-less presence of the equator, marked by ancestral solar-tracker wells that had allowed ritualized practices honoring the sun, many of today's practitioners of New-Age philosophy climb, as on a pilgrimage, to the forgotten hilltop, now considered as sacred site, in order to offer prayers for equilibrium, harmony, and peace. The same process is observed in the province of Cañar, where the ruins of Cojitambo, near Azogues, have drawn crowds of New Age believers in the mystic power of gold, quartz crystals, and grains of mercury, and who firmly believe in magnetic forces congruent with the tall stone wall of the north face of the outcrop and consider *Kuritambu* a sacred site of the *Kañary* nation. Also, in the northern province of *Imbabura*, ritualized respect for the sacred tree (*pinllu*) of the *Utawalu runakuna* or 'lechero' (*Euphorbia laurifolia)*, is practiced above the fort (*pukara*) of Reyloma (Fig. 5.2). The white sap exuded from the tree is believed to have healing properties, thus making this place one of the most hallowed indigenous sites of Ecuador (Sarmiento et al. 2008). These rites or *pagapu,* are associated with the

Fig. 5.2 The sacred tree of the *Utawalu runakuna* is an Euphorbiacea (*Euphorbia laurifolia*), known in Spanish as *lechero* and in Kichwa as *Pinllu*. Associated with the bodily fluids of reproduction in humans, the white latex is ceremoniously collected and used in medicinal applications. Connected with the fertility of the group, it receives the pagapu of byproducts of birth or even fetuses that are respectfully buried in the soil around it. It also receives, as payments by lovers, coins pressed onto its bark for prayers and wishes for good relations and prolific marriages. It is revered on top of the man-made fortress or *Pukara of Reyloma*, literally the hill of the king. Often referred to as *Yayitu* or *Taitiku*, it is the point of convergence of telluric forces of the *Imbakucha* watershed, being the highest point between the two tall mountains that frame the Otavalo-Cotacachi territory, between the male or *Taita Imbabura* volcano and the female or *Mama Kutakachi* volcano (Photo: Fausto Sarmiento)

fertility of newly weds, but also can include the burying of fetuses, placentas, and other matter associated with births or abortions in order to placate the fertility god (*wayakil*).

In Peru, several sites that were considered only as ruins in the past are (re)taking center stage as places of spiritual power, esoteric wisdom, or other sacral property. The best known example is *Machu Picchu*. This archaeological site was long held to be the ruins of *Inka* buildings hidden in the cloud forests, but *Machu Picchu* officially is recognized as a Historical Sanctuary, a sacred site in the *Inka* citadel of the isolated mountaintop. The sanctity of the site is highlighted by awe-inspiring vistas of the river below and the vertical cliffs. Other nearby sites include 'Choquequirao', 'Sacsahuaman' or 'Ollantaytambo'. Many sites of earlier origins, such as the *Chachapuya* site of Gran *Pajatén,* in the Rio Abiseo National Park, bring the notion of sacred constructions and ceremonial uses of the challenging relief of the cloud forest belt of the verdant Amazon. Even earlier, on the Pacific side of the northern *Supe* and *Casma* valleys, the oldest urban complex discovered in the *Caral* pyramid, calls the attention of researchers to the ancient tradition of adoration to mountain sacred sites (Moseley 2001).

In Argentina, traditional gathering places for ceremonial or festive purposes, such as in the 'Quebrada de Humahuaca' for carnival, retain the notion of holly spaces. *Umawaka* has always been a crossroads for many groups since antiquity, protected by *the pukara* of *Tilkara* and highlighted by the colored slopes of *Purmamarka.* Further to the south, in the 'valles calchaquies' of the Tucumán province, the 'Ruinas of Quilmes' stand as mute testament to the fleeing of the *Kilmi,* one group of the *Diaguita* people, to a restricted colony in an Indian reservation near Buenos Aires. Today, often branded by a popular beer name and the largest middle-class neighborhood namesake, few people are aware that the descendants of the *Diaguita* have (re)invigorated their identity and (re)claimed the ruins into a Sacred City of Quilmes. A similar situation is illustrated on the other side of the continental divide, where *Kichwa* communities in *Tukunci* and *Likan Antay* communities in *Chiu-chiu* and *Aiquina-turi* respectively, are linked spirituality to sacred sites. On the other side of the divide, the *Chango* people might reflect the ancestral tendency of petroglyph depictions that copy the larger geoglyphs of the *Nazca* culture with its lines; however, thousands of years earlier, ritualized mummification and the existence of sacralized places for ceremonial burial, such as in 'Chinchorro', illustrate the Andean notion of sanctified sites. Indeed, Latcham (1936) traces mummification through millennia from Chinchorro to the *Inka.*

Finally, the sacred dimension is also present in the *Mapuche,* who call themselves the "people of the earth" and whose language *Mapudungun* is still spoken amongst their communities; they continue to follow the medicine woman or *Machi* as spiritual guide for their well-being. *Rehue* or sacred trees often indicate the vicinity of the sacred dwelling (*ruca*) of the 'curandera'. Because of their long-standing conflict over land-rights, even from before the *Inka* conquest to the south, particularly after the 'Araucanization' of Patagonia, *Mapuche* leaders (*lonko*) remain protagonists in the conflict over indigenous land tenure rights.

5.6 Andean Hierophanies

> The bottom line for us is that all of our territory is sacred in the sense that it is deeply, powerfully imbued with spiritual reality. To take any particular location and call it "more sacred" makes little sense within a worldview where our interactions are with all of our environments, and all are sacred. (Randall Borman In Press)

It is difficult to describe Andean sacred sites from the perspective of Amerindian people, not only for the huge diversity of cultural backgrounds and philosophical and religious affiliations, but also because the notion of sacredness is highly intertwined with the notion of identity (Sarmiento et al. 2005). Unlike Judeo-Christian doctrine that puts people above the non-human entities of the planet with intangible existence after death in eternal heavenly existence or damnation, Amerindian doctrines rely on the notion that people are part of other tangible and intangible elements, of the web of life, in a process that cyclically converges towards new beginnings (*pachakutik*).

Borman (in press) calls attention to an overarching sentiment of a consecrated environment, the challenge for the revival of indigenous lifescapes relies mostly in the articulation of their sacredness with the overall Westernized beliefs via syncretism. Alternatively, the application of syncretic principles in daily chores and in yearly ceremonial observances of cleansing or divination is essential. In as much as the invasion of missionaries continues unabated towards indigenous territories, not only indigenous territories but also vernacular epistemographies are endangered by acculturation and secularization (Verschuuren et al. 2010). The fact that no artificial building is identified as the place to experience the manifestation of god, the communication with spiritual power does not convey the need of a constructed temple, thus there is no need for monumental architecture or a centralized power holding sanctuary in a citadel. For instance, mountains are abodes of gods and their telluric power is a manifestation of the need of respect, even reverence. Entering a sacred cove of the Cofan, where fish tree stumps survive centuries of exposure to the tropical forest of the headwaters of the Napo River in Ecuador, is equivalent to entering the holy basilica of the 'Virgen de Guadalupe', the largest Catholic sacred site in the region. A cursory search of sacred sites in Latin America, for instance, will include listing the most important sites of Catholicism and some Evangelical shrines and cathedrals.

5.7 Beyond Belief

> Sacred natural sites are increasingly recognized for making a positive contribution to the conservation and sustainable use of biodiversity globally. (Bas Verschuuren 2010, p. 3)

The Andes Mountains hold a diverse array of ecosystems and cultures in a mosaic of topographic and ethnographic palimpsest that pretends to unite a heterogeneous nature into a homogeneous cultural denominator. Many of the cultural landscapes of the region are considered Socio-Ecological Systems (SES) which function under global stressors of environmental change that must be carefully steered towards positive, sustainable, and transformative end points of development via Earth stewardship (Chapin et al. 2011). Sacred sites are thus generalized by ecoregion (tropandean, isthmian, equatorial, central, southern, Patagonian, Magellanic) and also by climate (humid, dry) or by political boundaries (country-based within the region, or provincial-based within a country) and are seen as ultimate socio-ecological transformative landscapes (Rhoades et al. 2008). In fact, there has never been a monolithic Andean identity, thus the appropriation of sacred spaces to turn them into places of sacred power has never been done in the same way in the region. However despite the milieu of choices and roots, a unifying tendency of radicular syncretism, or even a dendritic ritualization of pilgrimage, offerings, sacrifices or actual payments, are obviously present in the determination of Andean sacred sites (Sarmiento 2003). Biocultural heritage narratives of the present mountain communities require the affirmation of an indigenous sentiment of respect for the land and for the overall maternal homerange (*pachamama*) of their indigenous nations and the agreement to maintain traditional cultic practices amidst the powerful winds of change prompted by current society's ruling class. Many indigenous nations coped with this in colonial times with syncretism, but it remains to be seen if the proliferation of modern technologies employed to impose globalized ideologies will be detrimental to sacred natural sites.

References

Bastien JW (1978) Mountain/body metaphor in the Andes. Bulletin de l'Institut Français d'Etudes Andines 7(1–2):87–103

Bingham A, Roberts J (2010) South and Meso-American mythology A to Z, 2nd edn. Chelsea House Books, New York

Borman R (In Press) Whose sacred sites? Indigenous political use of sacred sites, mythology, and religion. In: Sarmiento F, Hitchner S (eds) Indigenous revival and sacred sites conservation in the Americas. Berghahn Books, New York

Borsdorf A (In Press) Structural changes in Latin American spirituality: an essay on the geography of religions. In: Sarmiento F, Hitchner S (eds) Indigenous revival and sacred sites conservation in the Americas. Berghahn Books, New York

Borsdorf A, Stadel C (2014) Die Anden. Springer, Heidelberg

Cachiguango LE, Domingo L, Porfirio E, Van Kessel J (2001) La Utopía Andina. Cuaderno de Investigación en Cultura y Tecnología Andina 17 IECTA, Iquique

Cánepa G (2008) The fluidity of ethnic identities in Peru. Centre for Research on Inequality, Human Security and Ethnicity, Oxford

Ceruti C (In Press) Frozen mummies and the archaeology of high Andean mountains in the construction of Andean identity. In: Sarmiento F, Hitchner S (eds) Indigenous revival and sacred sites conservation in the Americas. Berghahn Books, New York

Ceruti C, Reinhard J (2005) Sacred mountains, ceremonial sites, and human sacrifice among the Incas. Archaeoaston J Astron Cult. University of Texas Press, Austin

Chapin SF III, Pickett STA, Power ME et al (2011) Earth stewardship: a strategy of social-ecological transformation to reverse planetary degradation. J Environ Stud Sci 1:44–53

Don Segundo (2013) Personal communication. Shaman who practice in Ilumán, Otavalo

Gudynas E (2011) Buen vivir: today's tomorrow. Development 54(4):441–447

Gudynas E, Acosta A (2012) La Renovación de la Crítica al Desarrollo y el Buen Vivir como Alternativa. J Sustain Educ 3(1)

Harrison R (1989) Signs, songs, and memory in the Andes: translating Quechua language and culture. University of Texas Press, Austin

Kelly I (1965) Folk practices in North Mexico, Institute of Latin American studies. University of Texas Press, Austin

Kennedy M (2010) Editor's note. The Alpinist, 1 Apr
Kush R (1962) América profunda. Imprenta Hachette, Buenos Aires
Latcham RE (1936) Atacameño archaeology. Am Anthropol 38(4):609–619
Moseley ME (2001) The Incas and their ancestors: the archaeology of Peru. Thames & Hudson, London
Murra JV, Wachtel N, Revel J (eds) (1986) Anthropological history of Andean polities. Cambridge University Press, Cambridge
Naess A (1995) The shallow and the deep, long-range ecology movement: a summary. In: Drengson A, Inoue Y (eds) The deep ecology movement: an introductory anthology. North Atlantic Books, Berkeley, pp 3–11
Naveh Z, Liberman A, Sarmiento F et al (2002) Ecología de Paisajes. Teoría y Aplicación. Edición de estudiantes. [Landscape ecology: theory and application. Students' edition] Editorial Universitaria de Buenos Aires, EUDEBA, Buenos Aires
Odum E, Sarmiento F (1998) Ecología: El Puente Entre Ciencia y Sociedad. McGraw Hill/ Interamericana de Mexico, Buenos Aires
OGT (2009) Organización Indígena Gonawindúa Tayrona/Centro de Comunicaciones Zhigoneshi/ Casa indígena Santa Marta, Colombia
Pizzey J (1988) Sweat of the sun, tears of the moon: a Latin American journey. BBC documentary, London
Platt R (1986) Mirrors and maize: the concept of yanantin among the Macha of Bolivia. In: Murra JV, Wachtel N, Revel J (eds) Anthropological history of Andean polities. Cambridge University Press, Cambridge, pp 228–259
Randall R (1987) Return of the Pleiades. Nat Hist 96(6):42–53
Rhoades R, Zapata X, Aragundy J (2008) History, local perceptions, and social impacts of climate change and glacier retreat in the Ecuadorian Andes. In: Orlove B, Wiegandt E, Luckman B (eds) Darkening peaks: glacier retreat, science, and society. University of California Press, Berkeley, pp 216–225
Rössler M (2006) World heritage: linking nature and culture. In: Schaaf T, Lee C (eds) Conserving cultural and biological diversity: the role of sacred natural sites and cultural landscapes. UNESCO Division of Ecological and Earth Sciences, Paris, pp 15–16
Rozzi R (2012) South American environmental philosophy: ancestral Amerindian roots and emergent academic branches. Environ Ethics 34:343–366
Rozzi R (2013) Biocultural ethics: from biocultural homogenization toward biocultural conservation. In: Rozzi R, Pickett STA, Palmer C (eds) Linking ecology and ethics for a changing world: values, philosophy, and action. Springer, Dordrecht, pp 9–32
Rozzi R, Armesto JJ, Gutiérrez JR et al (2012) Integrating ecology and environmental ethics: earth stewardship in the southern end of the Americas. BioScience 62(3):226–236
Sallnow M (1987) Pilgrims of the Andes: regional cults in Cusco. Smithsonian Institution Press, Washington, DC
Sarmiento F (2000) Human impacts in man-aged tropandean landscapes: breaking mountain paradigms. Ambio 29(7):423–431
Sarmiento F (2002) Anthropogenic landscape change in highland Ecuador. Geogr Rev 92(2):213–234
Sarmiento F (2003) Protected landscapes in the Andean context: workshiping the sacred in nature and culture. In: Harmon D, Putney A (eds) The full value of parks. Rowman & Littlefield Publishing Group, Lanham, pp 239–249
Sarmiento FO (2012) Contesting Páramo: critical biogeography of the northern Andean Highlands. Kona Publishers, Higher Education Division, Matthews, p 134
Sarmiento FO (2013a) Lo Andino: integrating Stadel's views into the larger Andean identity paradox for sustainability. In: Borsdorf A (ed) Christopher Stadel Festschrift. Austrian Academy of Sciences, Innsbruck
Sarmiento FO (2013b) Ruinas Reificadas: El revivir indígena, los paisajes culturales y la conservación de sitios sagrados. Revista Parques 1:1–14

Sarmiento F, Hidalgo J (eds) (1999) Desarrollo Sustentable de Montañas: Entendiendo las interfaces ecológicas para la gestión de los paisajes culturales en los Andes. Memorias del III Simposio Internacional AMA. Editorial Corporación Editora Nacional, Quito, p 406

Sarmiento F, Rodríguez G, Argumedo A (2005) Cultural landscapes of the Andes: indigenous and colono culture, traditional knowledge and ethno-ecological heritage. In: Brown J, Mitchel N, Beresford M (eds) The protected landscape approach: linking nature, culture and community. World Conservation Union IUCN, Gland\Cambridge, p 262

Sarmiento F, Cotacachi C, Carter LE (2008) Sacred Imbakucha: intangibles in the conservation of cultural landscapes in Ecuador. In: Mallarach JM (ed) Cultural and spiritual values of protected landscapes, protected landscapes and seascapes, vol 2, IUCN and GTZ. Kaspareg Verlag, Heidelberg, pp 125–144

Urton G (1988) At the crossroads of the earth and the sky: an Andean cosmology. University of Texas Press, Austin

Vásquez GR (1998) The Ayllu. In: Apffel-Marglin F (ed) The spirit of regeneration: Andean cultures confronting western notions of development. Zed Books, London, pp 89–123

Verschuuren B (2010) Arguments for developing biocultural conservation approaches. In: Verschuuren B, Wild R, McNeely J et al (eds) Sacred natural sites: conserving nature and culture. Earthscan Publishing, Gland

Verschuuren B, Wild R, McNeely J et al (2010) Sacred natural sites: conserving nature and culture. Earthscan Publishing, Gland

Villafaña A (2009) Our ancestral view of territory: indigenous peoples from the snow-topped mountains of Santa Marta, Colombia. Organización Indígena Gonawindúa Tayrona

Webb HS (2012) Yanatin and Masintin in the Andean world: complementary dualism in modern Peru. University of New Mexico Press, Albuquerque

Webb HS (2013) Expanding western definitions of Shamanism: a conversation with Stephen Beyer, Stanley Kripnpner and Hillary S. Webb. Anthropol Conscious 24(1):57–75

Chapter 6
Spirituality and the *Pachamama* in the Andean Aymara Worldview

Vicenta Mamani-Bernabé

Abstract This chapter shows the importance of cultural values and the role of traditional religion in the daily life of the Aymara people of Bolivia and lifts up the special importance women have in continuing traditional culture. Aymara life is framed in rituals regarding the *Pachamama* or mother earth, planting and harvesting, especially potatoes, and is based on a close and communicative relationship with nature. The Andean worldview is based on male- female pairing because reciprocity, duality, and complementarity form the defining principle in which harmony and equilibrium are fundamental values. Likewise personhood is not granted to the individual but to the couple. Only when a young man and woman marry are they considered complete and are recognized as adults. Collective complementarity also is the basis of labor relations and forms of collective labor continue to be an important dimension of community life. This worldview, the values that sustain it and the traditional religious rituals, reinforce community solidarity and seek to protect the fields in order to respect the *Pachamama* as a living being, and to assure a good and abundant harvest for the Aymara communities.

Keywords Aymara • *Pachamama* • Potatoes • Reciprocity • Rituals

6.1 Introduction

The *Pachamama* is sacred life; to destroy it, is to destroy ourselves. As Llanque-Chana (1995, p. 29) sustains:

> The Aymara knows the earth as a fertile mother because life springs from her, whether it is animal or plant life; she herself is life that generates life, because only something that is alive is able to nourish life. For this reason the Aymara love, respect, venerate, and protect her with all their strength.

For the Aymara person or *jaqi*, the *Pachamama* is the reason for being. Since the *Pachamama* is a living being, she must be fed and nourished, if she is to

V. Mamani-Bernabé (✉)
Instituto Superior Ecuménico Andino de Teología (ISEAT), La Paz, Bolivia
e-mail: vicentamamani@yahoo.es

R. Rozzi et al. (eds.), *Earth Stewardship*, Ecology and Ethics 2,
DOI 10.1007/978-3-319-12133-8_6

produce well and provide good food. "Through rituals and religious acts," the *yatiri* (traditional priest) communicates and converses with the *Pachamama*, "seeking the welfare of community, economic, social and political life and that of the families as a single unity" (Huanca 1989, p. 12).

Land and territory are the principal sources for food. This has always been so, and nothing can replace them. For this reason the Aymara constantly are in contact with the *Pachamama*, dialoging with her as a mother that dialogues with her children. In so doing the Aymara people affirm that they are brothers and sisters of nature who care for her. As Lapiedra (1996, p. 50) explains: "coca and maize are alive, the hills and the valleys, the snow and the water, the thunder and lightning are also alive. As the source of life is the land and her spirit, the *Pachamama*, is called 'Mother Earth'. All this is honored in cultic adoration." Until the land reform of 1952, the land was left fallow every 5–7 years so that it could recover its vital energy. Since the mid-twentieth century, pressure due to increased population has made this practice difficult to continue.

This chapter is based on my life in the village of Ticohaya, which is located near Lake Titicaca on the Andean high plateau or altiplano of Bolivia. In west-central South America the Andes are at their widest and represent the most extensive area of high plateau on Earth, outside of Tibet. Fifty-five Aymara families live there in adobe and stone houses roofed with tin sheeting or *totora* reed combined with native grass. The community obtained electricity in 1995 and the following year houses were connected to potable water. The area is arid with little evident vegetation. Water often is scarce, and it is always cold at 3,800 m above sea level. The local economy is based on agriculture (potatoes, broad beans, barley) and animal husbandry (llamas, guinea pigs, sheep, chickens). Severe weather conditions such as drought, frost, and flooding, affect agricultural production often causing local hardship and forcing villagers, especially youth, to immigrate to La Paz.

The little town is governed by villagers who rotate administrative roles among themselves. The elderly are trusted advisors and are involved in all aspects of community decision-making. No one receives financial remuneration. Rather the various responsibilities are assigned by *sayañas* (families and their croplands). However, due to migration to the cities and the ever increasing influence of neoliberalism, this practice is being debilitated.

Women contribute the greater part to the family economy. They work the fields and manage the livestock. In the trade fairs they exchange through barter family-produced products with those produced in other villages. They participate in community collective labor activities, such as *ayni, waki*, and *mink'a*, that are organized to meet agricultural and social needs such as celebrations of marriage, baptism of a child, *rutuche* or first haircut, and construction and roofing of a new home. In spite of increasing individualism and neoliberal capitalism, these systems of solidarity and reciprocity continue to be practiced. Women have a key role in all of these activities and rituals (Lapiedra 1996, pp. 53–54).

6.2 Andean Symbolism and Interpretation for Planting

Careful observation of nature is an important aspect of daily life. The Aymara pay close attention to the *pacha* (time and space)—will rain come early or late, will it be a dry or wet year, when is it going to snow, freeze or hail?—given that the fruit of the earth is the vital source of family survival. For this reason, they communicate with nature as a living being. The Aymara have faith in what the *Pachamama* transmits.

Simón Yampara (1992, p. 160) explains:

> Forecasting the weather and ascertaining the right moment for planting are of capital importance for the Aymara. This is done by observing the planets and the stars, climatic phenomenon, flora and fauna; also through spiritual knowledge, the interpretation of dreams, reading coca leaves, and reciprocal relations with the *Achachilas* and the *Pachamama*. All these provide knowledge for organizing life in the *Ayllu* and for determining how to use the *sayañas, qallpas* and the *aynuqas*. It lets them foresee what the weather will be and thus prepare for agricultural production.

In Ticohaya, the elderly, especially the women, are mainly involved in observing and interpreting these natural signs. They watch closely certain plants that announce the beginning or postponing of the planting season. My maternal grandmother annually closely observed the little *uluypina* plant. If this little plant blooms early, it means that planting potato should be moved up but if it is late, then planting should be postponed. If, during its growth and flowering, a plant is eaten by a rat, the plants will be affected by freezing rain and if the leaves dry out, then there will be frost damage. If it flowered beautifully, then there would be a good harvest and so it always was.

Animal conduct also provides signs that help the Aymara know when to plant. Fox, toads, sheep, pigs, and birds give valuable information. If the fox has her pups in a low area, it means that place is good for planting. Likewise, if the place is high up, then planting should be there. Big toads appearing in the fields while the soil is being prepared for planting potatoes, is a sign of good luck and announces a good harvest. But little, skinny toads mean a poor harvest. Toads are considered symbolically as *juyra* or *isaplla* (the vital spirit of potatoes), so they must not be mistreated nor killed, lest they weep and prevent the production of good potato. Playful lambs and piglets indicate that there will be good rains. When the cows moo, rain is coming. When the *unqallas* and *quillwas* or Andean gulls, fly over places normally free of high water, it is said that there will be flooding. If large flocks of small birds called *khunu jamach'is* come around, snow is forecast. When the *jamach'i* or the *phichitanka* do not sing, then there will not be a good harvest. If *pankatayas* or beetles appear, or brown colored *k'isimiras* or ants, it means that there will be a period of drought. Likewise, when little rocks are found in the bird nests, sleet will be coming.

Other natural features are important for the information they provide. The Aymara always observe the phases of the moon for signs about how to proceed with agricultural work. The moon also indicates when women and the soil are most fertile. The sun too gives information. When it appears encircled by rainbow colors, sowing potatoes and *oca* must be stopped until the phenomenon passes. If the winter time sunset is very red and yellow, cold, freezing nights are predicted. This shows that it is time for preparing *chuño, tunta, uma kaya, juphi kaya*, freeze-dried products made from potatoes and *oca*, staples of the Aymara diet (Box 6.1).

Box 6.1. Ritual for Planting Potatoes

This ritual begins with the preparation of the soil for planting. I remember my parents coming to the field with the plow, the yoke and the oxen. They would take off their hats and sandals, make the sign of the cross, and pray softly to the *pachamama* asking permission to use the soil and to allow the yoked oxen to work without difficulty. This was accompanied by a brief rite, the *k'inthu*, to teach the oxen to work as a team. We carried a small bottle of alcohol which was sprinkled over the oxen and the soil that was to be plowed. At the same time some coca leaves along with lejía[1] or sugar were pushed into the soil and a prayer was said so that the oxen would be tranquil and encounter no difficulty. This rite still is practiced. After plowing the soil, animal manures are prepared. This has to be done over various days because the manures have to ferment so that they will be effective fertilizer and also so they will not be too heavy to carry.

On planting day, seed, oxen, plows, and the *sata manq'a* (food for the seed), are all brought to the field. The seed must be handled carefully because it is believed that the potato seed is pregnant and so must be treated delicately. It must not be stepped on nor thrown aside, because it might cry and go away.

Upon arrival at the field, the women, with the seed, arrange themselves on the soil in a circular fashion. The field's owner sits on the soil facing the sun with her companions around her. Once the potato seed is taken out of the woolen sacks and scattered upon a cloth, the women and the seed are encircled with the *wisqha* (rope made from llama wool) as protection assuring that the seed stays put and does not run away. The belief is that if the *wisqha* is not placed around them, *jathax sarxaspawa, jaltxaspawa* (the seed might go away).

After this, the women place in the center of the circle *tari*, a small woven cloth, along with coca leaves. In another circle they place sweets, colored wool, llama grease, and other required ingredients. Meanwhile, one of the

[1] In Aymara language lejía or *ñaka* is the name given to shrubs of the genus *Baccharis*. In vernacular Spanish it refers to the ashes of the plant, which are used to peel quinoa and corn (Villagrán and Castro 2004).

(continued)

Box 6.1. (continued)

men of the family plows a few furrows and then all gather around the seed for the rite of *k'inthu aysuña chuqi achuyañataki* to assure an early potato harvest. Immediately everyone removes their hats and sandals and pick up a few perfect coca leaves, prayfully invoking the seed, the *pachamama*, and the nearby *achachilas* to provide abundant fruit and not to let natural disasters such as frost, sleet, or flooding, befall the harvest. Absolutely everyone, even the children, must fulfill the *k'inthu* of coca leaves to assure *qhuya q'ara* or that everything is complete.

Simultaneously the mother of the family takes three large, good potatoes, and cuts them half. She stuffs them with coca leaves fixed in llama grease along with a little sugar or sweets, and aleluya and *sank'ayu* flowers. Then she wraps them with brightly colored or white llama wool, and kisses them reverently. These decorated and garlanded potatoes are called *qhuya achuyata, panqarayata, anakiri* (symbols of future flourishing and potato production) and will be planted either at the beginning or the end of planting, depending on the family custom. The ritual is concluded with the sharing of coca leaves and the *acullico*, the chewing of coca.

Then the women and girls purify themselves by anointing their hands with *llamp'u* or llama grease, so that they can touch the pregnant potato seed. They have to do this because they handle salt, pepper, and onions when they work in the kitchen. On this day you have to be careful not to make the girls weep because it is the same as making *espalla*—the potato's vital spirit—weep. Men do not handle the seed because, it is said, *chachax thä amparawa*—the hand of a male is cold and symbolizes ice and sleet. However women symbolically are considered *junt'u ampara*—warm handed, and so women must handle the seed. You must not peal the potato either because that would cause it go away.

In some villages at planting time the women build a fire of wood or cow dung and put coca leaves, candies and sweets and other things on it in order to scare away evil spirits and thus assure a good harvest. On one occasion in Copacabana I saw men and women, even their oxen, adorned with white roses as symbols of the flowering potato. It was heartwarming to see this.

Finally, after the sowing has been done, each person takes a coca leaf and gives it to the field's owner, or sometimes she alone takes the coca leaves and buries them in the ground. Following this brief rite, they all share the *sata manq'a* or common meal and then the helpers are paid with some potato or other product. To sum up, these rites serve as offerings to protect the fields in order to assure a good and abundant harvest (Fig. 6.1).

(continued)

Box 6.1. (continued)

Fig. 6.1 *Candelaria* ritual ensures abundant harvests and is celebrated every February 2 by women in potato fields with offerings of coca leaves and smoke (Mamani-Bernabé 2002). The Earth and its products, as well as women, are considered feminine because they are sources of life in the Aymara worldview (Photo by Vicenta Mamani-Bernabé)

6.3 Aymara Spirituality

The village of Ticohaya maintains its Aymara traditional customs and sacred rituals until today. In spite of 500 years of colonial invasion and Christian presence, the different Andean traditional sacred rituals continue to be practiced fervently, such as rites for planting and harvesting, marriage, roofing of a new house, a child's first haircut, among others. The village maintains numerous sacred places among the nearby hills, where the feminine divinity (*Pachamama*, Mother Earth) and the masculine divinity (*Achachilas*, hill or mountain) are celebrated to assure family and village well-being.

These practices are an essential dimension of the faith experience of the Aymara people. Such spirituality is part of their cultural identity and vice versa, because there is no cultural identity without spirituality, and no spirituality without cultural identity. For this reason, we are convinced that spirituality is an expression of life itself. Spirituality is born in the human heart, and there God is present, present as life itself, in all and everything, in the entire village.

Women have a key role in this spirituality. The Aymara woman coordinates the rituals for planting and harvesting potatoes. She keeps permanent contact with nature and speaks to the *Pachamama* and *Mama Ispalla* (protector spirit of agricultural products). When she does so, she is speaking with the God of life, and through speaking to God, is doing theology. Women know the importance of symbols and how to interpret them. Thus women live and practice their own experience of faith from deep within the Aymara culture itself.

6.4 Andean Worldview and the Aymara Concept of *Jaqi*

An important element of this traditional religiosity or spirituality is the concept of complementarity, the basis of the Andean worldview, as illustrated in Fig. 6.2.

Male- female pairing is the foundation of Aymara culture, because reciprocity, duality, and complementarity form the defining principle of the Andean worldview

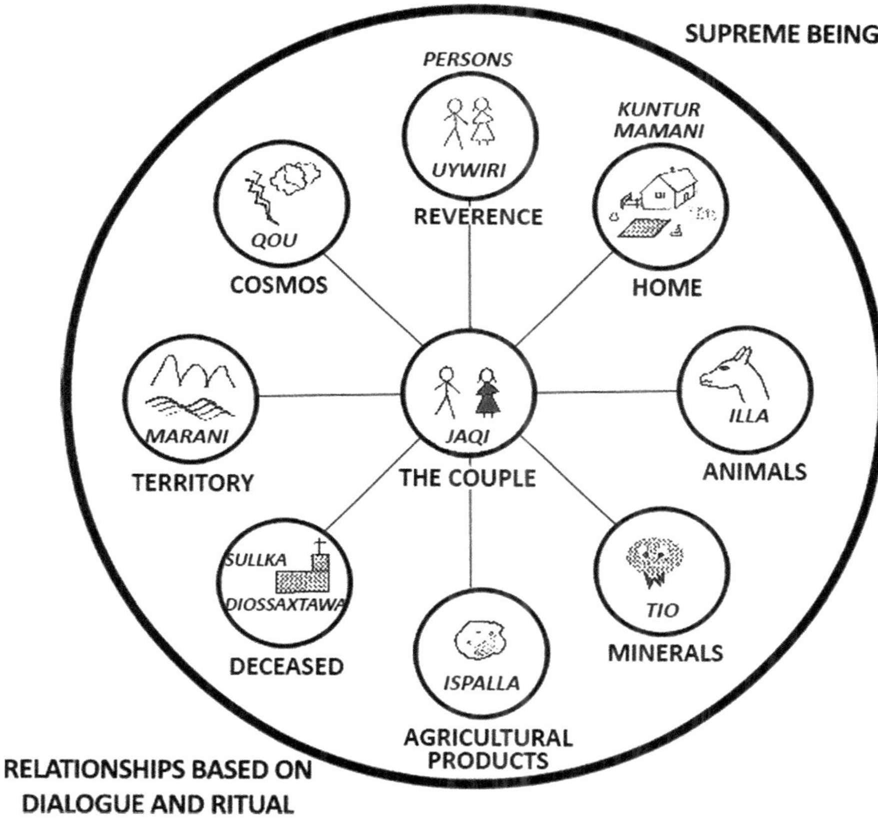

Fig. 6.2 The Aymara concept of person is multidimensional and complementary, especially between feminine and masculine, natural and supernatural. Each dimension of life is mirrored by a masculine or feminine supernatural protector or other spirit. The couple is paired with *Uywiri* or the spirit protector grandmother, while the spirit protector grandfather *Kuntur mamani* keeps the home safe; animals are enlivened by *Illa*, the feminine force of fecundity and the spirit of animals; the underworld of subsoils and minerals is governed by the masculine divinity *Tío*; *Ispalla* is the feminine divine protector of agricultural products especially potatoes and other tubers; upon dying, one joins the "minor Gods" *Sullka diosaxtawa* and thus has a supernatural presence in the community; the masculine *Marani* provide spiritual guidance and protection for communities and fields, deriving their power from the mountains; and *Qiju* protects the Aymara from inclement weather (Adapted from Mamani-Bernabé (2000))

Table 6.1 Examples of feminine-masculine complementarity in the Aymara worldview

Eco-social level of organization	Feminine	Masculine
Human persons	*Warmi* (woman)	*Chacha* (man)
	Tayka/Ipa (aunt)	*Awki/Tiyu* (uncle)
	Mama (mother)	*Tata* (father)
	Awicha (grandmother)	*Achachila* (grandfather)
Nature and animals	*Pachamama* (feminine divinity)	*Achachila* (masculine divinity)
	Qachu qala (males stone)	*Urqu qala* (female stone)
	Urqusuyo (masculine space)	*Qachusuyo* (feminine space)
Supernatural	*Pachamama* (mother earth)	*Achachila* (masculine hills and mountains)
	Phaxsimama (mother moon)	*Intitata* (father sun)
	Uywri (grandmother protector of the home)	*Kunturmamani* (grandfather protector of the home)
Social organization	Women to the left	Men to the right
Sacred places	*Wak'a awicha* (feminine place)	*Wak'a achachila* (masculine place)

in which harmony and equilibrium are fundamental values. This is manifested in traditional economic, socio-political, and religious practices. Everything is presented as complementarities. Nothing exists alone; the feminine and the masculine complement each other in multiple levels of socio-ecological organization (Table 6.1).

Table 6.1 shows only a few examples of the complementarity between feminine and masculine. Feminine-masculine complementarity is extended into all dimensions of life, including planting and harvesting, seed selection, traditional medicine, and religious functions. Even the "ritual of forgiveness" is complementary and communitarian. It consists in asking for forgiveness and then receiving it from all the men and women in order to re-establish equilibrium and harmony (Quispe 1996, p. 45). Complementarity in the Aymara mentality "includes not only men but also the feminine complement with all that exists on earth and in heaven, absolutely everything is interrelated through relationships of reciprocity" (Nina 1991, p. 172).

Complementarity, equilibrium, and reciprocity make up the formative paradigm of all socio-ecological relationships, including the human couple. In theory, neither male nor female is valued over the other. The relationship is one of mutual equality and the obligations of each correspond to those of the other. From an early age this model shows children how to be persons: *jaqi*. In Aymara worldview, *jaqi* exists in relationships of complementarity. The couple is the organizing center of life, and it exists in multiple dimensions that include both human and natural beings, among which there are traditional economic, socio-political and religious practices (Fig. 6.3).

Personhood is not granted to the individual but to the couple. Only when a young man and woman marry are they considered complete and are recognized as *jaqi-chachawarmi* or a couple. As a married couple they are considered to be one. Until they marry, they are considered minors. However, upon marriage

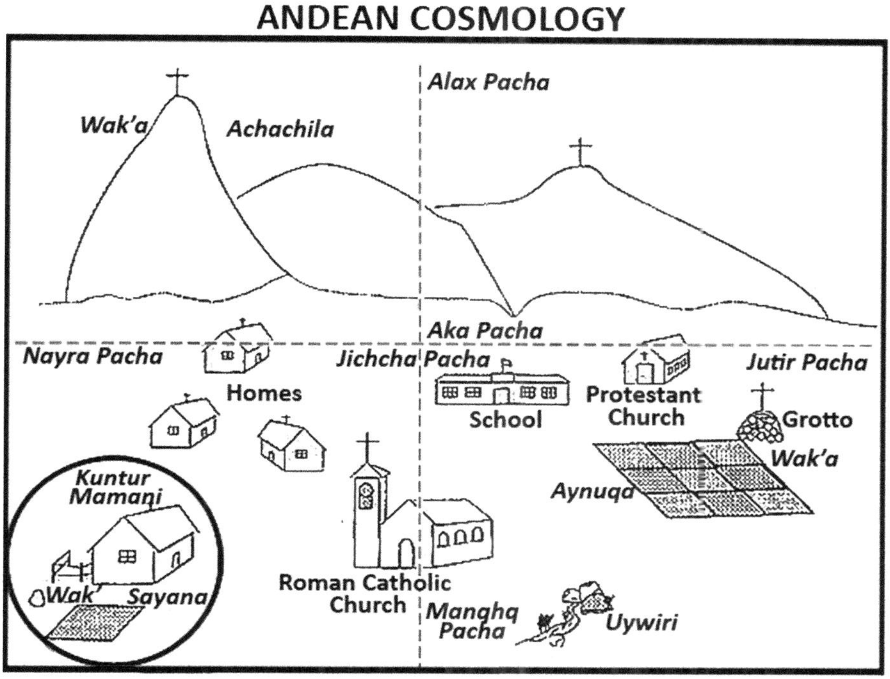

Fig. 6.3 The Aymara worldview is relational, including traditional economic, socio-political and religious practices among humans and natural beings. For example, scared places or *Wak'a* are found in the mountains (*Achachila*), the grotto (*calvario*), and garden (*Sayana*). *Pacha* refers to both space and time. *Alax Pacha* is space above Earth, it is a mysterious space where visible and invisible energies exist; *Aka Pacha* is the space where we live, planet Earth as a generous mother; *Manqha Pacha* is the space under the Earth, a sacred space where spirits dwell. *Nayra Pacha* is the past time, *Jicha Pacha* is the present or actual time in which we live, work, rest, and dream; *Jutir Pacha* refers to the future, a time that is only known to the *Pachamama* (Figure and caption adapted from Mamani-Bernabé (2000) and Mamani and Quispe (2007))

they become full members of the community and are accorded full rights, and given obligations.

> In the Aymara world the couple is central and constitutes the foundation of the community, because it is not just the union of two persons to guarantee physical reproduction, it means the recreation of the community itself. For this reason the word *jaqichasiña* (translated as matrimony) indicates something more: *jaqi*=person, and *chasiña*=to become; thus the approximate translation would be: to become or to be constituted persons. This is the meaning of matrimony seen and understood from the viewpoint of the Aymara and Quechua. (Argandoña 1996, p. 38)

Jaqi, in summary, is the complementarity of man and woman and implies completeness as persons. The single man or woman is *jaqi* but incomplete (Quispe 1995, p. 2). These are the values associated with *jaqi*. Unfortunately this model is more an ideal than reality because clearly there are contradictions between men and women in their daily lives and the ideal of male–female equilibrium and many other values of *jaqi* are being lost.

6.5 Collective Labor Practices

Collective complementarity also is the basis of labor relations and forms of collective labor continue to be an important dimension of community life. The *ayni* is an ancient means of mutual assistance based on reciprocity: "today for you, tomorrow for me."

The *mink'a* is solicited labor and frequently is practiced during the preparation of fields for planting, weeding, and harvesting. The *mink'a* is the person who helps another through direct assistance or by substituting them for certain tasks. In return the *mink'a* receives, as reciprocity, compensation in the form of products or money (Intipampa 1991, p. 195). When the *mink'a* is solicited for agricultural work, usually those who respond are women and relatives of the person in need.

The *waki-chikata* is associated labor. One family supplies the land for cultivation and another family provides the seed. The family that provides the seed works the field and cultivates the crops. The two families then divide the harvest equally between them. The *satxa-chikiña* consists in that a landowning family designates a certain number of furrows for the use of those persons who helped sow the owner's potatoes and *oca*. The helper sows her own seeds in the designated furrows. The *phawxa* sets aside a plot of land specifically for planting grains to benefit another, usually a recently married couple.

Various cultural moral values reinforce the former collective labor practices.

- **Hard work**, especially manual labor, is prized highly and laziness is frowned upon (Llanque-Chana 1990, p. 48).
- **Honesty and truthfulness** are founding values. Stealing, cheating, and deceiving are counterproductive to community solidarity required for survival, while being a "person faithful to their word"—*jaqix mayakiw parlaña*—makes for secure community contracts when help is needed.
- **Generosity and hospitality** not only provide for needs, they cement solidarity. Sharing with visitors or needy persons is a cultural norm. When a person passes by a potato field during harvest, the women working the fields always give the passerby some potatoes to take home. The person who receives the gift then kisses the potato because it is sacred, and is the Aymara's principal food. A traveling visitor never leaves a home without something, such as food, spices or coca. They always are invited to a meal or, at least, a snack. Many times the left-overs are given as a present for them to take home. Such generosity is understood as reciprocal because someday the visitor will have to return the favor and offer hospitality: you have to give hospitality to the traveler because someday you too will be a stranger in need. A wise Aymara saying dictates: *Jaya jaqix jilasawa kullakasawa*—the stranger is our brother

Fig. 6.4 Quechua men from Tarabuco, Bolivia dancing and playing the *zampoña* or reed pipes, a traditional musical instrument of the Aymara and Quechua. The earth is usually the center of such celebrations (Photo by Roy H. May Jr.)

and sister, so when they ask, they must not be denied. These various community practices of solidarity, reciprocity, and generosity continue to be practiced even as individualism and neoliberalism creep into traditional Aymara culture (Fig. 6.4).

6.6 Conclusion

This brief review shows the importance of cultural values and the role of traditional religion in the daily life of the Aymara. Among the positive, relevant values are reciprocity, solidarity, collective labor, respect for nature, neighbors, animals, village wisdom, equilibrium, family unity and others. Through a multiplicity of rituals Aymara people are able to converse with and enter into communion with nature: they speak with the mother earth, with agricultural products, with the hills and mountains, rivers, rocks, trees, and with protector spirits. All these answer and give signs for living. In this spirit we are called to commit ourselves to being little birds that announce planting, flourishing, and harvesting of Aymara culture and the Earth itself, the *Pachamama*.

References

Argandoña R (1996) Algunas consideraciones sobre inculturación. Fe y Pueblo 1

Huanca T (1989) El Yatiri en la Comunidad Aymara. CADA/HISBOL, La Paz

Intipampa C (1991) Opresión y Aculturación: la evangelización de los aymaras. CEPITA/HISBOL/ISETRA, La Paz

Lapiedra A (1996) Religiosidad popular y andina. In: Tamez E (ed) El rostro femenino de la teología. DEI, San José

Llanque-Chana D (1990) La cultura Aymara: destrucción o afirmación de identidad. IDEA/TAREA, Lima

Llanque-Chana D (1995) Ritos y espiritualidad aymara. ASET/IDEA/CTP, La Paz

Mamani-Bernabé V (2000) Identidad y Espiritualidad de la Mujer Aymara. Fundación Basilea, La Paz

Mamani-Bernabé V (2002) Ritos espirituales y prácticas comunitarias del Aymara. Privately published, La Paz

Mamani V, Quispe C (2007) Pacha. Editorial Verbo Divino, Cochabamba

Nina E (1991) La reciprocidad: ethos de la cultura aymara. Licencentiate thesis en philosophy. Universidad Católica Boliviana, Cochabamba

Quispe C (1995) Memoria de la IV jornada de reflexión de mujeres a nivel ecuménico de C.T.P. La Paz

Quispe C (1996) ¿Inculturación del Evangelio en los Andes? Fe y Pueblo 1

Villagrán C, Castro V (2004) Ciencia indígena de los Andes del norte de Chile. Editorial Universitaria, Santiago

Yampara-Huarachi S (1992) Economía comunitaria andina. In: van den Berg H y N Schiffers (eds) La cosmovisión aymara. UCB/HISBOL, La Paz

Chapter 7
Andean Llamas and Earth Stewardship

Roy H. May Jr.

Abstract Stewardship of the land often is used as a metaphor for environment ethics. However the term is problematic because of its origin in hierarchical social structure implying a master-servant relationship. Read from historical conditions of empire and technological prowess stewardship can lend itself to environmental exploitation. In contrast, read from the ancient Adam and Eve Hebrew myth and Sabbath tradition and non-Western peasant cultures such as the Aymara of Bolivia, stewardship signifies restraint, mutuality, and advocacy for the Earth.

Keywords Adam and Eve • Advocacy for justice • Bio-historical • Mutuality • Peasant societies • Sabbath

Many llamas roam the high plains of Bolivia and they leave a lot of dung around Andean villages, usually in specific places of their own choosing (Fig. 7.1). Llamas are culturally and economically important because they provide useful products for Andean peasants, including dung, which is a principal source of fertilizer.[1] They also figure prominently in the Andean symbolic universe. Llama representations are common in various artistic manifestations; llama fetuses often are buried under the threshold of newly constructed homes; llamas are sacrificed to assure people's well-being; and llama body parts are used in healing rituals.

Several years ago I visited an Aymara community south of La Paz, to consult with villagers about economic development projects. It was a beautiful day, the kind of day that made me understand why the Incas worshipped the sun, so we decided to hold our meeting outdoors. We spread out on the ground, continued our meeting, drinking coca tea and then having lunch. After a while it was evident that I was the only one who gave any importance to where we had chosen to sit: we were sprawled out in piles of dry llama dung.

[1] For a discussion of llamas and alpacas in rural Andean life, see Palacios-Ríos 1988; for information on agricultural production on the High Andean plains, see Mamani 1988.

R.H. May Jr. (✉)
Departamento Ecuménico de Investigaciones (DEI), San José, Costa Rica
e-mail: royhmay@gmail.com

R. Rozzi et al. (eds.), *Earth Stewardship*, Ecology and Ethics 2,
DOI 10.1007/978-3-319-12133-8_7

Fig. 7.1 A llama caravan transporting products to market near Lake Titicaca, Bolivia. Llamas have been used as beasts of burden in the Andes Mountains for hundreds of years, making possible extensive regional trade relations (Photo by Roy H. May Jr.)

My Aymara companions obviously had no problem at all sprawling out in llama dung, but their doing so was subversive of my Western cleanliness ethic. However for them, far from signifying something dirty, llama dung was part of the cycle of life, a feature of nature worthy of respect and care. Llama dung not only expressed life, it also symbolized future life through enriched soil for cultivation. Indeed, it seemed to me that they saw a connection between themselves and the dung. They knew themselves to be composed of the same organic material. Like the llamas that produced important dung, they too were creatures of the earth. My Andean companions were true stewards of the environment: they respected and cared for llama dung. Not only that, they had demonstrated many times during the half-millennium since the Spanish conquest that they were ever ready to defend vigorously llama dung and their own cultural integrity (see Cárdenas 1988; Stern 1987). Surely caring and respectful mutuality for the Earth and its defense, are the basis of environmental or Earth stewardship.

In recent years, "stewardship" has become a common way of talking about how people ought to relate to the earth. Although it echoes positively among many different social groups, the idea is especially important for Judaism and Christianity, religions that have had such an influence on the Western world. Indeed, using stewardship as a model for how humans ought to relate to nature is a Judeo-Christian contribution. However most of the discussion of stewardship has taken place in the world's affluent nations. Any idea "must be investigated in relation to the praxis out of which it comes," as José Míguez-Bonino (1975, p. 91) affirms. So, how might the idea be understood from the perspective of non-affluent, non-Western people and how could that perspective enrich an understanding of Earth stewardship? This chapter responds to these questions by discussing environmental or Earth stewardship using "peasant economy" as a hermeneutic and, therefore, proposes that, like the Aymara Indian--and Adam and Eve--, stewardship is best understood as caring and respectful mutuality for the earth and its defense.

7.1 Stewardship in Western Thought

By definition stewardship means taking care of and managing someone else's property. Its origin is in ancient English social structure (as well as that of the Ancient Near East, Rome, Greece and across feudal Europe) in which the household duties were performed by domestic servants who were guardians and managers of the property of the owners. Clare Palmer (1992, p. 77) explains, "The idea of stewardship originates in a society which is based on slavery or serfdom, and represents a despotic and autocratic form of government, a fact which is particularly clear when considering it in the feudal context." However the term became disassociated from its original context and became associated generally with managing other people's goods and as restraint on one's own use of those goods or property. As John Passmore (1974) indicates, Western traditions of stewardship and cooperation with nature are diverse, complex, and have changed over time. It was not until the late seventeenth century that the idea was applied to nature. Richard Bauckham (2011, p. 58) argues that it was first used as "a response to the growing sense of human control over nature" brought about by the Italian Renaissance and "the excessively anthropocentric Baconian view" of human domination, this in the framework of the cultural Christianity of the time. Human control of nature continued to be assumed, yet not in an unlimited fashion because, it was argued, "the world was not created solely for human benefit but for God's glory" (Bauckham 2011, p. 59). Although stewardship in this sense introduces the idea of restraint, it also argued for human control over chaotic nature (Bauckham 2011, p. 60). Humans were understood as separate from and above nature. For Michael Northcott (1996, p. 129), "the fundamental problem … is the implication that humans are effectively in control of nature" and sets up a master-servant relationship. Still, as Northcott (1996, p. 180) explains, "the concept of stewardship of nature is mobilised (sic) in the Western tradition from the Fathers to Benedict to refer to the just and gentle care of nature by humans." The problem is its later association with property rights thus turning stewardship into "a metaphor of human control and mastery over nature." In addition, Palmer (1992, pp. 72–73) argues that the idea became inseparably connected to money, as managing nature as a bank account for human enrichment.

Although stewardship in relation to nature is not a theological nor Biblical concept, but rather an idea used for theological construction and applied to certain Biblical texts, these interpretations were based mainly on readings of the creation stories in Genesis 1 and 2, especially Genesis 1 where human beings are given "dominion" over God's creation, that reflected the interpreter's own time. In historical context, the "praxis" of the time was that of emerging science, technology, and empire (seventeenth to the nineteenth centuries and the consolidation of modernity). These interpreters clearly saw themselves as having dominion over nature. The texts seemed to fit them well, even though some saw the need to place limits. Thus, as Palmer (1992, p. 82) explains, "These assumptions, which lie behind most uses of stewardship, demonstrate that stewardship is an anthropocentric ethic, which considers it to be better not only for human, but for the rest of the natural world, for nature to be managed and made fruitful by human standards." Stewardship, then, in the context of technological prowess (and expanding empire), understands the Earth in terms of its utility for (colonizing) people; it refers to the administration of nature to assure the needs of empire. Palmer (1992) thinks stewardship can never be disassociated from this meaning. David Ehrenfeld and Philip Bentley (2001, p. 132) rightly indicate the problem for today:

> When stewardship is corrupted by power in the absence of restraint, it becomes ecological tyranny and exploitation. This is the central problem of stewardship, a problem that has always existed but has become critical only with the rise of modern technology and its side effects, including overpopulation. With technology, humanity has achieved a power and a presence that is utterly subversive of the practice of stewardship. Modern theorists have despaired of finding noncoercive ways of resolving this tragic dilemma, and many environmentalists have condemned stewardship itself as an inherently unworkable concept.

Still, stewardship can have different meanings in different socioeconomic and political contexts; the material conditions in which stewardship is conceived inevitably shapes the concept itself. Thus different historical praxis—material conditions—can imbue stewardship with alternative meanings and can provide novel insights for interpreting the Biblical texts that often are argued as the conceptual origin of the idea of stewardship. The ancestral relationship between Aymara people and llamas, including llama dung as a metaphor for peasant societies, offers such an alternative interpretative framework.

7.2 Adam and Eve as Andean Peasants

The Ancient Near East (approximately fourth millennium BCE to the fourth century BCE) was a society of peasants and pastoralists (Wright 1990). Studies of peasant societies demonstrate several distinguishing characteristics (Chayanov 1966; Shanin 1971):

1. Production is motivated by, and oriented toward, the family unit, concerned with providing basic needs.
2. Market ties are weak; the major concern is not accumulation of capital, but rather equitable distribution within the family and village.
3. Labor is contributed by the family and the village through collective and reciprocal arrangements.
4. Land is the basis of livelihood, however land is not understood as private property in the capitalist sense, but as family or community property.

5. Existential identification is with the family, the land, and the village in such a way that the individual, family, and village form an indivisible whole.
6. The basic structure/control over land resources is vested collectively in the village.

Clearly the earth or land is fundamentally important to peasant societies. This is manifested in the rich symbolic representations and religious rituals and beliefs about the land that are common to peasant societies. These traditions emphasize the earth as the substance of human genesis and as the means that make life possible. Humanity is seen as being part of a networking of interrelationships binding together the earth and the human, in such a way that a good and just life is facilitated. Access to land is understood as a right that cannot be denied because the Earth, as the foundation of life, belongs to all living creatures (see Eliade and Sullivan 1987). Among the Quechua and Aymara of the Andes, the earth is *pachamama*, their mythical-religious concept of space. *Pacha* signifies the space of maximum security in the present, and is identified with the *ayllu* or traditional village with its homes, cultivated fields and common pasture lands. *Mama,* as feminine, maternalizes the *pacha* and is manifested in the earth. Thus the *pachamana* is fertile earth apt for cultivation, that nourishes and cares for humans and other creatures (Aguiló 1988).

The ancient Hebrews also were concerned for fertile earth. For them, land was a divine gift, an inheritance from Yahweh, to be respected and managed according to Sabbath. (See Brueggemann 2002; for an important study of land in the Hebrew Bible see Habel 1995). As "creator", God or Yahweh, is the "owner" of the land who establishes, through Sabbath, how the earth is to be lived. Yahweh, then, grants the Earth in usufruct to humanity as an inheritance. For the ancient Hebrews, the land was not property but rather the good that made life, as well as personal and cultural identity, possible.

"Inheritance" incorporates the idea that the earth itself is the substance of human genesis and that there is a symbiotic relationship between humans and the earth that is activated in mutual interaction. It also lifts up the social dimension of the earth and, therefore, was the basis of ancient Hebrew agrarian law that restricted how the land was to be used: land could not be bought or sold; part of the produce had to be left for the poor; the land had to rest, that is, be left fallow for certain periods of time, among other legal provisions governing the use of land. These were provisions required by Sabbath, the seventh day of creation according to Genesis 2: 2–3. Sabbath restricts the use of the Earth and subverts human efforts to control according to their own desires. To this restrictive end, the ancient Hebrew Scriptures contain numerous Sabbatical laws and regulations.

The proper Biblical text for basing the Judeo-Christian idea of stewardship as a conceptual ethical model for the earth-human relationship is the story of Adam and Eve (Gn 2: 4–25), the oldest of the Biblical creation stories.[2] This is because the text so clearly reflects peasant mentality and experience.

---

[2] Frequently Gn 1:1–2:4, the creation myth of 7 days, where "dominion" over the earth is granted to humanity, is cited as the model for Judeo-Christian stewardship. However this text is not properly about stewardship but rather about "public administration," although it introduces the key idea

Adam and Eve are Ancient Near Eastern peasants whose purpose in life is to cultivate and care for the land. The scene is a "garden"—such as the *chakra* of the Andes of South America or the *milpa* of Mesoamerica—where food is produced for the family by its own labor. Adam is placed there "to till it and to keep it" (Gn 2:15). The Hebrew word translated as "to till" or "to cultivate" literally is "to serve" while "to keep" in Hebrew means "to protect" or "to guard." So Adam is to cultivate and care for soil, to be its servant and to protect it. Indeed Yahweh made Adam because "there was no one to till the ground" (Gn 2:5). Adam himself is part of the soil. In an earlier verse we are told that Adam is made "from the dust of the ground" (Gn 2: 7). In the original Hebrew "dust of the ground" is *´adama*, humus or cultivable soil, top soil, the very substance of life. Adam, then, is a creature of the earth, named "soil." Adam also cares for animals and relates to them, indeed, is organically continuous with them. Not only are the animals made "out of the ground" like Adam (Gn 2:19), by giving them names Adam can develop a relationship; knowing a name opens the possibility of a relationship. Furthermore, as Biblical scholar George Ramsey (1988, pp. 34–35) explains, in naming the animals, Adam is "*discerning* something about the creatures—an essence which had already been established by God." Naming, as Ramsey says, is an act of discernment, not of domination. Adam, then, is tied to the soil and other creatures, first through creation from the earth, and secondly through toil, that is, caring for them. Finally Eve comes into Adam's life as "partner" (Gn 2:20; 3:20). Eve is "life"—here the Biblical text is a word play because in ancient Hebrew the word Eve resembles the word for life or living. Eve is the source of life, "the mother of all living" (Gn 3.20). In this ancient myth, "soil" and "life" are brought together integrally and intimately. So Adam and Eve are placed in this garden filled with fruit trees and even the "tree of the knowledge of good and evil" (Gn 2:16–17), a harmonious community typical of peasant utopic visions. The characteristics of this myth, molded to particular cultures, can be observed in peasant societies even today. In this sense, Adam and Eve are analogous to Andean peasants.

7.3 Andean Llama Stewardship

"Earth stewardship" is a metaphor that evokes a sense of responsibility and care beyond self-interest. It "recognizes value in the non-human creation other than its usefulness to humanity and gives humanity obligations to treat the nonhuman creation accordingly, while at the same time recognizing the unique degree of power over the rest of creation which human beings wield in modern times"

of Sabbath. Other possibly pertinent texts include Psalms 8 and, for contrasting anthropologies, Psalms 90, 103:14–16 and 144:3–4; Job 38–41 reflects another creation and anthropological tradition. The specifically Christian corpus includes Mt 25:45–51; 25:14–31 and twin texts in Luke, among others, that might be considered pertinent. A discussion of these texts requires more space than is available in this article.

(Bauckham 2011, pp. 60–61). The importance of stewardship is that it imposes limits on human conduct. Rooted in the Sabbath tradition of Judaism and the Adam and Eve peasant tradition of ancient Hebrew society, "restraint, noninterference, and humility were an integral part of the original Jewish concept of stewardship, regardless of corruptions that may have taken place subsequently, and these restraining virtues may yet prevail" (Ehrenfeld and Bentley 2001, p. 126). Its importance for ethics and thus stewardship is that it contains within itself normative language. Yet, as metaphor, content is plastic, not exhaustive and hardly literal. Still it is powerful because it stimulates the moral imagination. Certainly in many respects, stewardship is a flawed concept, given its historical origins in hierarchical social and economic structures, yet when understood from the ancient Hebrew tradition of Adam and Eve and Sabbath, and re-read from peasant economies such as the Aymara, new insights are to be had. Finally stewardship is about how humans are to interact with the earth.

Andean llama stewardship is practiced in a framework of mutuality. Andean ethics is based on reciprocity. So far from "domination" or "power over", this stewardship fosters a subject-subject relationship. That which is to be cared for is not an object but a subject worthy of respect. It nourishes "power with," a symbiotic empowering. Llamas, the dung they produce, and the peasants who spread it on their fields, are all in a kind of reciprocal partnership. Following Larson's discussion of environmental metaphors (2011, p. 119), stewardship, then, is a metaphor, that "[b]y emphasizing relationship… exemplif[ies] what has been called an ethic of partnership, as opposed to former ethics based on egocentrism, anthropocentrism, or even ecocentrism. This new ethic gives equal moral consideration to both the human and the nonhuman, thus balancing respect for biodiversity and cultural diversity." Andean llama stewardship is a kind of "ethics of care" that emphasizes relationship and the well-being of animals and people: peasant farmers take care of the llamas, the llamas take care of the peasant farmers. Following the meaning of cultivation as nurturing service, Earth stewardship nurtures a healthy earth. Leopold tells us that "[h]ealth is the capacity of the land for self-renewal." Paraphrasing him, we can say, "Stewardship is our effort to understand and preserve this capacity" (1949, p. 221).

7.3.1 Biohistorical Anthropology

Stewardship raises the question of anthropology. How should the human being be understood in this complex weaving called nature? Are humans to be considered a legitimate part of the natural order? What is their relationship to other living beings? These questions bear on the meaning of stewardship because it implies an anthropology.

Stewardship does not separate nature and society; furthermore, it understands humans as the artisans of history. In many respects it emphasizes the human situation, and that humans are not mere puppets of natural forces. Nor does stewardship downplay human needs and their right to use nature for survival. Such anthropology and concern ought to be maintained. Nevertheless, an exclusionary anthropology ought to be avoided in favor of an anthropology that overcomes the dualism characteristic of many environmental ethics that counter-pose "biocentrism" to "anthropocentrism." This can happen by understanding that the human being is not only biological but also cultural and lives historically; the human is an animal, but not just another animal.[3]

Rather, human beings are, as theologian Gordon Kaufman proposes, "biohistorical." "This way of conceiving the human emphasizes our deep embeddedness in the web of life on planet Earth while simultaneously attending to the significance of our radical distinctiveness as a form of life" (Kaufman 2004, p. 42; cp Rozzi 1997). Or, in the words of Ricardo Rozzi, we are "biocultural" (Rozzi 1997, 2001, 2012). "Biology" and "culture" are merged in humans and this union makes humans different from other forms of life. Human beings do not lose their importance, nor are other forms of life excluded or less appreciated. As in the Adam and Eve story, humans remember themselves as creatures of the earth who can relate to other living creatures. Among the Aymara, as well as other peasant societies, human need is not put aside nor underestimated. Indeed, much of the Earth-caring these societies demonstrate is in order to assure human welfare. Llama dung is respected precisely because it enriches the fertility of the soil that produces food. Yet it would be erroneous to consider the Aymara and other peasant societies as anthropocentric for whom llama dung is only of instrumental value. Their relationship to the Earth is not that of master-servant.

The idea of people as biohistorical or biocultural creatures offers a holistic framework for discussing the meaning of stewardship. It provides a conceptual basis for stewardship as symbiotic mutualism, thus undermining dualisms such as anthropocentrism versus biocentrism, instrumental versus intrinsic value, or nature versus culture.

7.3.2 Environmental Advocacy

Andean llama stewardship, however, is not passive but rather proactive in the defense of the Earth through concrete actions on its behalf. The many rebellions by Aymara and other Andean indigenous people demonstrate this. The Earth steward is Earth advocate. Alterity is the philosophical mode for understanding environmental issues. This view understands the Earth as "other", as the subaltern. This requires both an understanding of the modes and procedures that produce subalternity and political and other intervention to secure its liberation (Ortega 2011, p. 296). It denounces policies and activities that do not contribute to its health and announces ones that do. In this sense, stewardship is subversive of destructive policies and worldviews. Without Sabbath, stewardship becomes corrupted by power,

[3] For a discussion of human uniqueness in relationship to other animals, see Van Huyssteen 2006.

"associated with instrumentalist attitudes to nature which are linked with environmental exploitation..." (Northcott 1996, p. 180). Inevitably then, stewardship is political and conflictive as it challenges powerful interests.

Earth stewardship is advocacy for justice. Justice is concerned with power relations and the consequences for life-possibilities of those relationships. The Hebrew Sabbath tradition had much to say about justice. This tradition not only is concerned with the right relationship of humans to each other and their social groupings, it also calls for right relationships to the land itself. The Jubilee (Lv 25), as synthesis of the Sabbath tradition, proclaimed liberty to both human captives and to the land. It ordered the redress of wrongs committed against people and the Earth and the (re) establishment of a just situation. Redressing wrong is the essence of justice in this tradition. It is rooted in the Mesopotamian tradition of the right of the wronged to clamor for redress. According to this ancient custom, a person who had been wronged, whose "rights" were violated, could "clamor" to the king, who, in turn, was obligated to hear the complaint and to rectify the injustice. The Earth also "clamors" for redress. "A deep chesty bawl echoes from rimrock to rimrock, rolls down the mountain, and fades into the far blackness of the night. It is an outburst of wild defiant sorrow, and of contempt for all the adversities of the world," wrote Leopold (1949, p. 129). The Earth steward is obligated to hear the "chesty bawl" and to redress.

7.4 Conclusion

Reading stewardship from the praxis of peasant societies such as the Aymara provides a perspective that emphasizes mutuality, care and protection, and advocacy for the well-being of the Earth and its many forms of life. It urges humanity as a vital and legitimate participant in nature, as biocultural or biohistorical beings. This stewardship is not a master-servant one, but rather a community of beings together for the welfare of all. In this community even llama dung commands respect.

References

Aguiló F (1988) Tipología religiosa andina. Ensayo lingüistico y simbológico. Editorial Los Amigos del Libro, La Paz

Bauckham R (2011) Living with other creatures. Green exegesis and theology. Baylor University Press, Waco

Brueggemann W (2002) The land. Place as gift, promise, and challenge in biblical faith, 2nd edn. Augsburg Fortress Press, Minneapolis

Cárdenas VH (1988) La lucha de un pueblo. In: Albó X (ed) Raíces de América: El mundo Aymara. Alianza Editorial, Madrid

Chayanov AV (1966) The theory of peasant economy. Thorner D, Smith REF, Kerbloy BH (eds). Richard D Irwin, Homewood

Ehrenfeld D, Bently PJ (2001) Judaism and the practice of stewardship. In: Yaffe MD (ed) Judaism and environmental ethics. A reader. Lexington Books, Lanham\Oxford, UK

Eliade M, Sullivan L (1987) Land. In: Eliade M (ed) The encyclopedia of religion, vol 4. Macmillan Publishing, New York

Habel NC (1995) The land is mine. Six biblical land ideologies. Fortress Press, Minneapolis

Kaufman GD (2004) In the beginning… creativity. Fortress Press, Minneapolis

Larson B (2011) Metaphors for environmental sustainability. Redefining our relationship with nature. Yale University Press, New Haven

Leopold A (1949) A land county almanac and sketches here and there. Oxford University Press, New York

Mamani M (1988) Agricultura a los 4.000 metros. In: Albó X (ed) Raíces de América: El mundo Aymara. Alianza Editorial, Madrid

Míguez-Bonino J (1975) Doing theology in a revolutionary situation. Fortress Press, Philadelphia

Northcott M (1996) The environment and Christian ethics. Cambridge University Press, Cambridge

Ortega F (2011) Historical interpretation and ethics. Notes on the role of alterity. In: Cisneros AH, Premawardhana S (eds) Sharing values. A hermeneutics for global ethics. Globalethics.net, Geneva

Palacios-Ríos F (1988) Pastores de llamas y alpacas. In: Albó X (ed) Raíces de América: el mundo Aymara. Alianza Editorial, Madrid

Palmer C (1992) Stewardship: a case study in environmental ethics. In: Ball I, Goodal M, Palmer C et al (eds) The Earth beneath. A critical guide to green theology. SPCK, London

Passmore J (1974) Man's responsibility for nature. Ecological problems and western traditions. Charles Scribner's Sons, New York

Ramsey GW (1988) Is name-giving an act of domination in Genesis 2:23 and elsewhere? Cathol Biblic Q 50(1):24–35

Rozzi R (1997) Hacia una superación de la dicotomía biocentrismo-antropocentrismo. Ambiente y Desarrollo (September):2–11

Rozzi R (2001) Ética ambiental: raíces y ramas latinoamericanas. In: Primack R, Rozzi R, Feinsinger P et al (eds) Fundamentos de Conservación Biológica: Perspectivas Latinoamericanas. Fondo de Cultura Económica, México, pp 311–359

Rozzi R (2012) Biocultural ethics: the vital links between the inhabitants, their habits and regional habitats. Environ Ethics 34:27–50

Shanin T (ed) (1971) Peasants and peasant societies: selected readings. Penguin, Middlesex, UK

Stern S (ed) (1987) Resistance, rebellion, and consciousness in the Andean peasant world, 18th to 20th centuries. The University of Wisconsin Press, Madison

Van Huyssteen JW (2006) Alone in the world? Human uniqueness in science and theology. William B. Eerdmans Publishing, Grand Rapids\Cambridge, UK

Wright CJH (1990) God's people in god's land, family, land and property in the old testament. William B. Eerdmans Publishing\The Paternoster Press, Grand Rapids\Exeter

Chapter 8
Earth Stewardship and the Biocultural Ethic: Latin American Perspectives

Ricardo Rozzi

Abstract Latin America hosts a diversity of ecological worldviews and practices rooted in Amerindian cultures (e.g., Aymara, Quechua, U'wa, and Waorani) and schools of thought (e.g., geoculture, decoloniality, liberation philosophy and eco-theology) that have actual and potential value for Earth Stewardship. However, global discourses do not adequately include the diversity of languages and ethics rooted in the heterogeneous biocultural mosaic of Latin America and other regions. This is due in part to the limited *inter-linguistic* and *inter-cultural* dialogue among academics, educators, and policy makers that reside in different regions of the world. To contribute to solving this deficit, this chapter couples the conceptual frameworks of Earth stewardship and the biocultural ethic to foster: (i) inter-cultural dialogues and negotiations that fracture the current homogeneity of neoliberal global discourses through the acknowledgement and inclusion of the diversity of ecological worldviews, values, and languages, and (ii) forms of biocultural inter-species co-inhabitation embedded in the diversity of habitats and life habits. A basic principle of the biocultural ethic is that life *habits* are interrelated with the communities of *co-in-habitants* and their *habitats*. These "3Hs" of the biocultural ethic offer a conceptual framework that can be coupled with three terms that identify Earth Stewardship: the habitats of the *Earth*, the habit of *stewardship*, and the communities of co-inhabitants including the *stewards*. This coupling makes explicit the participation of diverse stewards. To better recognize the stewards' diversity is essential to identify their differential responsibility in the genesis of global environmental change, at the same time that to visualize and value a plethora of ways of conceiving and practicing Earth stewardship.

Keywords Biocultural • Environmental justice • Intercultural • Liberation philosophy • Traditional ecological knowledge

---

R. Rozzi (✉)
Department of Philosophy and Religion Studies, University of North Texas, Denton, TX, USA

Institute of Ecology and Biodiversity, Santiago, Chile

Universidad de Magallanes, Punta Arenas, Chile
e-mail: rozzi@unt.edu

R. Rozzi et al. (eds.), *Earth Stewardship*, Ecology and Ethics 2,
DOI 10.1007/978-3-319-12133-8_8

8.1 Naming the Diverse Earth Stewards

Earth is not only a biophysical entity, but it is also a word. A word that influences the way we understand and relate to the biophysical reality of the planet. Very often scientists forget the gravity of words, and focus their research on the biophysical reality. Conversely, philosophers often focus on examining the language of cultural reality, ignoring the biophysical realm. The biocultural ethic unites biological and physical realities with human cultural attainment. It provides Earth stewardship with a conceptual framework that integrates the biophysical and symbolic-linguistic realms of reality (Box 8.1), and explicitly integrates the concept of stewards. What is named exists in language and communication. Therefore, it can be included in analyses and evaluations. *Earth stewardship* is composed of two words, but the concept implies a third term: *stewards*. Naming these stewards allows us to better distinguish particular agents that have different types and degrees of impact and responsibility in causing the global environmental change we face today.

Box 8.1. Language: Human's Biocultural Lenses

Humans participate not only in the biophysical, but in the symbolic, cultural, and linguistic structures and processes of biocultural landscapes. Human perceptions and understanding of biological diversity are embedded in language, culture, and technology. The compound term *biocultural* makes explicit the role that the "cultural lenses" of any human "observer" (including scientists with their research methods, and conceptual taxonomies) have in shaping the construction and interpretation of biodiversity concepts. In turn, the ways humans perceive and understand biodiversity and their environment influence the ways humans inhabit ecosystems, and modify the structure, processes, and composition of living beings, from molecular to global scales. To illustrate this point, it is helpful to look at an example of two contrasting languages, Amazonian Waorani and English, regarding the way they refer to forest ecosystems.

The Waorani word *ömö* defines forests as *worlds inhabited by countless sentient beings*, who share with humans the same home, dispositions, values, and culture (Rival 2012). This *human-forest kinship* implicated in the word *ömö* stimulates the performance of rituals, and today it encourages Waorani people to protect their forests and oppose oil extraction in the Yasuní National Park (Sawyer 2004; Finer et al. 2009). In contrast, the English word *woodland*, implies that forest ecosystems are a *land of wood*. The focus on wood can lead to a further narrowing of mentality for understanding forest ecosystems because: (i) the existence of the many *non-woody beings* is excluded from language; (ii) *trees* may be interpreted *as mere resources*, for either fuel or building materials (Rozzi and Poole 2011). These contrasting definitions of forest ecosystems illustrate how concepts embedded in language influence both ecological practices (the ways in which humans transform other species and the environment), and ecological knowledge (the ways in which humans perceive other species and their environment) (cf. Rozzi 2001).

(continued)

Box 8.1. (continued)
By fostering an understanding of the multiple representations and classifications of biological diversity in various languages, this biocultural method can help constitute a new – global but regionally heterogeneous – covenant to sustain the *human-earth-system* (*sensu* Chapin et al. 2009). The need to de-construct and re-construct language, and to learn from the *ecologies of others* (*sensu* Descola 2013) is urgent for defending life (human and other-than-human) and fostering bioculturally diverse and complementary forms of Earth stewardship.

Chapin et al. (2011a, p. 44) point out that "a century ago, stewards were responsible for managing estates or for keeping order at public events. Today, the Earth is one global estate, and improved stewardship is vital for maintaining social order and for preserving life on Earth." In 2010, the Ecological Society of America (ESA) launched the "Earth Stewardship Initiative" to confront an environmental crisis that is now global in scope, rapidly worsening, and potentially catastrophic for human civilization. The ESA's Earth Stewardship Initiative provides a social–ecological framework for sustaining life in a rapidly changing world. The biocultural ethic's conceptual framework helps us to better understand that although the Earth is one global estate, there is a diversity of Earth stewards with their languages, cultures, social, and ecological practices that generate contrasting positive and negative socio-environmental impacts. Instead of making responsible the species *Homo sapiens*, in general, we should identify particular responsible agents (social groups, corporations, nations) of the current socio-environmental crisis. Unsustainable practices and agents that are detrimental to sustaining life need to be sanctioned and/or remedied; complementarily, more sustainable worldviews, forms of knowledge, values, economic, and ecological practices should be respected and eventually adapted as we develop new modes of Earth stewardship (Rozzi 2013).

The Earth Stewardship Initiative of the ESA (*sensu* Chapin et al. 2011a, b, and 2015 in this volume [Chap. 12]) aspires to contribute to a responsible administration of the planet. For this initiative to be respectful of the biophysical, linguistic, and cultural diversity of the planet, an inter-cultural and inter-regional dialogue is required. To contribute toward this aim, in this chapter I apply the conceptual and methodological frameworks of the biocultural ethic (*sensu* Rozzi 2012) to recognize and value the diversity of stewards, integrating their symbolic-linguistic and biophysical realities.

A basic principle of the biocultural ethic is that life *habits* are interrelated with the communities of *co-in-habitants* and their *habitats*. These "3Hs" of the biocultural ethic offer a conceptual framework that can be coupled with the three terms that identify Earth stewardship to better visualize the differential roles of diverse stewards: the habitats of the *Earth*, the habit of *stewardship*, and the communities of co-inhabitants (humans and other-than-humans) including the diverse *stewards* (Fig. 8.1). To do this analysis, I draw on Latin American traditions of environmental thought with a dual purpose of (1) examining little known concepts and practices of stewardship, and (2) developing a conceptual framework that can be used for similar analyses in other regions of the planet.

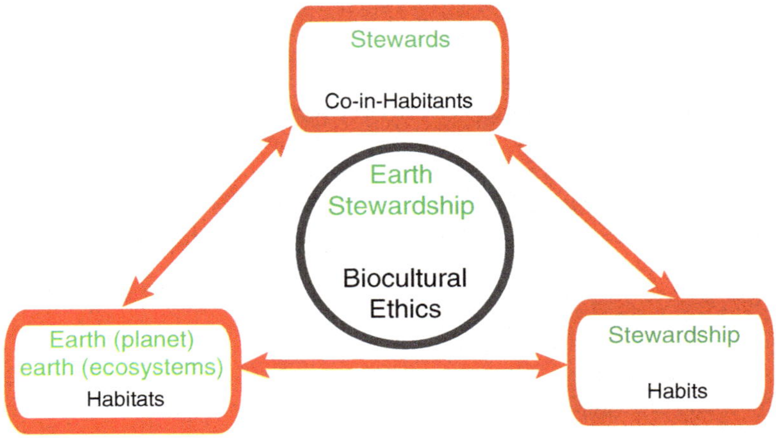

Fig. 8.1 The "3Hs" of the biocultural ethic coupled with the three core components identified for Earth Stewardship: habitat/*Earth*, habit/*stewardship*, and co-inhabitants/*stewards*

8.2 Amerindian, Scientific, and Pre-Socratic Perspectives on South American Co-inhabitation

The cultural and biogeographic identity of South America is marked by the presence of the Andes Cordillera, which crosses the continent north–south from Colombia to Chile. Soared over by the emblematic Andean Condor, this mountain range influences both (1) the symbolic-linguistic realm of the worldviews associated with environmental stewardship and philosophies, and (2) the biophysical realm of the heterogeneous mosaic of ecosystems in this continent.[1] According to the worldview of the pre-Incan civilization of Tiahuanaco, in ancestral times *Viracocha* (one of the most important deities for this primordial South American culture) emerged from Lake Titicaca in the heights of the Andes and created the sun with his radiant light, the rain and water with his tears, as well as the heavens, the stars, the humans and the other living beings that inhabited the region (Fig. 8.2).

This Andean cosmogony points out that humans share a common origin with all other-than-human beings. Viracocha is the source of both the biophysical entities and the order of the world; humans participate in both a cosmic community and a cosmic order. This Andean cosmology is similar to ancient Greek pre-Socratic cosmologies, which represent the philosophical roots of Western civilization. In the Quechua Andean

[1] The distinction of these two interwoven realms, the bio-physical and the symbolic-linguistic-cultural, is essential to the biocultural ethic (Rozzi 2012, 2013). Under this biocultural perspective, the term philosophy abandons its disciplinary character, which currently prevails in academia. Instead, I emphasize the plural character of philosophy, with its diversity of ways of understanding the natural world and of co-inhabiting in it, with in particular ecological and cultural contexts. The plural character of philosophy concurs with the conceptual framework developed by Raul Fornet-Betancourt for a Latin American intercultural philosophy (Fornet-Betancourt 1994).

Fig. 8.2 The iconic figure of *Viracocha* in the center of the Sun Gate in Tiahuanaco in the highlands of Bolivia was sculpted in stone 2,200 years ago. *Viracocha* is surrounded by 48 winged guardians, of which 32 have human faces and 16 have condor faces, illustrating how deities, humans, and nature have been and are still united in Amerindian worldviews and lives (Photograph Héctor Morales Deramond)

language, the name Viracocha is transcribed today as *Wairacocha* that means *waira* (wind) and *cocha* (lake, sea), or as *Ticci Vira Cocha Pachayachachic* that means the source of fire, earth, water, and air; i.e., the four pre-Socratic essential elements.

Viracocha also played an essential role in bringing order onto an originally chaotic world (Kusch 1962), a worldview reminiscent of Heraclitus's notion of *logos* or order (see Callicott 1994). Both ancestral cosmogonies –the Andean and the Heraclitean pre-Socratic–, in turn, have elements in common with current ecological scientific worldviews. The science of stoichiometry has determined that humans and all living beings are composed of the same major chemical elements. Moreover, planet Earth also shares the same basic chemical elements with the rest of the material bodies of the cosmos. Biogeochemical sciences have discovered ecological cycles of energy and nutrients in which humans participate, and geology and astronomy have disclosed dynamic exchanges of energy and chemical elements at the planetary scale (Schlesinger and Bernhardt 2013). The notion of *participation* embedded in these Andean, pre-Socratic, and scientific worldviews provides a solid foundation for both Earth stewardship and biocultural ethics. For Earth stewardship, the understanding that humans *participate* in the structure and order of the biosphere and the cosmos implies that appropriate forms of stewardship and governance need to adjust to such order, which is the condition of possibility for life. For biocultural ethics, the *ontological notion* of *participation* provides a foundation for the *ethical notion* of *co-inhabitation*.

The Andean backbone hosts the highest mountain peaks in the Americas and gives origin to an assemblage of vast and contrasting ecoregions that include the extensive Puna and Paramo in the high Andean Altiplano, the world's largest tropical forests in the Amazonian basin, the world's largest wetlands in the Pantanal, the widespread grasslands, savannas, and dry forests in the Gran Chaco, the world's driest habitat in the Atacama desert, and the most extensive area of Southern Hemisphere temperate and sub-Antarctic forests and fjords in the archipelagoes of southwestern South America (Olson et al. 2001; Spalding et al. 2007; Abell et al. 2008). In the twenty-first century, these ecoregions host the world's greatest diversity of plants and most animal groups, and are still inhabited by endemic cultures with their languages and worldviews (Guevara and Laborde 2008). For a planetary-scaled initiative, such as the ESA's Earth Stewardship Initiative, to be effective, however, it is essential that the richness and value of regions such as the intricate South American reservoir of biological and cultural diversity are better acknowledged by the discourses and policies that govern global society.

South American ecosystems play a critical role in the world's regulation of climate and conservation of biodiversity, and support a plethora of cultures with ancestral and contemporary ecological worldviews and sustainable practices. The value of these worldviews and practices for Earth stewardship has only recently begun to be considered by academics (Rozzi 2001). Neither do global discourses and environmental policies adequately include the diversity of languages and ethics rooted in the heterogeneous biocultural mosaic of South America and other continents. This is due in part to the limited *inter-linguistic* and *inter-cultural* dialogue among academics, educators, and policy makers that reside in different regions of the world (Li et al. 2015 in this volume [Chap. 13]). In order to contribute to solving this deficit, a specific purpose of this chapter is to provide conceptual elements for:

(i) facilitating inter-cultural dialogues and negotiations that acknowledge and include the diversity of ecological worldviews, values, and languages, thereby fracturing the current homogeneity of neoliberal global discourses and policies;
(ii) fostering forms of biocultural inter-species co-inhabitation embedded in the diversity of habitats and life habits.

As shown by practices associated with Earth Stewardship in South America (May Jr (2015a, b); Mamani Bernabé 2015 in this volume [Chaps. 6, 7, and 27]), biocultural inter-species co-inhabitation requires not only rational or verbal interactions but it also involves corporality, affection, and sharing everyday life experiences.[2] Llamas and potatoes, for example, are not mere natural resources but rather co-inhabitants that participate in rituals, farming and husbandry practices,

[2]The biocultural perspective of this essay shares central concepts with intercultural philosophy (*sensu* Fornet-Betancourt 1994, 1998). However, the biocultural ethic extends the moral community beyond the boundaries of the human species. The worldviews of contemporary ecological sciences and of Amerindian cultures support the concept of a community of life. This can be also considered as a moral community on the basis of the notions of kinship (based on evolutionary genealogies shared by humans and other living beings) and of co-inhabitation, embedded in the recurrent ecological and cultural interrelationships among human and other-than-human beings (Rozzi 2012, 2013). Recent ethnography of South America illustrates the creativity and agency of the other-than-human world, as well as the rich communication that takes place between human and other-than-human persons. The limits of extending personhood as a category of human-like subjectivity to non-humans has, however, been amply discussed (Rival 2012).

and everyday life of Aymara communities. The diversity of forms of ecological knowledge and practices rooted in Amerindian colonial and post-colonial cultural habits – which, in turn, are embedded in ancestral native habitats and contemporary anthropogenic habitats – offers today insights for stewardship and biocultural co-inhabitation both at local and planetary scales.

The complementarity of the forms of knowledge offered by Amerindian world-views, pre-Socratic philosophical foundations of Western civilization, and contemporary sciences is emphasized by the biocultural ethic. Ancient and modern forms of knowledge indicate that human beings participate in an ecological structure and order. Today, the ESA's Earth Stewardship Initiative can be built upon by incorporating the original meaning of economy: an administration of the *oikos* (or habitat in the terminology of the biocultural ethic), that understands and respects the ecological order of the *oikos* and the political order of human societies. Toward this aim, it is indispensable to reorient the current supremacy of prevailing neoliberal free-market policies. The essence of these policies has been to free themselves from restrictions for entrepreneurship and economic growth. As a consequence neoliberal policies override or often ignore both the ecological order and the political order, as we will discuss below.

8.3 Biocultural Roots of South American Environmental Philosophy

In mid twentieth century, some Latin American anthropologists and philosophers, such as Rodolfo Kusch in Argentina, forged pioneer studies that interrelated the biophysical reality of Meso- and South-American landscapes with the symbolic-linguistic reality of Amerindian cultures. These studies examine the intricate links between both realms of reality embedded in their fractured, dynamic, historical courses of colonialism. In his book *America Profunda* ("Deep America"), Kusch developed an approach that incorporates ways of understanding and inhabiting the landscapes rooted in pre- and post-colonial contexts that question the prevalence of theoretical models developed in academia (e.g., the Tragedy of the Commons as discussed by Kingsland 2015 in this volume [Chap. 2]): "as if" such theoretical models would have universal validity unaltered by local biocultural and historical contexts.

To counterbalance concentration on theoretical models and assumptions that become universal in academia, Kusch focused on dynamic forms of knowledge while researching the Incan legacies in peasant communities of Bolivia and northeast Argentina. Working at the University of Salta, he initiated a comparative ethno-philosophy that contextualizes supposedly "unalterable" universal notions, thereby enabling a better understanding of the diversity of local forms of knowledge and environmental thought. Kusch elaborated a *geocultural* perspective that considered both the cultures and the territories (Romero-Bedregal 2006). The integration of culture and territory is essential for an Earth stewardship, because concentration of land property and displacement of local poor communities is facilitated by prevailing development models (see May Jr (2015a, b); da Rocha and Possamai 2015; Viola and Basso 2015 in this volume [Chaps. 7, 24, 27, and 28]). These displacements often constitute social and ecological injustice that leads to the extinction of plant species traditionally cultivated (such as the hundreds of varieties of potato, chili, and many other plant species), of biocultural landscapes (such as the forest islands or *apete* created by the Kayapo people in the Amazonian region), and of cultural practices (such as exchanges of seeds among women of Aymara and Quechua communities inhabiting the Andean slopes) that form an integral part of the ecosystem dynamic in the Neotropics. Many ancestral ecological practices are currently alive among creole peasant and Amerindian communities that inhabit Latin America. These communities are traditional stewards of the land (see Mamani Bernabé (2015), in this book [Chap. 6]).

In the decade of the 1960s, the perspective of Kusch contrasted with the fact that forms of indigenous thought and life were ignored, even negated, by an academic philosophy dominated by an analytical-positivist school of thought. Counteracting this tendency, in *Geocultura y el Hombre Americano* ("Geoculture and the American Man"), Kusch (1976) introduced the term *geoculture*. With this term, South American geography ceases to be seen through a colonialist perspective, as a virgin territory to be conquered and used, and begins, instead, to be understood as a territory where cultural meanings are rooted. Kusch views the American continent as a place where an extended colonial Western culture coexists with Amerindian cultures, their ancestral memories, lifestyles, and thought patterns that have survived the colonial and postcolonial (or neocolonial) periods. Conflictive encounters between pre-Columbian peoples of the Andes and the Old World Europeans, Kush argues, established dialectical relationships between two polarized notions:

- *estar aquí* ("to be here" and "to be at"), which expresses the essence of what remains of the Amerindian cultures, and
- *ser alguien* ("to become someone"), which defines the attitudes of the European colonizers.

Since the arrival of the Spaniard conquistadors, the "New World" environments have been subjected to the prevailing colonial attitude of "possession of objects," which is established by individual self-centeredness focused on "becoming somebody in a future time" (*ser alguien*). This attitude contrasts with the customary Amerindian attitude of "participation and interaction with organisms," focused on present time, place and community (*estar aquí*). As emphasized by Mamani Bernabé (in this book [Chap. 6]), a person is fully-mature and virtuous through the cultivation of relationships. Kusch's conceptual framework converges with the approach of biocultural ethics because both aim to better understanding and valuing how *heterogeneous cultural habits* are interwoven with the *heterogeneous native, rural, and urban habitats* (Rozzi et al. 2008). Today, these biocultural relationships are disrupted as local communities are displaced, and native habitats are left open to accelerated processes of land-use changes, including large-scale mining and expansion of monocultures associated with a concentration of the ownership of land (Ceccon and Miramontes 1999; Neugebauer 2003; Tobasura 2006; Finer et al. 2008, 2009). The massive rural–urban migration that has taken place in Latin America since the 1950s has generated a loss of the *ancestral human stewards of the land*. The loss of the stewards of the land has led, in turn, to losses of biocultural life habits and native habitats, including their rich biocultural diversity, understood as vital communities of co-inhabitants (Rozzi 2013) (Box 8.2).

Box 8.2. Global Responsibility to Respect Amazonian Life

South American regions have been subjected to recurrent illegal deforestation and mining pollution, entailing multiple-scale processes that override the will of rural and indigenous populations and violate national laws (Ceccon and Miramontes 1999; Neugebauer 2003). The cases of the U'wa people in Colombia opposing oil companies and the decision by the Ecuadorian government to open the Yasuní National Park for oil exploitation, are emblematic.

The U'wa people inhabit the foothills and cloud forests of the Andes in northeast Colombia, and had almost no contact with the outside world until the 1960s (Fig. 8.3). In 1991, the oil company Oxy (Occidental) signed an exploration permit with the Ministry of the Environment (Tobasura-Acuña 2006). However the U'wa believe that oil is the blood of the mother Earth, and when faced with oil drilling against their wishes, in 1995 the U'wa threatened to commit collective suicide. Although in May 1998 Oxy announced that it was moving off of lands that were claimed under Colombian law by the U'wa (http://www.goldmanprize.org/1998/southcentralamerica), the disputes have continued for over a decade. The government militarized the zone and conflicts with the U'wa have been violent, including the murder of several U'wa children in 2000. Colombian environmental sociologist Isaías Tobasura Acuña (2006) concludes that the U'wa case demonstrates that the stated Colombian national environmental policy is continuously overridden by national and international economic power.

Fig. 8.3 An U'wa child holds a sacred shell, evoking the conception of his culture: "The U'wa territory is the heart of the world, run by the veins that feed the universe, if it is destroyed, then the world bleeds" (Photograph Terry Freitas, courtesy Project Underground, www.moles.org)

(continued)

Box 8.2. (continued)

In 2008, Ecuador's constitution was the first in the world to recognize legally enforceable rights of nature. In recent years, Ecuador's efforts to manage the Yasuní National Park – one of the most biologically diverse spots on Earth – have comprised innovative appeals to global responsibility through monetary compensations for not opening the park to oil exploration. However, in 2013 Ecuador withdrew its proposal to refrain from oil exploration in its Yasuní National Park because it had not received promised compensation from the world's industrialized nations (Espinosa 2013; Pellegrini et al. 2014). Ecuador's President Rafael Correa explained that "the Yasuní proposal was based on the principle of co-responsibility in the battle against climate change, but just 0.37 % of the target [US$ 2.3 billion] was provided by international donors…This failure of the international community touches on the wider issue of justice in the battle against climate change. What level of responsibility should be taken by the developed nations that have most contributed to the problem of climate change and are most able to tackle it? And what is the responsibility of the less developed nations? Clearly, a just solution would see the more developed nations bearing proportionally more of the responsibility" (in Falconi-Puig 2013).

Until 1956, the Yasuní region was entirely ancestral Waorani territory when first contacted. Now it is a complex mix of overlapping designations, and Waorani leaders are divided between those opposed to new oil development on their territory and those more inclined to negotiate with oil companies (Finder et al. 2009). The failure of Ecuador's innovative approach points to the ongoing tension between appeals to global responsibility and the values of the sustainability of life, human and other-than-human, and to national and global financial interests.

Conflicts such as the U'wa and Yasuní cases play out at once across local, national and global political contexts. On the one hand, they draw attention to transnational negotiations and distinct articulations of justice around planetary environmental sustainability. On the other, they demand from academics and policy makers a better understanding of the dynamic local forms of ecological knowledge. It is pressing to act, effectively addressing the complexity and multisided responsibility for implementing Earth Stewardship.

8.4 Liberation Philosophy and Decolonial Thinking

The perspectives of Kusch have provided one of the most important sources of inspiration for another Argentinean thinker at the end of the twentieth century, Walter Mignolo. In the 1990s Mignolo developed the notions of *border (boundary) thinking and epistemologies,* and *pluri-topical hermeneutics.* To critically examine the one-dimensionality established by the paradigm of conquest of the people and American nature, Mignolo has adopted key concepts from Enrique Dussel, who has led the liberation philosophy school of thought.

Latin American liberation philosophy involves two methodological moments: first, to liberate or free thinking from being encapsulated in colonizing conceptual frameworks (e.g., abstract Eurocentric ethics, economic models, or Christian credos); second, to reaffirm local forms of thought and material realities. Dussel seeks to overcome Eurocentrism and modernity, not simply by denying them, but also by "thinking from the perspective of the excluded other;" i.e., the impoverished communities of peasants, the colonized communities of indigenous people, the marginal communities of workers and urban citizens (Dussel 1996, p. 14). In his recent work, Dussel goes beyond the social domain to include ecological ethics. He criticizes formal Kantian ethics, discursive ethics, and utilitarian ethics to emphasize instead that:

> Having as our horizon the ecological destruction of the earth that is articulated concomitantly with misery, poverty, and the oppression of the majority of humanity (taking into account phenomena such as central and peripheral capitalism, racism, sexism, etc.), we must recover material references, since these "facts" can only be discovered critically via contrast (contradiction or non-compliance) to a positive material standard previously stated. For this reason, we need to reconstruct the truth of a material ethic –where ecological destruction and poverty are identified as ethical problems in themselves– and articulate it adequately to a formal morality –from which we can proceed consensually. (Dussel 2003, p. 32)

Dussel questions the hegemony of neoliberal capitalism in which the value of capital is ranked above the value of life. He demonstrates how this scale of values is in disagreement with the theological and philosophical texts that represent foundational traditions of belief systems and ethics in Western civilization. Consequently, Dussel argues that it is necessary to re-establish the right hierarchy of values: the value of life must be ranked above the value of capital. The land and humanity have "dignity;" only human-made products have "exchange value" or "economic value." Methodologically, he proposes that:

> Material ethics [of liberation] considers goods with use value to be wealth as such (objective goods of happiness, which is the subjective good). The political management of these public goods is the formal practical moment, which is public and consensual (democracy, for example). Hence, ecology and political economy speak first of the material level of ethics, but managed at the formal level of democracy or public morals. (Dussel 2003, p. 33)

In a similar manner, in his latest work Mignolo (2003a) has also gone beyond the purely social domain, extending it to the domain of life. His proposal of a *paradigm other* seeks to construct spaces of hope not only for human life, but also for all life forms. This extension of Mignolo's *paradigm other* is particularly pertinent for a Latin American Earth stewardship, because both the Amerindian cultures and the ecosystems, including their biodiversity, have been insensibly oppressed by the process of European conquest. In addition, after World War II the globalization of neoliberal market practices has led to accelerated processes of "biocultural phagocytosis,"[3] which has also oppressed the diversity of cultural and ethical traditions within Western civilization itself and has promoted a biocultural homogenization worldwide (Rozzi 2013).

Mignolo (1995) notes that Huntington's phrase "the West and the rest" expresses a model that should be overcome. This overcoming will occur when "the rest" emerge from, and in, its diversity. More than reproducing Western universal and abstract concepts, the alternative approach proposed by Mignolo constitutes a type of border thought that addresses the colonialism of Western epistemologies from the perspective of epistemological forces that have been relegated to subordinate forms of traditional, folkloric, religious, or emotional knowledge. Mignolo emphasizes the necessity of permitting expression of pluriversal epistemologies, histories, and local communities that today inhabit the borders or margins of globalization. This approach not only contributes to harmonious coexistence with diverse Amerindian people, but also with all groups whose histories are marked by colonialism and that "have lived or learned in their bodies the trauma, the unconscious lack of respect" (Mignolo 2003b, p. 20). As a vision for the future, he proposes that:

> boundary thinking is one of the possible ways toward a critical cosmopolitanism and a utopian horizon that helps us to construct a world where many worlds can fit. (Mignolo 2003b, p. 58)

Mignolo's critical optic opens a promising road for forms of Earth stewardship that could include diverse forms of life in a pluri-versal conception integrating people, ecosystems, and the other-than-human living beings with whom we co-inhabit. To forge his Latin American Modernity/Coloniality Research Program, Mignolo has collaborated with Arturo Escobar, who has elaborated a geopolitical perspective. Based on his work with Afro-American communities on the Pacific coast of his country of origin, Colombia, Escobar addresses problems of globalization and culture, gender, environment, and territory. In the Afro-American communities of tropical Colombia, he has found solid elements for ecological sustainability in the mythical and symbolic traditions related to specific ecosystems (Noguera 2012). These regional biocultural realities are, however, increasingly threatened by violence, poverty, and degradation of habitats in Latin America. Escobar (1996) opened his landmark book *The Invention of the Third World* by noting that "just a quick look at the biophysical, economic, and cultural landscapes of the Third World shows that the Project of Development is in crisis." Escobar makes an appealing call to inaugurate a *post-development era*. This call is especially relevant for a Latin American approach to Earth stewardship, because under the current model of development the original state of biocultural diversity

[3] By "biocultural phagocytosis" I refer to the appropriation and mercantilization of local cultures, their habitats, life habits, and communities of co-inhabitants.

and social well-being is being replaced by an accelerated process of biocultural homogenization and socio-ecological degradation.[4]

8.5 Eco-theology of Liberation

Broadening the perspective of Escobar, Brazilian liberation theologian Leonardo Boff (1995, p. 24) affirms that "today, in reality, it is not so much the development model that is in a state of crisis, but [more deeply] the model of society that dominates the world." Boff represents a major figure in Latin American environmental thought due to his original Christian Franciscan concepts, and to his arduous work dedicated to the communication of environmental problems and proposals to solve them.[5] In his landmark book *Ecology and Liberation: A New Paradigm*, Boff (1995, pp. 27–28) proposes a holistic, eco-social approach to environmental ethics, affirming that "the new model of society has to aim at a reconstruction of the social fabric, starting from the multiform potentiality of humankind and society." To articulate his holistic eco-theological proposal, Boff distinguishes seven complementary "pathways" or practices of ecology, which could be valuable for a holistic approach to Earth stewardship (Box 8.3).

Box 8.3. Leonardo Boff's Seven Paths of Ecology

Religious organizations in Latin America have linked environmental stewardship to concerns for equity by using the language of human rights. Much of the impetus for such efforts has come from church pastoral work among indigenous, Afro-Latino, and other minority cultural communities, supporting and advocating for their rights to territory, resources, and self-determination, often against government and corporate policies and interests. Such interventions by religious groups, aligned with popular struggle in efforts to combat the social, political, and economic injustices associated with marginalization and poverty, draw upon an established history of pastoral agency in Latin America, inspired by the Liberation Theology movement that emerged in the 1960s. One of the founders of liberation theology, Brazilian Leonardo Boff, has proposed seven peaceful "pathways" or practices of ecology to articulate a holistic eco-theological approach (Fig. 8.4).

(continued)

[4] The process of biocultural homogenization entails simultaneous and interdigitated losses of native biological and cultural diversity at local, regional, and global scales. This process leads to the disruption of the interrelationships between cultures and their land, and results in the massive replacement of native biota and cultures by cosmopolitan species, languages, and cultures (see Rozzi 2001, 2012, 2013).

[5] For example, Leonardo Boff had a leading role in the writing and divulgation of the Earth Charter, a declaration of fundamental ethical principles for building a just, sustainable, and peaceful global society in the twenty-first century. The Earth Charter involved a decade-long (1995–2005), worldwide, cross cultural dialogue on common goals and shared values, and the document has been further enhanced by its endorsement by over 4,500 organizations, including governments and international organizations. See http://www.earthcharterinaction.org/content, and Tucker (2015, in this volume [Chap. 26]).

Box 8.3. (continued)

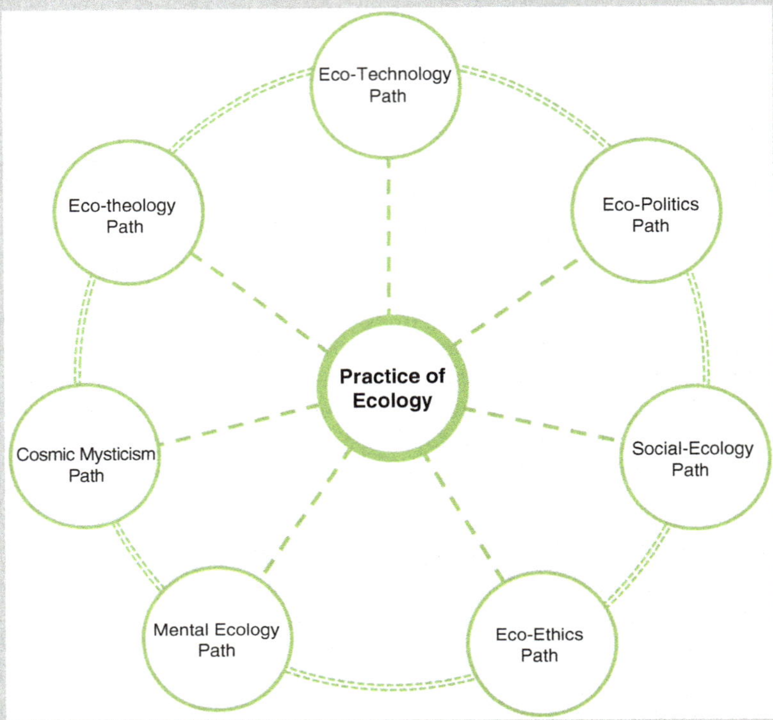

Fig. 8.4 Representation of the seven confluent and complementary pathways for a holistic practice of ecology, proposed by Brazilian, Franciscan, liberation theologian Leonardo Boff

1. The *Eco-technology Path* proposes that the technology that made the Earth bleed should also help to heal it. Economy should be reoriented toward the management of goods that are finite and necessary for human well-being.
2. The *Eco-politics Path* affirms that while human desire is structurally infinite, it should be confined by solidarity. Solidarity leads one to renounce things for the sake of the other, promoting governance and management for the common good. Boff evokes the eco-regional approach of Chico Mendes, the inspirational Brazilian rubber tapper and conservationist, and contrasts it with the socially and ecologically insensitive approach of economic megaprojects (see da Rocha and Possamai 2015, in this volume [Chap. 28]).
3. The *Social-ecology Path* calls for the transformation of the instrumental and mechanistic view that allows a few men and women, institutions, nations, and corporations to exploit without limits other persons, animals, plants, and minerals, thereby stripping them of their autonomy and intrinsic value, and reducing them to mere means to market ends. To counteract

(continued)

Box 8.3. (continued)

this trend, Boff proposes recovering the original meaning of economy: "the administration of the fair and modest means necessary for life and well-being. Rational application of scant income is the central activity of most households in the Third World.... [In this alternative] model of society, not only work but leisure, not only efficiency but gratuitousness, not only productivity but the absurd, playful dimension must be encouraged. Imagination, fantasy, utopia, dreams, emotions, symbolism, poetry, and religion have to be valued as much as production, organization, functionality, and rationality." (Boff 1995, pp. 19 and 28)

4. The *Eco-ethics Path* defines ethics as the "unlimited responsibility for everything that exists and lives. The supreme good is to be found in earthly and cosmic integrity. That does not amount merely to the common good of humanity, but includes the welfare of nature." (Boff 1995, pp. 29–30)
5. The *Mental Ecology Path* emphasizes the diversity of beings inhabiting not only nature but also ourselves, as images, symbols, and values. The water, plants, and animals that inhabit us are archetypes and figures filled with emotions. This understanding counteracts the modern fabrication of the "one-dimensional man."[6] Overcoming this one-dimensionality reintegrates the forces of reason with the multiple forces of the universe that are present in our impulses, visions, intuitions, dreams, and creativity.
6. The *Cosmic Mysticism Path* shows that spirituality and mysticism originate in the sacramental, symbolic, and affective reasoning that captures gratuitousness and the sense of communion among all beings.
7. The *Eco-theology Path*, based on a Christian-Franciscan panentheism, is rooted in tenderness as the main attitude in the encounter with other beings. This Christian praxis generates a cordial knowledge (*cordial*=from the heart) which does not distance itself from diverse realities. Instead it makes possible communion and friendship with them, as was done by St. Francis for whom the moon and the sun, water and fire, the birds and the herbs are our sisters and brothers with whom we share the same divine genealogy.

In the influential article "The Historical Roots of our Ecological Crisis," historian Lynn White Jr. (1967) criticized Western Christianity as "the most anthropocentric religion the world has seen" (1205). However, he concluded his essay proposing Saint Francis "as a patron saint of ecologists" (1207). Later, in 1979 Pope John Paul II formalized it, and in 1986 the World Wide Fund for Nature (WWF) organized in Assisi a meeting that generated The Assisi Declarations from Buddhist, Christian, Hindu, Jewish and Islamic relationships with nature and sacred duty to care for it (ARC 1986). Thirty years later, the Argentine-born Pope Francis took his name from Francis of Assisi, and is writing an encyclical

[6]Boff alludes to Herbert Marcuse's concept, and homonymous book *One-Dimensional Man: Studies in Ideology of Advanced Industrial Society* (Beacon Press: Boston, Massachusetts, 1964).

(continued)

Box 8.3. (continued)
on humanity's relationship with nature. In 2014, he expressed that "one of the greatest challenges of our time is to convert ourselves to a type of development that knows how to respect creation …when I look at America, also my own homeland [South America], so many forests, all cut, that have become lands… that can no longer give life. *This is our sin, exploiting the Earth and not allowing her to give us what she has within her*" (Pope Francis 2014). Saint Francis is not a frozen historical figure but continues to inspire a contemporary, holistic Earth stewardship, grounded in reality rather than dogma.

Starting from an initial impulse favoring the expression of multiple individual potentialities and those of diverse cultures and social groups, Boff's ethical turn toward an ecological ethics arises from the demand to "listen" to the other, to nature. In a recent interview,[7] Boff affirms that "not only the poor cry; also the lands cry, the waters cry, nature cries. Hence, we need an *eco-theology of liberation*" (see also May Jr 2015b, in this volume [Chap. 27]). In this attitude of listening among human and other-than-human beings "the decisive element in ethics is not what we want or what we seek to impose by force (thus creating various different moral standards), but what the same reality states and demands that everyone should heed and be in tune with it…. Human beings live ethically when they decide to stop placing themselves above all others, and decide instead to stand together with others" (Boff 1995, pp. 29–31).

A reconnection with the Earth as a whole, *a dignitas terrae*, demands a material and spiritual reconnection with both the exterior and the interior nature of each human being and society. Governed by market economy, global society tends to occupy the individual's attention with a flood of commercial messages and mundane demands that so assault a person that she or he cannot find their existential center. Another barrier to the dignity and emergence of healthy personhood is injustice in personal and social relations. Unjust processes are doubly inhumane. They force the oppressor to deny that the other (the oppressed) is like him or her, and even to dehumanize the oppressors themselves (to lose their own existential center). Only thus can an oppressor objectify and violate the integrity of the other. Oppression also triggers a process of dehumanization in the victims, a violence suffered through a negation in all aspects of their life, at table, at home, in school, and in the very core of human dignity. Based on this perspective that integrates psychoanalytic and liberation theology approaches, Boff broadens the spectrum of environmental ethics toward an environmental justice that includes poor and marginalized people around the world: the oppressed human beings side-by-side with the oppressed other-than-human beings (the plants, the animals, the running water, the oceans, the mountains, and so on).

In *Cry of the Earth, Cry of the Poor*, Boff (1997, p. 45) situates the concerns of social and political liberation within broader ecological frameworks: "without a minimum of social justice it is impossible to make ecological justice fully effective. The one involves the other." He inaugurates an ecotheology of liberation. According

[7] Unpublished interview by Ricardo Rozzi and Claudia Sepúlveda recorded in October 2008.

to this ecotheology, to achieve ecological justice it is necessary to overcome anthropocentrism and ethnocentrism. In addition, for his holistic approach, Boff demands that both the masculine and feminine be embraced, a position that echoes the philosophy of leading South American ecofeminists and liberation theologians, Ivone Gebara (Brazil) and Gladys Parentelli (Uruguay-Venezuela).

Based on their work with women living in urban poverty, Parentelli (1996) and Gebara (1999) have inaugurated a Latin American theology from the "optic" of women. Women as much as the poor are oppressed, hence poverty is not a gender-neutral category. Vicenta Mamani (2000), an Aymara woman of Bolivia, adds another layer of social oppression: being indigenous. Thus many women suffer triple oppression: gender, class, and race.

Gebara, Parentelli, Mamani and other Latin American ecofemenists show the complexity of socio-environmental problems. Methodologically they have directed their attention to the everyday life of women living in marginal neighborhoods. Moreover, this displacement of the poor is frequently associated with the destruction of their ancestral lands, farms, and working spaces – that is, their "habitats." During the last three decades, ecofeminists have created centers, networks, and periodical publications that explore the relationship between the oppression of women, indigenous people, and nature in Latin America (Ress 2006).

8.6 Earth Stewards and the Biocultural Ethic

The most severe social impacts associated with environmental degradation affect indigenous, peasant, coastal, and other rural and marginalized urban communities. In South America, poor communities (in monetary terms) are not the main agents but rather the main victims of environmental degradation.[8] The biocultural ethic affirms the value of the complex interrelationships among (i) the stewardship practices or *habits* of indigenous and rural women, (ii) their daily interactions with the

[8] The interpretation of poverty as a main cause of environmental degradation is still prevailing. However, many scholars and international organizations, including the United Nations World Commission on Environment and Development (WCED), have offered alternative evidence and approaches. In its landmark Burndtland Report, WCED (1987, p. 117) stated that "there has been a growing realization in national governments and multilateral institutions that it is impossible to separate economic development issues from environment issues; many forms of development erode the environmental resources upon which they must be based, and environmental degradation can undermine economic development. Poverty is a major *cause* and *effect* of global environmental problems. It is therefore futile to attempt to deal with environmental problems without a broader perspective that encompasses the factors underlying world poverty and international inequality." The Brundtland Report addressed in depth the disparities in income and ecological impact among countries, and documented that the countries with lower or middle income economies have 83 % of the world population, but only 21 % of the world's Gross Domestic Product (GDP). Conversely, the countries that are high-income oil exporters or have industrial market economies are inhabited by 17 % of the world population and accumulate 79 % of the world's GDP. The inequalities in income distribution are extreme in Latin American countries. For example, in Brazil, the wealthiest country of the region, the richest fifth of the population concentrates 68 % of the country's GDP, while the poorest fifth shares only 2 % of the national GDP (Baer and Maloney 1997).

land and the community of *co-inhabitants*, and (iii) the conservation and access to ancestral lands or *habitats*.

Ecofeminist analyses tend to agree with perspectives and studies of biological and cultural conservation in Latin America which demonstrate that women are key stewards of the land. Earth stewardship is not gender-neutral, neither is poverty. Indigenous and rural women have harvesting and farming habits that imply a rich empirical knowledge and close interactions with plants and other organisms that become companions or co-inhabitants in daily family life and other social interactions. Women acquire an experiential understanding about the need for conserving the integrity of and access to the habitats where they live and farm, obtain water and other goods. Hence they are not only stewards of the land but often custodians of it, and are leading resistance movements in its defense. Examples include the Afro-Latina *concheras* who defend the mangroves along the Pacific coasts of Colombia and Ecuador (Martínez-Alier 2001; Suárez and Ortiz 2006; Rozzi 2012); Rarámuri or Tarahumara women who defend the forests to assure continuous water supplies in the Sierra Madre of northwestern Mexico (Fig. 8.5, Rozzi 2001); and

Fig. 8.5 Tarahumara indigenous people from the Sierra Madre in northern Mexico marching to the city of Chihuahua to protest illegal deforestation that put at risk their water resources. Women marching with their babies through the streets of Guadalajara were graphically portrayed by US journalist Wesley Boxley in *The New York Times* on April 28, 1999, thereby contributing to stop the illegal deforestation, which ended three months afterwards (Photograph courtesy of Wesley Boxley)

Quechua women who trade complementary vegetal food products farmed or gathered at different altitudinal zones of the Peruvian Andean slopes (Box 8.4).

Box 8.4. Today's Challenges for Stewardship Habits by Quechua Women in Andean Habitats

The central Andes are one of the eight centers of the origin of agriculture, with the domestication of plants in this region dating back at least 8,000 years (NRC 1989). Today, however, indigenous Quechua communities are confronted with the expropriation of land and the resulting territorial displacement motivated mostly by the development of mining projects or of new agricultural practices that include the extensive use of commercial genetically modified varieties of potato. Quechua communities have repeatedly denounced the resulting marginalization of women who traditionally were responsible for the selection, storing, sowing, and harvesting of seeds and tubers of potatoes and other plants. The combined effects of exclusion from native habitats and the marginalization of women threaten the food security of peasant and indigenous communities whose health depends upon the exchange of edible vegetables from different agroecological zones.[9]

The biocultural ethic affirms that the links of specific life Habits with specific Habitats and communities of co-in-Habitants ought to be respected. Its "3Hs" descriptive and normative framework helps to better understand and value the delicate interrelationships among the Quechua *alimentary habits*, the mosaic of *Andean habitats*, and the human and other-than-human communities of *co-in-habitants*. Along the altitudinal gradient of the Andean Cordillera, each of the three primary altitudinal zones provides complementary plant foods for the human diet (Fig. 8.6):

1. In the *high Andean habitats*, or *Puna*, Quechua agricultural habits are rooted in the practice of growing tubers such as oca, isano, and hundreds of varieties of potatoes that are rich in carbohydrates and are co-inhabitants in rituals and everyday life (Mamani-Bernabé 2015, in this volume [Chap. 6]).
2. In the *intermediate-altitude habitats*, or *Andenes* in the sub-Andean terrace cultivation system, Quechua agricultural habits are based on quinoa and corn that provide grains rich in essential amino acids (Krogel 2006).
3. In the *low-land habitats*, including the *Yungas* and Amazonian rain forests, Quechua habits are grounded on planting and harvesting plenty of fruits that are rich in vitamins and coca leaves, which provides for chewing coca, an essential element of Quichua cultural identity (Allen 1981).

[9]Baseline information for this example is found in NCR (1989), Zimmerer (2003), Argumedo and Pimbert (2006), Primack et al. (2006), and Rolph and Obregón (2012).

(continued)

Box 8.4. (continued)

Fig. 8.6 View of the *Sallqantay* valley (*sallqa* = wild or invincible in Quechua language) in the *Willkapampa* mountain range in Peru. *Sallqantay* trail runs from Cuzco to Machu Picchu, the Inca sacred city. Along the trail it is possible to observe the agricultural landscape and harvested traditional crops, such as maize (*Zea mays*) and oca (*Oxalis tuberosa*) (Photograph J. Tomás Ibarra)

Healthy markets based on bartering and run by women from different altitudinal zones are interrupted by the territorial displacements of Quechua communities. Current displacements are caused by mining and other development projects, by global climate change, and by the substitutions of native varieties of plants by commercial and modern genetically modified varieties. This ecological and social disruption provokes:

(a) Losses of autonomy and capacity for self-determination of indigenous communities, due to the destruction and/or denial of access to their ancestral habitats and territories, which are essential for the continuity of their material and spiritual subsistence.
(b) Degradation of local economies and relations of reciprocity among the diverse human communities, and between these communities and regional ecosystems.
(c) Degradation of regional biological diversity; for example, of the more than 2,000 varieties of potatoes that have been traditionally cultivated in the high Andean slopes.

(continued)

Box 8.4. (continued)

(d) Losses of traditional ecological and cultural knowledge and practices; for example, the disappearance of local markets where women offer and exchange a wide variety of foods, provoking food insecurity leading to malnutrition, dependency, losses of autonomy and dignity.
(e) Immigration of Quechua women and their families toward marginal neighborhoods in cities where most frequently end up living in conditions of extreme poverty.

The clear interdependencies among the *life-habits* of human and other-than-human communities of *co-in-habitants* along Andean altitudinal gradients demonstrate that the conservation of *habitats* and access to them is a necessary condition for the autonomy, identity, dignity, and well-being of local communities. Assuring conditions that allow Quechua stewards to have access to their ancestral land constitute a challenge and a responsibility for an intercultural Earth stewardship endeavor involving all society.

The conservation of habitats and access to them is the condition of possibility for the autonomy, identity, dignity, continuity of habits, and well-being of local communities. The formal proposal of the biocultural ethic interrelates habits and habitats with the identities and well-being of the co-inhabitants, humans and other-than-humans. Consequently, the conservation of habitats and access to them by communities of co-inhabitants becomes an ethical imperative. The biocultural ethic demands that this imperative be incorporated into development policies as a matter of socio-environmental justice.

Once displaced from their traditional lands and ways of life, indigenous people, peasant, and fishermen communities often confront material and cultural misery in cities. In the marginal neighborhoods of metropolitan areas in Latin America, these displaced people frequently lack access to basic services, such as food, water, shelter, and sanitary conditions (Parentelli 1996; Gebara 1999; Rozzi 2001). Hence, they face extreme conditions of poverty that are rapidly expanding in the marginal neighborhoods of metropolitan areas in Latin America. At the same time, their ancestral lands lose their traditional stewards and local custodians, and become more vulnerable to large-scale, non-sustainable forms of exploitation. Box 8.4 illustrates the importance of conserving both the traditional habits and the regional habitats that have sustained the well-being of human and other-than-human communities in the Andean Cordillera.

I have proposed a *biocultural ethic* that aims to recover an integral understanding of the interrelationships among the cultural habits and the habitats where these habits take place (Rozzi 2012). I say recover, because these links have been largely ignored by modern dominant ethics that are centered on eurocentric human habits. However, pre-Socratic and other early Western philosophies (as well as ancestral Amerindian ecological worldviews) provide an ancient cultural foundation to support an ethic that affirms the value of the vital links between the life habits of the inhabitants and the habitats where these habits are practiced. Today, the value of these vital links is also confirmed by ecological and social sciences. The attention to the daily life of human communities and their biocultural landscapes – including ecosystems, historical, socio-political, and cultural settings – contributes to:

(i) discovering the inexhaustible biocultural diversity embedded in the spatial and temporal heterogeneity of the Latin American region, and
(ii) understanding how today these diverse human and other-than-human forms of life are threatened by development projects that are insensitive to their existence.

The Catalan (Spain) ecological economist Joan Martinez-Alier stresses that in Latin America conservation is far from being a luxury. On the contrary, the commitment and action in favor of conservation often springs from those communities who depend directly on natural resources to live. This conservation perspective is known as *environmentalism of the poor* (Martínez-Alier 2002).

Resistance movements and recurrent appeals to conservation made by local communities aim to maintain sustainable ecological practices rooted in regional biological and cultural diversity. To better understand the interrelationships between biological and cultural diversity, the Mexican ecologist Victor Toledo, founding editor of the journal *Etnoecologica,* promotes the study of the relationships between Amerindian cultures and nature.[10] Toledo has emphasized the necessity of developing hybrid disciplines that integrate the cultural, social, and ecological dimensions to enable communication and mutual respect among different socio-cultural actors (Toledo 2003; Toledo and Castillo 1999). The perspectives of diverse indigenous, peasant, and fisherman communities agree with those of ecologists and other researchers regarding the fact that levels of autonomy and social well-being are higher in areas where ecosystems and biodiversity have been protected (Rozzi and Feinsinger 2001). To enhance the understanding of this "win-win relationship" between the well-being of humans and of biotic communities and their ecosystems, and to better understand the value of their expression in local life histories, I have integrated ecological sciences and environmental ethics into the practice of *field environmental philosophy* (Rozzi et al. 2008; see Aguirre Sala 2015 in this volume [Chap. 15]). In this practice, students participate with philosophers, ecologists, and other researchers in long-term transdisciplinary projects of biocultural conservation. This *in situ* experience involves "face to face" encounters with co-inhabitants (human and other-than-human), their habits and habitats. With this methodology "biocultural diversity ceases to be a mere concept, and begins to be an experience of

[10] In Latin America, as in other regions of the world, ethnoecology has been essential to disclose the richness of Amerindian worldviews and the value of traditional ecological practices (Ulloa et al. 2001). This interdisciplinary field has involved fruitful collaborations between Latin American and international researchers. Indeed, the International Society of Ethnobiology was founded in Belem Brazil during the First International Congress of Ethnobiology in 1988. It involved an active collaboration between Brazilian, Latin American, and international researchers under the leadership Darrell Posey. In the 1990s, Victor Toledo's collaboration with U.S. ethnobotanist Janis Alcorn was essential to establish the journal Etnoecologica and to promote a conservation approach which attempts "to stabilize the traditional conservation ethics wherever it still exists, and improve the modern conservation ethic" (Alcorn 1993).

co-inhabitation with diverse living beings and life histories that regularly remain outside of areas considered in formal education and decision making" (Rozzi et al. 2008, p. 335). The field environmental philosophy methodological approach has allowed the incorporation of biocultural diversity into regional development policies, territorial planning, formal and non-formal education programs, including ecotourism (see Ogden et al. 2015 in this volume [Chap. 10]). In the context of academia and the ESA's Earth Stewardship Initiative, field environmental philosophy offers a methodology for students and researchers to integrate the theory and practice of ecology and ethics into intercultural, interdisciplinary, inter-institutional, and international forms of ecosystem co-management (Rozzi et al. 2012). In this way, field environmental philosophy provides a methodological basis for heterogeneous but articulated initiatives of Earth stewardship.

8.7 Concluding Remark

To a great extent the main challenge to an intercultural Earth stewardship is not to invent new paradigms but rather to allow the many traditions of stewardship to continue. Governed by a narrow neoliberal free-market economy, global society is blind to the beauty and refinement of traditions of environmental thought, ecological worldviews and practices, and forms of biocultural co-inhabitation that take place in Latin America and other overlooked regions of the world. By changing and enriching the language of global discourses and mindsets, Latin American philosophies contribute to broadening and modifying narrow economic mindsets and policies that are driving massive biocides and linguicides.

The recent establishment of indigenous networks and organizations, involving Amerindian people that had no contact with Western civilization prior to the 1950s, shows the accelerated dynamic and solidarity of resistance movements that include forms of Earth stewardship. Nourished by the collective work of ecologists, environmental philosophers, theologians, anthropologists, and other researchers together with fishermen, indigenous communities, farmers, government authorities, artists, journalists, and diverse members of society, who are collectively forging ethical guides, rooted in multiple modes of co-inhabiting in diverse biocultural landscapes, Earth stewardship could sprout with increasing strength today.

References

Abell R, Thieme ML, Revenga C et al (2008) Freshwater ecoregions of the world: a new map of biogeographic units for freshwater biodiversity conservation. BioScience 58:403–414

Aguirre Sala J (2015) Hermeneutics and field environmental philosophy: integrating ecological sciences and ethics into earth stewardship. In: Rozzi R, Chapin FS III, Callicott JB et al (eds) Earth stewardship: linking ecology and ethics in theory and practice. Springer, Dordrecht, pp 235–247

Alcorn J (1993) Indigenous people and conservation. Conserv Biol 7:424–426

Allen JC (1981) To be Quechua: the symbolism of coca chewing in highland Peru. Am Ethnol 8:157–171

Alliance of Religions and Conservation [ARC] (1986) The Assisi declarations. Messages on humanity and nature from Buddhism, Christianity, Hinduism, Islam and Judaism. ARC, Bath. http://www.arcworld.org/downloads/THE%20ASSISI%20DECLARATIONS.pdf. Accessed 17 Mar 2014

Argumedo A, Pimbert M (2006) Protecting indigenous knowledge against biopiracy in the Andes. International Institute for Environment and Development, London

Baer W, Maloney W (1997) Neoliberalism and income distribution in Latin America. World Dev 25:311–327

Boff L (1995) Ecology and liberation: a new paradigm. Orbis Books, New York

Boff L (1997) Cry of the earth, cry of the poor. Orbis Books, New York

Callicott JB (1994) Earth's insights: a multicultural survey of ecological ethics from the Mediterranean basin to the Australian outback. University of California Press, Berkley

Ceccon E, Miramontes O (1999) Mechanisms and social actors in the deforestation of the Brazilian Amazon. Interciencia 24:112–119

Chapin FS III, Kofinas GP, Folke C (eds) (2009) Principles of ecosystem stewardship: resilience-based natural resource management in a changing world. Springer, New York

Chapin FS III, Pickett STA, Power ME et al (2011a) Earth stewardship: a strategy for social-ecological transformation to reverse planetary degradation. J Environ Stud Sci 1:44–53

Chapin FS III, Power ME, Pickett STA et al (2011b) Earth stewardship: science for action to sustain the human-earth system. Ecosphere 2:89

Chapin FS, Pickett STA, Power ME et al (2015) Earth stewardship: an initiative by the Ecological Society of America to foster engagement to sustain planet Earth. In: Rozzi R, Pickett STA, Callicott JB et al (eds) Earth stewardship: linking ecology and ethics in theory and practice. Springer, Dordrecht, pp 173–194

Da Rocha FJR, Possamai FV (2015) Chico Mendes and Jose Lutzenberger: ecosystem management at multiple scales of government. In: Rozzi R, Chapin FS III, Callicott JB et al (eds) Earth stewardship: linking ecology and ethics in theory and practice. Springer, Dordrecht, pp 419–430

Descola P (2013) The ecology of others. Prickly Paradigm Press, Chicago

Dussel E (1996) The underside of modernity: Apel, Ricoeur, Rorty, Taylor and the philosophy of liberation (Translated and edited by Eduardo Mendieta). Humanity Books, New York, p 14

Dussel E (2003) Algunos Principios para una Ética Ecológica Material de Liberación (Relaciones entre la Vida en la Tierra y la Humanidad). In: Pixley J (ed) Por un mundo otro. Alternativas al mercado global. Quito, Ecuador, Comunidad Cristiana Mesoamericana y Consejo Latinoamericano de Iglesias, pp 29–44

Escobar A (1996) Encountering development: the making and unmaking of the third world. Princeton University Press, Princeton

Espinosa C (2013) The riddle of leaving the oil in the soil—Ecuador's Yasuní-ITT project from a discourse perspective. For Policy Econ 36:27–36

Falconi-Puig J (2013) The guardian (19 Sept 2013). UK

Finer M, Jenkins CN, Pimm SL, Keane B, Ross C (2008) Oil and gas projects in the western Amazon: threats to wilderness, biodiversity, and indigenous peoples. PLoS One 3:e2932

Finer M, Vijay V, Ponce F, Jenkins CN, Kahn TR (2009) Ecuador's Yasuni Biosphere Reserve: a brief modern history and conservation challenges. Environ Res Lett 4(3):034005

Fornet-Betancourt R (1994) Hacia una Filosofía Intercultural Latinoamericana. DEI, San José

Fornet-Betancourt R (1998) Supuestos filosóficos del diálogo intercultural. Utopia y Praxis Latinoamericana 5:51–64

Gebara I (1999) Longing for running water: ecofeminism and liberation. Augsburg Fortress Press, Minneapolis

Guevara S, Laborde J (2008) The landscape approach: designing new reserves for protection of biological and cultural diversity in Latin America. Environ Ethics 30(3):251–262

Kingsland SE (2015) Ecological science and practice: dialogues across cultures and disciplines. In: Rozzi R, Chapin FS III, Callicott JB et al (eds) Earth stewardship: linking ecology and ethics in theory and practice. Springer, Dordrecht, pp 17–26

Krogel AM (2006) *Ukhu Mankakuna*: culinary representations in Quechua cultural texts. Doctoral dissertation, comparative literature. University of Maryland, College Park

Kusch R (1962) America Profunda. Libreria Hachette, Buenos Aires
Kush R (1976) Geocultura y el Hombre Americano. Fernando García Cambeiro, Buenos Aires
Li B, Parr T, Rozzi R (2015) Geographical and thematic distribution of publications generated at the International Long-Term Ecological Research Network (ILTER) sites. In: Rozzi R, Chapin FS III, Callicott JB et al (eds) Earth stewardship: linking ecology and ethics in theory and practice. Springer, Dordrecht, pp 195–216
Mamani V (2000) Identidad y Espiritualidad de la Mujer Aymara. Fundación Basilea, La Paz
Mamani-Bernabé V (2015) Spirituality and the Pachamama in the Andean Aymara worldview. In: Rozzi R, Chapin FS III, Callicott JB et al (eds) Earth stewardship: linking ecology and ethics in theory and practice. Springer, Dordrecht, pp 65–76
Martinez-Alier J (2001) Ecological conflicts and valuation: mangroves versus shrimps in the late 1990s. Environ Plan C 19:715–716
Martinez-Alier J (2002) Environmentalism of the poor. Edward Elgar, Cheltenham
May Jr RH (2015a) Andean llamas and earth stewardship. In: Rozzi R, Chapin FS III, Callicott JB et al (eds) Earth stewardship: linking ecology and ethics in theory and practice. Springer, Dordrecht, pp 77–86
May Jr RH (2015b) Dorothy Stang: monkeys cry and the poor die, earth stewardship as liberation ecology. In: Rozzi R, Chapin FS III, Callicott JB et al (eds) Earth stewardship: linking ecology and ethics in theory and practice. Springer, Dordrecht, pp 407–418
Mignolo W (1995) The darker side of the renaissance: literacy, territoriality and colonization. University of Michigan Press, Ann Arbor
Mignolo W (2003a) Historias locales/Diseños globales. Ediciones Akal, Madrid
Mignolo W (2003b) The darker side of the renaissance: literacy, territoriality and colonization. University of Michigan Press, Ann Arbor
National Research Council [NRC] (1989) The lost crops of the Incas: little-known plants of the Andes with promise for worldwide cultivation. NRC, National Academy Press, Washington, DC
Neugebauer GP III (2003) Indigenous peoples as stakeholders: influencing resource-management decisions affecting indigenous community interests in Latin America. N Y Univ Law Rev 78:1227–1261
Noguera P (2012) Augusto Angel-Maya and environmental philosophy in Colombia. Environ Ethics 34:361–370
Ogden L, Heynen N, Oslender U et al (2015) The politics of earth stewardship in the uneven Anthropocene. In: Rozzi R, Chapin FS III, Callicott JB et al (eds) Earth stewardship: linking ecology and ethics in theory and practice. Springer, Dordrecht, pp 137–157
Olson DM et al (2001) Terrestrial ecoregions of the world: a new map of life on Earth. BioScience 51:933–938
Parentelli G (1996) Latin America's poor women. In: Ruether RR (ed) Women healing earth: third world women on ecology, feminism, and religion. Orbis Books, Maryknoll, pp 29–38
Pellegrini L, Arsela M, Falconíb F et al (2014) The demise of a new conservation and development policy? Exploring the tensions of the Yasuní ITT initiative. Extr Ind Soc 1:125–134
Pope Francis (2014) Exploitation of nature sin of our time. The Huffington Post USA: http://www.huffingtonpost.com/2014/07/05/pope-francis-nature-environment-sin-_n_5559631.html. Accessed 5 July 2014
Primack R, Rozzi R, Feinsinger P, Dirzo R, Massardo F (2006) Fundamentos de Conservación Biológica: Perspectivas Latinoamericanas. Fondo de Cultura Económica, México D.F
Ress MJ (2006) Ecofeminism in Latin America. Orbis Books, New York
Rival L (2012) The materiality of life: revisiting the anthropology of nature in Amazonia. Indiana 29:127–143
Rolph KS, Obregón Lázaro MF (2012) Quechua Oqrakashqa: the effects of mining consortia and globalization on local Quechua communities in the Peruvian Andes. Program on human rights, Center on Democracy, Development, and the Rule of Law Freeman Spogli Institute for International Studies, Stanford University, working paper 012, http://iis-db.stanford.edu/pubs/23774/Rolph-Lazaro_012.pdf

Romero Bedregal H (2006) Geocultura de Tarapacá: poiesis, autopoeisis y decisiones culturales. Revista Ciencias Sociales 16:16–37
Rozzi R (2001) Éticas ambientales latinoamericanas: raíces y ramas. In: Primack R, Rozzi R, Feinsinger P et al (eds) Fundamentos de Conservación Biológica: Perspectivas Latinoamericanas. Fondo de Cultura Económica, México, pp 311–362
Rozzi R (2012) Biocultural ethics: recovering the vital links between the inhabitants, their habits, and habitats. Environ Ethics 34:27–50
Rozzi R (2013) Biocultural ethics: from biocultural homogenization toward biocultural conservation. In: Rozzi R, Pickett STA, Palmer C, Armesto JJ, Callicott JB (eds) Linking ecology and ethics for a changing world: values, philosophy, and action. Springer, Dordrecht, pp 9–32
Rozzi R, Feinsinger P (2001) Desafíos para la conservación biológica en Latinoamérica. In: Primack R, Rozzi R, Feinsinger R, Dirzo F, Massardo F (eds) Fundamentos de Conservación Biológica: Perspectivas Latino-americanas. Fondo de Cultura Económica, México, pp 661–688
Rozzi R, Poole A (2011) Habitats-habits-inhabitants: a biocultural triad to promote sustainable cultures. In: Parodi O, Ayestaran I, Banse G (eds) Sustainable development: relationships to knowledge, ethics and culture. Karlsruher Studien Technik und Kultur Seroes, Kalsruher Institut für Technologie (KIT), Kalsruhe, pp 53–74
Rozzi R, Arango X, Massardo F et al (2008) Field environmental philosophy and biocultural conservation. Environ Ethics 30:325–336
Rozzi R, Armesto JJ, Gutiérrez J et al (2012) Integrating ecology and environmental ethics: earth stewardship in the southern end of the Americas. BioScience 62:226–236
Sawyer S (2004) Crude Chronicles: indigenous politics, multinational oil, and neoliberalism in Ecuador, American Encounters/Global Interactions Series. Durham Press, Durham
Schlesinger WH, Bernhardt ES (2013) Biogeochemistry: an analysis of global change. Elsevier, Waltham
Spalding MD, Fox HE, Allen GR et al (2007) Marine ecoregions of the world: a bioregionalization of coastal and shelf areas. BioScience 57:573–583
Suárez L, Ortiz D (2006) Producción de camarones y destrucción de manglares en Ecuador. In: Primack R, Rozzi R, Feinsinger P, Dirzo R, Massardo F (eds) Fundamentos de Conservación Biológica: Perspectivas Latinoamericanas, 2nd edn. Fondo de Cultura Económica, México, pp 195–197
Tobasura-Acuña I (2006) Ambientalismos y Ambientalistas: El Ambientalismo Criollo a Finales del Siglo XX. Universidad de Caldas, Colombia, Cuadernos de Investigación, no 21, Cali
Toledo T (2003) Ecología, Espiritualidad y Conocimiento: De la Sociedad del Riesgo a la Sociedad Sustentable. PNUMA, Mexico City
Toledo V, Castillo A (1999) La ecología en Latinoamérica: ocho tesis para una ciencia pertinente en una región de crisis. Interciencia 24:157–168
Tucker ME (2015) World religions, ethics, and the earth charter for a sustainable future. In: Rozzi R, Chapin FS III, Callicott JB et al (eds) Earth stewardship: linking ecology and ethics in theory and practice. Springer, Dordrecht, pp 395–405
Ulloa-Cubillos A, Campos C, Rubio H (2001) Manejo local por los Embera del Chocó colombiano. In: Primack R, Rozzi R, Feinsinger P et al (eds) Fundamentos de Conservación Biológica: Perspectivas Latinoamericanas. Fondo de Cultura Económica, México, pp 599–601
Viola E, Basso L (2015) Earth stewardship, climate change, and low carbon consciousness: reflections from Brazil and South America. In: Rozzi R, Chapin FS III, Callicott JB et al (eds) Earth stewardship: linking ecology and ethics in theory and practice. Springer, Dordrecht, pp 367–382
WCED (1987) Our common future. Oxford University Press, Oxford
White L (1967) The historical roots of our ecological crisis. publicado en Science 155:1203–1207
Zimmerer KS (2003) Geographies of seed networks for food plants (potato, ulluco) and approaches to agrobiodiversity conservation in the Andean countries. Soc Nat Resour An Int J 16:583–601

Chapter 9
Implications of the Biocultural Ethic for Earth Stewardship

Ricardo Rozzi

Abstract The biocultural ethic affirms the vital value of the links that have coevolved between specific life *habits, habitats*, and communities of *co-in-habitants* ("3Hs"). The conservation of habitats and access to them by communities of co-inhabitants is the condition of possibility for the continuity of their life; it becomes an ethical imperative that should be incorporated into development policies as a matter of eco-social justice. The conceptual framework of the biocultural ethic recognizes that there are numerous communities (inhabiting cities, rural, or remote areas) with cultural traditions that have ethical values centered in life, sustainable practices, and low environmental impact. It also recognizes agents that have values centered on short-term profit, non-sustainable practices, and disproportionately high environmental impact. Therefore, it would be technically and ethically right to define and enforce differential responsibilities among social groups, corporations, and nations that are contributing to the negative socio-environmental impacts that we face today. We have now reached a state of "plutonomy" that is dividing the world into two blocs: the wealthy 1 % of the world's population that owns 50 % of the world's wealth, and "the rest." To achieve Earth stewardship, this trend needs to be overcome by (i) changing the current regime of plutocracy towards one of more participatory democracy that ceases to be indifferent to the well-being of the majority of human and other-than-human living beings, (ii) reorienting the current habits of plutonomy, and its associated consumerism and land-grabbing practices, towards habits of stewardship, and (iii) broadening the prevailing perspective of ecosystem services toward an ethical concept of sustainable co-inhabitation. By more precisely identifying the diversity of Earth stewards, their languages, values, cultures, and practices in heterogeneous habitats of the planet, as well as the specific agents that are mostly responsible for current socio-environmental problems, the biocultural ethic can significantly contribute to orient clearer collaborative and supportive ways for a responsible and intercultural Earth stewardship.

R. Rozzi (✉)
Department of Philosophy and Religion Studies, University of North Texas, Denton, TX, USA

Institute of Ecology and Biodiversity, Santiago, Chile

Universidad de Magallanes, Punta Arenas, Chile
e-mail: rozzi@unt.edu

R. Rozzi et al. (eds.), *Earth Stewardship*, Ecology and Ethics 2,
DOI 10.1007/978-3-319-12133-8_9

Keywords Ecosystem services • Environmental justice • Global environmental change • Plutocracy • Sustainability

9.1 Introduction

The conceptual framework of the biocultural ethic links life habits with specific habitats and communities of co-in-habitants ("3Hs"), thereby emphasizing the great heterogeneity of the human species and its interrelationships with biodiversity (Rozzi 2012). Consequently it demands a change in language from a prevailing reference to the human species as a whole for causing "humanity's unsustainable environmental footprint" (*sensu* Hoekstra and Wiedmann 2014) toward a language that names and distinguishes specific human groups or individuals who have negative or have favorable environmental impacts. Complementarily, the biocultural ethic's conceptual framework discloses philosophical concepts and ecological practices for Earth stewardship that are still little known within academia, global discourses and decision making (Callicott 1994; Rozzi 2001; Berkes 2007, 2012). The previous chapters in this book offered a characterization of ecological worldviews and practices in Asia, South- and North-America. In this chapter I explore two sets of questions.

First, if there is a plethora of ways of conceiving and practicing Earth stewardship that have so much to offer to sustainability practices and global ecological discourses, such as those reviewed for Latin American schools of thought and living cultures (see Chap. 8 in this volume), why do they remain ignored? If there are so many cultural traditions and Earth stewards whose life habits imply a low ecological footprint and promote a sustainable life, why do we face a global environmental crisis today? To start answering these questions, I use the coupled Earth Stewardship/Biocultural Ethic's conceptual framework developed in Chap. 8 to examine: Who are the *stewards* that are most responsible for sustainable practices and who are the *human co-inhabitants* that are most responsible for the current excessive environmental footprints of humankind? Which *stewardship* and other *cultural habits* drive sustainable forms of co-inhabitation and which drive the largest unsustainable environmental footprints? In which locations or *habitats* do these unsustainable habits mostly take place?

The second set of questions explores what can the biocultural ethic, and more broadly environmental philosophy, contribute to the conceptual and practical framework of the Ecological Society of America's (ESA) Earth Stewardship Initiative (*sensu* Power and Chapin 2009; Chapin et al. 2011a, b, 2015, in this volume [Chap. 12]). Some concepts associated with Earth stewardship have elements in common with those proposed by the biocultural ethic. At the same time, some of the philosophical concepts of the biocultural ethic are incommensurable with those of the Earth Stewardship Initiative, and more importantly with those prevailing in today's global discourse. The identification of these incommensurable concepts enables a critical analysis of the prevailing global discourse of governance, while disclosing alternative ecological worldviews and practices of living cultures that can contribute to Earth stewardship. In this chapter I will analyze three core incommensurable concepts by using binary opposition to terms that, explicitly or implicitly, prevail in the global discourse: (i) democracy versus plutocracy, (ii) stewardship versus plutonomy, and (iii) biocultural co-inhabitation versus ecosystem services. Then, I will discuss the need to recover philosophical language and practices in order to foster intercultural dialogues, negotiations, and collaborations at multiple scales, with diverse stewards and languages, interacting in diverse local realities confronted with narrow economic prevailing global discourses, and forms of governance.

To conduct this analysis I will continue coupling the "3Hs" of the biocultural ethic with the three main components identified in Chap. 8 (Rozzi 2015, in this volume) as essential for an Earth Stewardship Initiative: Habitat/Earth, Habit/Stewardship, co-in-Habitants/Stewards. In addition, I will analyze this coupling from the perspectives of three families of ecological worldviews (involving traditions of ethic and philosophical thought, understood in a broad sense), that inform the biocultural ethic: (a) Amerindian ecological worldviews, (b) non-mainstream Western philosophies, and (c) contemporary ecological-evolutionary sciences. In turn, the biocultural ethic and Earth stewardship encompass a transdisciplinary endeavor (involving science, policy, economy, law, history, aesthetics, religion, ethics) that takes place at the interface of multiple institutions and practices. For this reason, the biocultural ethic incorporates an institutional, social-political, infrastructural-technological realm, in addition to the biophysical and symbolic-linguistic-cultural realms of reality analyzed in the previous chapter (Fig. 9.1).

9.2 Democracy Versus Plutocracy

Earth stewardship entails not only sciences but also governance (Steffen et al. 2011, p. 754). This adds a layer of complexity that limits, or modulates, the implementation of recommendations derived from the work of ecologists and other researchers committed to the Earth Stewardship Initiative. Regarding governance limitations for the implementation of an Earth stewardship as a "strategy for social–ecological transformation to reverse planetary degradation," (Chapin et al. 2011b, p. 44) former presidents and other ecologists of the Ecological Society of America (ESA) have critically observed that:

> Although the serious degradation of the Earth's system is widely recognized by the scientific community, governments are frequently reluctant to adopt policies that would radically reduce the rates of change and degradation, for fear of the *economic costs*. Aggressive actions that are taken now, however, are likely to be much less costly than the costs of failing to act (Stern 2007; NRC 2010). Institutional inertia and *cultural habits* are additional impediments to action. (Chapin et al. 2011b, p. 45; emphasis added)

For changing cultural habits, philosophy can make a valuable contribution: to clarify language and a cultural mentality embedded in it. Criticism such as that made by Chapin and collaborators suggest that ecological information is available, but decision making is governed ultimately by narrow economic interests. However, as they point out, even for economic reasons it would be wiser to include ecological information in governance decision making. The question is: how can we achieve institutional and cultural changes that make this possible?

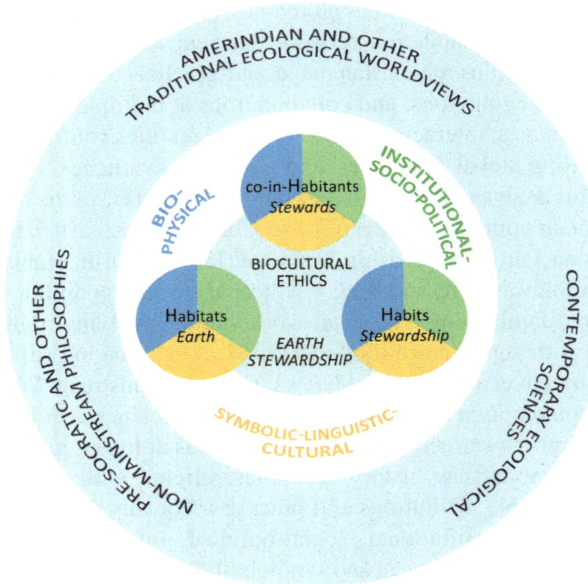

Fig. 9.1 The biocultural ethic affirms the vital value of the links that have coevolved between specific life Habits, Habitats, and communities of co-in-Habitants ("3Hs"). In the inner level of this figure, the core components of the biocultural ethic are matched with the core components identified for Earth stewardship: Habitat/Earth, Habit/Stewardship, co-in-Habitants/Earth Stewards (see Chap. 8 in this volume). The intermediate level illustrates that each of the core-components is constituted by biophysical dimensions (*blue*), symbolic-linguistic-cultural dimensions (*yellow*), and institutional-socio-political, infrastructural-technological dimensions (*green*). The green color, a blending of *blue and yellow*, indicates that the biocultural ethic requires that the particular biophysical and symbolic-linguistic–cultural dimensions are carefully considered by the institutional, policy, and infrastructure dimensions. The external *circle* is based on comparative philosophical analyses, which focus on three families of ecological worldviews that inform the biocultural ethic: (a) Amerindian and other non-Western ecological worldviews, (b) pre-Socratic and other non-mainstream Western philosophies, and (c) contemporary ecological sciences. The biocultural ethic fosters inter-cultural dialogues and practices among heterogeneous, rich cultural traditions and communities of Earth stewards (which are often overlooked in academia and socio-environmental policy-making). The *circular* forms in the figure indicate that both scientific and traditional ecological forms of knowledge and practices (including its multiple components and interactions) are dynamic; they have changed historically, and I emphasize that they can change today to orient forms of inter-cultural Earth stewardship

A first philosophical clarification is that: if decision making is governed ultimately by economic interests and the power of the wealthy, then the governance regime should be called plutocratic rather than democratic (cfr. Lutz et al. 2007; Freeland 2012). *Plutocracy* (Gr. *ploutos* = wealth; *kratos* = power or rule) defines a society ruled by the small minority of the wealthiest citizens, a form of *oligarchy* (Gr. *olígos* = few), while *democracy* (Gr. *dêmos* = people) defines a society ruled by the majority of people. Former U.S. president Abraham Lincoln concisely defined democracy as the "government of the people, *by* the people, *for* the people" (Lijphart 2012, p. 1; emphasis added). Democracy exhibits a variety of formal institutions and forms of organization, but all of them suppose consensual deliberation and decision making processes based on the best available information and arguments (see Dussel 2003). In this sense, a genuine democracy should attend to the concerns expressed by Chapin et al. (2011b) and other members of the ESA. However, this type of democratic process is often aborted, and democracy is reduced to a simple voting mechanism that ends up legitimizing the plutocratic regime (Borrero 2002; Winters 2011a). We can conclude that rather than by a democratic regime, today we are governed by a plutocratic regime (Winters 2011b).

Clarification of language and of a cultural mentality helps to diagnose problems that limit the implementation of environmental policies. The problem is not democracy, but the plutocratic regime of democracy. Colombian environmental lawyer José María Borrero, with reference to Latin America, has developed a critical evaluation of the current status of democracy, the participation of communities, types of dialogue, and levels of respect for social and economic differences. In his book *Imaginación Abolicionista* ("Abolitionist Imagination"), Borrero (2002, p. 129) states that in the collective mindset "political participation is increasingly deceptive, and becomes a trap." It gives the impression of broad popular participation, when in fact decisions are made by small but powerful economic minorities. Borrero (2002, p. 130) illustrates this deception with a graffiti found in the streets of Cali in Colombia that conjugates the verb "to participate" in the following manner:

Yo participo	I participate
Tú participas	You participate
Él participa	He participates
Ella participa	She participates
Nosotros (as) participamos	We participate
Vosotros (as) participáis	You (plural) participate
Ellos deciden	They decide

The conclusion of this graffiti is supported by cases such as the Yasuní Biosphere Reserve in Ecuador and the conflicts between the U'wa and Occidental Petroleum (Oxy) in Colombia (see Box 8.2 in this volume). Borrero's criticism is that democracy has been transformed into mere mechanisms for election and legitimization of governments. Democracy is reduced to competition among groups of elites, and citizens are treated as consumers for a political market. The tradition of citizen participation in Latin America is further discouraged by "administrative corruption, clientelism, bossism, and the lack of political and judicial security" (Borrero 2002, p. 131).

Borrero's conclusion concurs with the analysis developed for Brazil and South America. Viola and Basso (in this volume [Chap. 24]) caution that "when the plutocratic links reach such unbearable levels that corruption scandals become common, the resort to populism – another common discourse in the region – does not bring any relief." Plutocracy goes hand in hand with short-term thinking focused on quick profits that become more important than the ecological destruction of the Earth and humanity. After the Earth Summit Rio+20 held in Brazil in 2012, Viola and Basso lament that "short-term thinking in climate change consideration leads to South American failure to adopt the vanguard position that would be expected from a continent that leads in low carbon assets."[1] In North America, environmental lawyer William J. Snape, III has called on to the government of the United States to ratify the UN Convention on Biological Diversity (CBD, Snape 2010, see also Jamieson 2014). The United States is one of the only three countries worldwide that has not done so:

> What is missing in the US is any urgency to seek durable solutions to many of these problems. How this has come to be is a modern lesson in the power of oligarchical segments to take over political parties. In other words, old guard corporate users of the Earth's biological resources will not succumb lightly to new economic-ecologic paradigms that weaken their power. (Snape 2012, p. 3)

The evidence provided by ecological sciences as well as environmental law allows us to conclude that in order to achieve the ESA Earth Stewardship Initiative's central goal of "shaping of trajectories of change in coupled social–ecological systems at local-to-global scales to enhance ecosystem resilience and promote human well-being" (Chapin et al. 2011b, p. 45), *it is indispensable to change the current plutocratic regime toward a democratic one*.[2]

The clear distinction between plutocracy and democracy, and a reinforcement of the latter over the former form of governance, will enhance intercultural and interregional dialogues and negotiations at a planetary scale, which today are fostered by social networking, linked to communication and information technologies. This clarification and change in governance regime are necessary steps toward implementing Earth Stewardship and other international environmental initiatives (e.g., CBD) that better acknowledge and respect the linguistic and cultural diversity of communities, with their environmental and social interests, embedded in their ecological worldviews and practices.

9.3 Stewardship Versus Plutonomy

Will Steffen and collaborators (2011, p. 757) have lucidly argued that an effective Earth stewardship "can be built around scientifically developed boundaries for critical Earth System processes that must be observed for the Earth System to remain

---

[1] For example, in March 2008, the Brazilian House of Representatives passed a bill to change the law that governs forests. This change in legislation that could undermine authorities' power to halt deforestation was passed despite the established scientific facts that deforestation causes 15 % of global greenhouse gas emissions, and 75 % of Brazil's (Tollefson 2011).

[2] My conclusion concurs with Barry Commoner's concept of "ecodemocracy," which demands new social obligations to guide the course of both environmental improvement and economic development through democratic governance and make decisions that today are normally made on purely private economic grounds, such as profit maximization, by corporate managers. Commoner (1990) emphasized that the environment (whether local or planetary) is a sovereign social responsibility that takes precedence over the private interest in exploiting it.

within a Holocene-like state." However, current humanity's global ecological footprint exceeds Earth's annual biocapacity. Since 2007, humanity is annually consuming one-and-a-half times the biocapacity of the planet (Hoekstra and Wiedman 2014). A coupled Earth Stewardship/Biocultural Ethic's approach emphasizes that to amend this unsustainable path it is necessary to assess the:

(i) biophysical capacity of the planetary *habitat* to sustain life (Earth's carrying capacity),
(ii) cultural *habits* that influence human impact, and
(iii) human *co-inhabitants* that are most responsible for this impact.

In their article "The Anthropocene: From Global Change to Planetary Stewardship," Steffen et al. (2011) offer an analysis that is useful to assess differential responsibilities for the initiation of non-sustainable paths that have led to overshooting of the biocapacity of the planet. In terms of the spatial location (or habitat), they show that major ecological footprints have been generated in the Global North. In terms of the temporality, they affirm that ecological footprints have exponentially grown since the 1950s, at the "beginning of a second stage of the Anthropocene… after the Second World War – sometimes called the Great Acceleration" (Steffen et al. 2011, p. 755). They estimate human impact with reference to population and affluence (as an indicator for consumption). Since 1900 the world population has grown 3.9 times, from 1.8 to 7 billion (Table 9.1). During the same period, the world's Gross Domestic Product (GDP) has grown 30.6 times, from 1.8 to 55 trillion US dollars. This suggests that the increment in the rate of consumption is the most important factor in having surpassed the planet's biocapacity. Additionally, Steffen et al. (2011) state that while "developing" countries account for most of the population growth during the twentieth century and today contribute 5.9 billion to the 7 billion of the world population, "developed" countries (countries belonging to the OECD, Organization for Economic Cooperation and Development) are most responsible for the growth in GDP and today account for more than 75 % of the world's GDP. Based on these trends, they conclude that:

> Consumption in the OECD countries, rather than population growth in the rest of the world, has been the more important driver of change during the Great Acceleration, including the most recent decade. (Steffen et al. 2011, p. 757)

Table 9.1 World population and Gross Domestic Product (GDP) (Data provided by Steffen et al. (2011, p. 756) to show the "Great Acceleration" in growth that has taken place since 1950. Below, I added the rates of growth (GR) for Population and GDP, and the ratio between both)

Year	Population (billion)	GDP (trillion US $)	Population × GDP
1900	1.8	1.8	3.2
1950	2.5	5.3	13.3
2011	7.0	55.0	385.0
Interval	**Growth rate (GR-Pop)**	**Growth rate (GR-GDP)**	**GR-GDP/GR-Pop**
1900–1950	1.4	2.9	2.1
1950–2010	2.8	10.4	3.7
1900–2011	3.9	30.6	7.9

For a more precise diagnosis of the agents mainly responsible for global environmental change, our analysis of Steffen et al. (2011) conducted with the biocultural ethic's conceptual framework allows us to achieve three core conclusions:

(i) Regarding the temporal and spatial location (or *habitat*), the most dramatic change has taken place since the 1950s in the Northern Hemisphere.
(ii) Regarding the mechanism (or *cultural habit)*, the main responsible factor is the growth in affluence and consumption rates.
(iii) Regarding the social groups (or *human co-inhabitants*), societies of developed countries are those that have most contributed to global environmental change and, directly and indirectly, to the gestation of the Anthropocene.

With the conceptual framework of the biocultural ethic we are compelled to further distinguish among the levels of affluence by different sectors of the population. First, at a global scale, intercontinental analyses reveal great disparities in the proportions of the world population and total wealth (Table 9.2). North America and Europe are the richest continents; they concentrate 54 % of the global wealth, but have only 15 % of the world's population. On the poorer extreme, Africa has 10.7 % of the world population but only 1.5 % of the global wealth. Based on the data provided by Table 9.2 (see columns for the ratios *WW/WP* and *GDP/WP*), we can estimate that on average, a North American person participates 52 times more in the global wealth and 23 times more in the global GDP than an African person.

The figures of the intercontinental analysis are still too general, because within each continent and country there are also great inequalities and hence consumption differentiation. For example, in the United States of America the richest quintile (20 % of the population) possesses 84 % of the country's wealth, while the poorest quintile possesses only 0.1 % of this wealth (Norton and Ariely 2011). Furthermore, the richest 1 % of U.S. Americans holds nearly 50 % of the country's wealth. An equivalent concentration of wealth also is found at the world level.

The World Economic Forum (2013) reports that the richest 1 % of the world population owns 50 % of the world's wealth, i.e., *70 million people own US$ 115 trillion* (Table 9.3). In contrast, the poorest 50 % of the world population owns just

Table 9.2 Percentages of the world's population (WP), wealth (WW), and Gross Domestic Product (GDP), and WW/WP and GDP/WP ratios (Data based on Davies et al. (2007))

Continent	% World population (WP)	% World's wealth (WW)	% World GDP	Ratio WW/WP	Ratio GDP/WP
North America	5.2	27.1	23.9	5.2	4.6
Europe	9.6	26.4	22.8	2.7	2.4
Oceania & Others	3.1	3.7	5.4	1.2	1.7
Latin America	8.5	6.5	8.5	0.8	1.0
Asia	52.2	29.4	31.1	0.6	0.6
Middle East	9.9	5.1	5.7	0.5	0.6
Africa	10.7	1.5	2.4	0.1	0.2

Table 9.3 Wealth owned by different segments of the world's population (Data based on the World Economic Forum (2013))

Number of people	% World population	Wealth owned	% of the World's wealth	Average wealth owned per capita	Number of people that, combined, own the equivalent to one of the world's 85 richest people
85	0.000001 %	$2,300,000,000,000	1 %	$ 27,058,823,529	1
70,000,000	1.0 %	$115,000,000,000,000	50 %	$ 1,642,857	16,471
3500,000,000	50.0 %	$2,300,000,000,000	1 %	$ 657	41,176,471
6930,000,000	99.0 %	$115,000,000,000,000	50 %	$ 16,595	1,630,588
7000,000,000	100.0 %	$230,000,000,000,000	100 %	$32,857	823,529

1 % of the of the world's wealth, i.e., *3,500 million people own US$ 2.3 trillion.*[3] On average, each of the individuals belonging to the poorest 50 % of the world population owns US$ 657. Noticeably, the total amount owned by the bottom half of the world's population is the same as the amount owned by the richest 85 people in the world. On average, each of the 85 world's richest persons owns the same amount as owned by 41,176,471 people who belong to the bottom half the population (Table 9.3).

In summary, the data reported by the World Economic Forum (2013), Credit Suisse (2013), UNICEF (Ortiz and Cummins 2011), and the United Nations University and the World Institute for Development Economics Research (UNU-WIDER, Davies et al. 2007) show that just 0.000001 % of the world population owns the same amount of the world's wealth as 50 % of the world's population. Given this extreme concentration of wealth, the responsibility in terms of human consumption and impact cannot be presented in general terms of the human species *Homo sapiens* or *Humanity* in general. However, most publications analyze the problem in these terms. For instance, in the review article "Humanity's unsustainable environmental footprint" published by Hoekstra and Wiedmann (2014) in *Science* magazine, the authors conclude that:

> the various components of the environmental footprint of *humanity* must be reduced to remain within planetary boundaries. (Hoekstra and Wiedmann 2014, p. 1117; emphasis added)

Given the marked wealth gaps, it is technically misleading and ethically unjust to continue analyzing current challenges in terms of *humanity* in general, without defining differential responsibilities (Box 9.1). As philosopher and economist Amartya Sen (1997) has critically observed, the 1 % of the world's population is richer than ever, more powerful than ever, controlling the political and economic systems. The widening gap between the rich and non-rich has rapidly grown during the post-war Great Acceleration, and today we have reached a state of *plutonomy* where the majority of the wealth is controlled by an ever-shrinking minority, dividing humanity in two blocks: "the plutonomies, where economic growth is powered by and largely consumed by the wealthy few, and the rest" (Kapur et al. 2005, p. 1).

The economic growth of a plutonomic society becomes dependent on the fortunes of a wealthy minority (Box 9.1). However, as Canadian writer and politician Christine Freeland (2011) cautions, we are not merely living in a plutonomy, but a plutocracy, a form of oligarchy. The wealthy display "outsized political influence, narrowly self-interested motives, and a casual indifference to anyone outside their

[3] A similar figure is provided by Credit Suisse (2013), which reports a global wealth of $240.8 trillion. Share of wealth for the richest 1 % is 46 % (amounting to $110 trillion), and for the bottom half of the population is 0.71 % (amounting to $1.7 trillion). The United Nations Children's Fund (UNICEF, Ortiz and Cummins 2011, p. 12), and the United Nations University – World Institute for Development Economics Research (UNU-WIDER, Davies et al. 2007), offer complementary analyses whose global percentages are similar regarding wealth gaps at global scale.

Box 9.1. Mapping World Plutocracy

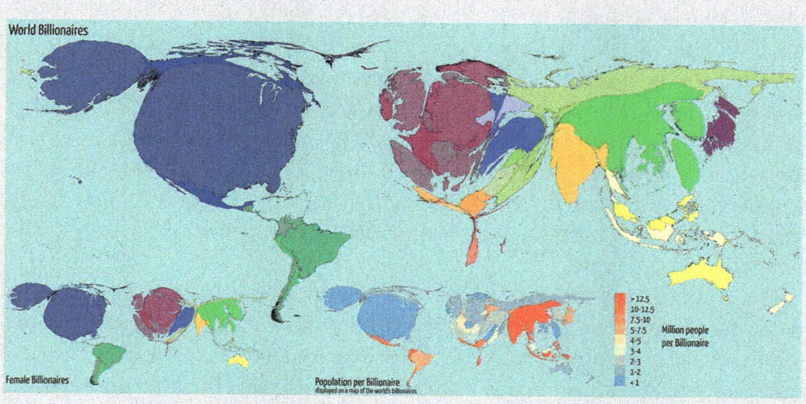

Fig. 9.2 The world's billionaires map (Figure reprinted from Hennig and Dorling (2013, p. 38), courtesy of John Wiley and Sons)

The wealth distribution shown in the above world map depicts the geographical area of each country multiplied by a factor associated with the country's number of billionaires as reported by Forbes magazine on March 7, 2012. Alaska looks very large in this map because it is transformed in one piece with the rest of the USA – although, it is likely to be much smaller in terms of its share of the billionaires within the country. According to Henning and Dorling (2013, p. 38), in 2012 "there were 1153 billionaires across the globe (this figure includes families, but excludes fortunes dispersed across large families where the average wealth per person is below a billion). The total wealth of the billionaires was US$3.7 trillion – as great as the annual gross domestic product of Germany. Top of this league table is the US with 424 billionaires, followed by Russia (96) and China (95)." Henning and Dorling (2013, p. 38), underline that "much of the wealth of billionaires is held offshore and their wealth is the tip of an iceberg of hard-to-tax personal assets. In a Tax Justice Network report, James Henry estimated the overall global offshore financial assets held by the world's richest to be between US$21 trillion and US$32 trillion (out of the total global wealth, estimated at US$231 trillion). Nearly half of these offshore assets are owned by the world's richest 91,000, just 0.001 % of the global population."

The map at the bottom left modifies the map on the top by considering only the number of women billionaires, who represent less that 10 % of the world's billionaires. Countries are shaded by the same colors in top and bottom left world maps to allow visual comparison between the sizes of total billionaires versus female billionaires. The map at the bottom right depicts the proportion of non-billionaire people per billionaire in each country; e.g., in the USA one

(continued)

Box 9.1. (continued)
billionaire can be found for every 740,000 people (dark blue), while in India one billionaire is found amongst every 26 million people (dark red).

In summary these three maps by Henning and Dorling (2013) depict: (i) *inter-hemispheric inequalities*, with a marked bias against the Southern Hemisphere, (ii) *gender inequalities*, with a marked bias against females, and (iii) *intra-continental* (and intra-country) *inequalities*, with the largest wealth gaps in Latin America (especially, Mexico, Colombia, and Chile), Africa, and Asia, where for each billionaire there are more than 12.5 million non-billionaire people.

For an Earth stewardship initiative, a main problem derived from an accumulation of power and wealth is its association with a current state of economic self-absorption and lack of socio-environmental responsibility. In 2005 three analysts at Citigroup, one of the major investment banking corporations, generated a report called "Plutonomy: Buying Luxury, Explaining Global Imbalances" (Kapur et al. 2005). They began by stating that today "the world is dividing into two blocks – the plutonomies, where economic growth is powered by and largely consumed by the wealthy few, and the rest" (Kapur et al. 2005, p. 1). They concluded their report by positing that:

> We hear so often about "the consumer." But when we examine the data, there is no such thing as "the consumer" in the U.S. or UK, or other plutonomy countries. There are rich consumers, and there are the rest. The rich are getting richer, we have contended, and they dominate consumption. As the rich have been getting richer, so too stocks associated with the rich have performed exceptionally well. Our Plutonomy Basket, generated returns of 17.8 % per annum, on average, from 1985. If Plutonomy continues, which we think it will, if income inequality is allowed to persist and widen, the plutonomy basket should continue to do very well. (Kapur et al. 2005, p. 30)

Kapur et al. (2005, 2006) have claimed that their plutonomy index outperforms the stock market. Noam Chomsky (2012) has critically analyzed how plutonomy does so by advancing the idea that money does not just represent a store of value, a medium of exchange and a unit of accounting, but also the power to claim the labor of others and natural resources in commodity form. In terms of the Citigroup analysts:

> In a plutonomy there is no such animal as "the U.S. consumer" or "the UK consumer", or indeed the "Russian consumer". There are rich consumers, few in number, but disproportionate in the gigantic slice of income and consumption they take. There are the rest, the "non-rich", the multitudinous many, but only accounting for surprisingly small bites of the national pie. Consensus analyses that do not tease out the profound impact of the plutonomy on spending power, debt loads, savings rates (and hence current account deficits), oil price impacts etc., i.e., focus on the "average" consumer are flawed from the start. (Kapur et al. 2005, p. 2)

(continued)

Box 9.1. (continued)

It is critical to note that plutonomy is indifferent to the *rest of humans* as well as to the *rest of non-human living beings*. A main socio-ecological problem is the association of the accumulation of wealth with unrestricted consumerism and a governance regime of indifference toward those who are irrelevant to plutonomies today, "the rest." A main ethical problem is that under plutonomic regimes, the value of capital is placed above the value of the life of "the rest" who represents the vast majority of human and non-human beings. In order to avoid the commodification of the labor of non-plutonomic humans and the life of other-than-human beings, it is necessary to change narrow economic discourses, structures, and policies that today override fundamental ethical values and ecological scientific understanding and advice, hindering the implementation of an urgently needed Earth stewardship.

own rarefied economic bubble" (Freeland 2011, p. 2). Plutonomy (from Greek *plouton* = wealth; *nomos* = rule or law) is a combination between plutocracy and economy, and Freeland critically states that:

> The rise of the new plutocracy is inextricably connected to two phenomena: the revolution in information technology and the liberalization of global trade. Individual nations have offered their own contributions to income inequality—financial deregulation and upper-bracket tax cuts in the United States; insider privatization in Russia; rent-seeking in regulated industries in India and Mexico. But the shared narrative is that, thanks to globalization and technological innovation, people, money, and ideas travel more freely today than ever before. (Freeland 2011, p. 14)

The unregulated free market has allowed some persons (individuals or corporate entities)[4] to accumulate unlimited wealth (Piketty 2014). The excessive accumulation of wealth and lack of limits on the free market and associated consumption

---

[4] A *legal person* is a subject of rights and obligations that exists, not as an individual but as an institution that is created by one or more individuals to fulfil a social objective, which may be for profit or not for profit. Hence, along with individual people there are also legal persons which are entities that the law accords and recognizes as having legal personality and, consequently, the ability to act as legal persons – that is, the capacity to acquire and to hold real estate of all kinds, to incur obligations and to engage in legal actions. In the case of the United States of America, *corporate personhood* is a legal concept in which a corporation may be recognized as an individual in the eyes of the law. This doctrine forms the basis for legal recognition that corporations, as groups of people, may hold and exercise certain rights under the common law and the U.S. Constitution. For example, corporations may contract with other parties and sue or be sued in court in the same way as natural persons or unincorporated associations of persons. Richard Watson (1992) concisely discusses the historical origin of corporate persons and the legal and moral implications for environmental ethics. He criticizes that: "Corporations are not responsible moral agents. They cannot reciprocate. They can have no primary rights because they cannot fulfill any duties. It is suspected that the concept of legal personhood for corporations is a device to allow actually responsible persons to escape punishment" (Watson 1992, p. 27).

rates create three types of difficulties to implement an effective Earth Stewardship: (a) ecological, (b) ethical, and (c) political problems.

(a) Regarding ecological problems, Steffen et al. (2011) have shown that a main cause of exceeding the limits of the biosphere is the increased consumption by the developed countries, consumption made possible by unlimited wealth accumulation – as if natural resources were unlimited and planetary boundaries were nonexistent. From the ecological perspective, it is problematic that the neoliberal free market regime does not sufficiently attend to the core concept of planetary boundaries (Chapin et al. 2011b; Steffen et al. 2011). The incorporation of the concept of *biophysical limits* (at the scale of the biosphere as a whole as well as of regional *habitats*) into economic and governance policies is a necessary condition for implementing Earth stewardship. The notion of limits has a long history in the concept of the economy of nature introduced by Linnaeus in the seventeenth century, and was extensively developed by ecological economics in the twentieth century. Limits on rates of consumption and accumulation of wealth challenge neoliberal free market theoretical assumptions and practices of production and consumption. Under the current neoliberal free market regime, risks and negative externalities (e.g., oil spills, and other forms of pollution and environmental damage) are often absorbed by communities of humans and other living beings, while monetary gains receive less taxation and are accumulated by persons (individuals or corporations) who commercially consume, use, and/or deteriorate "human and natural capital." This double standard involves not only economic problems, but also raises ethical and political issues.
(b) Ethically, the notion of limits has a long history in the philosophical roots of Western civilization, religious traditions, and Amerindian ecological worldviews. Aristotle develops an ethics based on the mid-way point; nothing in excess. This rule shares core concepts implicit in many religious traditions, including the Buddhist middle-way and the Christian values of solidarity and distributive equity. Furthermore, austerity, reciprocity and equality are values that are shared by the ecumenical community. Mary Evelyn Tucker (this volume [Chap. 27]) identifies six key "values for human-Earth flourishing" that are shared by world religions: reverence, respect, restraint, redistribution, responsibility, and renewal. In the tradition of Latin American liberation theology, Roy May (2002) and Guillermo Kerber (2011, p. 192) underline that "to regain a healthy relationship with all creation it is necessary to address, and not be indifferent to a world divided by extreme consumerism and starvation." Among Amerindians worldviews, equity and reciprocity among humans and nature are also core values for cultures such as the Quechua and the Aymara (see Sarmiento, in this volume [Chap. 5]; Mamani 2000, and in this volume [Chap. 6]). Additionally, the notion of limits and respect for others is a cornerstone of the ethical formulations substantiated in the ecological sciences, such as the land ethic of Aldo Leopold, who stated that "an ethic, ecologically, is a limitation on freedom action in the struggle for existence. An ethic, philosophically is a

differentiation of social from anti-social conduct. These are two definitions of one thing" (Leopold 1949, p. 202). The incorporation of the concept of *ethical limits* into cultural habits and socio-environmental policies is a second necessary condition for implementing Earth stewardship.

(c) In the political realm, the ethical imperative indicated by Aldo Leopold "to correct anti-social behaviors toward socially appropriate ones" coincides with a central concept of Aristotle. The ancient Greeks called *idiōtēs* people whose behavior put personal interests above the collective interests of the citizens of the Greek *polis* (or nation-state). Aristotle was relentless about the need to punish those *idiōtēs*, or idiots in order to sustain a democratic regime. Only if the idiots paid their fines, served their sentences, and corrected their unbalanced self-interested behavior, could they remain in the polis as citizens. If they did not, then the idiots were exiled. Aristotle affirmed that they should lose their citizenship because the *polis* could not be sustained in the presence of people taking only privileges but not respecting their obligations as citizens. The restoration of the judicial system capacity to sanction exacerbated, self-absorbed, individualism (such as the *idiōtēs* by Aristotle) is a third necessary condition for implementing an Earth stewardship.

Under a plutocratic regime (national and international), nation-states and citizens often do not have the ability to sanction violators of environmental, economic and social laws. Colombian sociologist Isaías Tabasura-Acuña (2006) discussed this problem in the case of the conflict between the U'wa and Oxy (see Box 8.2, in this volume), and many other Latin American and other regional cases could be mentioned. To enforce penalties on those that cause environmental and social damage, it is necessary to change the plutocratic regime. In turn, the change from plutocracy to democracy would favor the enforcement of national and international environmental regulations, as well as agreements of co-responsibility for the management of hotspots of biological and cultural diversity that are critical for the sustainability of life at local and global scales (see Chaps. 2, 3, 6, 7, 8, 27, and 28 in this volume). The biocultural ethic extends the community of citizens beyond the Aristotelian *polis*, and the modern nation-state, to include all human beings, involving diverse genders, languages, and human societies, as well as considering the well-being of all other living beings that constitute communities of co-inhabitants.

Through our analysis of *stewardship* versus *plutonomy*, we can conclude that Western philosophical and theological traditions, Amerindian ecological knowledge and practices – ancestral and contemporary – as well as ecological sciences provide a basis for restoring the concept of limits to the prevailing global economic system. This is essential in order to overcome the current indifference of plutonomy to ecological, social, and ethical boundaries within which economic activity unfolds. In consequence, *to open novel biocultural pathways toward Earth stewardship and sustainable co-inhabitation, it is essential that the prevailing economic system be amended so that it ceases to be indifferent to the well-being of the majority of human and other-than-human living beings.*

9.4 Biocultural Co-inhabitation Versus Ecosystem Services

The notions of stewardship and co-inhabitation have relevant ethical and ontological differences. Stewardship, as conceived by the initiatives of Earth and Planetary stewardship (Chapin et al. 2011a, b; Steffen et al. 2011), is based on a notion of ecosystem services where human subjects administer goods and services of ecosystem objects and processes. Consequently, the only subjects (active agents with their own interests) are humans (see Naeem 2013). Biodiversity and ecosystems are viewed as passive objects without intentionality or interests. Under the prevailing perspectives of ecosystem services, these objects are managed with a utilitarian ethics, to produce the greatest good for the greatest number of people, and for the longest time. This utilitarian ethics has a long and influential history in the philosophy of conservation and rational use of resources inaugurated by Gifford Pinchot at the beginning of the twentieth century (see Norton 1991). Later, at the end of the twentieth century, it also became the central school of ethics for the concept of sustainable development envisioned by the Bruntland Commission report, *Our Common Future* (WCED 1987). The utilitarian ethics that has inspired Pinchot and Bruntland supposes an ontological split between human-subjects and nature-objects that has a long history in Western philosophy (see Morin 1990). As environmental philosopher Irene Klaver underlines:

> The dualism between *subject* and *object* has been pervasive, deeply imbedded in Western thought, and at the root of a variety of interlocking dualisms, such as *activity* (or agency) versus *passivity*, resonating in *culture* versus *nature*. A dualistic mindset comes with a value attribution, with an implied sense of *superiority* (culture, agency) versus *inferiority* (nature, passivity) and hence an implied legitimation for use, domination and exploitation. The *inert material* or *natural object* is waiting for the *human intentional subject* to do something with it. It became the basis for a *Western conception of passive nature, ready to be used by culture.* (Klaver 2013, p. 93, emphasis added)

In contrast to utilitarian ethics, the concept of co-inhabitation proposed by the biocultural ethic is based on an ontology that considers all living beings as active subjects with their own interests (see Rozzi 2013, pp. 26–28). Recent scientific discoveries have determined that even invertebrates have the capacity to feel pain and stress (Horvath et al. 2013). These invertebrates actively seek and build their own habitats (Contador et al. 2014), and exhibit behaviors that seek pleasure and avoid pain (Barras 2007). Contemporary sciences provide an avalanche of evidence supporting the continuity of biological nature between humans and all living beings. Therefore, it is more appropriate to conceive living beings as a community of active subjects with their own interests with whom we co-inhabit – and not merely as "natural resources" that we rationally manage to only get goods and services.

The ontology of the biocultural ethic has ancient roots in Western philosophy. Aristotle considered all living beings as having a soul. Soul (Lat. *anima*) means spirit, and spirit (Lat. *spiritus*) means breath. According to Aristotle, plants and animals (humans and other-than-humans) have a vegetative soul; that is, all living beings breathe, grow, and reproduce. The Aristotelian view is consistent with the scientific theory of the unity of life. In the nineteenth century it was discovered that all living beings are made of cells, and during the twentieth century it was demonstrated that all living beings, including humans, share a fundamental genetic basis. These Western philosophical and scientific worldviews offer a conceptual framework to understand the implications of Amerindian concepts such as *Pachamama* (see Mamani-Bernabé 2015 in this volume [Chap. 6]) not as folk curiosities, but as worldviews consistent with cutting-edge scientific knowledge. Like the Aymara worldview contained in the concept of *Pachamama*, Western philosophical and scientific worldviews enable us to consider the community of living beings as a community of active subjects with their own interests. Comparative analyses of Amerindian, philosophical, and scientific forms of ecological knowledge generate a congruent and complementary understanding that invites us to revise the dualism between human-subjects and ecosystem-objects established by a utilitarian ethics that prevails in the logic of ecosystem services. Modification of this dualism could extend the concepts and practices of Earth Stewardship towards forms of intercultural dialogue and interspecific co-inhabitation. This biocultural modification would enlarge the human community of stewards participating in Earth stewardship practices, as well as broaden the community of human and other-than-human co-inhabitants considered in the analyses of life well-being.

9.5 Concluding Remarks

In an era of rapid socio-environmental change, it is technically misleading and ethically unjust to ascribe responsibility to humanity in general. The biocultural ethic's conceptual framework contributes to an Earth stewardship initiative by more precisely identifying the diversity of Earth stewards as well as the specific agents that are mostly responsible for current socio-environmental problems and by demonstrating the need to question, clarify, and change language, governance regimes, and life habits in order to effect cultural transformations. Framed in the tradition of liberation philosophy (see Chap. 8 in this volume), the biocultural ethic involves two methodological steps: (a) to *liberate* diverse forms of thinking from being encapsulated by colonizing global conceptual frameworks; (b) to *reaffirm* languages, forms of thought, ethics, and cultures that are marginalized from global discourses and media.

Regarding the first methodological step of the biocultural ethic, it is critical to transform the state of indifference toward the diversity of life and cultures that prevails in global discourses today. Governed by a plutocratic regime, global discourses are centered on a free market culture. In this chapter I have highlighted the distinctions between democracy and plutocracy, stewardship and plutonomy to better understand the current state of absorption in a consumerist culture not as a trend that is inherent to "human nature" (as it is often portrayed), but as a particular and recent cultural trend in human history. To achieve Earth stewardship, this trend needs to be overcome because it alienates global society from complex, multifaceted, dimensions of human culture and other-than-human life.

Regarding the second methodological step, the biocultural ethic provides a conceptual and methodological framework to discover the richness of ecological values and forms of knowledge grounded in the worldviews of Amerindian cultures, Western philosophical and scientific traditions of thought, as well as everyday practices of urban and rural organizations and socio-cultural groups, which are essential for an Earth stewardship initiative that is inclusive and effectively incorporates socio-environmental justice into it. In order to achieve the recovery of understanding and valuation of biocultural diversity, in this chapter I have highlighted the need for a change of language to more precisely name and identify particular (1) Earth habitats at planetary and ecosystem scales, (2) habits of stewardship or co-inhabitation, and (3) Earth stewards or co-inhabitants.

(1) *Regarding the Earth habitats,* the biocultural ethic's conceptual framework clarifies that the main drivers of the Anthropocene have accelerated since the 1950s, and have mainly originated in the Northern Hemisphere. The impact, however, reaches worldwide. Several chapters of this book document a diversity of active Earth stewards who oppose this trend. However, these Earth stewards face growing challenges for maintaining their stewardship habits in their traditional places or habitats in the Northern and the Southern hemispheres. Today, transnational and national economic actors are acquiring 'empty' lands, often in distant countries, which can serve as sources of alternative energy production (primarily biofuels), food crops, mineral deposits (new and old), and reservoirs of environmental services (Borras et al. 2011, 2012). In their article "The Anthropocene: From Global Change to Planetary Stewardship," Steffen et al. (2011, p. 739) point out that "the new economic giants of Asia move to secure food resources in non-Asian territories;" therefore, land grabbing represents a rapidly growing twenty-first century driver of social-environmental problems. Social scientists have criticized land grabbing as a form of neocolonialism:

> Some of this *land has been cleared of existing inhabitants* and users but not yet put into production; in many cases buyers and investors are simply preparing for the next global crisis (Borras et al. 2011, p. 209; emphasis added)

Land grabbing and other forms of concentration of land ownership are a major driver for the rapid rates of rural–urban migration in Africa, Asia, and Latin America, since the mid-twentieth century (Fig. 9.3). For the native habitats, this migration causes a loss of ancestral human stewards or custodians of the land. For the displaced people, this migration causes a loss of everyday contact with their communities of co-inhabitants and diverse life habits. In the cities, displaced people frequently lose their autonomy and lack access to basic services, such as food, water, shelter, and sanitary conditions. They face extreme poverty conditions that are rapidly expanding in the marginal neighborhoods of metropolitan areas. To confront these policies that imply social and environmental injustice, the 3Hs formal proposal of the biocultural ethic is grounded in the notion of *ethos* as habitat. Then, the biocultural ethic links the habitats with the life habits and the identity, autonomy, and well-being of the co-inhabitants (humans and other-than-humans). The conservation of habitats and access to them by communities of co-inhabitants is the condition of possibility for the continuity of their life, and becomes an ethical imperative that should be incorporated into development policies as a matter of eco-social justice. Consequently, the conservation of habitats and access to them by communities of co-inhabitants provide a basis for indexes of sustainability and well-being that broaden the current emphasis on GDP and monetary indicators to measure the nations' success or levels of poverty (see also Kubiszewski et al. 2013; Costanza et al. 2014).

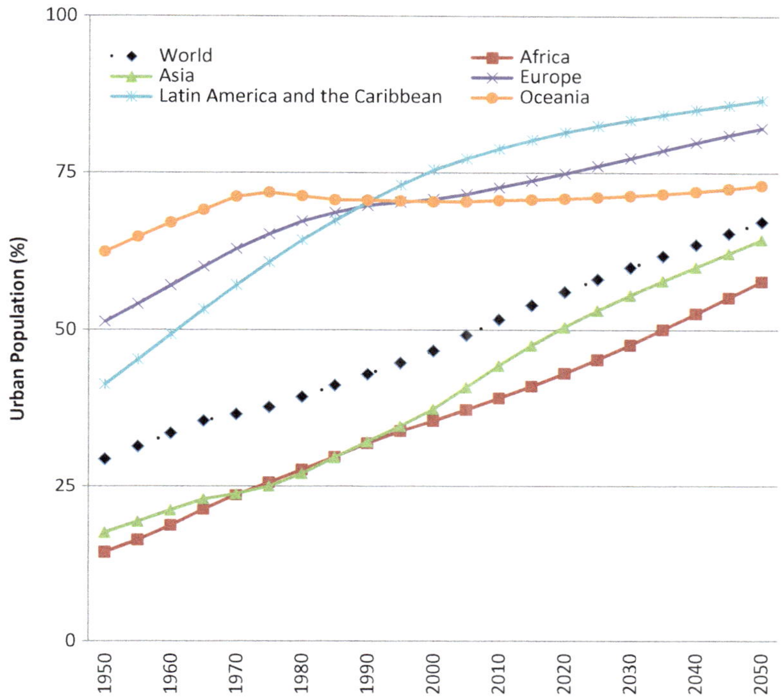

Fig. 9.3 Relative percentages of urban populations in Africa, Asia, Europe, North- and South-America, Oceania, and the world since 1950, including estimated percentages until 2050 (Data from Heilig (2012))

In summary, on the one hand, to conserve and have access to a habitat is the condition of possibility of exercising the role of steward of the land, or Earth. On the other hand, it is not ethically acceptable to accumulate territory and not properly administer the land to the interest of the community of co-inhabitants.

It is important to understand that the habitat includes not only its biophysical dimension (the biosphere at a global scale, sensu Vernadsky; see Huggett 1999), but also its cultural and symbolic-linguistic dimensions (the logosphere at a global scale, sensu Krauss 2007), and its socio-political, institutional, and technical dimensions (the technosphere, sensu Naveh and Lieberman 1990). Changes in one dimension imply changes in the other dimensions (see Fig. 9.1). The concentration of wealth and ownership of the habitats generates a replacement of very diverse life habits and communities of co-inhabitants by a few plutonomic, consumerist habits involving the well-being of a minor fraction of the co-inhabitants. This process leads to a non-sustainable and unjust process of biocultural homogenization, which oppresses the majority of human and other-than-human co-inhabitants (Rozzi 2013). Fortunately, given that history is not linear, but instead it is dynamic and complex, global society is not condemned to continue its path towards biocultural homogenization. Today, a greater precision in the language used to identify the diversity of Earth stewards, practices of land stewardship in heterogeneous habitats of the planet, as well as the specific agents that are mostly responsible for current socio-environmental problems, can significantly contribute to orient clearer collaborative and supportive ways for a responsible and inter-cultural Earth stewardship.

(2) *Regarding stewardship and other cultural habits,* the biocultural ethic's conceptual framework adds clarity to the Earth Stewardship Initiative in two complementary directions: (a) better understanding and valuing a plethora of sustainable ecological worldviews and practices, and (b) better identifying that the main *cultural habit* driving global environmental impact is the growth in consumption rates and affluence. In its current style and magnitude, the consumerist habit has a very recent history (triggered after World War II), and is affordable to only a small fraction of the world population (Ortiz and Cummins 2011). Religious, philosophical, and Amerindian ethics criticize this *consumerist habit,* because it does not contribute to a flourishing life of those who have too much nor of those who have too little. Those who are irrelevant to plutonomies today, "the rest," represent the vast majority of human and other-than-human beings, and they are not passive victims or objects; instead, they are active subjects with beauty, creativity, dignity, and solidarity. To transform reductionist, individualist, and selfish behaviors and values embedded in prevailing, hegemonic, narrow economic discourses, the biocultural ethic fosters inter-cultural dialogues and practices, based on partnerships among the majority of overlooked, heterogeneous, rich cultural traditions and communities of Earth stewards. Toward this aim philosophers can act as translators and initiators. In terms of environmental philosopher Irene Klaver:

> Translating various concerns along multiple perspectives opens up new situations and affords us the freedom of ongoing new beginnings. It is crucial to an understanding of the various viewpoints, positions, places and experiences of others. Environmental philosophy enlarges the category of the "other" beyond human beings. It enlarges ethics in the direction of ethos, resonating with "habitat," "inhabitants," and "habits" (Rozzi et al. 2008). It questions certain mentalities and provokes and evokes different modes of knowledge and experience, to enhance cultural imagination into environmental imagination. (Klaver 2013, p. 91)

Philosophers contribute to "pluralizing" human natures. This plural understanding of human natures fosters intercultural forms of Earth stewardship at multiple scales by including the diversity of Earth stewards, their cultural habits and languages, interacting in complex and often non-linear ways in the context of diverse local realities confronted with increasingly prevailing global discourses and forms of governance. The biocultural ethic recovers the archaic meaning of the Greek term ethos, and interprets it ecologically in terms of "habitats" and "habits" of communities of human and other-than-human co-inhabitants (Rozzi 2013). By conducting comparative ethical analyses of (i) pre-Socratic and other non-mainstream Western philosophies, (ii) Amerindian and other non-Western ecological worldviews, and (iii) contemporary ecological-evolutionary sciences, it introduces into Earth stewardship an intercultural philosophical language that broadens the prevailing spectrum of normative ethics that emphasize utilitarianism and deontology, or more recently virtue ethics (see Bina and Vaz 2011; Jax et al. 2013). The biocultural ethic asserts values, virtues, and forms of ecological knowledge that are complementary to those preponderant ethical schools. For example, it is interesting to note that the pre-Socratic notion of ethos by Heraclitus resonates with the Andean Amerindian notion of *Pacha* (see Rozzi 2015 [Chap. 8] in this volume). For the Aymara culture the *Pacha* [= cosmos] is all what exists; everything is of the *Pacha*, and all is in the *Pacha* (Mamani and Quispe 2007, p. 21). *Pacha* encompasses time and space, and *Aka Pacha* (the space here and now) or planet Earth welcomes all living beings that inhabit it (Mamani and Quispe 2007, p. 13). The Aymara worldview understands cosmos as a totality in which humans participate, and co-inhabit with other beings. Aymara ethics does not accept the notion of "the rest;" the *Pacha* includes all beings. It implies an ethics that is congruent with a tradition of virtue ethics, which is oriented toward the flourishing of each living being according to its talents. If the harmony of co-inhabitation is ruptured, Aymara ethics demands reconciliation in order to restore equity to the *Pacha*. With a complementary perspective based on a scientific understanding, Chapin et al. (2011b, p. 52) state that "given the pace of environmental deterioration and the increased recognition that this path is untenable, society should seize the opportunity to reorient its relationship to the biosphere." As much as the Aymara perspective, the scientific ecological worldview that sustains the ESA's Earth Stewardship Initiative is interested in the sustainability of the biosphere and human well-being.

Intercultural comparisons disclose differences and commonalities; intercultural dialogues build partnerships for bioculturally diverse but synergic forms of Earth stewardship that are informed by ancient philosophical schools of thought and forms of traditional ecological knowledge, as well as by cutting-edge sciences. The comparative analysis conducted with the biocultural ethic's lens discovers that the Heraclitean notion of ethos, Aristotelian ethics (and the neo-Aristotelian concept of good life, eudaimonia), Amerindian ecological worldviews and contemporary evolutionary, ecological and biogeochemical sciences provide an understanding that transcends the dichotomy between human-subjects and natural-objects (or passive resources to be used). This biocultural understanding demands and requires an ethic of responsibility and reciprocity, where the provision of services should flow from ecosystems to humans and also from human to human and other co-inhabitants of ecosystems.

(3) *Regarding the stewards or human co-inhabitants*, not all humans are equally responsible for generating the Anthropocene, and having surpassed the biocapacity of planet Earth. Today's degree of responsibility is associated with a degree of accumulation of wealth and power. In addition to the degree of responsibility associated with wealth, it is necessary to distinguish between socio-environmentally responsible stewards and irresponsible agents with an unbalanced self-interested behavior (*idiōtēs* in terms of Aristotle), who are indifferent to the collective interests of citizens. However, it is necessary to not only have a more precise diagnosis of the agents mainly responsible for environmental changes at local and global scales. It is also essential to foster stewardship habits and build on the capacity to aspire to a broad arrange of values and practices that favor the flourishing of the life of each member of the community of co-inhabitants. Complimentarily, it is also necessary to recover the capacity to sanction the agents that cause major negative environmental impacts.

For implementing an Earth stewardship it is indispensable to overcome the current state of impunity in which *idiōtēs* gain power; instead, nation-states and citizens should recover their capacity to enforce laws, and sanction their violations. It is not Mankind or the human species as a whole that is responsible for causing the Anthropocene and the current unsustainable environmental footprints, as it has been mostly portrayed for over a century. However, it is the whole humanity and community of life who is in peril due to the actions of a few specific agents, who need to be reoriented. To achieve Earth stewardship, omitting this specification in the diagnosis of global environmental change would be a mistake as serious as a physician that treats a patient with an infectious disease and blames microorganisms in general for this disease, instead of identifying the specific organisms that are actually responsible for the infection. As Aldo Leopold (1949, 258) stated, "health is the capacity of the land for self-renewal. Conservation is our effort to understand and preserve this capacity." A biocultural approach to Earth stewardship helps to achieve a better diagnosis of specific threats and a better identification of opportunities that already exist in many communities for conserving the health of the land and the people.

Acknowledgments This chapter benefited from discussions at the Departamento Ecumenénico de Investigaciones (San José, Costa Rica), especially with Roy H. May Jr., and Francisca Massardo. I thank Irene Klaver, Kelli P. Moses, Eugene C. Hargrove, and Shaun Russell for their constructive comments on the manuscript, and Paula Viano and Paola Vezzani for their artistic help in the preparation of Fig. 9.1. The National Science Foundation (Project SES-10581630), and the grants PO5-002 ICM and PFB-23 CONICYT awarded to the Institute of Ecology and Biodiversity, Chile, provided valuable support.

References

Barras C (2007) Crustacean 'pain response' gives food for thought. New Sci 196:14

Berkes F (2007) Community-based conservation in a globalized world. Proc Natl Acad Sci U S A 104:15188–15193

Berkes F (2012) Sacred ecology. Routledge, New York

Bina O, Vaz SG (2011) Humans, environment and economies: from vicious relationships to virtuous responsibility. Ecol Econ 72:170–178

Borras SM, Hall R, Scoones J, White B, Wolford W (2011) Towards a better understanding of global land grabbing. J Peasant Stud 38:209–216

Borras SM, Franco JC, Gómez S, Kay C, Spoor M (2012) Land grabbing in Latin America and the Caribbean. J Peasant Stud 39:845–872

Borrero JM (2002) Imaginación Abolicionista. PNUMA/CELA, Cali

Callicott JB (1994) Earth's insights: a survey of ecological ethics from the Mediterranean basin to the Australian outback. University of California Press, Berkeley

Chapin FS, Power ME, Pickett STA et al (2011a) Earth stewardship: science for action to sustain the human-earth system. Ecosphere 2:89

Chapin FS, Pickett STA, Power ME et al (2011b) Earth stewardship: a strategy for social-ecological transformation to reverse planetary degradation. J Environ Stud Sci 1:44–53

Chapin FS, Pickett STA, Power ME et al (2015) Earth stewardship: an initiative by the Ecological Society of America to foster engagement to sustain planet Earth. In: Rozzi R, Pickett STA, Callicott JB et al (eds) Earth stewardship: linking ecology and ethics in theory and practice. Springer, Dordrecht, pp 173–194

Chomsky N (2012) A rebellious world or a new dark age? Page 5 of 8 http://www.tomdispatch.com/dialogs/print/?

Commoner B (1990) Making peace with the planet. Pantheon Books, New York

Contador T, Kennedy J, Ojeda J, Feinsinger P, Rozzi R (2014) Ciclos de vida de insectos dulceacuícolas y cambio climático global en la ecorregión subantártica de Magallanes: investigaciones ecológicas a largo plazo en el Parque Etnobotánico Omora, Reserva de Biosfera Cabo de Hornos (55°S). Bosque 35(3):429–437

Costanza R, Kubiszewski I, Giovannini E et al (2014) Time to leave GDP behind. Nature 505:283–285

Credit Suisse (2013) Global wealth report 2013. Credit Suisse, Zurich. https://publications.credit-suisse.com/tasks/render/file/?fileID=BCDB1364-A105-0560-1332EC9100FF5C83

Davies JB, Sandström S, Shorrocks A et al (2007) Estimating the level and distribution of global household wealth. UNU-WIDER research paper 2007/77

Dussel E (2003) Algunos Principios para una Ética Ecológica Material de Liberación (Relaciones entre la Vida en la Tierra y la Humanidad). In: Pixley J (ed) Por un mundo otro. Alternativas al mercado global. Quito, Ecuador, Comunidad Cristiana Mesoamericana y Consejo Latinoamericano de Iglesias, pp 29–44

Freeland C (2011) The rise of the new global elite. The Atlantic 307(1):44–55

Freeland C (2012) Plutocrats: the rise of the new global super-rich and the fall of everyone else. Penguin, New York

Heilig GK (2012) World urbanization prospects the 2011 revision. United Nations, Department of Economic and Social Affairs (DESA), Population Division, Population Estimates and Projections Section, New York

Hennig BD, Dorling D (2013) The world's billionaires. Polit Insight 4(1):38

Hoekstra AY, Wiedmann TO (2014) Humanity's unsustainable environmental footprint. Science 344:1114–1117

Horvath K, Angeletti D, Nascetti G et al (2013) Invertebrate welfare: an overlooked issue. Ann Ist Super. Sanità 49 Roma Jan/Mar 2013, http://dx.doi.org/10.4415/ANN_13_01_04

Huggett RJ (1999) Ecosphere, biosphere, or Gaia? What to call the global ecosystem. Glob Ecol Biogeogr 8:425–431

Jamieson D (2014) Reason in a dark time: why the struggle against climate change failed. Oxford University Press, New York

Jax K, Barton DN, Chan K, de Groot R, Doyle U, Eser U, Wichmann S (2013) Ecosystem services and ethics. Ecol Econ 93:260–268

Kapur A, Macleod N, Singh N (2005) Plutonomy: buying luxury, explaining global imbalances. Citigroup, equity strategy, industry note, 16 Oct 2005

Kapur A, Macleod N, Singh N (2006) Revisiting plutonomy: the rich getting richer. Citigroup, Equity Strategy, Industry Note, 5 Mar 2006

Kerber G (2011) Ecoteología y cambio climático: perspectivas ecuménicas. Cuadernos de Teología XXX:185–194

Klaver IJ (2013) Environment imagination situation. In: Rozzi R, Pickett STA, Palmer C et al (eds) Linking ecology and ethics for a changing world: values, philosophy, and action. Springer, Dordrecht, pp 85–105

Krauss M (2007) Mass language extinction and documentation: the race against time. In: Miyaoka O, Sakiyama O, Krauss ME (eds) The vanishing languages of the pacific rim. Oxford University Press, New York, pp 3–24

Kubiszewski I, Costanza R, Franco C et al (2013) Beyond GDP: measuring and achieving global genuine progress. Ecol Econ 93:57–68

Leopold A (1949) A sand county almanac and sketches here and there. Oxford University Press, New York

Lijphart A (2012) Patterns of democracy: government forms and performance in thirty-six countries. Yale University Press, New Haven

Lutz C, Nonini D, Holland D (2007) Local democracy under siege: activism, public interests, and private politics. NYU Press, New York

Mamani V (2000) Identidad y Espiritualidad de la Mujer Aymara. Fundación Basilea, La Paz

Mamani-Bernabé V (2015) Spirituality and the Pachamama in the Andean Aymara worldview. In:Rozzi R, Chapin FS III, Callicott JB et al (eds) Earth stewardship: linking ecology and ethics in theory and practice. Springer, Dordrecht, pp 65–76
Mamani V, Quispe C (2007) Pacha. Editorial Verbo Divino, Cochabamba
May R (2002) Ética y Medio Ambiente. DEI, San José
Morin E (1990) Introducción al pensamiento complejo. Editorial Gedisa, Barcelona
Naeem S (2013) Ecosystem services: is a planet servicing one species likely to function? In: Rozzi R, Pickett STA, Palmer C et al (eds) Linking ecology and ethics for a changing world: values, philosophy, and action, ecology and ethics. Springer, Dordrecht, pp 303–321
Naveh Z, Lieberman AS (1990) Landscape ecology. Springer Science & Business Media, New York
Norton BG (1991) Thoreau's insect analogies: or, why environmentalists hate mainstream economists. Environ Ethics 13:235–251
Norton MI, Ariely D (2011) Building a better America – one wealth quintile at a time. Perspect Psychol Sci 6(1):9–12
NRC (2010) America's climate choices: adapting to the impacts of climate change. National Academies Press, Washington
Ortiz I, Cummins M (2011) Global inequality: beyond the bottom billion – a rapid review of income distribution in 141 countries. United Nations Children's Fund (UNICEF), New York
Piketty T (2014) Capital in the twenty-first century. Harvard University Press, Cambridge, MA
Power ME, Chapin FS III (2009) Planetary stewardship. Front Ecol Environ 7:399
Rozzi R (2001) Éticas ambientales latinoamericanas: raíces y ramas. In: Primack R, Rozzi R, Feinsinger P et al (eds) Fundamentos de Conservación Biológica: Perspectivas Latinoamericanas. Fondo de Cultura Económica, México, pp 311–362
Rozzi R (2013) Biocultural ethics: from biocultural homogenization toward biocultural conservation. In: Rozzi R, Pickett STA, Palmer C et al (eds) Linking ecology and ethics for a changing world: values, philosophy, and action. Springer, Dordrecht, pp 9–32
Rozzi R (2015) Earth stewardship and biocultural ethics: Latin American perspectives. In: Rozzi R, Chapin FS III, Callicott JB et al (eds) Earth stewardship: linking ecology and ethics in theory and practice. Springer, Dordrecht, pp 87–112
Rozzi R, Arango X, Massardo F et al (2008) Field environmental philosophy and biocultural conservation. Environ Ethics 30:325–336
Sen A (1997) On economic inequality. Clarendon Press\Oxford University Press, Oxford\New York
Snape WJ III (2010) Joining the convention on biological diversity: a legal and scientific overview of why the United States must wake up. Sustain Dev Law Policy 10:6–8
Snape WJ (2012) Why everyone loses from US boycott of the UN biodiversity agreement. China Dialogue (19 Oct 2012). https://www.chinadialogue.net/article/show/single/en/5197. Accessed 30 Mar 2014
Steffen W, Jansson Å, Deutsch L et al (2011) The Anthropocene: from global change to planetary stewardship. Ambio 40:739–761
Stern N (2007) The economics of climate change: the Stern review. Cambridge University Press, Cambridge
Tabasura-Acuña I (2006) Ambientalismos y Ambientalistas: El Ambientalismo Criollo a Finales del Siglo XX, Cuadernos de Investigación, no 21. Universidad de Caldas, Cali
Tollefson J (2011) Brazil revisits forest code. Nature 476:259–260
Watson RA (1992) Self-consciousness and the rights of nonhuman animals and nature. In: Hargrove E (ed) The animal rights/environmental ethics debate. SUNY Press, New York, pp 1–36
Winters JA (2011a) Democracy and oligarchy. Am Interest VII(2):18–27
Winters JA (2011b) Oligarchy. Cambridge University Press, New York
World Commission on Environment and Development [WCED] (1987) Our common future. Oxford University Press, Oxford
World Economic Forum (2013) Outlook on the global agenda 2014. World Economic Forum, Geneva. http://www3.weforum.org/docs/WEF_GAC_GlobalAgendaOutlook_2014.pdf

Chapter 10
The Politics of Earth Stewardship in the Uneven Anthropocene

Laura Ogden, Nik Heynen, Ulrich Oslender, Paige West, Karim-Aly Kassam, Paul Robbins, Francisca Massardo, and Ricardo Rozzi

Abstract The Anthropocene is not only an epoch of anthropogenic dominance of the Earth's ecosystems, but also an epoch characterized by new forms of environmental governance, institutions, and uneven development. Following the literature in political ecology, we are calling these new forms of environmental governance, "global assemblages." A key argument from a political ecological perspective is that socio-ecological changes historically disproportionately impact communities in the Global South, and minority and low-income communities in the Global North. While global assemblages are powerful mechanisms of socio-ecological change, we

---

This paper is based on discussions that emerged out of the Ecological Society of America's Earth Stewardship Workshop on June 18–19, 2012 in Chevy Chase, M.D., funded by the National Science Foundation. We are grateful to Sue Silver and the editorial staff at *Frontiers in Ecology and the Environment* for allowing us to publish a modified version of "Global assemblages, resilience, and Earth Stewardship in the Anthropocene" (Ogden et al. 2013). Some research for this paper is supported by the National Science Foundation under grants DEB-0823293 and DEB-1237517.

L. Ogden (✉)
Department of Anthropology, Dartmouth College, Hanover, NH, USA
e-mail: laura.a.ogden@dartmouth.edu

N. Heynen
Department of Geography, University of Georgia, Athens, GA, USA
e-mail: nheynen@uga.edu

U. Oslender
Department of Global and Sociocultural Studies, Florida International University, Miami, FL, USA
e-mail: uoslende@flu.edu

P. West
Department of Anthropology, Barnard College and Columbia University, New York, NY, USA
e-mail: cw2031@columbia.edu

K.-A. Kassam
Department of Natural Resources, College of Agriculture and LifeSciences, Cornell University, 122 Fernow Hall, Ithaca, NY 14853-3001, USA
e-mail: ksk28@cornell.edu

R. Rozzi et al. (eds.), *Earth Stewardship*, Ecology and Ethics 2,
DOI 10.1007/978-3-319-12133-8_10

demonstrate the ways transnational networks of grassroots organizations can challenge their negative social and environmental impacts, and thus foster socio-ecological resiliency.

Keywords Global assemblages • Political ecology • Uneven development

10.1 Introduction

At the beginning of the twenty-first century, Crutzen and Stoermer (2000) proposed that we are now in the Anthropocene, a geologic epoch characterized by human dominance over the Earth's ecosystems. For many, the Anthropocene has become an important framework for thinking about the processes and consequences of global environmental change, particularly global climate change, widespread species extinctions, and the erosion of the "global life support system" (Steffen et al. 2011; Viola and Basso 2015 in this volume [Chap. 24]). For others, the Anthropocene is the affirmation of the long-held proposition that nature and society coproduce each other, and that capitalism has become the dominate logic of this coproduction process (see Dussel 2003; Swyngedouw 2013). We write this chapter to stimulate a conversation with our colleagues in ecology, philosophy, and other disciplines about Earth stewardship in the Anthropocene.[1] We do so by discussing how some social theorists are thinking about the Anthropocene's emergent properties of scale-defying governance, with particular attention to the role social and economic inequalities play in the transformation of the world's ecosystems, communities, and more broadly, global approaches to Earth stewardship. In this chapter,

[1] During the last decade, the Ecological Society of America has initiated a broad Earth Stewardship platform (Chapin et al. 2015 this volume [Chap. 12]). This platform includes a call for a more action-oriented science that is oriented toward understanding pathways to sustainability (Chapin et al. 2011, see also Sayre et al. 2013).

P. Robbins
Nelson Institute for Environmental Studies, University of Wisconsin-Madison, Madison, WI, USA
e-mail: director@nelson.wisc.edu

F. Massardo
Universidad de Magallanes, Puerto Williams, Chile

Institute of Ecology and Biodiversity, Santiago, Chile
e-mail: francisca.massardo@umag.cl; massardorozzi@yahoo.com

R. Rozzi
Department of Philosophy and Religion Studies, University of North Texas, Denton, TX, USA

Institute of Ecology and Biodiversity, Santiago, Chile

Universidad de Magallanes, Punta Arenas, Chile
e-mail: rozzi@unt.edu

we offer two conceptual frameworks that contribute to an ethics of Earth stewardship.

First, we argue that the Anthropocene, as a lens for understanding global environmental change, makes certain processes of change visible, while blinding us to others. We examine how insights from political ecology help us to see the Anthropocene as an uneven process of global change. A political ecological approach suggests that particular agents (social groups, corporations, nations) have caused the global environmental change we face today, instead of blaming a generalized species, *Homo sapiens*. Unsustainable practices that are detrimental to life (in all its human and non-human complexities) need to be sanctioned and/or remedied. Complementarily, more sustainable worldviews, forms of knowledge, values, economic and ecological practices should be respected and eventually adapted as we develop new models of Earth stewardship (Rozzi 2013, p. 10).

Second, we introduce the concept of the "global assemblage," a framework adopted widely in the social sciences (Collier and Ong 2005; Sassen 2006). Many scholars in the social sciences and humanities draw upon "assemblage theory," a theoretical approach indebted to the relational philosophy of Donna Haraway (2008), Bruno Latour (1993, 2004), and Gilles Deleuze and Félix Guattari (1987; DeLanda 2006). While distinct, this scholarship shares an understanding of world making, or life, as a dynamic process of becoming beings in the world through changing constellations of humans, other species, technologies, and institutions.

Like many of our colleagues in ecology, we, as anthropologists, geographers, and philosophers, share research approaches that focus on a range of different kinds of socio-ecological relationships that are embedded at different spatial and geopolitical scales within communities, ecosystems, biomes, and the biosphere (cfr. Wu 2013). While these methodological and conceptual lenses provide a rich analytical framework for understanding the complexities of social and environmental change, we suggest that Earth stewardship requires more deliberate inclusion of conceptual approaches that help us to understand how the "local" articulates with and is transformed by economic and cultural globalization and global climate change.

When Arthur Tansley (1935) developed his ideas about ecosystems, he suggested that they were not simply comprised of "natural" dynamics, but also human made dynamics. In so doing, he laid the foundations for understanding the Anthropocene, forcefully arguing for a new conceptual apparatus for ecology:

> We cannot confine ourselves to the so-called "natural" entities and ignore the processes and expressions of vegetation now so abundantly provided by man [sic]. Such a course is not scientifically sound, … The "natural" entities and the anthropogenic derivates alike must be analyzed in terms of the most appropriate concepts we can find (1935: 304).

Following Tansley and others, political ecologists have developed approaches that analyze the complex ways "natural" entities are transformed and contested through changing social contexts. However insightful, Tansley and early political ecologists were writing at a time before globalization and global institutions began to transform ecological processes and functions to the extent we all recognize today. In this chapter, we suggest the Anthropocene is an epoch constituted by processes of socio-ecological change that are no longer localized, as they were for most of human history. We suggest that "global assemblages," as a conceptual framework, provides a sophisticated multi-scalar approach for analyzing these changes. In this chapter, we demonstrate how diverse forms of global assemblages drive these changes—with some forms facilitating and other forms hindering socio-ecological resiliency. Our argument is that we only will be in a position to fully understand and respond to the Anthropocene's challenges by acknowledging:

1. the diversity of human societies and cultures that are part of increasingly connected world, and
2. the unbalanced power of contemporary global socio-ecological relations.

10.2 The "Uneven" Anthropocene

Though there have been important debates about the "start" of the Anthropocene, most scholars locate its origins with European industrialization (Zalasiewicz et al. 2010). Not only did the Industrial Age herald practices of production that have led to widespread degradation of the Earth's ecological systems, it also created profound transformations in social, economic, and political relations. For most of human history, our subsistence strategies (such as foraging, hunting, small-scale agriculture) were predicated on local cultural expectations regarding the use and meaning of the material world (plants, animals, land, water) and, importantly, by social obligations regarding the distribution of resources among members of a community, as well as, for many peoples, social obligations to the material world (see chapters by Kingsland and Mamani-Bernabé in this volume [Chaps. 2 and 6]). Key to the Anthropocene's shift has been a decoupling of societal obligations to other fellow humans and nature, or Earth stewardship, from practices and ideologies of societal reproduction. This decoupling has generated inequalities not only among different societies and cultures, but stratification within societies and cultures (producing racial, gender, economic, and generational divisions), and inequalities among human and non-human life.

These shifts in social-ecological relations are historically constituted and unevenly deployed. In other words, there is a politics to the Anthropocene that is often unrecognized in our discussions of "global environmental change." American anthropologist James Ferguson (1990) famously called the international development industry an "anti-politics machine." He argued that development efforts treat social and environmental inequalities (such as food and water scarcity, loss of ancestral lands, environmental degradation) as "problems" best resolved through the application of modern technical solutions (such as techniques for intensified agricultural production to alleviate food insecurity). Ferguson's point is that this apolitical framing masks the legacies of colonialism and ongoing global economic exploitation. As Ferguson makes clear, development projects fail, in part, because apolitical, technical approaches often reinforce the economic and structural inequalities that have created the problems in the first place (Ferguson 1990; see also Escobar 1995).

We need to resist a similar "anti-politics" that blinds us to the asymmetrical economic and ecological relations of the post-industrial world. Doing so will help us understand the unevenness of the "anthropogenic." As Nathan Sayre (2012, p. 59) asks, "who caused which changes, and what impacts on whom?" The "uneven anthropogenic" is visible at multiple scales: from cross-national analyses of the relationship between environmental degradation and economic development, to more nuanced examinations of the dynamics and drivers of local socio-ecological changes.

10.3 The Uneven Anthropogenic

The anthropologist Eric Wolf (1982) was particularly interested in understanding the ways in which industrialization gave rise to a global economic system that relied on cheap labor and raw materials from the Global South. As Wolf and others have demonstrated (see Smith 2008), the reach of industrial capitalism transformed the livelihood strategies of peasants, horticulturalists, and pastoralists throughout the world, and in many cases, promoted the overuse of resources.[2] Political ecologists have sought to understand the role of the modern economic system in creating "ecological distribution conflicts," such as conflicts over access and control of land and resources (Escobar 2008; May Jr 2015 in this volume [Chap. 27]). Political ecology, including the influential work of Wolf and others, was shaped by systems approaches, such as World Systems Theory (Wallerstein 1974), that conceptualized the connections between the Global South and North as historically constituted by uneven political economic relations.

Recently, Steffen and colleagues analyzed broad patterns of global change associated with transformations in the global economy since the Industrial Revolution (2011). In particular, numerous social and economic indicators demonstrate that rates of consumption, production, and population growth have accelerated dramatically in the past 50 years (Steffen et al. 2011, p. 742). The authors go on to carefully demonstrate the correlations between these accelerated rates of change and the continued degradation of the functioning and structure of the Earth's systems. Importantly, the authors make the point that this "accelerated" rate of global change was disproportionately driven by consumption patterns in the Global

[2] For the most part, the "politics" of political ecology has concerned itself with the means by which people exert control over other people, as well as the environmental transformations (deforestation, desertification, for example) spurred by these material processes (Blaikie 1985; Blaikie and Brookfield 1987). Paige West has defined political ecology as "a sophisticated contemporary theory of accumulation by dispossession and the vast effects of this ongoing process" (2012, p. 30; see also Biersack and Greenberg 2006; Neumann 2005; Paulson and Gezon 2005; Peet and Watts 2004). This scholarship has produced critical appraisals of the symbolic and material absorption of other beings within capitalism and other arenas of socioeconomic power—including through discursive regimes, practices of governance, and contests over resources and the equitable distribution of environmental risk.

North *even* in the context of increased population growth in the rest of the world (Steffen et al. 2011, p. 746).

The Anthropocene is an epoch of the Earth where nations with the highest rates of consumption also have the lowest levels of environmental degradation within their borders (Leichenko and Solecki 2005). As incomes rise in countries, environmental risks (such as air and water pollution) initially increase, then subsequently fall. There are several related reasons for this paradox, as Rinku Roy Chowdhury and Emilio Moran (2012) elaborate. For instance, wealthier countries are able to "outsource" the production of goods that degrade the environment to other nations (HDR 2011), and have the resources for environmental protection and restoration activities within their own nations.[3]

Much of our understanding about the relationship between environmental change and social equity is derived from macro-scale indexes of wealth (such as Gross Domestic Product) or quality of life (such as the Human Development Index). While these indicators are insightful, they miss the heterogeneous patterns that reveal differential vulnerabilities to environmental hazards, including within nations in the Global North (Ravallion 2007). How individuals, institutions, and social groups experience these vulnerabilities is structured by a complex and dynamic set of historical, economic, cultural, and political conditions. Classic studies within the environmental justice literature have documented how toxic waste dumps and facilities are disproportionately sited near places where ethnic and racial minorities live in the United States (Bullard 1994; UCCCRJ 1987; Cole 1994). Other research has shown that as cities change, these past patterns of residential segregation may produce unexpected outcomes—such as how the legacy of "white flight" in Baltimore has created new patterns of spatial segregation and access to environmental amenities such as parks (Boone et al. 2009). As Robbins's work on the political ecology of lawns in the United States has shown (Box 10.1), wealth and education do not always correlate with healthier local environments.

The Anthropocene is also an epoch of the Earth where new forms of governance are creating these uneven processes of social and environmental change. Moreover, these forms of governance, which we are calling "global assemblages," often transcend the boundaries and power of the nation state. Global assemblages are socio-ecological constellations that include multinational corporations, development initiatives, media networks, trade agreements, political treaties and other forms of governance, lending organizations, and non-governmental organizations (see, for example, Tsing 2005). As Sassen describes (2008), key institutions and agents within global assemblages, whether the International Monetary Fund (IMF), or multinational corporations, serve as "instruments," that drive socio-ecological changes associated with our world today. Many of these global assemblages are guided by neoliberal trade agreements that leave communities and ecosystems in the Global South less resilient to socio-ecological change, as Daly and Goodland (1994) demonstrate. These global assemblages should be understood as new, transnational forms of socio-ecological governance.

[3] For examples of how global markets negatively impact regions of Latin America see Rozzi and Feinsinger (2001), and Rozzi (2012).

Box 10.1

Fig. 10.1 American lawns complicate expectations regarding Earth stewardship (Photo by L. Ogden)

The turf grass lawn is one of the largest and fastest growing landscapes in the United States. It is also a landscape that requires the application of chemicals, nutrients, and water on a scale rivaling commercial agriculture. Paul Robbins' (2007) research on lawns illustrates the complexities of understanding the drivers and risks associated with environmental degradation in the United States. Contrary to expectations about "green" behavior, Robbins' research shows that homeowners who apply lawn chemicals are disproportionately better educated and have higher incomes than homeowners who do not. Surprisingly, these same "chemical appliers" do so even though they also are more likely to believe that lawn chemicals have a negative effect on local water supplies. For these homeowners, the rewards of a green lawn, specifically as a way of protecting capital investment, outweighs fears about the environmental harm and human health consequences of lawn chemicals. In effect, wealthier and better-educated Americans live in toxic landscapes of their own making. That said, Robbins demonstrates how homeowner anxieties about property values, community standards, lawn care industries, input manufacturers, developers, municipalities, and even turf grass itself drives the production and maintenance of this unique urban ecology.

In many cases, poorer countries articulate with these global assemblages through the export of natural resources. Countries dependent the upon export of primary commodities (such as agricultural goods and minerals) tend to experience persistent levels of social inequalities and poverty (UNCTD 2004) and are extremely vulnerable to price fluctuations, market consolidations, environmental hazards (floods, pests, etc.), as well as changing demands for these commodities. Yet the ways in which global commodity markets transform people and places vary considerably, as Box 10.2 demonstrates.

Box 10.2. Coffee in Papua New Guinea

Fig. 10.2 Exotic images, such as this one, are often used to sell coffee from Papua New Guinea (Photo by L. Ogden)

After petroleum, coffee is the second most frequently traded commodity on world markets. Coffee cultivation has profoundly shaped global economies, transformed tropical mountain ecosystems, and has redefined how both consumers and producers live in the world. For instance, in Papua New Guinea (PNG) coffee production has been part of social and ecological life since the

(continued)

Box 10.2 (continued)
early colonial period. Today, one in three people in PNG is connected to the coffee industry, illustrating the way in which demand for commodities can define local socio-environmental relations even in places that are often thought of as on the edges of the global market economy. Paige West (2012) has examined the movement of coffee from indigenous producers in PNG to consumers around the world. Her research with the Gimi peoples, who grow coffee in PNG's highlands, shows how eager they are to expand their business and social relationships with the buyers, processors, and exporters, as well as with consumers in cities such as Hamburg, Sydney, and London. At the same time, West shows how the "market" for specialty coffee misrepresents the Gimi, using images of primitivity and poverty to sell coffee. By implying that the "backwardness" of PNG impedes economic development, these images obscure the structural relations and global political economy that actually cause poverty in PNG. Coffee producers, in PNG, make about .15 cents per hour on specialty coffee that sells for over $12.00 per pound at Starbucks, exemplifying the asymmetrical relations that constitute the Anthropocene's global economic system.

As the United Nation's last *Human Development Report* makes clear, climate change driven by fossil fuel consumption in the world's wealthiest nations poses the greatest challenge to achieving environmental and social equity in the world (HDR 2011). The Anthropocene's poorer nations have contributed less than 1 % of the cumulative emissions that are driving climate change (Steffen et al. 2011, p. 746). At the same time, it is the 1.3 billion people who rely on natural resources for their income and subsistence (such as through export agriculture, forest products, and fishing) that are the least resilient to climate-driven environmental change. For example, societies in resource rich Arctic regions contributed little to the causes of climate change, yet they are among the first to observe and respond to its impacts (Crate and Nuttall 2009; Krupnik et al. 2004), as illustrated in Box 10.3. Women in poorer countries, who are disproportionately involved in subsistence farming, gathering of forest goods, and water collection, are even more vulnerable to the environmental impacts of climate change (HDR 2011; see Mamani-Bernabé 2015 in this volume [Chap. 6]).

These socio-ecological inequalities are a result of what Leichenko and O'Brien (2008) have called a "double exposure" to global environmental change and processes of globalization associated with market-driven drivers of change. Leichenko and O'Brien's double exposure framework directs our attention to the non-linear interactions and feedbacks of these two "transformative" processes of change, as further illustrated in Box 10.4. Importantly, they demonstrate how groups most vulnerable to the impacts of global environmental change often simultaneously experience the negative impacts of globalization. In other words, the interactions between these processes contribute to "growing inequalities, increasing vulnerabilities, and accelerated and unsustainable rates of change" (Leichenko and O'Brien 2008, p. 9).

Box 10.3. Northern Indigenous People, Climate Change and Food Security

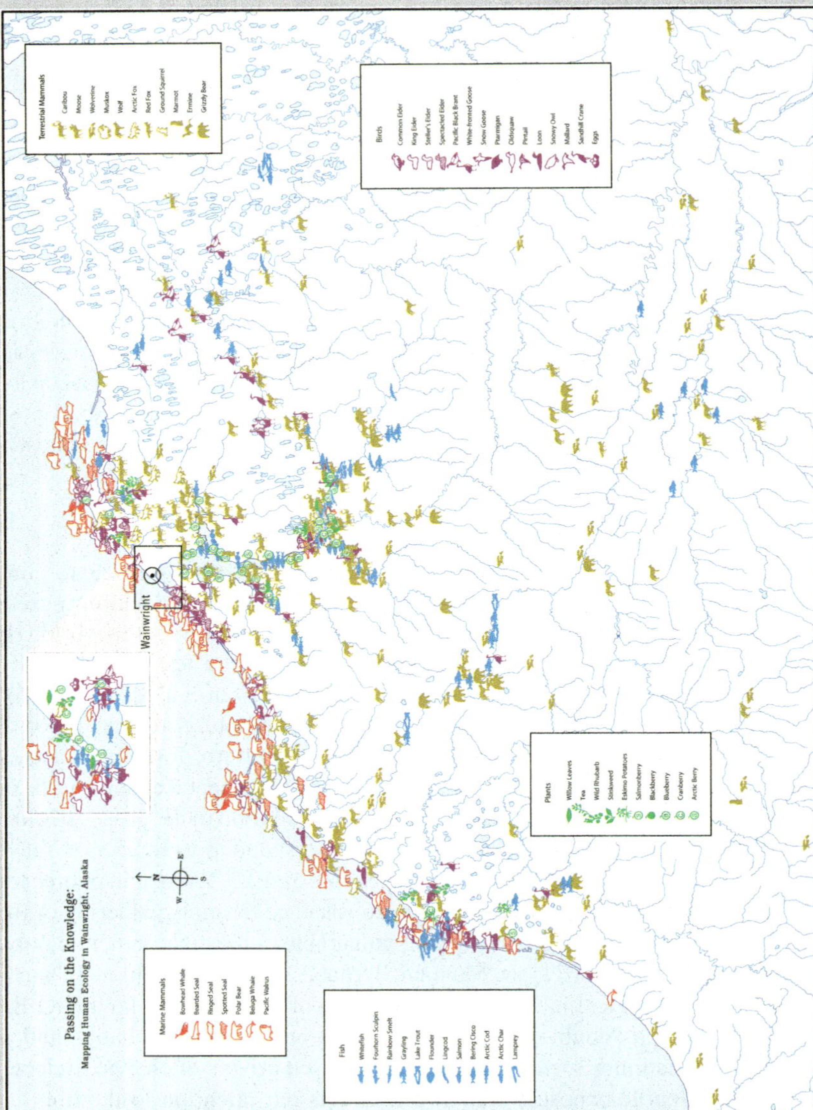

Fig. 10.3 Map of Iñupiat human-ecological relations illustrating how subsistence is fundamentally dependent on biodiversity

(continued)

Box 10.3 (continued)

Subsistence activities of the Iñupiat of Wainwright on the North Slope of Alaska at the Chukchi Sea intimately connect them to a diversity of life in their habitat (Kassam 2009). The residents maintain ecological relations with many sea mammals to meet their nutritional needs despite risks associated with travel on open water and sea ice (Fuller and George 1999; Ivie and Schneider 1988; Kassam and The Wainwright Traditional Council 2001; Luton 1986; Nelson 1969, 1982). The Iñupiat have therefore developed significant context-dependent knowledge of sea-ice and methods for interacting with it. Climate change is leading to increasing uncertainty in patterns of sea-ice formation, challenging the predictive capacity of Iñupiat knowledge of sea-ice. Therefore, climate change impacts make subsistence activities potentially dangerous, because the harvests of marine mammals require calm winds and strong ice for safe travel. The risk that climate change impacts may overwhelm these communities is increased by other chronic stressors, including legacies of colonialism and economic imperialism that constrain local economies. Put tersely, climate change is an additional layer of complexity on already existing inequities (Kassam et al. 2011).

Box 10.4. Palm Oil Production: A Complex Global Assemblage

Fig. 10.4 Pacific coast settlements in Colombia where locals resist oil palm cultivation (Photo by U. Oslender)

(continued)

Box 10.4 (continued)
Motivated by concerns over loss of biodiversity from unregulated timber extraction and gold mining, in 1993 Colombia passed Law 70, which granted land rights to Afro-Colombian communities living in the tropical rainforests of the Pacific coast lowlands. With this law, some five million hectares of lands were to be passed into communal land ownership, an acknowledgement of the communities' role in preserving this fragile ecosystem for hundreds of years. Social activism in these communities was decisive to this landmark achievement.

Today, however, this conservationist rationale has been all but abandoned. As Oslender (2008) shows in his research, powerful oil palm and gold mining multi-national corporations are colluding with illegal armed groups to displace local residents from their lands in order to gain access to the rich resource base. Targeted killings of activists and massacres of entire communities have led hundreds of thousands of *campesinos* to flee their lands since the mid-1990s.

Palm oil production has been aided by national and international organizations, such as the United Nations, who credit Colombia's oil palm industry for playing an important role in climate change mitigation. Its principal product—palm oil—is converted into biodiesel, considered an important resource for reducing carbon emissions into the atmosphere. As Oslender's work shows, palm oil companies mobilize environmental discourses to expand oil palm cultivation, while Afro-Colombian peasant farmers and fisherfolk (who are considered "guardians" of the region's rich biodiversity) continue to be displaced by the thousands. In the battle of these competing environmental interests, community resilience and conservation continue to be jeopardized (Oslender 2008).

As we have briefly described, the Anthropocene is characterized by unequal processes of global connection and governance (economic, political, social, and technological), as well as patterns of uneven development. Moreover, conceptual apparatus, such as the global assemblage, provides a multi-scalar approach to examining these new forms of socio-ecological governance in ways that our traditional attention to the local (community or ecosystem) only partially reveals.

10.4 Repoliticizing the Anthropocene

While national governments and supra-national organizations, such as the United Nations, have shown an increasing interest in sustainability and socio-ecological resiliency, grassroots groups have tended to be the most vocal proponents of protecting the environment and promoting local autonomy in the process. What is very clear: these social movements emerge out of global networks of activists and offer important insights into alternative forms of socio-ecological governance in the Anthropocene. An example from Ogden's research site in Tierra del Fuego, Chile, serves to illustrate the ways in which social movements can contest global assemblages to foster environmental conservation. In 1993, the Trillium Corporation, out of Bellingham, Washington, purchased 400,000 ha of sub-Antarctic rain forest at the southern end of Isla Grande, the largest island in the archipelago. Trillium's *Rio Condor* project, as they named the operation, was considered a model of sustainable forestry and corporate responsibility (Ginn 2005). Though the company had all the necessary legal permits and funding to move forward on *Rio Condor*, Chilean environmental activists were able to use the media to cast considerable doubt about the true environmental impacts of the project and, ultimately, use the Chilean courts to slow the project's implementation and drain the project's financial resources (Klepeis and Laris 2006). These environmental activists, as Ogden's fieldwork reveals, organized against the *Rio Condor* project by collaborating with anti-Trillium activists in Bellingham, Washington. This international alliance was effective in defending the forests of Tierra del Fuego, in part, because supporters of *Rio Condor* were less effective in engaging relevant communities and organizations. Despite the sound science published in extensive reports that were delivered to the regional government and published in peer-reviewed journals (Arroyo et al. 1996), project supporters did not sufficiently involve the regional government and community in the design or decision-making process (Rozzi et al. 2006).[4] Ultimately, *Rio Condor* went bankrupt and, in the process of debt liquidation, the Goldman, Sachs, & Co. investment firm acquired the forests and transferred them to the New York-based Wildlife Conservation Society for protection.

In another example, Colin McFarlane describes the global reach of the "Slum/Shack Dwellers International" (SDI), an urban housing rights organization based in Mumbai, India (2009, p. 563). SDI, unlike a traditional development organization, does not have a centralized, hierarchical structure. Instead, as McFarlane demonstrates, SDI-modeled housing rights organizations have emerged in 20 different countries, from Cape Town to Phnom Penh, through global networks of social activists. While modeled on SDI's activities in Mumbai, which entail demonstrating home building in the city's informal settlements, each node of the SDI global network is highly localized in terms of aesthetics, knowledge, materials, and the organizational infrastructure of the groups.

Transnational activist networks of this kind are increasingly important in the struggle to protect local environments from unsustainable extraction and exploitation practices—for example, in the case of gold mining and agro-industrial monoculture in Colombia's tropical rainforests, as shown by Arturo Escobar (2008) and Ulrich Oslender (2008). Some of these struggles have led to remarkable achievements and a significant rethinking of humankind's relationship with nature. In 2008,

[4] The outcome of this large-scale project involving the Chilean National Academy of Science and nearly 100 researchers who attempted to establish sustainable forestry and biological reserves in Tierra del Fuego was unexpected, and shows the limitations of purely technical scientific approaches in conservation.

Ecuador issued a new constitution that has garnered international attention because of its pioneering treatment of the rights of nature. On par with human rights, nature is endowed, constitutionally, with the right to be protected and to be treated with respect. In this conceptualization, nature is no longer seen as an inert object for humans to appropriate. According to Escobar (Forthcoming), nature's inclusion in the Ecuadorian Constitution is based on an ecological worldview in which all beings exist in relation to others. As Escobar explains:

> To endow Nature with rights means to shift from a conception of nature as object to be exploited to one in which Nature is seen as subject; indeed, in this conception the idea of rights of Nature is intimately linked with the humans' right to exist. This notion implies an expanded ecological notion of the self, which, unlike the liberal notion, sees the self as deeply inter-connected with all other living beings and, ultimately, with the planet as a whole (Forthcoming, 66).

What these examples illustrate, as do many others, is the power of grassroots global movements, in collaboration with other institutions, to foster resilience in sites vulnerable to economic globalization and global environmental change.

We are suggesting that Earth stewardship in the Anthropocene requires an engagement with new forms of environmental politics. It also requires paying attention to multiple forms of environmental knowledge and to transdisciplinary approaches that reach beyond the standard players in collaborative conservation efforts. For example, at the southern end of the Americas the collaboration among artists, artisans from the indigenous community, philosophers, journalists, teachers, the tourism sector, and scientists have generated novel education, and conservation practices, and decision making (Box 10.5). As another example, the NSF-funded Coweeta Long-Term Ecological Research (LTER) Project has been conducting ground-breaking ecological research for over 30 years in the eastern deciduous forests of the southern Appalachian Mountains. As a long-term collaboration between the University of Georgia (UGA) and the United States Department of Agriculture (USDA) Forest Service, it has shown that producing scientific knowledge about complex, place-based, ecological issues in a manner that makes information relevant to local policy makers and citizens alike presents extreme challenges. At the same time, democratically co-producing ecological knowledge will become more and more important within the era of uneven anthropogenic development. Coweeta LTER researchers have thus initiated the "Coweeta Listening Project" (CLP) to address the long-standing difficulties of scientific knowledge production being done in democratic and "user-friendly" ways that better help inform community members in making the difficult ecological decisions they face amidst rampant exurban development.

The model for the CLP is explicitly based on the "spiral model of learning" (Fig. 10.6). This model comes from the traditions of community based participatory research, and the pedagogic traditions of John Dewey, Paulo Freire, and Myles Horton, among others. Through the creation of the CLP, Coweeta scientists have worked toward the ideal that dialogic participation and interaction is necessary to create the kind of reflexive socio-ecological science that can help transcend the limitations of "expert only" science, and move toward more evenly distributed and

Box 10.5. Changing Lenses to Assess Biodiversity: Ecotourism with a Hand-Lens in Cape Horn

Fig. 10.5 Ecotourism with a Hand-Lens in the Magellanic sub-Antarctic ecoregion, Chile (Photos by Adam Wilson)

Today a vast diversity of living beings and human values are invisible, because they do not have a place in the narrow worldview that governs global society. The limitations of a worldview -the lenses through which we view the world- become evident when confronting *otherness*. As an example, global standard biodiversity assessments based on vascular plants had considered floristic diversity of southwestern South America to be poor. However, long-term botanical fieldwork in the region disclosed its *floristic*

(continued)

Box 10.5 (continued)
otherness: non-vascular plants had a greater diversity than vascular plant species. Moreover, the sub-Antarctic Magellanic ecoregion hosts >5 % of the world's species of non-vascular plants. This stimulated the research team at Omora Park to "change the lenses to assess biodiversity richness" and focus on non-vascular plants for defining conservation priorities in high latitude ecoregions (Rozzi et al. 2008). This change made visible an idiosyncratic biodiversity, and provided an argument to create the UNESCO Cape Horn Biosphere Reserve in 2005. For the first time, a protected area was designated based on the diversity of mosses, tiny organisms that had rarely been perceived and valued in the international conservation community. This led to a change in the language referring to mosses, and to an awareness of interspecific co-inhabitation. Thus, the "miniature forests of Cape Horn" metaphor was composed to communicate that little organisms are co-inhabitants rather than mere "natural resources." The field activity of *ecotourism with a hand-lens* (EHL) was invented to help citizens and decision makers discover the beauty, diversity and ecological importance of a flora that regularly goes unnoticed (Goffinet et al. 2012). By observing *floristic otherness*, global citizens recover the awareness of the vital pulse that is common to mosses, humans, and all living beings. From the south of the world, EHL summons ethical, aesthetical, and ecological values that broaden the narrow economic lens that prevails in the relationship of global society with nature.

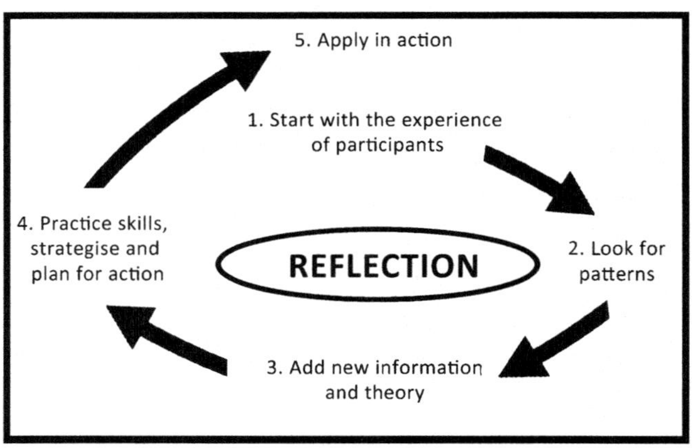

Fig. 10.6 The spiral model of learning (Redrawn from Arnold et al. 1991)

egalitarian forms of knowledge. The Anthropocene casts a bright light of the politics of knowledge production. The CLP's approach is working to create broader approaches within "big science" to better connect ecology through broader ideas of what Earth stewardship can be.

10.5 Implications for an Earth Stewardship Initiative

Here we propose that the Anthropocene is not only an era of anthropogenic change. Instead we suggest that the Anthropocene's changes arise out of new processes linked to a diversity of global assemblages. This reframing allows us to develop new approaches for more holistically engaging with a broader and necessarily more complicated articulation of Earth stewardship. Here we suggest that analyzing global assemblages, as a new form of socio-ecological governance, allows us to better articulate and understand how processes of uneven development often disproportionately impact vulnerable communities and environments, making them less resilient to global environmental change. At the same time, we see how grassroots social movements facilitate socio-ecological resilience even in contexts of "double exposure," as Leichenko and O'Brien describe.

From this new approach, Earth stewardship requires a willingness to recognize the politics inherent to the Anthropocene. In other words, it requires us to grapple with the complex ways our global connections create and maintain social and environmental inequalities. Taking these politics seriously, necessitates asking who produces what kind of socio-ecological configurations and for whom? Or in other words, this more holistic ethic is about formulating socio-ecological perspectives that are radically democratic in terms of organizing the interconnected processes through which ecologies we inhabit (humans and non-humans) come to be and evolve into the future.

This more robust version of an *anthropocenic* logic, if applied to ecological research and thinking, would be transformative and would necessitate the inclusion of a whole range of alternative approaches to the *status quo*. It would situate as a central pillar the view that processes of metabolic change are never socially or ecologically neutral; that socio-ecological conditions under which particular trajectories global environmental change occur sometimes destabilizes the coherence of *some* people living in *some* places, while the "sustainability" of *some* people living in *some other* places might be improved by those same conditions. In sum, a broader political-ecological perspective on Earth stewardship in the Anthropocene exposes the fundamentally uneven development of the global environmental system and amplifies the sorts of socio-ecological alternatives necessary for science and practice to do better at ameliorating environmental injustices the world over and sustainability for some and ecological calamity for others.

While not tending toward an unproductive apocalyptic sensibility, the uneven socio-ecological processes sound an important warning about the need to seek alternative forms of thinking about, and action toward, the world around us. To this end, Gibson-Graham and Roelvink suggest "responding to the challenges of the Anthropocene is not simply about human beings finding a technological or normative fix that will control and restore the earth. It is about human beings being transformed by the world in which we find ourselves—or, to put it in more reciprocal terms, it is about the Earth's future being transformed through a living process of inter-being" (2010, p. 322). Following Ecuador's lead in spelling out the rights of nature might be a step in this direction, though Ecuador's recent permitting of increased oil drilling within the Yasumi National Park, a UNESCO biosphere reserve, suggests how fragile nature's rights may be.

References

Arroyo MT, Donoso C, Murúa RE et al (1996) Toward an ecologically sustainable forestry project. Protecting biodiversity and ecosystem processes in the Río Cóndor project, Tierra del Fuego. DID Universidad de Chile, Santiago

Biersack A, Greenberg JG (2006) Reimagining political ecology. Duke University Press, Durham

Blaikie PM (1985) The political economy of soil erosion in developing countries. Longman, London

Blaikie P, Brookfield P (1987) Land degradation and society. Methuen, London

Boone CG, Buckley GL, Grove MJ et al (2009) Parks and people: an environmental justice inquiry in Baltimore, Maryland. Ann Assoc Am Geogr 99(4):767–787

Bullard RD (1994) Overcoming racism in environmental decisionmaking. Environ Sci Policy Sustain Dev 4(1):10–44

Chapin FS III, Pickett STA, Power ME et al (2015) Earth stewardship: an initiative by the Ecological Society of America to Foster Engagement to Sustain Planet Earth. In: Rozzi R, Chapin FS III, Callicott JB et al (eds) Earth stewardship: linking ecology and ethics in theory and practice. Springer, Dordrecht, pp 173–194

Chapin FS III, Power ME, Pickett STA et al (2011). Earth stewardship: science for action to sustain the human-earth system. Ecosphere 2:art89. 10.1890/ES11-00166.1

Cole LW (1994) Civil rights, environmental justice and the EPA: the brief history of administrative complaints under Title VI of the Civil Rights Act of 1964. J Environ Law Litigation 9:309–398

Collier JS, Ong A (2005) Global assemblages, anthropological problems. In: Ong A, Collier JS (eds) Global assemblages: technology, politics, and ethics as anthropological problems. Blackwell, Malden, pp 3–21

Crate SA, Nuttall M (2009) Anthropology and climate change: from encounters to actions. Left Coast Press, Walnut Creek

Crutzen PJ, Stoermer EF (2000) The "Anthropocene". Glob Chang Newsl 41:17–18

Daly H, Goodland R (1994) An ecological-economic assessment of deregulation of international commerce under GATT. Ecol Econ 9:73–92

DeLanda M (2006) A new philosophy of society: assemblage theory and social complexity. Continuum, London

Deleuze G, Guatarri F (1987) A thousand plateaus: capitalism and schizophrenia (trans: Massumi B). University of Minnesota Press, Minneapolis

Dussel E (2003) Algunos principios para una ética ecológica material de liberación (Relaciones entre la vida en la tierra y la humanidad). In: Pixley J (ed) Por un mundo otro. Consejo Latinoamericano de Iglesias, Quito, pp 29–44

Escobar A (1995) Encountering development: the making and unmaking of the Third World. Princeton University Press, Princeton
Escobar A (2008) Territories of difference: place, movements, life. Duke University Press, Durham
Escobar A (unpublished manuscript, forthcoming) Notes on the Ontology of Design
Ferguson J (1990) The anti-politics machine: 'development', depoliticization and bureaucratic power in Lesotho. Cambridge University Press, Cambridge, UK
Fuller AS, George JC (1999) Evaluation of subsistence harvest data from the North Slope Borough 1993 census for eight North Slope villages: for the calendar year 1992. Department of Wildlife Management, Barrow, North Slope Borough
Gibson-Graham JK, Roelvink G (2010) An economic ethics for the Anthropocene. In: Castree N, Wright M, Larner W et al (eds) The point is to change it: geographies of hope and survival in an age of crisis. Blackwell, Malden
Ginn W (2005) Investing in nature: case studies in land conservation in collaboration with business. Island Press, Washington, DC
Goffinet B, Rozzi R, Lewis L et al (2012) The Miniature Forests of Cape Horn: Eco-Tourism with a Hand-lens ("Los Bosques en Miniatura del Cabo de Hornos: Ecoturismo con Lupa"), Bilingual English-Spanish editionth edn. UNT Press–Ediciones Universidad de Magallanes, Denton/Punta Arenas
Haraway DJ (2008) When species meet. University of Minnesota Press, Minneapolis
HDR (Human Development Report) (2011) Sustainability and equity: a better future for all. United Nations Human Development Programme, New York
Ivie P, Schneider W (1988) Wainwright: land use values through time in the Wainwright area. North Slope Borough and the Anthropology and Historic Preservation Section of the Cooperative Park Studies Unit, University of Alaska, Fairbanks
Kassam KA (2009) Biocultural diversity and indigenous ways of knowing: human ecology in the Arctic. University of Calgary Press, Calgary
Kassam KA, The Wainwright Traditional Council (2001) Passing on the knowledge: mapping human ecology in wainwright, Alaska. Arctic Institute of North America, Calgary
Kassam K-AS, Baumflek M, Ruelle M, Wilson N (2011) Human ecology of vulnerability, resilience and adaptation: case studies of climate change from high latitudes and altitudes. In: Kheradmand H (ed) Climate change – socioeconomic effects. InTech. doi:10.5772/24054. ISBN: 978-953-307-411-5. Available from: http://www.intechopen.com/books/climate-change-socioeconomic-effects/human-ecology-of-vulnerability-resilience-and-adaptation-case-studies-of-climate-change-from-high-la
Klepeis P, Laris P (2006) Contesting sustainable development in Tierra del Fuego. Geoforum 37:505–518
Krupnik I, Huntington H, Koonooka C et al (eds) (2004) Watching ice and weather our way: Sikumengllu Eslamengllu Esghapalleghput. Arctic Studies Center, Smithsonian Institution, Washington, DC
Latour B (1993) We have never been modern (trans: Porter C). Harvard University Press, Cambridge, MA
Latour B (2004) Politics of nature: how to bring the sciences into democracy (trans: Porter C). Harvard University Press, Cambridge, MA
Leichenko RM, Solecki WD (2005) Exporting the American dream: the globalization of suburban consumption landscapes. Reg Stud 39(2):241–253
Leichenko RM, O'Brien KL (2008) Environmental change and globalization: double exposures. Oxford University Press, New York
Luton HH (1986) Wainwright, Alaska: the making of Inupiaq cultural continuity in a time of change. Ph.D. dissertation, University of Michigan, Ann Arbor
Mamani-Bernabé V (2015) Spirituality and the Pachamama in the Andean Aymara Worldview. In: Rozzi R, Chapin FS III, Callicott JB et al (eds) Earth stewardship: linking ecology and ethics in theory and practice. Springer, Dordrecht, pp 65–76
May Jr RH (2015) Dorothy Stang: monkeys cry and the poor die, Earth stewardship as liberation ecology. In: Rozzi R, Chapin FS III, Callicott JB et al (eds) Earth stewardship: linking ecology and ethics in theory and practice. Springer, Dordrecht, pp 407–418

McFarlane C (2009) Translocal assemblages: space, power and social movements. Geoforum 40:561–567
Nelson R (1969) Hunters of the northern ice. University of Chicago Press, Chicago
Nelson R (1982) Harvest of the sea: coastal subsistence in modern Wainwright, a report for the North Slope Borough's Coastal Management Program. North Slope Borough, Barrow
Neumann RP (2005) Making political ecology. Routledge, New York
Ogden L, Heynan N, Oslender U et al (2013) Global assemblages, resilience and Earth stewardship in the Anthropocene. Front Ecol Environ 11(7):341–347
Oslender U (2008) Another history of violence: the production of 'geographies of terror' in Colombia's Pacific coast region. LA Perspect 35:77–102
Paulson S, Gezon LL (eds) (2005) Political ecology across spaces, scales and social groups. Rutgers University Press, New Brunswick
Peet R, Watts M (eds) (2004) Liberation ecologies: environment, development, social movements, 2nd edn. Routledge, New York
Ravallion M (2007) Looking beyond averages in the trade and poverty debate. In: Nissanke M, Thorbecke E (eds) The impact of globalization on the world's poor. Palgrave, New York
Robbins P (2007) Lawn people: how grasses, weeds, and chemicals make us who we are. Temple University Press, Philadelphia
Roy Chowdhury R, Moran EF (2012) Turning the curve: a critical review of Kuznets approaches. Appl Geogr 32:3–11
Rozzi R (2012) Biocultural ethics: the vital links between the inhabitants, their habits and regional habitats. Environ Ethics 34:27–50
Rozzi R (2013) Biocultural ethics: from biocultural homogenization toward biocultural conservation. In: Rozzi R, Pickett STA, Palmer C (eds) Linking ecology and ethics for a changing World: values, philosophy, and action. Springer, Dordrecht, pp 9–32
Rozzi R, Feinsinger P (2001) Desafíos para la Conservación Biológica en Latinoamérica. In: Primack R, Rozzi R, Feinsinger P et al (eds) Fundamentos de Conservación Biológica: Perspectivas Latinoamericanas. Fondo de Cultura Económica, México, pp 661–688
Rozzi R, Massardo F, Anderson C, Heidinger K, Silander J (2006) Ten principles for biocultural conservation at the southern tip of the Americas: the approach of the Omora Ethnobotanical Park. Ecol Soc 11(1):43. http://www.ecologyandsociety.org/vol11/iss1/art43/
Rozzi R, Armesto J, Goffinet B et al (2008) Changing lenses to assess biodiversity: patterns of species richness in sub-Antarctic plants and implications for global conservation. Front Ecol Environ 6:131–137
Sassen S (2006) Territory, authority, rights: from medieval to global assemblages. Princeton University Press, Princeton
Sassen S (2008) Neither global nor national: novel assemblages of territory, authority and rights. Ethics Glob Politics 1:61–79
Sayre N (2012) The politics of the anthropogenic. Annu Rev Anthropol 41:57–70
Sayre NF, Kelty R, Simmons M et al (2013) Invitation to Earth stewardship. Front Ecol Environ 11:339
Smith N (2008) Uneven development: nature, capital, and the production of space. University of Georgia Press, Athens
Steffen WÅ, Persson DL, Zalasiewicz J et al (2011) The Anthropocene: from global change to planetary stewardship. Ambio 40:739–761
Swyngedouw E (2013) Anthropocenic politicization: from the politics of the environment to politicizing environments. In: Hedrén J, Bradley K (eds) Green utopianism: politics, practices and perspectives. Routledge, London/New York
Tansley AG (1935) The use and abuse of vegetational terms and concepts. Ecology 16:284–307
Tsing AL (2005) Friction: an ethnography of global connection. Princeton University Press, Princeton
UCCCRJ (United Church of Christ Commission for Racial Justice) (1987) Toxic waste and race in the United States: a national report on the racial and socioeconomic characteristics of communities with hazardous wastes sites. Public Access, New York

UNCTD (United Nations Conference on Trade and Development) (2004) The least developed countries report. United Nations Publication, New York

Viola E, Basso L (2015) Earth stewardship, climate change, and low carbon consciousness: reflections from Brazil and South America. In: Rozzi R, Chapin FS III, Callicott JB et al (eds) Earth stewardship: linking ecology and ethics in theory and practice. Springer, Dordrecht, pp 367–382

Wallerstein I (1974) The modern world-system: capitalist agriculture and the origins of the European world-economy in the sixteen century. Academic, New York

West P (2012) From modern production to imagined primitive: the social world of coffee from Papua New Guinea. Duke University Press, Durham

Wolf E (1982) Europe and the people without history. University of California Press, Berkeley

Wu J (2013) Hierarchy theory: an overview. In: Rozzi R, Pickett STA, Palmer C et al (eds) Linking ecology and ethics for a changing World: values, philosophy, and action. Springer, Dordrecht/Heidelberg/New York/London, pp 281–302

Zalasiewicz J, Williams M, Steffen W et al (2010) The new world of the Anthropocene. Environ Sci Technol 44(7):2228–2231

Part II
Integrating Stewardship Across Disciplines and Scales

Chapter 11
The Centennial Return of Stewardship to the Ecological Society of America

J. Baird Callicott

Abstract In 1917, Victor Shelford, a founder and the first president of the Ecological Society of America (ESA), formed and chaired the ESA's Committee for the Preservation of Natural Conditions (CPNC), an advocacy arm of the organization. Representing the Wilderness Society (WS), Aldo Leopold's efforts to form an alliance between the two organizations were spurned by Shelford because the WS encouraged the recreational use of wilderness reserves, thus altering their "primeval" condition. Seized by positivist zealotry, in 1945 the ESA prohibited members of the CPNC from advocacy activities and the committee disbanded in 1946. Shelford formed the private Ecologists' Union in 1946, which became The Nature Conservancy in 1950. In response to an intensifying global ecological crisis, the ESA has latterly returned to the policy-oriented advocacy under the banner of "Earth Stewardship" that characterized it during its first three decades." In addition to the values of its first president, Victor Shelford, the concept of Earth stewardship also expresses the spiritual values that its thirty-second president, Aldo Leopold, found in the scientific study of nature. The concept of Earth stewardship also returns ecology to its first organismic paradigm, for the Earth as a whole, Gaia, exhibits the defining characteristics of a "superorganism."

Keywords Advocacy • Leopold • Recreation • Superorganism • Shelford • Wilderness

The Ecological Society of America (ESA) was established in 1915. Victor Shelford was its first president, serving for the year 1916. Under Shelford's leadership, the Committee for the Preservation of Natural Conditions for Ecological Study was formed in 1917, chaired by Shelford himself (Anonymous 1917). In a paper titled "Preserves of Natural Conditions," Shelford reveals both the scientific and the philosophical rationales for the preservation of natural conditions. He also reveals himself to be an active advocate of preservation: "The first work [of the Committee] was to make a list ['of preserved or preservable areas'] and when this had made some

J.B. Callicott (✉)
Department of Philosophy and Religion Studies, University of North Texas, Denton, TX, USA
e-mail: callicott@unt.edu

R. Rozzi et al. (eds.), *Earth Stewardship*, Ecology and Ethics 2,
DOI 10.1007/978-3-319-12133-8_11

progress, *to agitate* for the reservation of such important areas as demanded immediate attention" (Shelford 1929, p. 37 emphasis added).

After about a decade of committee work, the "list" was published as *The Naturalist's Guide to the Americas* (Shelford 1926). And the agitation was stepped up, directed at federal and state governmental agencies, especially the National Park Service of the U. S. Department of Interior and the National Forest Service of the U. S. Department of Agriculture—both to acquire more "examples of primeval nature" and to restrict access to these "First Class Nature Sanctuaries" as "Research Reserves" to which "admission is by permit only" (Shelford 1929, p. 242). The Committee had representatives in every state who marshaled conservation and preservation organizations to conduct letter-writing campaigns addressed to politicians and bureaucrats urging the preservation of tracts of land deemed examples of primeval nature (Kinchy 2006; Shelford 1943).

The last three words (which later seem to have been dropped) of the original name of the Committee indicate the principal scientific rationale for the preservation of "natural conditions": Such conditions are the only proper objects of ecological study. Shelford prefers to make this clear by quoting other authors. According Francis B. Sumner, "The science of ecology depends upon *undisturbed* patches of *nature* as its 'material'" (Shelford 1929, p. 41 emphasis added). Joseph Grinnel had argued that "the *original balance*" of the "plant and animal life of the national parks . . . should be maintained" for ecological study among other reasons (Shelford 1929, p. 38 emphasis added). And Robert Griggs, also a member of the ESA, had argued for "the desirability of parks and reservations which preserve something of the original conditions *before the advent of man*" (Shelford 1929, p. 39 emphasis added).

In the first quarter of the twentieth century it was commonly believed that the "advent of man" in the Americas followed their "discovery" in 1492 by an Italian navigator commissioned by the Spanish Crown. Before then, "nature" in the Americas was "undisturbed" and remained in a "balanced" condition that was "original." Disturbance, in short, was assumed to be exogenous, anthropogenic, and relatively recent. What about the inconvenient truth that the Americas were long inhabited by indigenous peoples prior to their so-called "discovery"? To save the nature-primeval story on which their new science was erected, early ecologists had to downplay the environmental impact of pre-Columbian American Indians to the point of reducing them to the status of native wildlife. Shelford (1933a, p. 241) is explicit about such institutionalized racism:

> Primitive man, who could not remove the forest or exterminate the animals, is properly called a part of nature. At the time of the discovery of America, a scattered population of Indians had locally modified the vegetation, but had not destroyed any of the vegetation types. However, most of the areas which are now available for reservation as nature sanctuaries or nature reserves were probably not much affected by these primitive men. This is the argument for leaving them out of the picture.

Such were the assumptions lying behind ecology's first paradigm set forth by the first dean of American ecology, Frederic E. Clements; and during the first half of the twentieth century, ecology was dominated by the Clementsian paradigm. Clements (1905) thought that the objects of ecological study were what might be called third-order organisms or, as they actually came to be called, "superorganisms." The first organisms—first-order organisms—were single-celled. Through close symbiotic association, single-celled organisms evolved into multi-celled organisms—second-order organisms. Likewise, through close symbiotic association, multi-celled organisms evolved into third-order organisms—superorganisms. Until the invention of the microscope, we could not perceive single-celled organisms—because they are too small—nor did we even know that they existed. Neither do we perceive superorganisms, as organisms, because they are too big. The invention of ecology, however, provides a conceptual, if not a physical, lens by means of which they may be revealed and studied. In response to skeptics such as Henry A. Gleason (1926), "Clements is quoted [by whom is not specified] as saying that biologists present at the evolution of multicellular from unicellular organisms would have denied that they *were* organisms because they were *different*" (Tansley 1935, p. 305, n. 5).

In any case, by this conceptual device—this paradigm—Clements was able to organize and subdivide the science ecology by analogy with organismal biology. Taxonomic ecology would identify "types" of superorganisms, such as piñon-juniper and post-oak cross-timber forests, long- and short-grass prairies, sphagnum-tamarack bogs and tupelo-cypress swamps. Ecological ontogeny would trace how—after catastrophic, usually anthropogenic disturbance—such superorganisms return to their "adult" or "climax" condition through the process of succession, which was the specialty that Clements (1916) made his own. Physiological ecology would study the functions of the various components of such superorganisms—how tree-roots hold soil, how bacteria and fungi reduce detritus to minerals ready to be taken up again by plants, how predators prevent the irruption of prey populations, and so on. As are all organisms, superorganisms were conceived to be closed, homeostatic, and self-regulating. Civilized human beings were regarded as external to them and the principal source of their disturbance. Thus the need to preserve natural conditions for ecological study free of human interference. For there to remain only humanly disturbed superorganisms for ecologists to study would be as if there remained only mutilated organisms for organismal biologists to study.

Note that the "pristine myth", as geographer William Denevan (1992) styles it, privileges *American* ecology. The so-called "Old World" had been humanized from time immemorial. Clements's European contemporaries—such as Eugenius Warming, a Dane, and Oscar Drude, a German—could not have shared such a parochial assumption. Perhaps for this reason, as well as others, British ecologist Arthur G. Tansley (1935) criticized and rejected the climatic-climax concept in favor of multiple climax types, such as "fire climaxes" and even "mowing climaxes." Tansley also criticized and rejected Clements's superorganism paradigm in ecology and introduced the ecosystem paradigm to replace it. Tansley's rejection of Clementsian organicism is, however, too often too greatly exaggerated—due in large part to the misrepresentation of ecosystem ecology by historian Donald Worster (1994). Tansley (1920, 1926, 1935) repeatedly declares that what he eventually dubbed "ecosystems" were, albeit not literally organisms, nevertheless "quasi-organisms." Further, he also surmised that those quasi-organismic ecosystems which exhibited the greatest degree of stability and dynamic equilibrium had evolved by natural selection (Tansley 1935). So Tansley hardly put the kibosh to superoganismic thinking in the science of ecology.

Primarily an animal ecologist, Shelford's lasting Clementsian affinities were so ingrained that he could comfortably team with Clements, primarily a plant ecologist, to write a unified ecology text titled *Bio-Ecology* (Clements and Sheford 1939). The dean of mid-twentieth-century ecology Eugene P. Odum (1969) fully returned ecosystem ecology to its Clemensian roots by attributing even more sophisticated and subtle equilbria to "mature" ecosystems, such as a ratio of 1 between biomass production and respiration and between nutrient uptake and release. Indeed, Odum (1969, p. 596) tellingly alludes to Clements when he writes, "As viewed here, ecological succession involves the development of ecosystems; it has many parallels in the developmental biology of organisms. . . ." As Robert P. McIntosh (1985, p. 81) wryly remarks, "superorganisms are not easily killed." Like Shelford, Odum had strong preservationist sympathies, which show through his writings. He was a gentle advocate of "landscape law" and "environmental rights" as well as human rights; and he expressed a "need to start teaching the principles of ecosystem in the third grade" (Odum 1969), p. 269. But, unlike Shelford, Odum was no agitator for specific preservationist legislation and regulation.

That Shelford declined to join forces with the Wilderness Society (WS) seems to be a historical mystery, given that one pillar of his agitational strategy was to enlist the membership of various conservation and preservation organizations in letter-writing campaigns on behalf of some item on the Committee's wish list of preservable areas. The mystery deepens when we learn that, in 1940, none other than Aldo Leopold was tasked by Robert Sterling Yard, then the WS's president, with forming a WS-ESA alliance (Warren 2008). Shelford repeatedly rebuffed Leopold's entreaties.

Based in a study of Leopold's correspondence, Juliann Lutz Warren (2008) concludes that Shelford's rebuff of Leopold's WS-ESA affiliation efforts came down to what Shelford perceived to be a conflict of interest between the two organizations: the WS was interested in the *use* of wilderness set-asides for the recreational activities of its members, while the ESA was interested in preserving "Nature Sanctuaries" as "Research Reserves," from which the non-professional public was excluded. And at that time, wilderness recreation was a far cry from the "leave-no-trace" wilderness ethic of the present day. Wilderness recreation was then conceived to be the exercise of "woodcraft"—cutting down saplings to build rough shelters, hunting, fishing, traveling by pack train—generally living off the land (Turner 2002). And all that, of course, would constitute anthropogenic disturbance and spoil the putative natural condition of wildlands and thus ruin them for purposes of ecological study. Interestingly, a secondary difference was spatial scale: the WS was interested in large tracts of land, "big enough," as Leopold (1921, p. 719) first put it, "to absorb a two week's [mule] pack trip"; the ESA was then interested in "undisturbed patches of nature" as its "material" for study.

Leopold's role as matchmaker for the would-be groom (the WS) in pursuit of a marriage with the prospective bride (the ESA) led him to develop a scientific rationale for wilderness preservation (Warren 2008). In "Wilderness as a Land

Laboratory," Leopold (1941, p. 3) "set forth the need of wilderness as a base-datum for problems of land-health." Leopold, by then, had relocated from the Southwest to the Midwest, founded the field of wildlife management, and become a professor at a land grant institution (the University of Wisconsin). In that capacity he was doing "extension work" with farmers and other rural landowners in an effort to foster the growing of wild "crops" of "game" for purposes of sport hunting. Leopold, to his credit, did not betray his commitment to land *use* and disingenuously claim that he and fellow members of the WS were no less interested in pure ecological study than Shelford and fellow members of the ESA. Rather, land use, such as horticulture, ranching, and, yes, manipulative wildlife management, needed a "base-datum"—a control, land in the pink of health—against which to assess the success or failure of various experimental techniques of land management. Leopold was, in effect, elaborating a point that Shelford (1933b, p. 535) himself had already adumbrated: "There has been so much interference with natural processes in the form of 'control' of this and that organism that the student of 'wild life' management"—that would be Leopold—"who would seek a scientific basis for more scientific treatment of the animals in his charge, is left without guiding principles or reliable information and will continue thus until preservation measures . . . are put into effect in as many nature reserves as possible." Leopold's envisioned base-data wilderness areas would map neatly on to the distinct types of natural areas that the Committee had listed. As Leopold (1941, p. 3) put it, "One cannot study the physiology of Montana in the Amazon; each biotic province needs its own wilderness for comparative studies of used and abused land."

With this article, Leopold was addressing two audiences: members of the WS and Victor Shelford. That he had Shelford personally in mind is suggested not only by picking up Shelford's own suggestion that wildlife management needed a sound scientific basis if it were to succeed in its control efforts, but by the heavy emphasis that Leopold (1941, p. 3) gives to Clementsian ecological organicism: "There are two organisms in which the unconscious automatic processes of self-renewal have been supplemented by conscious interference and control. One of these is man himself (medicine and public health). The other is land (agriculture and conservation)." Leopold is not here shamelessly pandering to Shelford's theoretical commitments. No, Leopold was beginning to lose confidence in the essentially mechanistic "factors" approach that he had set out in *Game Management*, which was not working out on the ground as he had theorized in the abstract (Flader 1974, Meine 2010). He was beginning to think that soil, plant, animal, and climate interactions were so complex as to warrant portraying them as more like the interactions among the parts of an organism than those among the parts of a machine. Leopold (1939, p. 727) recorded his own experience as a wildlife manager in "A Biotic View of Land," an address to the joint meeting of the ESA and the Society of American Foresters:

> The emergence of ecology has placed the economic biologist in a peculiar dilemma: with one hand he points out the accumulated findings of his search for utility, or lack of utility, in this or that species; with the other he lifts the veil from a biota so complex, so conditioned by interwoven cooperations and competitions, that no man can say where utility begins or ends.

Shelford's main problem with the WS was its promotion of rough-rider wilderness recreation, which he viewed with something of the same contempt that present-day leave-no-trace wilderness adventurers view snow-mobile and dirt-bike off-road recreation. And that was indeed so big and so real a problem, Leopold realized, as to be a deal breaker. Leopold thus had to lead the members of the wilderness society toward a more refined form of wilderness recreation and to convince Shelford that such a change was afoot. Leopold (1941, p. 3) ends "Wilderness as a Land Laboratory" with these three sentences: "All wilderness areas, no matter how small or imperfect have a large value to science. The important thing is to realize that recreation is not their only or even their principal utility. In fact, the boundary between recreation and science, like the boundaries between park and forest, animal and plant, tame and wild, exists only in the imperfections of the human mind." With the first sentence Leopold addresses the scale issue—small preserves are also good and should not be snubbed by WS members. With the second, he downgrades the importance of recreation for future WS policy.

The third sentence cryptically suggests that recreation and science might not be altogether distinct activities. In two essays, "Conservation Esthetic" and "Wildlife in American Culture"—one antedating "Land Laboratory," one following it, both of which eventually found their way into *A Sand County Almanac*—Leopold (1938, 1943) envisions wildlife research and husbandry to become the ultimate form of outdoor recreation or sport. In his initial pitch to the ESA, Leopold put it this way:

> The [Wilderness] Society as now constituted is interested mainly in wilderness *recreation*. Another group, the Ecological Society, is interested mainly in wilderness *study*. There is little or no cooperation between the two groups, though both need the same changes in public policy.
> What needs to be done, I think, is to persuade both groups that wilderness recreation is destined to become more "studious," and wilderness studies more and more appreciative of esthetics, i.e., recreation. Therefore the two groups should get acquainted (*fide* Warren 2008, p. 98).

Note that Leopold here indicates that the membership of the ESA could also stand some refinement. A hallmark of Leopold's worldview is that embedded within the sciences of evolutionary biology and ecology—both in the practice of those sciences and in what those sciences reveal about nature—is a vast reservoir of aesthetic and spiritual potential, which most scientists fail to actualize. Just at the moment that he was attempting to ally the WS and the ESA, Leopold indulged himself in a little public fit of frustration at the insensitivity of science and scientists to the quasi-religious nature of their vocation and the deeply moving beauty of their revelations. Addressing his fellow-members of the Wildlife Society at its annual meeting, Leopold (1940, p. 338, emphasis added) said point blank, "We are not scientists. We disqualify ourselves at the outset for professing loyalty to and affection for a thing: wildlife. A scientist in *the old sense* may have no loyalties except to abstractions, no affections except for his own kind." He went on with palpable pique:

> The definitions of science written by, let us say, the National Academy, deal almost exclusively with the creation and exercise of power. But what about the creation and exercise of wonder, of respect for workmanship in nature? I see hints of such dissent, *even* in the

> writings of the scientifically elect—Fraser Darling, for example. Of course, we have always had such writers (David, Isaiah, John Muir) but they were not scientifically elect; they were poets. Is Fraser Darling *only* a poet? . . .
> There are straws which indicate that this *senseless barrier* between science and art may one day blow away and that wildlife ecology, if not wildlife management, may help do the blowing.

Strong stuff. Heartfelt. And typical. Any alliance between the WS and the ESA would have to involve some growth on both sides. Wilderness recreation would have to become more studious; ecological study would have to become more axiological—more frankly imbued with value and religious sensibility. One might say that for Leopold (1940, p. 338) the "dramas of ecology and evolution" were the liturgy and rites of a new paganism. He lamented that the model scientist of his day—the heyday of Logical Positivism in the philosophy of science—had "become as callous as an undertaker to the mysteries at which he officiates" (Leopold 1938, p. 107). That, Leopold hoped, would soon change.

A change did come to the ESA, but not the one Leopold envisioned. Indeed, the positivist temperament seized the ESA with a vengeance. Throughout the 1920s and 1930s, led by the Committee for the Preservation of Natural Conditions, the ESA was a vigorous player in national politics. For example, it agitated for the establishment of the Glacier Bay National Monument; vigorously opposed predator control in Yellowstone National Park and the diversion of water from Yellowstone Lake for commercial use; and it actively opposed opening Organ Pipe National Monument to prospecting and mining (Kinchy 2006). From the outset there were members of the ESA who were uncomfortable with advocacy and political agitation, in the interest of scientific objectivity and credibility; others, who were employees of government agencies, were forbidden by law from lobbying elected officials (Kinchy 2006). By the 1940s the tide began to turn against Shelford and the Committee within the ESA (Kinchy 2006).

According to Kinchy (2006), World War II was the storm surge that eventually swept the Committee away. The war effort and the need for resource extraction overtopped nature preservation when push came to shove. Scientists in other fields patriotically put themselves in service to the US government—and at the same time, with urgent government sponsorship and funding, made rapid scientific advances—while the posture of the ESA had been largely adversarial in its relations with the government. The Executive Committee of the ESA became increasingly disaffected with the preservation committee and more and more attracted to the posture of other professional scientific organizations, which stood for a relationship of science to politics as informing and advising and not as agitating and advocating policy and legislation. But Shelford remained adamant and uncompromising—and as obstreperous as ever. Robert Griggs, an early ally of Shelford's, led the campaign against advocacy on the part of the ESA. He served the Society as president in 1944 and put the fate of the preservation committee to a vote of the membership. In 1945, the constitution of the ESA was amended preventing the preservation committee from advocating policy positions. In 1946, the members of the preservation committee voted to disband.

Shelford responded by forming the independent Ecologists' Union (EU)—which was joined not only by a sizable number of ESA members, but was also open to non-ecologists (Kinchy 2006). As the membership of the EU swelled, its overlap with the ESA membership was attenuated. By 1950, the EU transformed itself into The Nature Conservancy (TNC). There is an irony in this history. The EU was formed in order to free members of the ESA to *continue* independently agitating and advocating for governmental policy and legislation that would protect and preserve natural conditions. The *goals* of the TNC remained the same, but the TNC abandoned this *method* for another: raising money and purchasing the properties wanted for protection and preservation—to hell with the government, so to speak. A further historical irony is the election of Aldo Leopold as president of the ESA for the year of 1947—and no one was more surprised by that than he.

However hard to kill, superorganismism had pretty much died out by the mid-1970s, supplanted by a neo-Gleasonian paradigm in ecology (Pickett and Ostfeld 1995). Although aggressive Shelfordian agitation and advocacy has not been re-institutionalized in the ESA, growing alarm at planetary-scaled environmental change, especially global climate change and the down-scale impacts it entrains, has moved the ESA to embrace an environmental ethic under the rubric of "stewardship"—first "ecosystem stewardship" (Chapin et al. 2009) then "planetary stewardship" (Power and Chapin 2009) and finally "Earth stewardship" (Chapin et al. 2011). Three past presidents of the ESA, Mary E. Power (2009–10), F. Stuart Chapin III (2010–11), and Steward T. A. Pickett (2011–12) have led the ESA's stewardship initiative, the culmination of its renewed social engagement, begun in the last decade of the previous century as the Sustainable Biosphere Initiative, under the leadership of Lubchenco (et al. 1991), also a past president of the ESA (1992–93).

Chapin, Pickett, Power, and three other authors begin their initiative-defining tract with a historical contextualization of "stewardship" and how it might be adapted to present environmental concerns: "A century ago, stewards were responsible for managing estates or keeping order at public events. Today the Earth is one global estate and improved stewardship is vital for maintaining social order and for preserving life on Earth" (Chapin et al. 2011, p. 44). But back in the day of stewards, estates had proprietary lords and many believe that the Earth also has a proprietary Lord. In response to a sharp critique of the Judeo-Christian worldview by Lynn White Jr. (1967), Christian theologians developed a Stewardship Environmental Ethic (Barr 1972, Black 1970).

White had based his critique on narrow but powerful textual evidence—Genesis 1:26–28—in which God *creates humans in His own image*, gives humans *dominion over the other creatures*, and commands humans to *subdue the Earth*. But, respond the Stewardarians, prior to these verses, God declares the pre-human creation to be "good"—that is, he invested it with what environmental ethicists call intrinsic value. That humans alone are created in the image of God, they go on to note, is a double-edged sword. True, by that token, humans have a unique right to use the Earth and all its other creatures for meeting their own needs, but they also have a unique responsibility not to destroy other species (the "kinds" that God had declared to be good) or undermine the health (ecosystem functionality) and beauty of the whole creation. This interpretation of "dominion," the Stewardarians point out, is confirmed in Genesis 2:15 where it is written that God put Adam (the first human) into the Garden of Eden (Nature) "to dress it and to keep it."

The ESA's current presidential stewardship initiative, with its religious overtones, returns the ESA to some of the most important extra-scientific values present at its founding. Note that Shelford, the ESA's first president, consistently characterizes "First Class Research Reserves" as "Nature *Sanctuaries*," endowing them with a kind of holy aura. Leopold, the ESA's thirty-second president, carries forward the science of ecology's perennial reverential romance with Nature, but less in a Judeo-Christian than in a neo-pagan modality.

Further, while the superorganismic paradigm has been long abandoned at traditional scales of ecological study, it too has been revived at the planetary scale.

Ecosystems were once thought to be (a) closed; (b) self-regulating; (c) tending toward a single point of stable equilibrium, (d) through determinant and invariant successional pathways; (e) with disturbances as exceptional and exogenous events, (f) most of which were due to external human activities (Pickett and Ostfeld 1995). Now they are thought to be (a) open to nutrients, pollutants, and motile organisms; (b) subject to external (often distant) as well as internal regulatory regimes; (c) capable of tending toward multiple domains of ecological attraction and (d) manifest directionless and endless successional change; (e) to incorporate natural disturbances some of which constitute disturbance regimes; (f) to also incorporate human influences everywhere for millennia; and (g) to have indeterminate spatial and temporal boundaries set by the scientists who investigate them (Pickett and Ostfeld 1995).

At the scale of the single, integrated *planetary ecosystem*, the Earth's biosphere exhibits many of the characteristics of the erstwhile organismic ecosystems in the tradition of Clements, Shelford, Leopold, and Odum (Margulis 1998). The Earth is readily and unambiguously bounded; its atmosphere excludes much (otherwise lethal) incident ultraviolent radiation and burns up small meteorites before they reach the surface—so it is closed; its magnetic field channels the charged particles of cosmic radiation and the solar wind to the poles and thus prevents the atmosphere from being stripped away; the chemical composition of the atmosphere and the oceans is internally regulated, as is Earth's average temperature; and until relatively recently (from the beginning of the Industrial Revolution), humans have had little effect on these planetary-scaled ecological processes—the chemical composition of the atmosphere and oceans, the ozone membrane in the stratosphere, global average temperature, global biogeochemical cycles (Zalasiewicz and Williams 2012).

And for nearly half a century the biosphere—the planetary ecosystem—has been swathed in a religious aura, having been baptized in the name of a Greek goddess, Gaia, by James Lovelock (1972).

References

Anonymous (1917) Committee on preservation of natural conditions. Bull Ecol Soc Am 1(5):1
Barr J (1972) Man and nature: the ecological controversy and the Old Testament. Bull J Rylands Libr 55:9–32
Black J (1970) The dominion of man: the search for ecological responsibility. Edinburgh University Press, Edinburgh
Chapin FS III, Carpenter SR, Kofinas GP et al (2009) Ecosystem stewardship: sustainability strategies for a rapidly changing planet. Trends Ecol Evol 25:241–249
Chapin FS III, Pickett STA, Power ME et al (2011) Earth stewardship: a strategy for social-ecological transformation to reverse planetary degradation. J Environ Stud Sci 1:44–53
Clements FE (1905) Research methods in ecology. University Publishing Company, Lincoln
Clements FE (1916) Plant succession: an analysis of the development of vegetation, vol 242. Carnegie Institution, Washington, DC
Clements FE, Sheford VE (1939) Bio-ecology. Wiley, New York
Denevan WM (1992) The pristine myth: the landscape of the Americas in 1492. Ann Assoc Am Geogr 83:396–385
Flader SL (1974) Thinking like a mountain: Aldo Leopold and the evolution of an ecological attitude toward deer, wolves, and forests. University of Wisconsin Press, Madison
Gleason HA (1926) The individualistic concept of the plant association. Bull Torrey Bot Club 53:7–26
Kinchy AJ (2006) On the borders of post-war ecology: struggles over the ecological society of America's preservation committee, 1917–1946. Sci Cult 15:23–44
Leopold A (1921) The wilderness and its place in forest recreational policy. J For 19:718–721
Leopold A (1938) Conservation esthetic. Bird Lore 40:101–109
Leopold A (1939) A biotic view of land. J For 37:727–730
Leopold A (1940) The state of the profession. J Wildl Manag 4:338–341
Leopold A (1941) Wilderness as a land laboratory. Living Wilderness 6:3
Leopold A (1943) Wildlife in American culture. J Wildl Manag 7:1–6
Lovelock J (1972) Gaia as seen through the atmosphere. Atmos Environ 6:579–580
Lubchenco J, Olson AM, Brubaker LB et al (1991) The sustainable biosphere initiative: an ecological research agenda. Ecology 72:371–412
Margulis L (1998) Symbiotic planet: a new look at evolution. Basic Books, New York
McIntosh RP (1985) Background of ecology: concept and theory. Cambridge University Press, New York
Meine C (2010) Aldo Leopold: his life and work. University of Wisconsin Press, Madison
Odum EP (1969) The strategy of ecosystem development. Science 164:262–270
Pickett STA, Ostfeld RS (1995) The shifting paradigm in ecology. In: Knight RL, Bates SF (eds) A new century for natural resource management. Island Press, Washington, DC, pp 261–277
Power ME, Chapin FS (2009) Planetary stewardship. Front Ecol Environ 7:399
Shelford VE (ed) (1926) The Naturalist's guide to the Americas. Williams and Wilkins, Baltimore
Shelford VE (1929) Preserves of natural conditions. Trans Ill State Acad Sci 13:37–58
Shelford VE (1933a) The preservation of natural biotic communities. Ecology 14:240–245
Shelford VE (1933b) Conservation versus preservation. Science 77:535
Shelford VE (1943) Twenty-five year effort at saving nature for scientific purposes. Science 98:280–281
Tansley AG (1920) The classification of vegetation and the concept of development. J Ecol 22:118–144
Tansley AG (1926) Succession: the concept and its values. In: Duggar MB (ed) Proceedings of the international congress of plant sciences, 16–23 Aug 1926. George Banta, Menasha, pp 677–686
Tansley AG (1935) The use and abuse of vegetative concepts and terms. Ecology 16:318–307

Turner JM (2002) From woodcraft to "leave-no-trace": wilderness, consumerism, and environmentalism in twentieth-century America. Environ Hist 7:162–487

Warren JL (2008) Science, recreation, and Leopold's quest for a durable scale. In: Nelson MP, Callicott JB (eds) The wilderness debate rages on: continuing the great new wilderness debate. University of Georgia Press, Athens, pp 97–118

White LW Jr (1967) The historical roots of our ecologic crisis. Science 155:1203–1207

Worster D (1994) Nature's economy: a history of ecological ideas, 3rd edn. Cambridge University Press, New York

Zalasiewicz J, Williams M (2012) The Goldilocks planet: the four billion year story of Earth's climate. Oxford University Press, New York

Chapter 12
Earth Stewardship: An Initiative by the Ecological Society of America to Foster Engagement to Sustain Planet Earth

F. Stuart Chapin III, S.T.A. Pickett, Mary E. Power, Scott L. Collins, Jill S. Baron, David W. Inouye, and Monica G. Turner

Abstract The Ecological Society of America (ESA) has responded to the growing commitment among ecologists to make their science relevant to society through a series of concerted efforts, including the Sustainable Biosphere Initiative (1991), scientific assessment of ecosystem management (1996), ESA's vision for the future (2003), Rapid Response Teams that respond to environmental crises (2005), and the Earth Stewardship Initiative (2009). During the past 25 years, ESA launched five new journals, largely reflecting the expansion of scholarship linking ecology with broader societal issues. The goal of the Earth Stewardship Initiative is to raise awareness and to explore ways for ecologists and other scientists to contribute more effectively to the sustainability of our planet. This has occurred through four

F.S. Chapin III (✉)
Institute of Arctic Biology, University of Alaska Fairbanks, Fairbanks, AK, USA
e-mail: terry.chapin@alaska.edu

S.T.A. Pickett
Cary Institute of Ecosystem Studies, Millbrook, NY, USA
e-mail: picketts@caryinstitute.org

M.E. Power
Department of Integrative Biology, University of California, Berkeley, CA, USA
e-mail: mepower@berkeley.edu

S.L. Collins
Department of Biology, University of New Mexico, Albuquerque, NM, USA
e-mail: scollins@sevilleta.unm.edu

J.S. Baron
Fort Collins Science Center, U.S. Geological Survey, Fort Collins, CO, USA
e-mail: Jill_baron@usgs.gov

D.W. Inouye
Department of Biology, University of Maryland, College Park, MD, USA
e-mail: Inouye@umd.edu

M.G. Turner
Department of Zoology, University of Wisconsin, Madison, WI 53706, USA
e-mail: turnermg@wisc.edu

R. Rozzi et al. (eds.), *Earth Stewardship*, Ecology and Ethics 2,
DOI 10.1007/978-3-319-12133-8_12

approaches: (1) articulation of the stewardship concept in ESA publications and Website, (2) selection of meeting themes and symposia, (3) engagement of ESA sections in implementing the initiative, and (4) outreach beyond ecology through collaborations and demonstration projects. Collaborations include societies and groups of Earth and social scientists, practitioners and policy makers, religious and business leaders, federal agencies, and artists and writers. The Earth Stewardship Initiative is a work in progress, so next steps likely include continued nurturing of these emerging collaborations, advancing the development of sustainability and stewardship theory, improving communication of stewardship science, and identifying opportunities for scientists and civil society to take actions that move the Earth toward a more sustainable trajectory.

Keywords Earth Stewardship Initiative • Ecological Society of America • Interdisciplinary integration • Practitioner Engagement • Sustainability

12.1 Introduction

Societies around the world are anxious to meet the needs of their growing human populations and to satisfy their rising aspirations. Human desires for high quality of life, material comfort, and consumption-based lifestyles are now shared around the world. Response to these pressures relies on industrial processes and global trade, which together are greatly expanding the human capacity to disrupt the biosphere. Growth in these human capacities has led to the global decline in biodiversity and other benefits that society receives from ecosystems (MEA 2005). These impacts have accelerated over the last 60 years (Steffen et al. 2004) and may now be approaching or exceeding the limits of ecologically tolerable environmental change (Ellis and Ramankutty 2008; Foley et al. 2005; Rockström et al. 2009).

Although the serious degradation of the Earth System is widely recognized by the scientific community, governments are frequently reluctant to adopt policies that would radically reduce the rates of change and degradation, for fear of economic repercussions. Aggressive actions that are taken now, however, are likely to be much less costly than the price of failing to act promptly (NRC 2010; Stern 2007). However, it is not only governments that seem constrained from acting. Individuals may not see the relevance of the status of the Earth's ecological processes to their lives and may therefore be tone deaf to their own responsibilities for the health of the Earth System (Hargrove 2015 in this volume [Chap. 20]).

Given the pace of environmental deterioration and the increased recognition that this path is unsustainable, society in all its aspects must seize the opportunity to reorient its relationship to the biosphere (DeFries et al. 2012) and ask what do humans owe to nature and to future generations? The scientific community has worked to develop the science needed for a more sustainable relationship between society and the planet (Lubchenco et al. 1991; MEA 2005) and to assess the rates, causes, and consequences of human pressure on the environment (IPCC 2014;

Melillo et al. 2014). Civil society, including individual citizens, businesses, religious and non-governmental organizations, communities, and tribes, have sought to apply this understanding to reduce society's impacts on the environment, but these efforts have so far been insufficient to stem the tide of degradation of Earth's life-support system. A broader, ethically framed approach is needed to move forward. We believe the concept of stewardship provides a compelling framework to move beyond what science can accomplish on its own.

In 2009, the Ecological Society of America (ESA) launched an initiative in Earth Stewardship to raise awareness and to explore ways that ecologists and other scientists could increase their effectiveness in shifting the planet toward a more sustainable trajectory. This parallels the Planetary Stewardship Initiative developed internationally as part of scientific planning for Future Earth (Steffen et al. 2011). We define Earth Stewardship as *a strategy to shape the trajectories of change in coupled social-ecological systems to foster ecosystem resilience and human well-being*. It builds on sustainability science (Clark and Dickson 2003; Kates et al. 2001; Matson 2009; Turner et al. 2003) and explores approaches to apply this science to urgent problems facing society and the biosphere (Chapin and Fernandez 2013).

Stewardship, according to the Merriam Webster dictionary, means "the activity and job of protecting and being responsible for something" (http://www.merriam-webster.com/dictionary/stewardship). The word is an old one, dating from the fifteenth century. According to the Online Etymology Dictionary (etymonline.com), it combines the idea of a house or hall (*stig*), such as on an estate or large farm, with the concept of a guard (*weard*). Thus, a steward is one who is entrusted with the care of a household. Responsibility in a deep and participatory sense is suggested by stewardship. However, it also implies that the task is undertaken on behalf of someone else or a larger entity (May Jr (2015) in this volume [Chap. 7]). In English and Scottish use, it can also apply to the care of a large political jurisdiction. The term has more recently come to mean provisioning of ships, and by extension, events, trains, or airplanes.

The original meaning, focusing on households, seems quite appropriate for an environmental application. A household associated with an area of land would include related and unrelated persons and would keep and maintain animals, woodlots, and gardens. The sense of responsibility and careful guardianship would attend the stewardship of a household. Consider that the terms "ecology" and "economics" also come from a formulation based on Greek that includes the idea of the household – of nature in this case. Ecology is the study of the household of nature, and economics relates to its management. Stewardship of Earth acknowledges that humans are members of the household of nature and that they bear responsibility to care attentively for this household.

The concept of Earth Stewardship, although rooted in religious thought (Conradie 2006; Hargrove 2015 in this volume [Chap. 20]; Kearns and Keller 2007), is a broadly ethical idea that does not rely on any one religious tradition in its call for responsibility to and membership in the larger Earth system and community. Indeed, its inclusiveness is suggested by similarity to principles underlying efforts as different as U.S. environmental policy, strategies for sustainability in developing nations (UN 2010; WCED 1987), and adaptive ecosystem management (Chapin et al. 2009; Christensen et al. 1996; Szaro et al. 1999). The concept of stewardship is familiar to the general public and has essentially the same meaning in lay terms as we intend in its scientific usage. Its goals are thus widely accepted by scientists, policy makers, and civil society, although their application inevitably raises contentious issues regarding tradeoffs (Clark and Levin 2010). The familiarity of the term stewardship facilitates communication with the larger civil society, although its diverse connotations can be problematic in some quarters (Hargrove 2015 in this volume [Chap. 20]), just as with "sustainability".

12.2 Evolution of ESA's Stewardship Approach

Since ESA's founding in 1915, the society has sought to provide leadership in both cutting-edge science and its application to environmental issues. Early leaders such as Victor Shelford and William Cooper played important roles in establishing National Parks and other areas for conservation. Eugene Odum advocated passionately throughout his career for the protection of Earth's endangered life-support systems (Odum 1989). However, tension between "basic" and "applied" research caused a group of ecologists to split away from ESA and form The Nature Conservancy in 1951 to pursue issues of explicit societal relevance, leaving ESA as the home for "basic" scientific ecology (Callicott (2015) in this volume [Chap. 11]).

Beginning in the late 1980s, ESA developed a research agenda for ecology. Under the leadership of five successive ESA presidents (1988–1992), the society came together to establish the Sustainable Biosphere Initiative (SBI), whose goal was to "define the role of ecological science in the wise management of Earth's resources and the management of Earth's life support system" (Lubchenco 2012; Lubchenco et al. 1991). The SBI identified three research priorities requiring particular attention in addressing global environmental problems: *global change*, *biodiversity loss*, and *sustainable ecological systems*. An important contribution of the SBI was the recognition of tight coupling between human activities and ecological processes on an increasingly human-dominated planet, with an emphasis on the application of ecological science to address these issues.

There were several important outcomes of the SBI. Membership in ESA broadly embraced the SBI's commitment to research that bridged basic and applied ecological science to contribute to the wise management of Earth's resources. As part of this commitment, ESA established an SBI office 1992 in Washington, D.C. to facilitate access to national government and relevant agencies and to inform government more effectively about the ecological repercussions of its policies. ESA established a policy office in 1983, which developed an education program in 1998 that subsequently branched off as an independent education office in 2003. The SBI office became the ESA science office in 1997. Together these offices foster the development of societally relevant ecological science and its application to policy and education. An ad-hoc committee was formed by ESA to assess the scientific underpinnings of ecosystem management, which took a holistic approach toward managing ecosystems and strongly emphasized sustainability (Christensen et al. 1996). In 2003, some 15 years after the SBI was launched, the ESA Ecological Visions Committee engaged in a second visioning exercise to assess the fit of ESA's activities to its goals and mission (Palmer et al. 2004; Palmer et al. 2005). Key points derived from this exercise were the need to acknowledge the extent of the human footprint globally and to use ecological knowledge as a solution-based science to improve ecosystem services and human well-being.

This more recent visioning process led to two significant outcomes. One recommendation was for the establishment in 2005 of Rapid Response Teams, a group of ecologists who are knowledgeable about ecological issues of societal relevance and are committed to respond rapidly when this knowledge is needed to inform government actions or issue media statements. This team of about 50 experts serves as panelists in briefings for congressional staff, provides expert testimony to Congress, analyzes the likely ecological consequences of proposed changes to environmental regulations, and provides scientific feedback for news stories. A second recommendation from the Visioning Committee was the establishment of a center that would link ecologists, other researchers, managers and policy makers for communicating and implementing ecological science for solutions. The National Socio-Environmental Synthesis Center (SESYNC; http://www.sesync.org/), funded by the National Science Foundation, directly addresses this recommendation. Projects at SESYNC focus on actionable science that can inform decisions within government, business, and households to improve the implementation of public policies and inform environmental planning.

ESA's commitment to stewardship is also reflected in the history of its journals. In 1991 it undertook publication of a new journal, *Ecological Applications*, which is concerned broadly with the applications of ecological science to environmental problems. It publishes papers that develop scientific principles to support environmental decision-making, as well as papers that discuss the application of ecological concepts to environmental issues, policy, and management. *Ecological Applications* is intended to be accessible to both scholars and practitioners. More recent ESA journals show an increasing commitment to societal issues: *Frontiers in Ecology and the Environment* (started in 2003), *Ecosphere* (started in 2010), and *Ecosystem Health and Sustainability* published jointly with the Ecological Society of China (started in 2015). All demonstrate this commitment. The series *Issues in Ecology* (started in 1997) report the consensus of scientific experts on specific issues related to the environment, using commonly understood language. Its intended audience includes decision-makers at all levels for whom an objective presentation of the underlying science will increase the likelihood of ecologically-informed decisions. Many of the numbers of the series *Issues in Ecology* are available not only in English, but also in Spanish.

Parallel to ESA's efforts, the National Academy of Sciences brought together scholars from a variety of natural and social sciences to advance societally relevant "sustainability science" (Clark and Dickson 2003; Kates et al. 2001; Matson 2009;

NRC 1999), whose goal is to "promote human well-being while conserving the life support systems of the planet" (Clark and Levin 2010). In 2004, ESA initiated a Sustainability Science Award to recognize authors who have made the greatest contribution to sustainability science through the integration of ecological and social sciences.

ESA's Earth Stewardship Initiative developed over several years reflecting the commitment of several ESA presidents and a broad spectrum of ESA members (Chapin et al. 2011; Power and Chapin 2009). Most significantly, the Earth Stewardship Initiative coincided with increased engagement and commitment to action by ESA's student section, one of the society's largest sections, clearly indicating the desire of the next generation of ecologists to address important environmental challenges. The Earth Stewardship Initiative builds upon the research agendas of the SBI and sustainability science with an emphasis on applying this understanding to help shape a more sustainable pathway for Earth as a social-ecological system. There are numerous ways to shape pathways of change toward a more sustainable future, including building the science as advocated by SBI and the Ecological Visions Committee, engaging the public and practitioners, communicating more effectively with the public and with policy makers, and conducting research that explicitly includes efforts to shape a more sustainable future. Box 12.1 illustrates some of these approaches, and the following sections describe ESA's efforts to engage ecologists and a broader range of scientists and practitioners in meeting the needs for a more sustainable future of our planet.

Box 12.1: Examples of Stewardship Applications

SEEDS Campus BioBlitz Campaign

BioBlitz is a community engagement exercise developed by ESA's Applied Ecology Section to acquaint local residents with the biodiversity in their neighborhoods. It is a quick comprehensive inventory of local biodiversity that typically requires both professional scientists with ecological and taxonomic expertise and resident volunteers to search for and collect local species of flora and fauna. It has been an effective approach to engagement and communication between ESA members and underserved communities in cities where ESA holds its annual meetings (Fig. 12.1). ESA's Strategies for Ecology Education, Diversity and Sustainability (SEEDS) Program expanded the use of BioBlitzes by organizing BioBlitzes in communities associated with local campus chapters, using an informational document they developed. SEEDS students find that a BioBlitz helps raise community awareness of the diversity of living organisms in their neighborhood and the ecosystem services they provide. Goals of the BioBlitz program include promoting environmental programs on campuses and their surrounding communities, engaging volunteers in citizen science, providing a vehicle for both informal and formal environmental education, creation of databases of local species, and stimulating political awareness about biodiversity and environmental degradation.

(continued)

Box 12.1 (continued)

Fig. 12.1 BioBlitz collaboration between ESA students and local community members

ESA Graduate Student Response to the BP Oil Spill of 2010

In response to the British Petroleum (BP) oil rig explosion and fire of April 2010, Student Section chair Rob Salguero-Gomez and chair-elect Jorge Ramos harnessed the enthusiasm, energy, networking skills, and commitment to the environment of ESA's student membership. They assembled metadata from the work of ecologists, both ESA members and others, documenting pre-spill conditions in estuaries, shorelines, and marine environments in the affected states along the Gulf Coast. Mark Stromberg of the University of California Natural Reserve System shared database software developed by the Organization for Biological Field Stations, which was subsequently tweaked by ESA web-developers. Student section leaders and ESA SEEDS students assembled an ESA database on research and researchers with relevant pre-spill information and shared this with research institutions, agencies, and local universities working on spill assessment and recovery. Through listservs and social networks, ecologists and other scientists learned about the effort and emailed datasets and photographs to the ESA's Student Section. Jorge and Rob collated the information, made it available via ESA's website to resource managers in the affected Gulf Coast states. ESA Student Section leaders and Public Affairs staff also distributed a compilation of state-specific links for opportunities to volunteer with clean-up and rescue of oiled wildlife (http://www.esa.org/esablog/research/conservation/taking-action-what-is-being-done-and-what-you-can-do-for-the-gulf/) (Ramos et al. 2012).

(continued)

Box 12.1 (continued)

Ranching, Local Ecological Knowledge, and the Stewardship of Public Lands

After decades of controversy over grazing and fire, ranching families, conservation groups, agency officials, and engaged citizens are finding ways to link sustainable grazing with conservation in prairie grasslands of the Southwestern US. Sustainable grazing can preserve open space and wildlife habitat, allow oversight of exploding recreation, and motivate restoration of degraded lands and watersheds (Sayre 2005; Silbert et al. 2007). These outcomes, however, depend critically on the knowledge of local ecosystems held by multi-generational ranching families, particularly during this era of rapid environmental change. Two efforts in the Grand Canyon region have enhanced stewardship of the social-ecological systems on ranches and our public lands. In the early 1990s, two ranching families joined with former critics in the environmental community to form the Diablo Trust, a collaborative management group sponsoring monitoring research that informs ranch practices, conservation projects, and policy reform (Muñoz-Erickson et al. 2009; Sisk 2010). On the North Rim of the Grand Canyon, another collaborative effort came together when the Grand Canyon Trust, a leading conservation organization, purchased the historic Kane and Two-mile Ranches to reform the livestock business from within, linking ranching with overarching commitments to ecosystem restoration and biodiversity conservation across 380,000 ha of public land (Sisk et al. 2010). These collaborations moved controversy out of the courtroom and into the use of evidenced-based science to improve stewardship of public lands and resources.

Salmon, Cyanobacteria, and Watershed Stewardship in Northwestern California

In 2011, people living along the Eel River in northwestern California, concerned about diminishing flows, recovery of salmonids, and a rash of toxic algal blooms, formed the Eel River Recovery Project (ERRP) (Fig. 12.2). Like many rivers of the western US, the Eel historically supported iconic Pacific salmon populations (Yoshiyama and Moyle 2010). Juvenile salmonids thrive when their invertebrate prey are fueled by edible algae (particularly diatoms). These diatoms and their macro-algal hosts, which act as substrates that vastly increase diatom surface area, can colonize in rivers and dominate when summer flows connect and flush channel habitats. However, when drought and/or human water extraction decrease the flows of river waters, these edible algal assemblages can become overgrown by cyanobacteria, some of which are toxic. Summer water extraction has recently been greatly exacerbated by burgeoning marijuana cultivation. ERRP volunteers, tribal members from the Eel and Klamath basins, and researchers (ecologists and

(continued)

Box 12.1 (continued)

phycologists) at the Angelo Coast Range Reserve have teamed together to: (1) share algal identification skills, so local residents can distinguish the "good, the bad, and the structural" algae (Fig. 12.3), and (2) partner in basin-scale surveillance to track changes in salmonids, algae and channel environments under climatic and human-induced drought. The Eel River Critical Zone Observatory (http://criticalzone.org/eel/), which hosts scientists studying the effects on stream flow of geology, topography, vegetation cover, human activities and climate in these steep forested basins, promotes exchange among scientists, ERRP volunteers (http://www.eelriverrecovery.org/algal_foray), and other citizens and tribal members concerned about rivers along the California North Coast. The collaboration of researchers and citizen scientists and tribal members in watching, analyzing, interpreting, and forecasting flow-driven changes in river ecosystems will guide practices that could enhance resilience under drought for this vulnerable but important coastal landscape (Power et al. 2015).

Fig. 12.2 Floating cyanobacterial mats in the Eel River. These are incubators where diatoms and filamentous green algae die, and cyanobacterial propagules are "spawned," floating down the river to colonize other places

(continued)

Box 12.1 (continued)

Fig. 12.3 Identifying "the good, the bad, and the slimy" taxa in collected algae

12.3 Engaging Ecologists in Stewardship

Both the SBI and the Earth Stewardship Initiative were initially proposed to ESA members with some trepidation, given ESA's history of reluctance to address the link between science and policy, which may have reflected a fear that this could lead to advocacy, such that the credibility or objectivity of the science would be jeopardized (Lubchenco 2012; Callicott (2015) in this volume [Chap. 11]). However, both initiatives came to be widely supported by ESA membership, particularly by younger members. Both initiatives represent an expansion of ESA's goals from a focus on communication of ecological science among members to "raising public awareness and ensuring the appropriate use of ecological science in environmental decision making" (http://www.esa.org/esa/). ESA has explored and promoted the Earth Stewardship Initiative among ecologists largely through four approaches:

1. articulation of the Earth Stewardship Initiative concept in ESA publications (Chapin et al. 2011; Power and Chapin 2009; Sayre et al. 2013) and Website (http://www.esa.org/esa/?page_id=2157),

2. selection of meeting themes and symposia (Box 12.2),
3. engagement of ESA sections to implement the initiative more broadly, and
4. outreach beyond ecology through collaborations and demonstration projects.

Box 12.2: ESA Meeting Themes (in Bold) and Examples of Stewardship-Related Symposia Since Launching of the Earth Stewardship Initiative

2010: **Global warming: The legacy of our past, the challenge for our future**

Environmental scientists as effective advocates: Above the din but in the fray
Planetary stewardship and the MAHB
Climate and justice: Exploring equity through land, water, and culture
Global warming, smallholder agriculture, and environmental justice: Making critical connections
Contributions of citizen science to our understanding of ecological responses to climate change

2011: **Earth Stewardship: Preserving and enhancing Earth's life-support systems**

Earth stewardship: Defining the scientific challenges and opportunities
Building a global sense of place, responsibility and stewardship
How we manage our share of Planet Earth
Thirty years of Earth Stewardship research: Long-term matters
Stewardship of urban systems: Socio-ecology, governance, and equity in the ULTRA network
Micro-managing the planet: Integrating microbial ecology and Earth Stewardship
A natural history initiative for ecology, stewardship, and sustainability
Revolutionary ecology: Defining and conducting stewardship and action as ecologists and global citizens
Integrating evolution into policy: Improved science-based decision-making for environmental stewardship
Warfare ecology: Impacts of conflict on environmental security and stewardship
Global perspectives of Earth Stewardship

2012: **Life on Earth: Preserving, utilizing, and sustaining our ecosystems**

Interacting with practitioners to facilitate Earth Stewardship
Human behavior and sustainability: Addressing barriers to change
Revolutionary ecology: The role of diversity in unleashing ecology's potential to improve environmental conditions and societal welfare
Translational ecology: Forging effective links between knowledge and action

(continued)

Box 12.2 (continued)

The new grand challenge for ecology: Sustaining agriculture while promoting environmental justice

Ecological consequences of multiple changes in Asia and their implications to global sustainability

Grappling with intangibles: Bringing cultural ecosystem services into decision-making

The evolving role of environmental scientists in informing ecosystem policy and management

Conservation in a globalizing world

Commodifying nature: The scientific basis for ecosystem service valuation in environmental decision making

2013: **Sustainable pathways: Learning from the past and shaping the future**

Resilience, disturbance and long-term environmental change: Integrating paleoecology into conservation and management in the Anthropocene

Can ethics and justice pave a sustainable pathway for human ecosystems

Ecology across borders: International, national, and cultural challenges of managing species internationally

Ecological sustainability in a telecoupled world

Past, present and future design of infrastructures for a resilient society

The ecology-policy interface: Perspectives on student engagement

2014: **From oceans to Mountains: It's all ecology**

Ecosystem stewardship through traditional resource and environmental management: Indigenous management models from around the globe

Use-inspired ecological research that moves knowledge to action

The view from the trenches: Perspectives and advice from scientists engaged in science, policy and advocacy

What can ecologists learn from communities: A dialogue on Earth Stewardship from the dual perspectives of communities engaged in ecology and ecologists engaged in communities

Ecological design and planning for ecologists: Applying Earth Stewardship

Engaging with business and industry to advance Earth Stewardship: Business and biodiversity

Sustainable sourcing of food products: Social-ecological perspectives of constraints and opportunities for sustainable food production strategies

Green cities: Ecology and design in urban landscapes

Understanding and managing ecological resilience to natural disasters in a changing environment

Mitigating impacts to ecosystem services: Approaches, assumptions, and advances

(continued)

Box 12.2 (continued)

From studying to shaping: A design charette bridging site analysis to conceptual design

Analysis of the ecological dimensions in general public energy education programs of major justice, faith-based, indigenous, and environmental organizations: Energizing a future role for ecologists

Promoting urban sustainability via linkages among stewardship, urban yards, biodiversity, and ecosystem services

The student section of ESA has been most active and innovative in exploring ways to incorporate Earth Stewardship into their section activities. Five ESA student members summarized some of the ways that graduate students and their university departments could individually and collectively be more effectively engaged in Earth Stewardship (Colón-Rivera et al. 2013). In addition, the student section has been a reasoned and effective advocate for "action ecology," an expansion of ecological science into the realm of research that directly supports decision-making and policy (Bonilla et al. 2012; Rivera et al. 2010). They have done this, for example, by sponsoring symposia on this topic (sometimes under the label of "Revolutionary Ecology"; Marshall et al. 2011) at several recent ESA annual meetings. They were instrumental in organizing an initiative to assess ecosystem services in response to the British Petroleum oil spill in the Gulf of Mexico (Ramos et al. 2012) and have participated actively since 2008 in BioBlitzes that engage residents in documenting local biodiversity (Box 12.1). ESA graduate students have been consistent, active participants in congressional staff visits in Washington. For example, in April 2014, five graduate students visited congressional offices to explain the value of ecological science to the nation and to press for continued support for scientific research (http://www.esa.org/newsletter/eiaSpring14.html).

The extent of engagement of other ESA sections in the Earth Stewardship Initiative has been variable. In general, the sections that focus explicitly on human-nature interactions have been consistently active and account for much of the current implementation of Earth Stewardship within ESA. For example, the Human Ecology Section has regularly organized symposia at annual meetings and has served as the interface between ESA and its international counterpart—the Society for Human Ecology. The Environmental Justice Section has also organized symposia and played an active outreach role by engaging environmental groups associated with various communities of faith and by organizing a speakers bureau, as described in the next section. The Traditional Knowledge Section has regularly met with local tribes in the region of each ESA annual meeting to increase the awareness of ESA members of the indigenous heritage of the US, and on occasions also with indigenous people from other countries, to foster engagement of indigenous peoples in local and global ecological and environmental issues. About half of the ESA Sections (including Agroecology, Applied Ecology, Aquatic Ecology, Asian Ecology, Education, Environmental Justice, Long-term Studies, Microbial Ecology,

Natural History, Paleoecology, Policy, Rangeland Ecology, and Urban Ecosystem Ecology) have also organized symposia at annual meetings that explore the societal relevance of their subdisciplines in an Earth Stewardship context.

Since the launching of the Earth Stewardship Initiative, there has been a gradual increase in the number of ESA sections actively involved in the initiative. During the past 5 years, topics of symposia, which are generally co-sponsored by multiple ESA sections, have gradually evolved from conceptualization to implementation to evaluation of Earth Stewardship approaches (Box 12.2). In general, the involvement of ESA sections has broadened the leadership and intellectual framework of the Earth Stewardship Initiative and has led to more diverse pathways for engagement of ESA members in its implementation.

The 2014 meeting included a demonstration project for the application of ecosystem stewardship and other aspects of ecology: "Cities that work for people and ecosystems." Using the American River Parkway that runs through downtown Sacramento CA, the project demonstrates how ecological research, working at the intersection between ecological science and urban design, can monitor and adjust management practices using ecological principles, in order to work toward sustainability goals.

ESA's Public Affairs office sponsors or co-sponsors congressional briefings on topics relevant to the Earth Stewardship Initiative, taking advantage of its Washington, D.C.-based policy office and the expertise represented by its members. Recent briefings have included topics such as water resources, climate-change impacts and adaptation, and improvement of flood management. Field trips and exhibits targeting policy makers are another way that ESA tries to broaden its impact. The ESA Office of Science Programs focuses its activities on advancing ecological science, but also on projects that link ecological research and management communities to more effectively integrate ecological science into decision-making and education. Its third category of activities focuses on solutions for sustainability, through a series of activities that examine and articulate the intellectual foundations for a new sustainability science. Since 2008 the Education and Diversity Programs Office has coordinated workshops, webinars, and speaking tours to promote the future of continental-scale science and education primarily to undergraduate institutions and underrepresented audiences in ecology. Its project on the Future of Environmental Decisions also included graduate students.

12.4 Moving Beyond Ecology

Recognizing that Earth Stewardship must be much broader than ecology, ESA began a series of efforts to collaborate with other disciplines and practices. This began with a symposium on scientific foundations of Earth Stewardship organized jointly with physical scientists at the 2010 annual meeting of the American Geophysical Union (AGU). This symposium highlighted readily implementable opportunities for biophysical collaborations to address Earth Stewardship. One such initiative, led by AGU in collaboration with several academic societies, explores the challenge of communicating climate change (AGU 2013). ESA organized a series of informal meetings with leaders of (1) various social-science societies, (2) various societies representing practitioners (e.g., planners and engineers), (3) various federal agencies, and (4) various religious groups in the hopes that ESA might collaborate with these groups to develop jointly the concept of Earth Stewardship or a suite of compatible concepts that would engage a range of disciplines and practices in shifting the planet toward a more sustainable trajectory.

These conversations led to a workshop of natural and social scientists, practitioners, and religious scholars in 2012. The workshop brought together representatives from academia, federal agencies, religious organizations, business, and planning/design organizations to discuss building strategic interdisciplinary partnerships to foster sustainability. During the workshop participants identified challenges to implementing Earth Stewardship, along with possible solutions and novel ways to collaborate across sectors and disciplines. The special issue of *Frontiers in Ecology and the Environment* resulting from the workshop (2013, Vol 11, issue 7) contained a series of papers about diverse stewardship issues, each co-authored by scholars and practitioners from multiple disciplines and led by a non-ecologist. The goal of the workshop was to develop a more inclusive integrated framework for Earth Stewardship that would facilitate collaborative engagement across multiple disciplines and practices.

The participation of urban designers and engineers in the 2012 workshop and the issue of *Frontiers* described above symbolized the importance of interacting with professions that are engaged in the front lines of shaping the world in which we live. Sustainable or ecological approaches are becoming increasingly important to urban designers, regional planners, civil engineers, and those interested in restoring ecosystems that are embedded in urban territories. The fact that most of the world's human residents already live in cities or other places classified as urban suggests that the various practitioners of urban design and planning will play important roles in promoting Earth Stewardship. Consequently, ecologists must engage with these professions in order to: (1) help shape the urban designs, rather than study the outcomes after the fact; and (2) learn how to engage better with the real estate industry, the developer community, and those who write and enforce zoning and building regulations. Working with urban designers can help insert ecological principles and knowledge into the process of urban, suburban, and rural "place making," and may help formulate new procedures and regulations that are more attuned to the ecological processes that must be maintained or restored in sustainable urban areas (Felson et al. 2013; Felson and Pickett 2005; Pickett et al. 2013; Steiner et al. 2013). Professional societies such as the American Planning Association, the American Society of Landscape Architects, the Associated Collegiate Schools of Planning, and the Association of Collegiate Schools of Architecture are examples of practitioner organizations through which mutually beneficial pursuit of Earth stewardship may exist. In 2013 and 2014, ecologists engaged with landscape architects in symposia at the American Society for Landscape Architecture annual meeting to offer examples of how to incorporate ecological science in landscape and urban design, not just in the design phase, but throughout the life of the built landscape in order to move toward sustainability goals. This joint ESA/ASLA effort is repeated at ESA annual meetings, building a community from both societies determined to work together to achieve lasting provision of ecological services.

In their 2010 meeting with ESA, leaders of eight Judeo-Christian groups expressed concern about sustainability and an interest in exploring ways to collaborate with ESA to foster Earth Stewardship. Unlike the meeting of social scientists, the religious leaders had explicit suggestions about how this might be done. They felt, in general, that they had no ready access to the environmental science community, which they felt looked down on religious groups. They questioned whether environmental advocacy groups would be unbiased sources of scientific information. They suggested three concrete steps: (1) preparing fact sheets or short YouTube-type videos on issues that would be of concern to the religious community, (2) initiating a speakers' bureau that was co-trained by ecologists and by religious leaders to speak effectively to religious audiences, if invited to do so, and (3) an open letter from scientific and religious leaders to the religious community summarizing their common concern about the future. They emphasized that more progress would be made by focusing on issues of common concern (e.g., Earth Stewardship) than on issues that had a history of divisiveness (e.g., evolution). They also emphasized that issues of social and environmental justice would be of greater interest to religious groups than issues of environment. These conversations resulted in the development of a speakers' bureau led by ESA member Greg Hitzhusen (http://www.esa.org/enjustice2/projects/faith-communities/).

ESA reached out to the business community in 2013 and continues to work toward lasting relations with business leaders around the world. Businesses are among the largest agents of environmental degradation in the world. This offers tremendous opportunities for companies to become agents for positive change. A growing number of companies around the world realize they can galvanize the global business community to create a sustainable future for business, society, and the environment. The first workshop held in 2013 (standing room only) brought together sustainability officers from large corporations with ecologists to address how the science of ecology can be put to use by corporations such as 3M and Weyerhauser in meeting their sustainability goals. The ESA workshop was followed by a meeting that included several ESA members at PricewaterhouseCoopers (PwC) in London during the British Ecological Society Centennial Celebration in 2013 to explore how the science community can communicate more effectively with senior business leaders on sustainability issues. In 2014 a panel of business representatives convened to deepen the conversation between ESA members and business leaders, with a focus on businesses and biodiversity. Topics that remain to be explored include how business and industry view the need for biodiversity, what kinds of ecological information will enable businesses and industries to achieve sustainability goals that help preserve biodiversity, and what are the avenues for building collaborations between ecologists and businesses to protect biodiversity and the services it provides?

ESA is developing partnerships with public relations firms to help train ecologists in the art of effective communication with business leaders and has begun to develop a speakers' bureau of ecologists with these skills. We hope to deepen our ties with public relations companies who can help spread the word regarding Earth Stewardship. These discussions and the above-mentioned Demonstration Project not only serve to expand the conversation of Earth Stewardship to audiences with real ability to enact lasting positive change in environmental practices, but they also identify career paths and opportunities for ecologists with businesses and organizations that are trying to meet sustainability goals of economy, environment, and equity.

In addition to outreach to communities of faith and business, ESA is developing collaborations via the arts and humanities. Currently, this effort is being led by the Long-Term Ecological Research Network via Ecological Reflections (http://www.ecologicalreflections.com/), an effort to link environmental science with the arts and humanities (Goralnik et al. (2015) in this volume [Chap. 16]). This effort led to environmental art exhibits at the 2012 and 2013 ESA Annual Meetings as well as temporary exhibits of environmental art at the National Science Foundation headquarters in Ballston, Virginia in 2012 and 2013. The goal of this collaboration is to connect environmental science and Earth Stewardship to the general public through the languages of the arts and humanities. Similarly, the 14th Cary Conference brought together philosophers, ethicists, religious scholars, and ecologists to explore the linkages among values, philosophy, and action and to explore a new framework for conversations about how to motivate and implement actions toward sustainability (Rozzi et al. 2013). That conference was an important steppingstone toward the present volume (see Introduction to this volume).

12.5 The Future of Stewardship at ESA

The growing interest in Earth Stewardship from the leadership and membership of ESA bodes well for future involvement of the Society in this area. Continued effort is clearly warranted; indeed, we consider it urgent. The wide range of scales at which stewardship can be approached allows individuals to be involved in a variety of ways and to identify activities that resonate personally. A spatially small scale, such as a local park, a backyard, or the area designated for a BioBlitz (see Box 12.1) can motivate some individuals, while others may find regional or global scales more compelling. The existence of many environmental organizations focused on watersheds, ranging in size from small neighborhood watersheds to the Chesapeake Bay watershed that encompasses six states plus the District of Columbia, exemplifies the range of scales at which a particular disciplinary approach to stewardship can be applied (Kingsland (2015) in this volume [Chap. 2]). ESA can continue to encourage involvement across a wide range of scales. Here, we highlight several directions that seem important and tractable.

12.5.1 Building Stronger Partnerships

Contemporary environmental challenges go well beyond science alone. ESA must continue to build strong partnerships with people and institutions that can effect change, finding key areas of commonality that reflect shared goals and making sure that ecological science is at the table. As with any ecosystem, particular components or linkages within the system may be highly influential, and identifying keystone institutions and leverage points is important. Linkages with other groups must broaden to include greater representation from the business community and politicians. An "us vs. them" attitude will not serve the goals of Earth Stewardship well, and many leaders are keenly interested in sustaining resources in their local environment. Actions that enhance sustainability may be good for the bottom line. Throughout the country, business and engineering schools are developing new degree programs and certificates in sustainability, and ESA could cultivate partnerships with such programs. The business community will remain influential, and technology will surely play a role in addressing stewardship issues. Developers should be encouraged to collaborate with ecologists during the early phases of land-development projects so that subsequent ecological problems (and litigation) might be minimized. Ecologists are not generally well schooled in how to develop such partnerships and engage effectively; ESA should assist its membership in developing these critical skills.

ESA can also encourage more interaction with specialized interest groups, such as societies devoted to fish and game species that are working to preserve or improve habitat for their particular species. For example, there are now some large organizations focused on conservation of trout and other salmonids, elk, deer, turkey, quail, and waterfowl. These organizations reflect the broader recognition of stewardship in society at large, although there are often tradeoffs among competing interests of different groups.

12.5.2 Science Communication

ESA should continue to enhance its leadership in science communication. The challenges of communicating ecological science within civil society remain profound, especially when some sectors of society consider scientific data to carry only the weight of an opinion. An ecologically literate citizenship is essential for achieving the goals of the Earth Stewardship Initiative. Thus, ESA must continue to help our members become more effective at communicating what we do, what we know, and most importantly, why it matters. ESA might develop more widespread communication training programs, perhaps modeled on the successful Leopold Fellows Program, targeted especially for graduate students and non-academic scientists that are not eligible for the Leopold Fellows Program. The ability to anticipate and use new communications media effectively will be key for these efforts. Earth

Stewardship requires ecological literacy, and ecologists must be better at understanding their audiences in order to enter into dialogues that will result in more effective communication with the public at large. By partnering with other groups and engaging our younger scientists in the planning effort, ESA could make a major contribution to Earth Stewardship by directly enhancing the professional preparation of early-career ecologists.

12.5.3 Leading Theory Development in Sustainability Science

ESA members can also contribute to the theoretical basis for sustainability science. Historically ecologists have developed theory that integrates classical ecology with theory from evolutionary biology, molecular biology, geophysical sciences, etc. We are in early stages of integrating ecological theory with theory from various social sciences (Collins et al. 2011; Matson 2009) and currently lack a thoroughly developed theory for sustainability science. ESA can provide leadership to go beyond thinking of stewardship as "applied sustainability science" and rather to understand when and why (or why not) scientific understanding is effective in moving toward more sustainable pathways at various scales. Action ecology, such as ideas developed by the ESA student section, and discussions with practitioners need to become part of the learning loop for developing broader theory. Theory must be applied and tested against real societal and ecological problems. This remains a formidable challenge, but one that ESA is well positioned to nurture, perhaps by encouraging ESA sections to tackle relevant issues and by emphasizing sustainability theory in different venues during annual meetings.

12.5.4 Encouraging Personal Involvement

Ecologists can engage directly in stewardship activities that emerge from their research programs. There are many examples of academic scientists who have felt compelled to focus their efforts on conserving the species and habitat they study, after realizing that the subjects of their studies are rapidly disappearing. For example, the Golden-Lion Tamarin, an endemic primate in Brazil, is now the only primate species to have been upgraded in terms of its endangered species status, following prodigious efforts by researchers who spent most of their careers studying them (Kierulff et al. 2012). In other cases, scientists have advocated strongly for habitat connectivity on regional scales or for sustaining a key resource, such as fresh water, or for reducing pollution. These constitute another avenue by which current and future ESA members could become involved in Earth Stewardship activities that are personally important to them. Workshops at the annual meeting might include training in best practices for members to pursue stewardship related to their research.

Given the successes documented from previous ESA efforts, future ESA Presidents will likely choose to sharpen the Society's focus on Earth Stewardship in different ways. Recent discussions with other professional societies whose expertise is related to stewardship have documented broad common interests that can be developed in the future. A recent effort by the ESA and the British Ecological Society to foster regular discussions among leaders of all the world's ecological societies will provide an opportunity to interest a global audience of ecologists.

The changes that we ecologists have seen in less than a generation include remarkable advances in technology (e.g., computing power, global positioning systems, geographic information systems, sensor networks), rapid changes in global climate, a blossoming of quantitative analytical techniques, an explosion of information with the digital revolution, and a great increase in cross-disciplinary and international collaborations. The kinds of science that can be done have changed, and the training of new generations of ecologists must change accordingly. Amidst all these changes to our field, the natural world is also changing at an unprecedented rate. This set of circumstances puts ESA at a critical juncture where we have the opportunity to train future generations of ecologists to work effectively in a world that is fundamentally different from the one in which we grew up. Further, ESA must intensify efforts to partner with a wider range of institutions and become more active participants in problem-solving, recognizing that compromise is often necessary. Having realized these challenges and begun to respond, ESA must continue to embrace them.

Acknowledgments We thank the staff, governing board, and members of the Ecological Society of America, especially Executive Director Katherine McCarter, for their engagement in crafting and implementing ESA's Earth Stewardship Initiative. We also thank Baird Callicott, Cliff Duke, Katherine McCarter, and Teresa Mourad for their careful review of this chapter.

References

AGU (2013) Helping audiences receive climate-science information: a message framework for scientists. Unpublished report

Bonilla NO, Scholl J, Armstrong M et al (2012) Ecological science and public policy: an intersection of action ecology. Bull Ecol Soc Am 93:340–345

Callicott JB (2015) The centennial return of stewardship to the Ecological Society of America. In: Rozzi R, Chapin FS III, Callicott JB et al (eds) Earth stewardship: linking ecology and ethics in theory and practice. Springer, Dordrecht, pp 161–171

Chapin FS III, Fernandez E (2013) Proactive ecology for the Anthropocene. Elementa. doi:10.12952/journal.elementa.000013

Chapin FS III, Power ME, Pickett STA et al (2011) Earth stewardship: science for action to sustain the human-earth system. Ecosphere 2(8):art89 doi:10.1890/ES11-00166.1

Chapin FS III, Kofinas GP, Folke C (eds) (2009) Principles of ecosystem stewardship: resilience-based natural resource management in a changing World. Springer, New York

Christensen NL, Bartuska AM, Brown JH et al (1996) The report of the Ecological Society of America committee on the scientific basis for ecosystem management. Ecol Appl 6:665–691

Clark WC, Dickson NM (2003) Sustainability science: the emerging research program. Proc Natl Acad Sci U S A 100(14):8059–8061

Clark WC, Levin SA (2010) Toward a science of sustainability. Princeton Environmental Institute, Princeton

Collins SL, Carpenter SR, Swinton SM et al (2011) An integrated conceptual framework for long-term social-ecological research. Front Ecol Environ 9:351–357

Colón-Rivera RJ, Marshall K, Soto-Santiago FJ et al (2013) Moving forward: fostering the next generation of Earth stewards in the STEM disciplines. Front Ecol Environ 11:383–391

Conradie E (2006) Christianity and ecological theology: resources for further research. Sun Press, Stellenbosch

DeFries R, Ellis E, Chapin FS III et al (2012) Plantetary opportunities: a social contract for global change science to contribute to a sustainable future. BioScience 62(6):603–606

Ellis EC, Ramankutty N (2008) Putting people on the map: anthropogenic biomes of the world. Front Ecol Environ 6(8):439–447

Felson AJ, Pickett STA (2005) Designed experiments: new approaches to studying urban ecosystems. Front Ecol Environ 3:549–556

Felson AJ, Bradford MA, Terway TM (2013) Promoting Earth stewardship through urban design experiments. Front Ecol Environ 11:362–367

Foley JA, DeFries R, Asner GP et al (2005) Global consequences of land use. Science 309:570–574

Goralnik L, Nelson MP, Ryan L et al (2015) Arts and humanities efforts in the US Long-Term Ecological Research (LTER) network: understanding perceived values and challenges. In: Rozzi R, Chapin FS III, Callicott JB et al (eds) Earth stewardship: linking ecology and ethics in theory and practice. Springer, Dordrecht, pp 249–268

Hargrove E (2015) Stewardship versus citizenship. In: Rozzi R, Chapin FS III, Callicott JB et al (eds) Earth stewardship: linking ecology and ethics in theory and practice. Springer, Dordrecht, pp 315–323

IPCC (2014) Climate change 2014: impacts, adaptation, and vulnerability. Working group II contribution to the IPCC 5th assessment report. Cambridge University Press, Cambridge

Kates RW, Clark WC, Corell RW et al (2001) Sustainability science. Science 292(5517):641–642

Kearns L, Keller C (2007) Eco-spirit: religions and philosophy for the Earth. Fordham University Press, New York

Kierulff MCM, Ruiz-Miranda CR, de Oliveira PP et al (2012) The Golden lion tamarin *Leontopithecus rosalia*: a conservation success story. Int Zoo Yearb 46(1):36–45

Kingsland SE (2015) Ecological science and practice: dialogues across cultures and disciplines. In: Rozzi R, Chapin FS III, Callicott JB et al (eds) Earth stewardship: linking ecology and ethics in theory and practice. Springer, Dordrecht, pp 17–26

Lubchenco J (2012) Reflections on the sustainable biosphere initiative. Bull Ecol Soc Am 93:260–267

Lubchenco J, Olson AM, Brubaker LB et al (1991) The sustainable biosphere initiative: an ecological research agenda. Ecology 72:371–412

Marshall K, Hamlin J, Armstrong M et al (2011) Science for a social revolution: ecologists entering the realm of action. Bull Ecol Soc Am 92:241–243

Matson PA (2009) The sustainability transition. Issues Sci Technol 25(4):39–42

May Jr RH (2015) Andean llamas and earth stewardship. In: Rozzi R, Chapin FS III, Callicott JB (eds) Earth stewardship: linking ecology and ethics in theory and practice. Springer, Dordrecht, pp 77–86

MEA (2005) Ecosystems and human well-being: current status and trends, vol 1. Cambridge University Press, Cambridge, UK

Melillo JM, Richmond TC, Yohe GW (eds) (2014) Climate change impacts in the United States: the third National Climate Assessment. U.S. Global Change Research Program, Washington, DC

Muñoz-Erickson T, Aguilar-Gonzáles B, Loeser MR et al (2009) A framework to evaluate ecological and social outcomes of collaborative management: lessons from implementation with a northern Arizona collaborative group. Environ Manage 45:132–144

NRC (1999) Our common journey: a transition toward sustainability. National Academies Press, Washington, DC

NRC (2010) America's climate choices: adapting to the impacts of climate change. National Academies Press, Washington, DC
Odum EP (1989) Ecology and our endangered life-support systems. Sinauer Associates, Sunderland
Palmer M, Bernhardt E, Chornesky E et al (2004) Ecology for a crowded planet. Science 304:1251–1252
Palmer MA, Bernhardt ES, Chornesky EA et al (2005) Ecological science and sustainability for the 21st century. Front Ecol Environ 3:4–11
Pickett STA, Cadenasso ML, McGrath M (eds) (2013) Resilience in ecology and urban design: linking theory and practice for sustainable cities. Springer, New York
Power ME, Bouma-Gregson K, Higgins P, Carlson SM (2015) The thirsty eel: summer and winter flow thresholds that tilt the eel river of northwestern California from salmon-supporting to cyanobacterially degraded states. Copeia 103(1) (in press)
Power ME, Chapin FS III (2009) Planetary stewardship. Front Ecol Environ 7(8):399
Ramos J, Lymn N, Salguero-Gomez R et al (2012) Ecological Society of America's initiatives and contributions during the Deepwater Horizon oil spill. Bull Ecol Soc Am 93:115–116
Rivera N, Calderon-Ayala J, Calle L et al (2010) Multidisciplinary and multimedia approaches to action-oriented ecology. Bull Ecol Soc Am 91:313–316
Rockström J, Steffen W, Noone K et al (2009) A safe operating space for humanity. Nature 461:472–475
Rozzi R, Pickett STA, Palmer C et al (eds) (2013) Linking ecology and ethics for a changing World: values, philosophy, and action. Springer, Dordrecht/Heidelberg/New York/London
Sayre N (2005) Working wilderness: the Malpai Borderlands Group and the future of the western range. Rio Nuevo Press, Tucson
Sayre NF, Kelty R, Simmons M et al (2013) Invitation to earth stewardship. Front Ecol Environ 11:339
Silbert S, Chandler G, Nabhan GP (eds) (2007) Five ways to value working landscapes of the West. Center for Sustainable Environments, Northern Arizona University, Flagstaff
Sisk TD (2010) Ranching, local ecological knowledge, and the stewardship of public lands. In: Power ME, Chapin FS III (eds) Planetary stewardship in a changing world: paths toward resilience and sustainability. Bull Ecol Soc Am 91(2):150–152
Sisk TD, Albano CM, Aumack EA et al (2010) Integrating restoration and conservation objectives at the landscape scale: the Kane and Two-mile Ranch project. In: van Riper C III, Wakeling BW, Sisk TD (eds) The Colorado Plateau IV: shaping conservation through science and management. University of Arizona Press, Tucson, pp 45–66
Steffen WL, Sanderson A, Tyson PD et al (2004) Global change and the Earth system: a planet under pressure. Springer, New York
Steffen W, Jansson Å, Deutsch L et al (2011) The Anthropocene: from global change to planetary stewardship. Ambio. doi:10.1007/s13280-011-0185-x
Steiner F, Simmons M, Gallagher M et al (2013) The ecological imperative for environmental design and planning. Front Ecol Environ 11(7):355–361
Stern N (2007) The economics of climate change: the Stern review. Cambridge University Press, Cambridge, UK
Szaro RC, Johnson ND, Sexton WT et al (eds) (1999) Ecological stewardship: a common reference for ecosystem management. Elsevier, Oxford
Turner BL II, Kasperson RE, Matson PA et al (2003) A framework for vulnerability analysis in sustainability science. Proc Natl Acad Sci U S A 100(14):8074–8079
UN (2010) The millennium development goals report 2010. United Nations, New York
WCED (World Commission on Environment and Development) (1987) Our common future. Oxford University Press, Oxford
Yoshiyama RM, Moyle P (2010) Historical review of Eel River Anadromous Salmonids, with emphasis on Chinook salmon, Coho salmon and steelhead. Report for California Trout. 1–107, Center for Watershed Sciences, University of California, Davis

Chapter 13
Geographical and Thematic Distribution of Publications Generated at the International Long-Term Ecological Research Network (ILTER) Sites

Ben Li, Terry Parr, and Ricardo Rozzi

Abstract The International Long-Term Ecological Research (ILTER) network is currently unmatched by other global networks in its ability to coordinate and collaborate on long-term ecological research and monitoring at a planetary scale. This offers an ideal research, information, and infrastructural platform for the Earth Stewardship initiative. However, to achieve an effective synergy between ILTER and Earth Stewardship it is critical to overcome problematic geographical and conceptual gaps in ILTER Research. To quantify these gaps we produced a new database of scholarly and grey literature generated at long-term ecological or socio-ecological research (LTER) sites worldwide. We assessed: (1) the geographical origin of LTER researchers; (2) the geographical regions where these researchers conduct their studies; (3) which thematic areas are investigated in LTER research, and to what extent do they include concepts associated with Earth Stewardship; (4) in which venues are LTER research outputs published. Regarding the production of knowledge at ILTER, we found a marked *Northern Hemispherism*: > 90 % of the ILTER publications are generated by researchers from the Northern Hemisphere. Furthermore, 89 % of ILTER publications are generated by researchers associated with LTER networks in the North Temperate region (23° N – 66° N). Regarding conceptual gaps, < 0.5 % of ILTER

B. Li (✉)
Department of Information Processing Science, University of Oulu, Oulu, Finland
e-mail: banjili@gmail.com

T. Parr
Lancaster Environment Centre, Centre for Ecology and Hydrology, Lancaster, UK
e-mail: twp@ceh.ac.uk

R. Rozzi
Department of Philosophy and Religion Studies, University of North Texas, Denton, TX, USA

Institute of Ecology and Biodiversity, Santiago, Chile

Universidad de Magallanes, Punta Arenas, Chile
e-mail: rozzi@unt.edu

R. Rozzi et al. (eds.), *Earth Stewardship*, Ecology and Ethics 2,
DOI 10.1007/978-3-319-12133-8_13

publications are included in social sciences databases. Noticeably, however, > 99 % of all ILTER publications in the arts and the humanities are generated by researchers working in the South Temperate region (23°N – 66°N), especially Chile. Additionally, in Southern Hemisphere LTER networks research themes associated with Earth Stewardship were the most represented. Our concise analysis aims to call attention to the fact that opportunities exist for greater collaboration and complementarity in research across the ILTER Network. The southern regions can significantly add to the integration of social, ethical, and artistic dimensions to transdisciplinary socio-ecological research at ILTER, providing an intercultural and participatory foundation for Earth Stewardship.

Keywords Earth Stewardship • Ethics • Knowledge production • Long-Term Ecological Research (LTER) • Research outputs

13.1 World Distribution of ILTER Sites and Research Themes

The International Long-Term Ecological Research (ILTER) network consists of approximately 40 national-scale long-term ecological research (LTER) networks, including prospective network members. National LTER networks conduct site-based research and monitoring in a variety of ecosystems and geographies. ILTER's work addresses international ecological and socio-ecological problems through collaborative question- and problem-driven research, as well as data collection and sharing (ILTER Network n.d.). Although the formal ILTER Network was established only two decades ago in 1993, many member networks and sites have been conducting long-term monitoring and research during prior decades.

Our ability to conscientiously interact with the world is limited by our direct sources of ecological knowledge. However, a relatively small portion of the world, encompassing a limited portion of the ecological and cultural diversity of the planet is included in longer ecological studies. Distribution and availability of ecological knowledge directly affects an Earth Stewardship initiative (see Chapin et al. 2015 in this volume [Chap. 12]). Formal scientific publication or data availability is one benchmark by which knowledge is accepted into scientific (Christensen et al. 1996) and policy (Turnhout et al. 2007) communities. However, not all ecological knowledge, e.g., traditional ecological knowledge (Huntington 2000), might be suited for the predominant scientific publication forums. The thematic and geographic distribution of those forums potentially shape and reflect available ecological knowledge and interests.

In spite of its limitations, the ILTER network's ability to coordinate and collaborate on long-term ecological comprehensive research and monitoring is currently unmatched by other networks (see Maass and Equihua 2015 in this volume [Chap. 14]). Our chapter complements previous research concerning the geographic

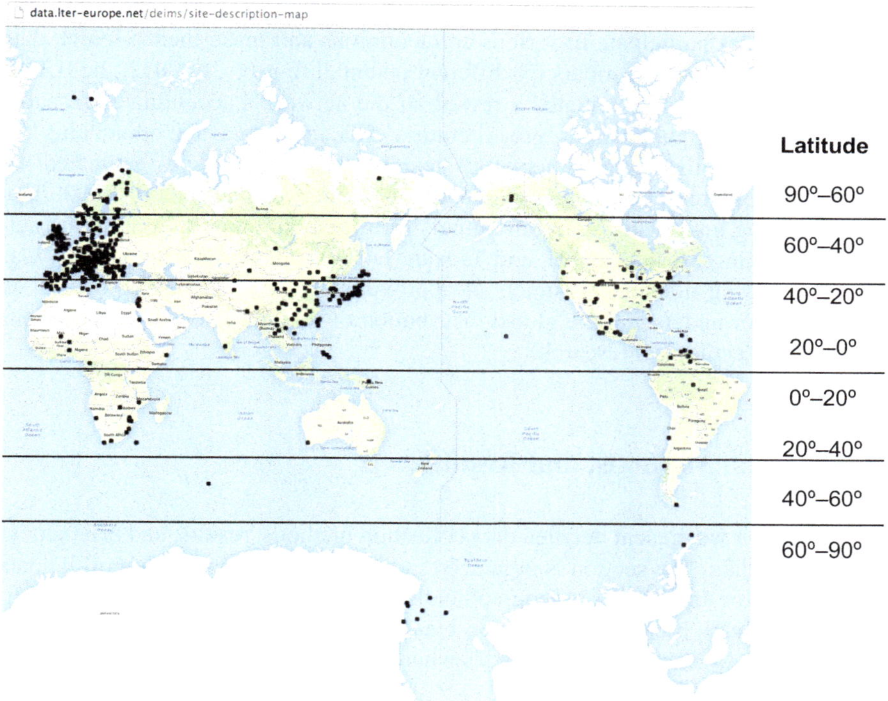

Fig. 13.1 Official map of ILTER sites as of May 2014 (Based on a screenshot from: http://data.lter-europe.net/deims/site-description-map, with latitudes overlaid from Rozzi et al. (2012))

distribution of ILTER sites. Rozzi et al. (2012) found a marked *Northern Hemispherism:* of the 543 ILTER sites distributed in 44 countries, 509 sites (93.7 %) are located in the Northern Hemisphere, while only 34 sites (6.3 %) are located in the Southern Hemisphere. Figure 13.1 shows that the majority of ILTER sites are concentrated in:

(a) the Northern Hemisphere, and
(b) within relatively small terrestrial areas of the Northern Hemisphere, mostly in Europe, and Japan.

Regarding the thematic distribution of research conducted at ILTER sites, Rozzi et al. (2012) stated that most research was purely ecological, and when it was socio-ecological it focused on socio-economic themes. They cautioned that this *economicism* was problematic because it left out aesthetic, ethical and multicultural that core attributes of socio-ecological systems.

In order to quantitatively assess thematic and geographical distributions of ILTER publications, this chapter draws on a newly compiled bibliography of research outputs from the International Long-Term Ecological Research (ILTER)

Network. Within the ILTER Network, national and supranational LTER networks and local nodes participate in various collaborations and make their research data and results available in formats for different potential re-users. In 2012, the ILTER management committee initiated a review of the network's accumulated research outputs, including publications, grey literature, data, meta-data, and other items. An initial survey of the research network's member network websites estimated the number of research output items produced by the network at 30,000–40,000, not fully counting un-collated research outputs from several major national-scale networks including Mexico, Israel, and Taiwan. For this chapter, a new analysis was conducted using titles and abstracts of approximately 30,000 research outputs in order to better understand the global distribution of research themes and locations of long-term ecological research.[1]

13.2 Data, Methods, and Results

In this section we present detailed data collection methods, results, and brief analysis of those data. The section is organized in four parts: (1) description of the data source used for this study, (2) geographic distribution of ILTER research and publications or more broadly research outputs, (3) thematic distribution of ILTER research, and (4) analysis of the venues where research outputs are published.

13.2.1 Methods and Data Source

In 2013, with the ILTER network we began to compile an accumulated bibliography of all LTER research outputs generated by its member networks. Since ILTER lacks a network-wide standard for materials eligible to be classified as research outputs, the kinds of references gathered varied among member networks and their sites. Research outputs included data and meta-data descriptions of data, patents, scholarly articles, book chapters, theses and dissertations, popular news articles, edited volumes, commissioned reports, poster and presentation abstracts, meeting and workshop proceedings, compendia, and other materials compiled by regional, national, and local LTER networks and sites.

The timespans covered by member network bibliographies also differ. For example, while the US LTER includes scholarly publications, dissertations, and theses dating from the late 1970s shortly before the US LTER Network's formal initiation, the Taiwan Forestry Research Network (TFRI, part of the national Taiwan Ecological Research Network) includes in its bibliography those kinds of items plus patents

[1] This dataset should be considered only as an initial attempt to collect a bibliography of ILTER work, and is subject to revisions and omissions as detailed later.

and commissioned industrial work dating from the 1960s, when the TFRI's annual reports first included extensive bibliographies.

In total, over 30,000 research outputs and over 30,000 meta-data[2] outputs were collected from over 30 of the approximately 40 ILTER networks, spanning approximately 40 years of research. The networks from which no bibliographic information was obtained consist of networks that are inactive (e.g., Canada and some networks in Eastern Europe and Africa) or recently established (e.g., Philippines and Malaysia). While some regional and national networks actively maintained comprehensive bibliographies of their own research outputs, others maintained bibliographies at the sub-regional or site levels.

The set of ILTER research sites is not identical to the combined sets of research sites under each of the regional or national networks. Furthermore, non-ILTER research is conducted at many ILTER research sites. Consequently, the 30,000 research outputs collected include research outputs produced at research sites and by individuals affiliated with national LTER networks, but which may not be formally part of the ILTER Network. Inclusion of such research outputs from outside the formal core of the ILTER network is consistent with the inclusion of networks that have in their bibliographies work initiated or published before the formal establishment of the ILTER network in 1994.

All available abstracts from meta-data outputs and over 5,400 abstracts from other research outputs also were collected into the same database, containing among others the following columns:

$$Author(s), Title, Year, Publication\ Name, Keywords, LTER\ Network(s), Abstract$$

In cases of multiple authorship, a single publication may appear in the bibliography of more than one LTER network. However, each publication only is counted once. Texts of the research output and meta-data titles and abstracts were automatically deconstructed into one-to-three word long alphabetized N-grams (Cavnar and Trenkle 1994) of Porter Stemmed (1980) words, excluding stop word such as "of", "is", and "the".[3] Plausible place-names were initially identified as those containing

[2] Meta-data are searchable data about data. In LTER, a meta-data record about a data set might include time and location of data collection, methods used, species and geographies involved, etc. Many LTER networks (also) publish their data and meta-data in a Global Biodiversity Information Facility repository or other repositories.

[3] For example, the title "Geographical and Thematic Analysis of Publications Generated at ILTER Sites" would be processed first into: "geograph*", "themat*", "analysi*", "public*", "gener*", "ILTER*", "site*", "geograph* themat*", "analysi* themat*", "analysi* public*", "gener*, public*", "analysi* geograph* themat*", "analysi* themat* public*", "analysi* gener* public*", "gener* ILTER* site*". Each N-gram was considered to be a plausible concept discussed in the research outputs. Other concepts included "disturb*" (capturing "disturbance", "disturbed", etc.), "chang* environment*" (capturing "changing environments", "environmental change", etc.), and "chang* impact*" (capturing "impact of change", "changes impact", etc.).

at least one capital letter in the first position of each word.[4] Three main analyses were conducted using place-name and concept N-grams as input data: place-names, research topics, and publication venues.

13.2.2 Where Are ILTER Researchers Based, Which Regions Do They Study?

Plausible place-names matched the names of a political geographic unit (including countries, autonomous regions, and major sub-national states) or a major geographical feature (such as the Andes, the Arctic, or the Pacific Ocean). The automatically coded and uncoded data were then inspected manually. Plausible place-names that appeared five[5] or more times in the data were given manual coding rules (e.g., place-names ending in "-shan" were coded as occurring in China since "-shan" is a common Romanization of the Chinese word for mountain).

A single title may include more than one place-name (such as "Kruger National Park, South Africa"). No attempts were made to identify any hierarchical or other relationships among such place-names. Errors of automatic coding were culled by adding manual coding rules (e.g., excluding matches based on the n-gram "Rio" alone, which matched many rivers in Latin and South America and parts of Europe). A small number of endemic species, such as the Adelie penguin endemic to Antarctica, were also used to geo-locate publications. From over 60,000 plausible place-names, over 11,000 place-names were coded from 10,228 publication titles. The vast majority of capitalized words in titles not accurately identifiable as place-names were excluded from the place-name analysis. Over 90 different countries and regions were identified from titles and abstracts in this way.

The geographic origin of researcher and the geographic areas that are studied by researchers were both coded into one of the following six geographic zones (A-F) (Fig. 13.2):

A = *Arctic* (> 66° N), north of the Arctic Circle;
B = *North Temperate* (66° N – 23° N), south of the Arctic Circle and north of the Tropic of Cancer;
C = *North Equator* (23 °N – 0 °), south of the Tropic of Cancer and north of the Equator;
D = *South Equator* (0 ° – 23° S), south of the Equator and north of the Tropic of Capricorn;

[4] For example, N-grams including "Antarctic", "Cascade Mountains", and "Wisconsin United States" were identified as plausible place-names. These plausible place-names are the basis of further analysis.

[5] The lower limit of five is arbitrarily chosen, but reasonable in light of other place-names and kinds of place-names that appear dozens or hundreds of times. Frequent non-place-names included any word that appeared at the beginning of the title, such as "Assessing" and "The", along with genus names.

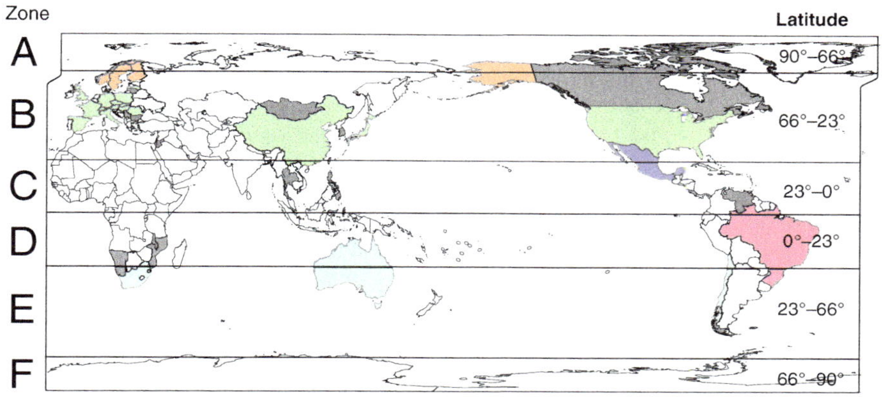

Fig. 13.2 World map representing LTER networks with research outputs and data included in this chapter (*colored areas*). The *grey* areas represent LTER networks whose research outputs and data were not accessible. The *white* areas lack national-scale LTER networks. The geographical (latitudinal) zones are the following: *A = Arctic* (> 66° N); *B = North Temperate* (66° N – 23° N); *C = North Equator* (23° N – 0°); *D = South Equator* (0° – 23° S); *E = South Temperate* (23°S – 66°S); *F = Antarctic* (> 66° S). For countries included in geographical zones A to F see Table 13.1

E = *South Temperate* (23° S – 66° S), south of the Tropic of Capricorn and north of the Antarctic Circle
F = *Antarctic* (>66°S), south of the Antarctic Circle.

For each publication or other research output, the allocation to a geographic zone was based on latitude of the place-names (when available), or based on country or biome information if no more specific place-name was identified. The list and number of LTER networks per geographic zones and countries are given in Table 13.1.[6] The origin of researchers was identified based on the location of the national LTER network from which the titles of the publications were obtained. For example, a research output listed by the US LTER concerning Antarctica would be coded as: *Researcher's Origin* = Zone B; *Research Subject* = Zone F.

Regarding which geographic zones are being studied at ILTER sites, the number of publications of articles and other research outputs excluding meta-data is similar in the Northern (56.6 %) and the Southern (43.4 %) hemispheres (Table 13.2). However, the production of meta-data is markedly concentrated in the Northern Hemisphere (95.3 %). The ratio of research outputs to meta-data (RO/M-D) is noticeably contrasting between two hemispheres: In the Northern Hemisphere the RO/M-D is 38.7 times greater than in the Southern Hemisphere. [7]

[6] There is no code for a 'global' zone, because among ILTER publications only few papers included research at a global scale.

[7] Caution should be exercised in interpreting this ratio because the generation and use of meta-data in the production of research outputs is not well characterized within LTER, and because indexed meta-data may itself refer to other sets of meta-data that have as yet uncharacterized extents.

Table 13.1 Distribution of national-scale LTER networks affiliated with ILTER

Zone	Number/+ *repeated country*	Relative percentage (%)	National LTER networks
A = *Arctic* (> 66°N)	3/+ *3*	5	Finland, Norway[a], Sweden/+ *Alaska (US), Germany-Norway Arctic Ocean*[b]
B = *Temperate North* (66°N – 23°N)	26	63	Austria, Bulgaria, Canada, Czech Republic, France, Germany, Hungary, Israel, Italy, Japan, Jordan, Latvia, Lithuania, Mongolia, People's Republic of China, Poland, Portugal, Romania, Serbia, Slovakia, Slovenia, South Korea, Spain, Switzerland, United Kingdom, United States of America
C = *North Equator* (23°N – 0°)	6	15	Costa Rica, Mexico, Philippines, Republic of China (Taiwan), Thailand, Venezuela
D = *South Equator* (0° – 23°S)	3	7	Brazil, Malawi, Mozambique
E = *Temperate South* (23°S – 66°S)	4	10	Australia, Chile, Namibia, South Africa
F = *Antarctic* (> 66°S)	0/+ *1*		*/+ Palmer Antarctica LTER (US) and McMurdo Dry Valleys LTER (US)*
Total	**42/+ *3***	**100**	

[a]Norway is a prospective ILTER member
[b]The results of this collaboration are attributed to a distinct network by LTER Europe

Table 13.2 Geographic zones studied by ILTER in terms of published articles and meta-data

Zone	Research outputs (RO)		Meta-data (M-D)		(RO/M-D) ratio
	N	Rel. (%)	N	Rel. (%)	
A (*Arctic*)	561	5.3	1,310	5.0	0.4
B (*Temperate North*)	4,615	43.7	23,263	89.6	0.2
C (*North Equator*)	801	7.6	185	0.7	4.3
Subtotal Northern Hemisphere	**5,977**	**56.6**	**24,758**	**95.3**	**0.2**
D (*South Equator*)	422	4.0	685	2.6	0.6
E (*Temperate South*)	3,510	33.2	438	1.7	8.0
F (*Antarctica*)	660	6.2	86	0.3	7.7
Subtotal Southern Hemisphere	**4,592**	**43.4**	**1,209**	**4.7**	**3.8**
Total	**10,569**	**100.0**	**25,967**	**100.0**	**0.4**

Within each hemisphere, research outputs and meta-data are concentrated in temperate zones (B and E; Table 13.2). Combined, temperate zones of the Northern and Southern hemispheres account for 76.9 % of the research outputs, and 91.3 % of the meta-data produced by ILTER sites. Equatorial regions account for only

11.6 % of the research outputs and 3.3 % of the meta-data produced by ILTER sites (C and D; Table 13.2).

Incorporating the geographical origin of ILTER researchers reveals a similar pattern: the North Temperate region (Zone B) concentrates most published outputs overall. Published outputs excluding meta-data (Fig. 13.3), and meta-data (Fig. 13.4) also show that authors of research outputs and meta-data in all regions write primarily about their own zone. Noticeably, most of the outputs concerning research on the equatorial zones are produced by researchers residing in Zones B and D (*North and South Temperate*).

No meta-data contributions came from Zone A (*Arctic*), and all meta-data concerning Zone F (*Antarctic*) were published by Zone B (*North Temperate*) (Fig. 13.4). A high number of items originating from Zones C and D were not codeable with respect to their subject zones since their titles did not clearly specify place-names (e.g., "census", "development"). Place-names that occurred a small number of times (<5) are not included.[8] These place-names can be interpreted in the LTER context to mean that there are ILTER sites about which there are not yet coherent bodies of published research.

13.2.3 Which Thematic Areas Do ILTER Researchers Study?

This section examines the geographic distribution in terms of LTER research topics: What are the thematic contrasts among regions? To answer this question, the data source is the same as in Sect. 13.2.2, and the thematic concepts were classified into eight categories (Table 13.3). The classifications reflect major themes of the ILTER Strategic Plan (ILTER Network 2006) and stewardship themes.

For each of the categories of research concept, researchers from LTER networks in the North Temperate region (Zone B) generated more than 75 % of the total publications (Fig. 13.5). For all categories, researchers from LTER networks in the Southern Hemisphere have generated approximately 10 % of the publications,

[8] Regarding our methodology it is important to note that the lack of detailed coding of infrequent place-names is not detrimental to the scale of analysis conducted with this method because of its low numbers. In Figs. 13.3 and 13.4, and Table 13.2 each article title may contain more than one place-name, and some place-names may represent more than one geographical location (e.g., the municipality of China in the Mexican state of Nuevo León, the People's Republic of China, and the Republic of China, etc.). Figure 13.3 counts the number of relationships between research networks (known from their network homes) and the geographic zone investigated (inferred from place-names in article/data titles). Table 13.2 counts the number of times an identifying place-name occurs in each of the zones in articles and data. The number of research outputs/meta-data reported for a zone in Table 13.3 is equal to or lower than the sum of the number of articles/meta-data where that zone is the right side of Fig. 13.3. Table 13.2 counts a small number of research outputs/meta-data not counted in Fig. 13.3, namely those having an uncoded researcher origin Zone due to inadequate meta-data. Finally, it is also important to note that titles of meta-data contained more N-grams about methods and theoretical approaches than did titles of publications.

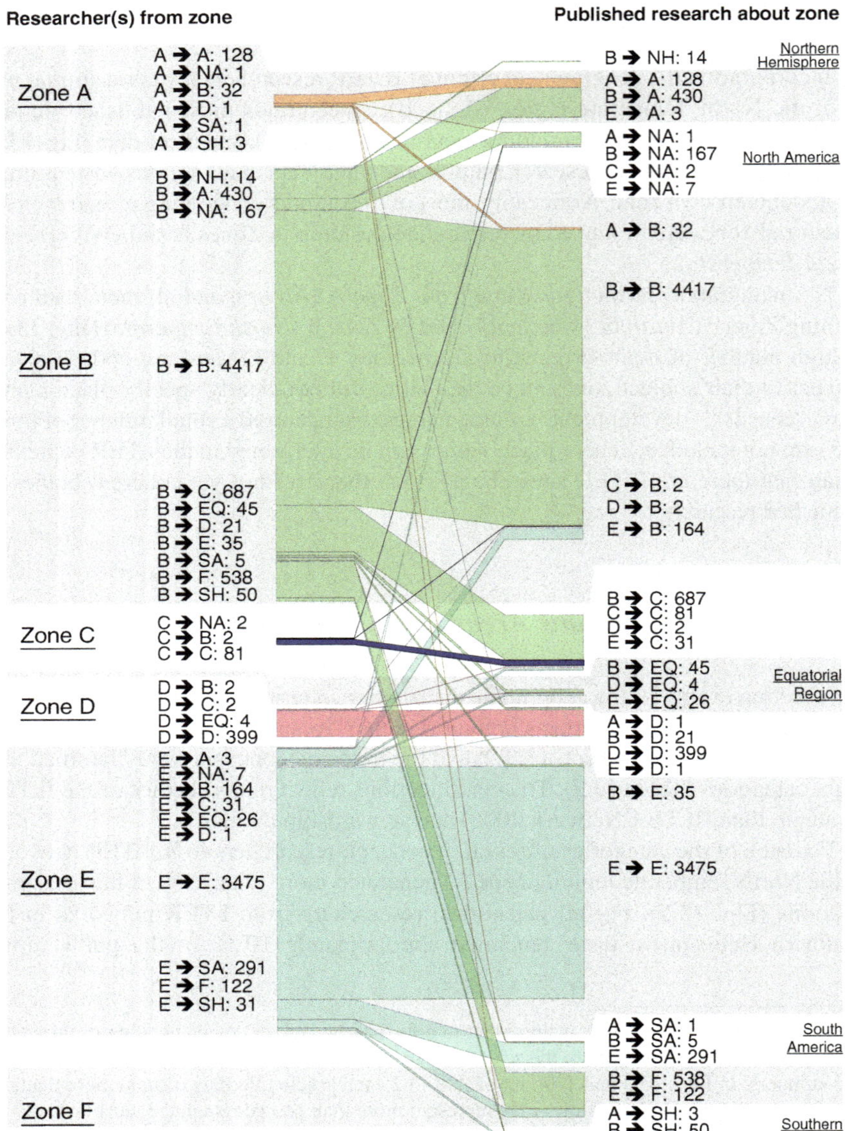

Fig. 13.3 Directions of LTER research considering the geographical origin of LTER researchers (*left column*) and the geographical region covered in the research outputs (*right column*) in terms publications, excluding meta-data. Vertical heights are proportional to the number of research originating from, or about, a Zone. For latitudinal ranges and countries included in zones A to F see Table 13.1. The following zones are given where a research output provides no more detailed geographic information: *NH* Northern Hemisphere, *EQ* Equatorial, Includes Africa, Tropics; *SH* Southern Hemisphere, *NA* North America, *SA* South America, *AF* Africa. Colors of the Zones match those given in Fig. 13.2

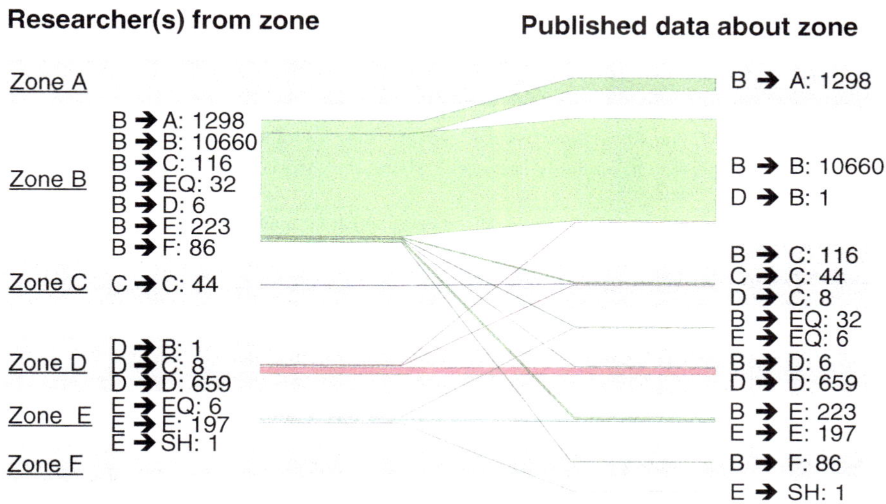

Fig. 13.4 Directions of LTER research considering the geographical origin of LTER researchers (*left column*) and the geographical region covered in the research outputs (*right column*) in terms publications, considering meta-data only. For latitudinal ranges and countries included in Zones A to F see Table 13.1; for color matches of the Zones see Fig. 13.2

mostly in the South Temperate region (Zone E). Therefore, in most categories the proportion of Northern/Southern Hemisphere ILTER publications is 9:1. In addition, the equatorial zones C and D are the least represented, accounting for less than 5 % of the publications in all categories.[9]

Regarding the thematic areas, *management* and *stewardship* are the categories that include higher numbers of ILTER publications; both have more than 5,000 research outputs (Fig. 13.5). *Location*, *methods*, and *monitoring* include more than 3,000 research outputs, and *scale* more than 2,000. *Event* and *LTER* are the least represented research themes, each including a total of less than 2,000 research outputs.

[9] It is important to note that Zone C is not fully represented in the data included in Figs. 13.5 and 13.6 because the data for Mexico LTER are incomplete, and several Asian LTER networks' databases are still in the early stages of work. Also, not all zones had a network with publications about concepts that were shared by more than seven other networks. (Seven networks as a cutoff is based on the proposition that ILTER Network-wide research should be defined as that which could draw on work from each of the continents. It is also based on the practical consideration that the other 99% of the approximately 10,000 possible concepts not represented here is too vast to code reliably into relevant categories.)

Table 13.3 Concepts included in each research thematic category

Event = Things that happen to the biophysical world (event*, disturb*, storm*, damag*, acidifi*, extrem*)
Location = Spatial (kinds of) location and processes {local*, catchment*, air*, hydrolog*, sediment*, stream*, fauna*, wetland*, aquat*, adapt*, state*, ground*, stress*, arctic*, alien*}
LTER = Network research, synthesis {network*, shortterm*, workshop*, longterm* studi*, lter*, ilter*, integr*, review*}
Methods = Theories about and measures of socioecological systems {case* studi*, evid*, flow*, precipit*, map*, regim*, concept*, methodolog*, techniqu*, commun* composit*, detect*, chang* climat*}
Management = Concepts and policies concerning human actions on the world {theori*, implic*, project*, establish*, budget*, load*, perspect*, remov*, reduct*, problem*, health*, histor*, vulner*, pressur*, uncertainti*, reconstruct*, chang* environment*, challeng*, promot*}
Monitoring = Using measures of the world over time to understand change {natur* regener*, stabl*, shift*, learn*, sens*, remot* sens*, satellit*, forest* monitor*, recoveri*, paramet*, regener*, consequ*, eutroph*, assess*, monitor*, carbon* flux*}
Scale = Understanding the world across locations {biodivers*, biospher*, food* web*, gradient*, complex*, global*, transfer*, fluctuat*}
Stewardship = Human interventions on the world {predict*, strategi*, risk*, futur*, biospher* reserv*, human*, emiss*, appli*, air* pollut*, crop*, artifici*, rural*, plan*, design*, polici*, district*, framework*, farm*, area* protect*, forestri*, programm*, implement*, social*, govern*, scheme*, optim*, agricultur* landscap*, econom*, activ* human*, dam*, ecolog* impact*, chang* impact*, conserv* natur*}

Categories are based on root words common to publication titles and abstracts published by at least eight national LTER networks (=20 % of all ILTER member networks), with the exception of the *LTER* category which includes the concept *LTER* appearing in only publications of seven national LTER networks. This table lists the concepts in each category. The categories as constructed here as mutually exclusive, and for convenience of analysis. Doubtless, there are many other useful ways to categorize and interpret this data

Regarding the distribution of research themes within each of the LTER networks, it is salient that *stewardship* is the most represented research area in the Southern Hemisphere. It includes more than 40 % and 20 % of the publications generated by South Equator (Zone D) and Temperate (Zone E) zones, respectively (Fig. 13.6). In the Northern Hemisphere, stewardship is also well represented in North Temperate (Zone B). In this zone, management and stewardship combined account for 40 % of the publications. At ILTER sites in the North Temperate Zone, broad scale research represents less than 10 % of the research outputs. In addition, the Arctic (Zone A) is the geographical region that is most concentrated on local topics, having more than 35 % of its research outputs focused on location. Hence, broad scale research is better resented in the Southern Hemisphere LTER networks where it accounts for more than 10 % of the research outputs in zones D and E (Fig. 13.6).

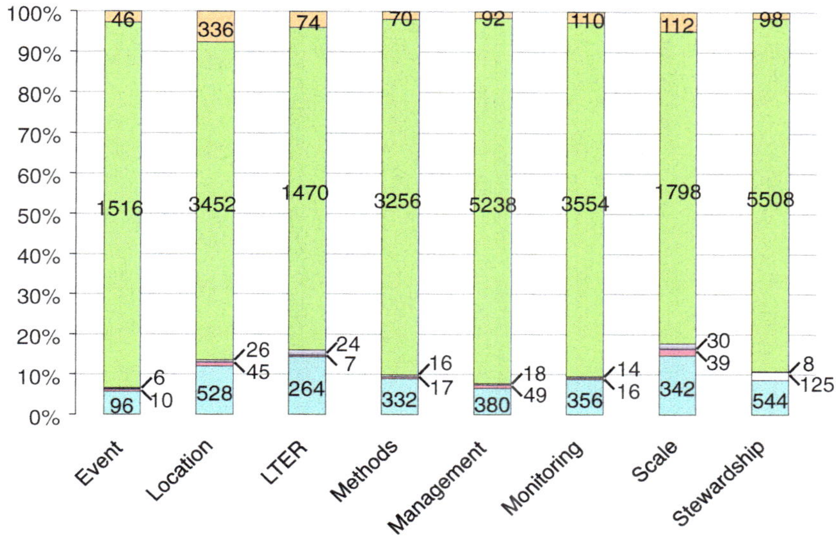

Fig. 13.5 Absolute numbers (in the bars) and relative percentage (indicated by Y axis) of ILTER publications for each of the defined research concepts contributed by contributed by ILTER researchers from each of the geographical zones (A to E). Geographical Zone F is not included because all researchers in Antarctica are from other parts of the world. For latitudinal ranges and countries included in geographical zones A to E see Table 13.1; for color matches of the Zones see Fig. 13.2

13.2.4 From Which Geographical Region and in Which Venues Are LTER Researchers Publishing?

A first of level of analysis was conducted based on the ISI's Web of Knowledge database service. Publication venues were identified by automatically matching ISI's Master Journal List (http://ip-science.thomsonreuters.com/mjl/) with publication names from the collected ILTER bibliographies.[10] The majority (89 %) of publication venues is generated by researchers based in the North Temperate region (in Zone B) (Table 13.4). Zone B together with zone A (Arctic) account for 90 % of the ISI-ILTER publications. Adding Zone C (North Equator), the proportion of

[10] It is important to note data quality issues. They included: typos and inconsistent spelling and use of publication names in national- and regional-scale bibliographies, lack of DOIs, and lack of public availability of some documents listed in bibliographies. These issues existed in bibliographies from both small and large networks regardless of geographic location. Furthermore, Asian and non-Latin journal names presented an additional challenge since they are not well represented in the ISI Master Journal List. The impact is clear from the ISI/non-ISI ratios for CERN and Brazil national-scale networks, which both listed many publications in Chinese and Portuguese publication venues, respectively (see Fig. 13.7).

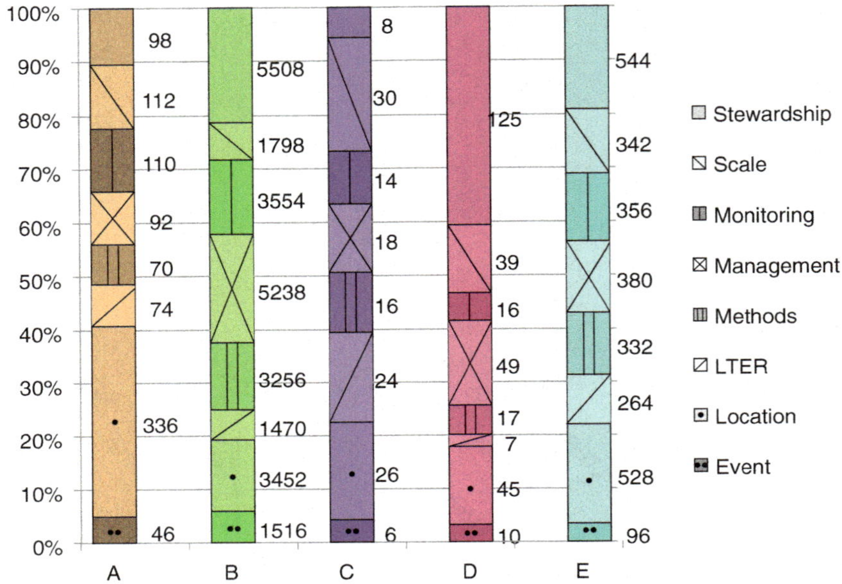

Fig. 13.6 Number and relative percentage each of the defined research concepts within the accumulated number of LTER publications produced by researchers based at each of the geographical zones, A to E. Geographical Zone F is not included because all researchers in Antarctica are from other parts of the world. For latitudinal ranges and countries included in geographical Zones A to E see Table 13.1

Table 13.4 ISI titles published from each Zone. For latitudinal ranges and countries included in Zones A to E see Table 13.1

Zone	Number of ISI titles	Fraction of ISI titles (%)
A	112	0.64
B	15,734	89.33
C	706	4.01
D	500	2.84
E	561	3.19
TOTAL	17,613	100.00

ILTER publications generated in the Northern Hemisphere accounts for 94 % of the world's total. Therefore, for ILTER ISI publications the Northern/Southern Hemispheres ratio is even greater than 9:1.

Most LTER research outputs listed on LTER bibliographies are not published in ISI journals (Fig. 13.7).[11] Notably, networks in regions with numerous local

[11] It is important to note that there are an unknown number of LTER research outputs that are not listed in bibliographies, and the national and site-level bibliographies themselves are often inconsistent in what they report as publications.

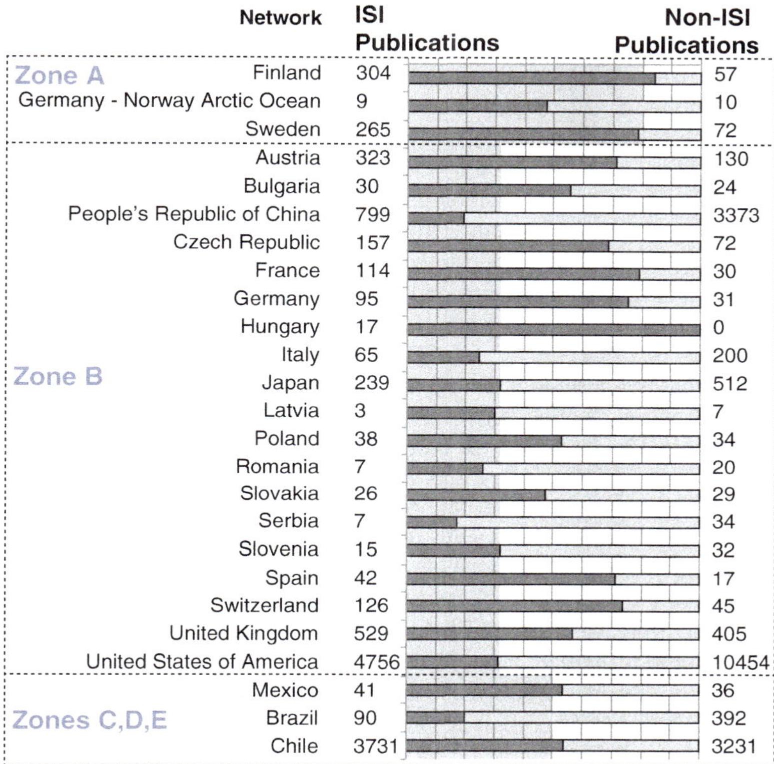

Fig. 13.7 Number and relative percentage of ISI and Non-ISI publications produced by researchers of each of the LTER networks associated with ILTER. For latitudinal ranges and countries included in zones A to E see Table 13.1

language publication venues such as Japan, China, and those in Eastern Europe appear to publish relatively less in ISI journals. Networks in Zone A (*Arctic*) have the highest proportion of their publications in ISI journals. This may indicate careful targeting of publication, and/or success in getting local publication titles listed in ISI. In Zone B (*North Temperate*), European countries tend to have more ISI than non-ISI publications. In contrast the US, Japan, and specially China have larger numbers of non-ISI than ISI publications. In the US many of the non-ISI publications include theses and dissertations. China, in turn, includes many publications in national non-ISI venues.

Regarding Zones B, C, and D, it is noticeable that Mexico and Chile produce more ISI than non-ISI publications. The opposite is true for Brazil. In the analyses summarized by Fig. 13.7 it is important to note that there probably significant data missing due to incomplete bibliographies available in national-scale networks in Zones C, D, and E. Networks included in these zones have not made network-wide bibliographies readily available. For example, Red Mex-LTER in Mexico includes

11 research sites and approximately 200 active and interested researcher members, yet had only compiled an internal list of fewer than 100 ILTER research outputs as of early 2014. TFRI in Taiwan has a history of over 50 years of intensive production of numerous scholarly and other publications, each compiled in annual reports, which has accumulated a vast bibliography that is incompletely digitized. Collectively, these gaps highlight broader problems recognizing and sharing these networks' outputs as bodies of ecological knowledge within the formal scientific publication model. These gaps also highlight different priorities among national-scale LTER networks with respect to how ecological knowledge is to be accessed. For example, TFRI also maintains an extensive physical library of pre-war long-term Japanese ecological research that is currently only accessible and searchable in person.

Core databases for ecological and socio-ecological sciences were used to analyze the distribution of publication interests for each geographical zone.[12] In 11 of the 14 bibliographic databases, the North Temperate region (Zone B) accounts for over 50 % of all ILTER publications (Table 13.5). The only three databases that have more publications generated in other ILTER geographic regions are: BIOSIS Reviews Reports and Meetings with over 50 % of the publications generated in North Equator (Zone C); Arts & Humanities Citation Index and Current Contents Arts & Humanities with over 99 % of the publications generated in South Temperate (Zone E). Therefore, in the fields of the arts and humanities the Northern/Southern Hemisphere ratio is 0.1/9.9.

The scarcity of publications in social sciences, engineering, and medicine databases is noteworthy. Only 63 ILTER publications were found in the Social Sciences Citation Index, and 60 in the Current Contents – Social Sciences & Behavioral Sciences. Combined, these two databases account for less than 0.2 % of all ILTER publications. Five large databases of natural sciences (*Science Citation Index*, *Social Sciences Citation Index*, *BIOSIS Previews*, *Current Contents – Agriculture, Biology & Environmental Sciences*, and *Zoological Record*) concentrate 90 % of all ILTER publications.

13.3 Discussion and Implications for Earth Stewardship

The data presented here confirm both the geographic and the conceptual biases in ILTER research. A *Northern Hemispherism* is quantitatively demonstrated by a Northern/Southern Hemispheres ratio greater than 9:1 in ILTER ISI publications, and an even higher ratio for meta-data. Furthermore, within the Northern Hemisphere, the production of knowledge is concentrated in the Temperate region (Zone B), which includes the US, Western Europe, and North-East Asia.

[12] For each ISI publication attributed to a zone, the ISI index in which that publication appears is counted. Note that some publications appear in more than one ISI index.

Table 13.5 Number and relative percentage of articles published in core databases for ecological and socio-ecological sciences generated by ILTER researchers from each geographical zone. For latitudinal ranges and countries included in geographical Zones A to E see Table 13.1. Geographical Zone F (Antarctica) is excluded because no research is a resident of the Antarctic region

	Zone A		Zone B		Zone C		Zone D		Zone E		Total	
Index	*N*	%	*N*	%	*N*	%	*N*	%	*N*	%	*N*	%
Arts & Humanities Citation Index	0	*0.0*	6	0.6	0	0.0	0	0.0	1,059	99.4	**1,065**	100
BIOSIS Previews	482	*5.1*	6,397	67.8	32	0.3	87	0.9	2,435	25.8	**9,433**	100
BIOSIS Reviews Reports and Meetings	0	*0.0*	30	47.6	32	50.8	0	0.0	1	1.6	**63**	100
Current Contents - Agriculture, Biology & Environmental Sciences	469	*5.3*	6,338	71.4	29	0.3	81	0.9	1,955	22.0	**8,872**	100
Current Contents - Arts & Humanities	0	*0.0*	6	0.6	0	0.0	0	0.0	1,059	99.4	**1,065**	100
Current Contents - Clinical Medicine	0	*0.0*	5	71.4	0	0.0	0	0.0	2	28.6	**7**	100
Current Contents - Engineering, Computing & Technology	6	*2.3*	254	96.9	0	0.0	0	0.0	2	0.8	**262**	100
Current Contents - Life Sciences	61	*4.8*	807	64.0	1	0.1	6	0.5	385	30.6	**1,260**	100
Current Contents - Physical, Chemical & Earth Sciences	118	*7.9*	1,262	84.9	2	0.1	8	0.5	96	6.5	**1,486**	100

(continued)

Table 13.5 (continued)

Index	Zone A		Zone B		Zone C		Zone D		Zone E		Total	
	N	%	*N*	%	*N*	%	*N*	%	*N*	%	*N*	%
Current Contents - Social & Behavioral Sciences	4	6.7	48	80.0	1	1.7	0	0.0	7	11.7	**60**	100
Science Citation Index	569	5.6	7228	71.3	24	0.2	71	0.7	2248	22.2	**10,140**	100
Science Citation Index Expanded	569	5.6	7,228	71.1	35	0.3	87	0.9	2,248	22.1	**10,167**	100
Social Sciences Citation Index	4	6.3	50	79.4	1	1.6	0	0.0	8	12.7	**63**	100
Zoological Record	431	5.2	5,272	63.2	33	0.4	79	0.9	2,526	30.3	**8,341**	100
Total	**2,713**	**5.2**	**34,931**	**66.8**	**190**	**0.4**	**419**	**0.8**	**14,031**	**26.8**	**52,284**	**100**

Consequently, information and perspectives expressed in the published literature may be less sensitive to socio-ecological phenomena and concepts grounded in other regions.

The geographic bias is also grounded in the uneven distribution of ILTER sites around the globe. As illustrated in Fig. 13.2, currently vast zones in the African, Asian and Latin American continents lack LTER networks. Consequently, current long-term ecological and socio-ecological research programs are missing some of the world's most diverse countries and regions in terms of both biological and cultural diversity.

Regarding cultural diversity, it is critical to note that standards and embedded concepts employed by LTER networks to obtain and share data, and to collaborate, do not facilitate sharing or reuse of data and underlying theories that lack a tabular or matrix representation of discreet values (Li 2014). This is clearly seen in discussions about "long-term" data sets and observations valued for their large quantitative size in time or geography, and in the information infrastructures that make such data desirable, describable, achievable, and sharable. Difficult to fit into that model are interview transcripts, images of interactions among human and natural communities, or models of such interactions. In particular, LTER's EML standard to describe ecological data encodes a bureaucratic hierarchical understanding of ecological knowledge production and prioritizes attribution rather than stewardship of data (Li 2013). Nature is thereby framed in service of largely individual knowledge discovery, rather than in terms of stewardship, advocacy, or responsibility for the underlying life processes and relationships.

The conceptual bias is expressed in the scarcity of publications in the social sciences. The recent call to implement long-term socio-ecological research in the ILTER network (Maass and Equihua 2015 in this volume [Chap. 14]) will have to address the fact that less than 0.5 % of ILTER publications are indexed in social sciences bibliographic databases. However, it is promising that the South Temperate region (Zone E), especially Chile, is leading the publications in the humanities and arts, accounting for over 99 % of ILTER publications in these thematic areas. Additionally, the Chilean LTSER network is generating methodologies to integrate ecological sciences and environmental ethics that can be adapted by LTER programs in other regions (Rozzi et al. 2008; Aguirre Sala 2015 in this volume [Chap. 15]).

During the last decade Northern Hemisphere LTSER networks in the US (Redman and Miller 2015 in this volume [Chap. 17]), Europe (Singh et al. 2013), and Japan (Shibata 2015 in this volume [Chap. 3]) have called attention to the need to incorporate social dimensions of ecological research into ILTER. To achieve this goal, we need to consider how research infrastructures might emerge and be adapted to suit those needs. Presently, LTER networks across the world have largely adopted and adapted the US LTER's infrastructure for meta-data, and with it notions of what is or is not to be considered valid forms of research. For example, the Kepler workflow engine is becoming increasingly optimized to handle large anonymous sensor networks, and offers little value to handling interview transcripts. The GBIF data and meta-data repository and data standard, used by LTER networks worldwide, was passed over for official adoption by the ILTER in favor of the infrastructure developed by the US LTER.

For an Earth Stewardship initiative, it is relevant to consider the extent to which ethnographic methods that focus on individuals, sites, or individual networks research can be incorporated. This type of research has been underrepresented in Northern Hemisphere long-term socio-ecological (LTSER) networks, which have focused on socio-economic variables (Rozzi et al. 2012, p. 303). South American and Asian socio-ecological research initiatives highlight the relevance of traditional ecological knowledge, as well as ethical, aesthetic, and spiritual values (see chapters by Shibata, Gao, Sarmiento, Mamani-Bernabé, Rozzi, Aguirre Sala, May Jr, in this volume [Chaps. 3, 4, 5, 6, 8, 9, 15, 27]). Overall, the language and practice of long-term ecological data favors quantitative measures of single parameters. Interestingly, today local forms of ecological knowledge are beginning to appear on the conceptual radar of ILTER.

Technological advances in the LTER networks, such as automated sensor networks, present a bias toward the Northern Hemisphere. In the Southern Hemisphere, LTER networks face unique and great challenges in terms of costs to build and maintain capital-intensive infrastructures. Additionally, novel technological infrastructure, which monitors nature in more automated ways, increases the distance between humans and nature. However, an Earth Stewardship initiative requires a social engagement and the participation of researchers from diverse regions and cultures. Therefore, it is necessary to also develop novel participatory models to promote an inclusive intercultural approach to LTER research.

As demonstrated by our analyses, currently it is not possible to interrogate directly the knowledge superstructure that the various LTER infrastructures have built collectively. As we advance toward that goal, the role played by ILTER infrastructure, the geographical and conceptual constraints in the production of knowledge at ILTER cannot remain invisible. The marked geopolitical biases in the knowledge production at ILTER suggests that it is urgent to better balance the inclusion of quantitative and qualitative forms of knowledge from different regions and cultural traditions. Epistemologically, the inclusion of broader geographical areas and qualitative research will broaden the spectrum of ecological forms of knowledge. Ethically, it will broaden the spectrum of values and the participation of local and regional communities. Our concise analysis aims to call attention to the fact that ILTER research outputs could better represent the multiplicity of existing ecological worldviews in order to avoid excluding diverse stakeholder communities to Earth Stewardship, and enhance intercultural and interregional dialogues and collaborations in this planetary initiative.

Acknowledgements This chapter was made possible by a project of the International Long-Term Ecological Research (ILTER) Network, in conjunction with FinLTSER, the THULE Institute, and the Academy of Finland, SYKE – the Finnish Environment Institute, the Centre for Ecology and Hydrology (UK), and the support of the Institute of Ecology and Biodiversity (IEB-Chile).

References

Aguirre Sala J (2015) Hermeneutics and field environmental philosophy: integrating ecological sciences and ethics into earth stewardship. In: Rozzi R, Chapin FS III, Callicott JB et al (eds) Earth stewardship: linking ecology and ethics in theory and practice. Springer, Dordrecht, pp 235–247

Cavnar WB, Trenkle JM (1994) N-gram based text categorization. In: Proceedings of SDAIR-94, 3rd annual symposium on document analysis and information retrieval

Chapin FS III, Pickett STA, Power ME et al (2015) Earth stewardship: an initiative by the ecological society of America to foster engagement to sustain planet earth. In: Rozzi R, Chapin FS III, Callicott JB et al (eds) Earth stewardship: linking ecology and ethics in theory and practice. Springer, Dordrecht, pp 173–194

Christensen NL, Bartuska AM, Brown JH, Carpenter S et al (1996) The report of the Ecological Society of America committee on the scientific basis for ecosystem management. Ecol Appl 6(3):665–691

Gao S (2015) Aesthetic and moral appreciation of nature in philosophical traditions in China. In: Rozzi R, Chapin FS III, Callicott JB et al (eds) Earth stewardship: linking ecology and ethics in theory and practice. Springer, Dordrecht, pp 39–48

Huntington HP (2000) Using traditional ecological knowledge in science: methods and applications. Ecol Appl 10(5):1270–1274

ILTER Network (2006) International long-term ecological research network final draft strategic plan. Published by the ILTER Network. http://www.ilternet.edu/about/key-documents/FINAL-DRAFT-STRATEGIC-PLAN-Aug2006.pdf

ILTER Network (n.d.) http://ilternet.edu/

Li B (2013) Searching for theory in metadata. iConference 2013 Proceedings, pp 377–388. doi:10.9776/13216

Li B (2014) Sustainable value and generativity in the Ecological Metadata Language (EML) platform: toward new knowledge and investigations. Proceedings of HICSS 2014

Mamani-Bernabé V (2015) Spirituality and the Pachamama in the Andean Aymara worldview. In: Rozzi R, Chapin FS III, Callicott JB et al (eds) Earth stewardship: linking ecology and ethics in theory and practice. Springer, Dordrecht, pp 65–76

Maass M, Equihua M (2015) Earth stewardship, socioecosystems, the need for a transdisciplinary approach and the role of the international long term ecological research network (ILTER). In: Rozzi R, Chapin FS III, Callicott JB et al (eds) Earth stewardship: linking ecology and ethics in theory and practice. Springer, Dordrecht, pp 217–233

May Jr R (2015) Dorothy stang: monkeys cry and the poor die, earth stewardship as liberation ecology. In: Rozzi R, Chapin FS III, Callicott JB et al (eds) Earth stewardship: linking ecology and ethics in theory and practice. Springer, Dordrecht, pp 407–418

Porter MF (1980) An algorithm for suffix stripping. Proc Natl Acad Sci U S A 14(3):130–137

Redman CL, Miller TR (2015) The technosphere and earth stewardship. In: Rozzi R, Chapin FS III, Callicott JB et al (eds) Earth stewardship: linking ecology and ethics in theory and practice. Springer, Dordrecht, pp 269–279

Rozzi R (2015a) Earth stewardship and the biocultural ethic: Latin American perspectives. In: Rozzi R, Chapin FS III, Callicott JB et al (eds) Earth stewardship: linking ecology and ethics in theory and practice. Springer, Dordrecht, pp 87–112

Rozzi R (2015b) Implications of the biocultural ethic for Earth Stewardship. In: Rozzi R, Chapin FS III, Callicott JB et al (eds) Earth stewardship: linking ecology and ethics in theory and practice. Springer, Dordrecht, pp 113–136

Rozzi R, Arango X, Massardo F, Anderson C, Heidinger K, Moses K (2008) Field environmental philosophy and biocultural conservation: the Omora Ethnobotanical Park educational program. Environ Ethics 30(3):325–336 [en español: Filosofía ambiental de campo y conservación biocultural: el programa educativo del Parque Etnobotánico Omora. Environ Ethics 30(S3):5–128]

Rozzi R, Armesto JJ, Gutiérrez JR, Massardo F et al (2012) Integrating ecology and environmental ethics: earth stewardship in the southern end of the Americas. BioScience 62(3):226–236

Sarmiento F (2015) The antlers of a trilemma: rediscovering Andean sacred sites. In: Rozzi R, Chapin FS III, Callicott JB et al (eds) Earth stewardship: linking ecology and ethics in theory and practice. Springer, Dordrecht, pp 49–64

Shibata H (2015) Biogeochemistry and traditional ecological knowledge and practices in Japan. In: Rozzi R, Chapin FS III, Callicott JB et al (eds) Earth stewardship: linking ecology and ethics in theory and practice. Springer, Dordrecht, pp 27–38

Singh SJ, Haberl H, Schmid M, Mirtl M, Chertow M (2013) Long term socio-ecological research. Studies in society: nature interactions across spatial and temporal scales. Springer, New York

Turnhout E, Hisschemöller M, Eijsackers H (2007) Ecological indicators: between the two fires of science and policy. Ecol Indic 7(2):215–228

Chapter 14
Earth Stewardship, Socioecosystems, the Need for a Transdisciplinary Approach and the Role of the International Long Term Ecological Research Network (ILTER)

Manuel Maass and Miguel Equihua

Abstract The way we see ourselves and understand the world we live in guides and determines the types of solutions we are designing and implementing to deal with our global change problems. System thinking is helping us to recognize humanity as complex, self-organized, multi-level, and highly integrated socio-bio-physical entities that we refer to as socioecosystems. This new ontological paradigm requires new epistemological tools, and transdisciplinary research is inducing changes in different aspects of our scientific endeavor, including: the philosophical approach we use to observe our world; the level of commitment we put in our scientific work; the extent and scope we envision in our research goals; the geographical scale and context in which we focus our case-studies; the type of collaboration we engage in with other scientists; and the institutional arrangements we construct to accomplish our research efforts. The International Long Term Ecological Research Network (ILTER) includes national-level networks of scientists engaged and committed to conducting long-term and site-based ecological and socio-economic research and monitoring, with a strong interest in capacity building. ILTER members have expertise in the collection, management, and analysis of long-term environmental data and, together, they are responsible for creating and maintaining a large number of unique long-term datasets. ILTER has been a natural partner for global initiatives dealing with environmental issues, and many members of its community have been participating in these international programs. We should not underestimate the urgency, nor the level of commitment, required to foster worldwide socioecosystem research with a transdisciplinary approach, which are essential for the success of the sustainable Earth Stewardship initiative.

M. Maass (✉)
Centre for Ecosystem Research (CIEco), Universidad Nacional Autónoma de México (UNAM), Morelia, Michoacan, Mexico
e-mail: maass@cieco.unam.mx

M. Equihua
Instituto de Ecología, A.C. (INECOL), Xalapa, Veracruz, Mexico
e-mail: miguel.equihua@inecol.mx

R. Rozzi et al. (eds.), *Earth Stewardship*, Ecology and Ethics 2,
DOI 10.1007/978-3-319-12133-8_14

Keywords Capacity building • Coupled socio-ecological systems • Epistemological paradigm • Long-term ecological research • Socioecosystems • Transdisciplinary research

14.1 Introduction

The extent of our current environmental crises has reached planetary proportions, clearly shown in a variety of challenges collectively known as "global change" (Vitousek 1992; Steffen et al. 2004). It includes not only climate change, loss of biodiversity, soil erosion of arable land, and stratospheric ozone depletion, but other problems less mentioned in the mass media, but equally important, such as ocean acidification and disruption of the global N and P biogeochemical cycling (Rockström et al. 2010). All of these are interconnected in various ways. Among these earth-scale environmental problems, land use change particularly is relevant since it is not only the main cause of biodiversity loss, it also embodies the transformation of natural ecosystems and thus the processes eroding earth's life support system (Ehrlich and Ehrlich 1991).

Society's development highly depends on the benefits it obtains from nature (Daily et al. 1997). In order to get these ecological services, humans interact and transform their local ecosystems. These local transformations became regional with human expansion currently, and have reached global proportions (Kates and Paris 2003). Sustainability has been proposed as the goal of societal development in response to this severe environmental crisis (ICSU 2010; Spangenberg 2011). Global-level problems require global-level solutions, an idea that is embedded in the Earth Stewardship concept (Chapin et al. 2011). As Power and Chapin (2009) state:

> Planetary stewardship requires that decision makers and stakeholders be well-informed about how global change is likely to affect households, resources, livelihoods, and quality of life. They must also learn how local actions and reactions to change could feed back to influence the trajectory of planetary change. To provide this information, ecologists must redouble their efforts to understand and forecast ecosystem changes across multiple scales.

An important initiative within the global research arena is the International Long Term Ecological Research Network, known by its acronym ILTER (www.ilternet.com). Since its creation, ILTER has grown at an average rate of two countries and 30 sites per year, reaching now 37 national networks and embracing nearly 600 academic groups anchored in specific sites over the five continents and committed to conducting scientific research spanning decades (Gosz 1996; Parr 2013).

In the following lines we will describe the type of changes that are already occurring in the scientific sector to deal with this global-scale environmental crisis. Also we will identify changes that we need to foster and speed up in order to advance towards an earth-level stewardship process, and will lift up the role of ILTER in this endeavor.

14.2 The Complex Nature of Socioecosystem: A New Ontological Paradigm

Global change not only refers to changes at global scales, but changes associated with human activities. The extent of the human impact on earth has been so deep that some authors are calling the current times the Anthropocene as a new geologic era (Crutzen and Stoermer 2000). We know that humans are not the only organisms capable of transforming their environment at global scales. The appearance of photosynthetic cyanobacteria transformed the oxygen-free atmosphere into an oxidizing one, which dramatically changed the composition of life forms on Earth billions of years ago. However, humans are the only species that has been conscious about its global effect on the environment, and with technological means to do it in a much faster manner.

Being conscious and able to generate technology is generating ecological drawbacks, but these human characteristics are also our best tools to deal with those environmental problems. In fact, most organisms do not think about their environmental problems. They just react to them using their natural arsenal encoded in their genes, and, through a Darwinian evolutionary process, the best momentary solutions are selected in each generation and transmitted to their descendants. Humans, instead, have the capability of thinking about their environmental problems. Through knowledge generation and technological development humans make a conscious attempt to deal with those challenges, and this is what Earth Stewardship is all about. Our best solutions are incorporated into our cultural legacy and transmitted not only from one generation to the following, but also to other humans of the same generation in other places in a more horizontal fashion. As Callicott (2007) has pointed out, this conscious and horizontal evolution in humans, somewhat of a Lamarckian type, is many times faster than Darwinian evolution, giving to humans a peculiar character that differentiates us from the rest of living organisms. This more conscious evolution of humans highly depends on the way we see and understand the world, and guides and determines the types of solutions we design and implement to deal with our environmental problems. Therefore, the way we see the world is critical for the solution of our environmental problems.

System thinking has produced a profound change in the way we appreciate and understand our world (Ackoff 1999; ICSU 2010). The ecosystem concept brought fresh air to our perception and comprehension of life phenomena at levels higher than individual species (Golley 1993; Maass and Martínez-Yrízar 1990; Kaya et al. 1999). What remains controversial, though, is the conceptual place of humans in nature.

Physicists state that life is just another type of organized star dust. Although it is true that all living organisms are made of atoms following the laws of nature, biologists have show that life, in comparison to most entities in the universe, has the particularity of being able to store information in genes. This ability of store and reproduce genetically encoded information, generates new and different entities, built from already existing ones, without the need to start from zero every time it deteriorates as a result of interaction with other components of the system or just thermodynamic decay. This biological evolution is a much faster process than physical-chemical evolution.

Likewise, some biologists believe that humans can be conceptualized as just another type of biological species. Although it is also true that humans are biological entities that store, reproduce, and transfer genetic information, they also store and encode information in the form of a symbolic language with a highly complex syntactic structure (Maass 2012). This ability of humans to store and share cultural information allows them to generate knowledge and develop technology in a progressive way without precedence in the history of life on our planet (Ehrlich 2002).

In the same way as living nature is a biological-physical-chemical phenomenon, human nature is a socio-cultural-biological-physical-chemical phenomenon (Maass 2012). And in the same way living organisms cannot exist without their physical-chemical matrix, humans cannot live without their ecosystem matrix (O'Neill 2001). From a system perspective an ecosystem is the result of living and non-living entities interacting in time and space at different hierarchical scales (Odum 1953, 1969). Ecosystems are as small as a drop of water (or even smaller as a group of bacteria interacting in a corner of a cell wall), or as large as the entire planet. From the same system perspective, a **socioecosystem** is the result of humans and ecosystems interacting in time and space at different hierarchical scales. Socioecosystems are as small as a farmer with his family interacting with his agricultural piece of land, and as large as the entire planet (and beyond, if we consider the satellites, the International Space Station, and other human made space crafts visiting the Moon, Mars, and other planets).

We see humans as embedded in socioecosystems, recognizing their sociocultural-biological-physical nature. The recognition of this complex, multi-level and highly integrated socio-bio-physical entities, require new epistemological frameworks to properly study and deal with them.

14.3 The Epistemological Paradigm of Transdisciplinary Research: A Must for the Study of Socioecosystems Required for an Earth Stewardship Initiative

Scientific research has evolved in its attempt to deal with this new ontological paradigm, which implies the study of these highly coupled socio-ecological systems, or "socioecosystems", as we like to call them. Changes occurred as early as the middle of the last century when the systems approach appeared in the scientific arena (Bertalanffy 1950). However, these changes have gained important momentum in the last 20 years. This shift in the way we do science has happened in different aspects of our scientific endeavor, including: the philosophical approach we use to observe our world; the level of commitment we put in our scientific work; the extent and scope we envision in our research goals; the geographical scale and context in which we focus our case-studies; the type of collaboration we engage in with other

Table 14.1 Aspects in the way science is changing ("*from...*" to "*a more...*") in order to deal with **socioecosystem** research and in its quest for earth stewardship towards sustainability

CURRENT PARADIGM	→	+ NEW TRANSDISCIPLINARY PARADIGM
Change in philosophical approach		
Reductionist	→	+ Holistic
Analysis	→	+ Synthesis
Rational (Cartesian)	→	+ Empirical (phenomenology)
Change in commitment		
Current generation concern	→	+ Future generation concern
Curiosity driven	→	+ Result based research
Understanding	→	+ Managing ecosystems
Change in scope		
Disciplinary	→	+ Interdisciplinary
Process oriented	→	+ System oriented
Short term	→	+ Long term
Change in geographical scale of focus		
Local	→	+ Global
National	→	+ International
North-north	→	+ North-south
Indoors	→	+ Outdoors
Change in type of collaboration		
Competition	→	+ Cooperation
Individual	→	+ Collective
Teamwork	→	+ Network
Disciplinary	→	+ Transdisciplinary
Change in institutional arrangements		
Peer review	→	+ Society review
Vertical governance	→	+ Horizontal
Institute	→	+ Meta-institute

scientists; and the institutional arrangements we develop to accomplish our research efforts (Table 14.1). We will describe briefly all these changes that constitute a whole new epistemological paradigm of science for the study of socioecosystems.

14.3.1 Philosophical Approach

System thinking is a relatively new philosophical approach to observe nature. This approach has been able to comprehend the hierarchical character of nature, show the limitations of the analytical approach to studying its complexity, and demonstrates the importance of stepping back to ponder the whole and to identify the emerging properties of that whole, which is "more than the sum of its parts". With a more phenomenological approach, we can recognize a world in which reality expresses itself, in contrast to a strict Cartesian view, which starts from doubting the existence of reality itself. As Sokolowski (2012) explains:

> Phenomenology is the study of human experience and of the ways things present themselves to us in and through such experience. (…) Phenomenology is a significant philosophical movement because it deals so well with the problem of appearances. (…) [I] n its classical form, [it] insists that parts are only understood against the background of appropriate wholes, that manifolds of appearance harbor identities, and that absences make no sense except as played off against the presences that can be achieved through them.

Moreover, phenomenology, since its inception by Husserl (1913), has opened what can be named a correlational view, in which any kind of reality or "world" (a "noema" in phenomenological terms) can only be understood in its mutual relationship with subjective lived processes ("noesis" in phenomenological terms) in which it is given or experienced. Thus, humans can only be understood against the proper (socioeco)system in which they live, and correlatively, this socioecosystem should be understood as a correlate of human life and intentions (Hopkins 2010).

14.3.2 Commitment

Sustainability originally was stated as a trans-generational issue, i.e., *"how can we develop, as a society, without putting at risk the development of future generations?"* Therefore sustainability science has incorporated a commitment to future generations. However, on the face of the magnitude of the problem and the urgency of scientists to supply the solutions society is demanding to deal with global change, science also is moving from just a "curiosity driven approach", to a more "problem oriented" and a more "result based" research. Still driven by curiosity; however, it recognizes that curiosity alone is not enough to understand how the world works. It is also necessary to promote the incorporation of this understanding into public policy (Vaughan et al. 2007). And furthermore, it is important for policies we design to be implementable and functional. This requires evaluating whether or not the socioecosystem is really going in the direction it was expected and that damage to the environmental life support system is being avoided. Frequently this can be done following an "adaptive management" approach, when suitable options are available (Holling 1978), but there are cases when it is not possible given that there are no management options available that prevent serious damage to the environmental life support system.

14.3.3 Scope

In our efforts to study and understand how socioecosystems are structured and work, the extent and scope of our research approach has increased. Within the reductionist approach, scientists interested in functional aspects tend to specialize on particular process, and by studying the same processes under different settings or contexts, a better understanding of the process is achieved. Within a system approach, the strategy shifts to a focus on a particular system and examines different processes within it, developing an understanding of the whole. Under this system approach, there is a need for multi and interdisciplinary efforts in which several disciplines interact to understand the complexity of socioecosystems. In the same way as "problem oriented" science does not kill "curiosity driven" science, interdisciplinary approaches do not replace disciplinary efforts. The difference between *multi*-disciplinary research and *inter*-disciplinary research is the level of interactions among the disciplinary efforts. In the latter, the interdisciplinary group identifies and defines the problems, and the level of interaction among disciplines demands common conceptual frameworks and stronger communication skills (García 1994).

14.3.4 Scale of Focus

A major shift in science as a result of incorporating the system approach, is the recognition of needing multiple level of analysis to cope with the hierarchical nature of systems. The study of socioecosystems is not the exception. The need for long-term research has been identified since the last century, and there are very good examples of studies conducted for decades long before formal research programs were established to foster long-term endeavors (Swank and Crossley 1988). As we mentioned, and we will further discuss below, the establishment of the United States LTER network in the 1980s (Gosz et al. 2010) and the International LTER network 10 years later (Gosz 1996), have been important advances to this change of research scope in science (Parr 2013). Socioecosystems research not only requires a shift in time scale, but also implies a shift in spatial scales. Socio-ecological processes take place in multiple spatial scales and the shift from strictly local research to a more regional and global scope is crucial for an earth stewardship undertaking. Cultural diversity in a particular region is aligned with the local biodiversity (Toledo 1995, 2001). Therefore, the great ecosystem diversity found on earth has produced a large diversity of socioecosystems, as well as an enormous variation in the ways humans see, interact with, and transform their natural environment. It is very important to recognize and consider all these variations in human expressions to truly understand their impact in the earth socioecosystem. However, it is essential to recognize the deep contrast in the amount of economic resources allocated to science between north and south. In order to overcome these gaps in multiescalar research and geographical representativeness, research has become more international. However a stronger effort should be placed on moving from dominant north-north collaboration, to more north–south and south-south collaborations.

14.3.5 Collaboration

The complexity of socioecosystems is forcing scientist to engage in collaborative work. As we suggested above, interdisciplinary research does not mean converting ourselves into generalists. What it means is the need of collaboration with scientists from other disciplines (social scientists and natural scientists working together on a common problem). And because of the regional and global scope, these collective efforts have transcended our traditional local disciplinary institutions, inducing the creation of networks of teams at different scales. Even more, the socioecosystem paradigm is stimulating the development of a truly transdisciplinary approach, in which the intelligence behind the understanding of our world cannot come only from the scientific research (Spangenberg 2011), but also from knowledge acquired in a more empirical way, sometimes over hundreds of generations (Toledo 1995; Rozzi et al. 2008; Rozzi 2010). Following this new approach, research is conducted in collaboration with other sectors of society directly involved in the particular problem that is the object of study. Research tools and approaches like "co-design", "participatory monitoring" and "citizen science" have been developed to incorporate local and traditional knowledge into the research process (Burgos et al. 2013).

14.3.6 Institutional Arrangements

All of the above are pressuring scientific institutions interested in transdisciplinary research to find new and creative arrangements to accomplish the task. A difficult aspect is how research performance should be evaluated. Peer review is very important to assure the rigor of the research, but it is not enough if we accept the commitment to cross the line from "curiosity driven" research all the way to "solution base" research. Under these new conditions, other sector of society involved in the enquiry subject should participate in the evaluation process to assure the *pertinence of the study*, since they are experts in the matter (Spangenberg 2011). Another complicated aspect of collaborative research, in which many groups and institutions are involved, are "author's rights" and "governance" issues. Sharing data protocols and multi-authored documents are becoming important aspects within socioecosystem research. Polycentric governance approach in which multiple governing bodies interact to make and enforce rules within a specific arena or location, have been suggested to deal with this multi-level and nested institutions (Simonsen et al. 2014). Most academic institutions are big and old, with enormous inertia. It has been very difficult to move them toward new administrative arrangements. One way to overcome the need for interaction between scientists of different disciplines and sectors without dramatically changing the current administrative arrangement has been the creation of meta-institutes. They consist in a particular arrangement, in which the associated researchers, belonging to different institutions and thematically and geographically separated, collaborate on a regular basis with the help of new information technologies and communication protocols.

14.4 The Role of ILTER in the Earth Stewardship Initiative

Moving from the current dominant disciplinary science to a more interdisciplinary approach is a requirement for the study of global change and its consequences for society. Seeking ways to deal with this challenge has been present in the admonitions and efforts of several international research organizations endeavoring to effectively prevent environmental deterioration on Earth. Such are the cases of the International Geosphere and Biosphere Program (IGBP) and its social counterpart, the International Human Dimensions of Global Environmental Change (IHDP) Program. They have documented the magnitude of the problem, the urgency of taking actions, and the need of long-term research and monitoring to understand the causes and consequences of global change. Other initiatives focused on more specific aspects such as the Millennium Ecosystem Assessment (MA) launched to evaluate the state of ecosystem services at regional and global scales, and their importance for the human wellbeing. This initiative not only documented the fragility of our life support system, but also the severity of knowledge fragmentation and the difficulties of the world scientific system to conduct interdisciplinary research (Norgaard 2008). As we will describe next, the International Long Term Ecological Research Network (ILTER) is also engage in this effort to conduct socioecosystem research for a sustainable earth stewardship (Maass and Equihua 2014).

14.4.1 Vision and Mission

ILTER envisions a world in which science helps to prevent and to solve environmental and socio-ecological problems. ILTER contributes to solving international ecological and socio-economic problems through question and problem-driven research, with a unique ability to design collaborative, site-based projects, compare data from a global network of sites, and detect global trends (www.ilternet.edu). Most ILTER members are national or regional networks of scientists engaged in long-term, site-based ecological research and monitoring. They have expertise in the collection, management, and analysis of long-term environmental data. Together they are responsible for creating and maintaining a large number of unique long-term datasets (Parr 2013). ILTER is a natural partner to global initiatives, and many members of its community have been participating in these international programs.

14.4.2 From LTER to LTSER

There has been a natural evolution of scientific groups of ecologists interested in long-term research to move from strictly ecological research (LTER) to a more socio-ecological research (LTSER; Fig. 14.1). One of the main objectives of LTER, 30 years ago, was to fill up the knowledge gap created by the established scientific

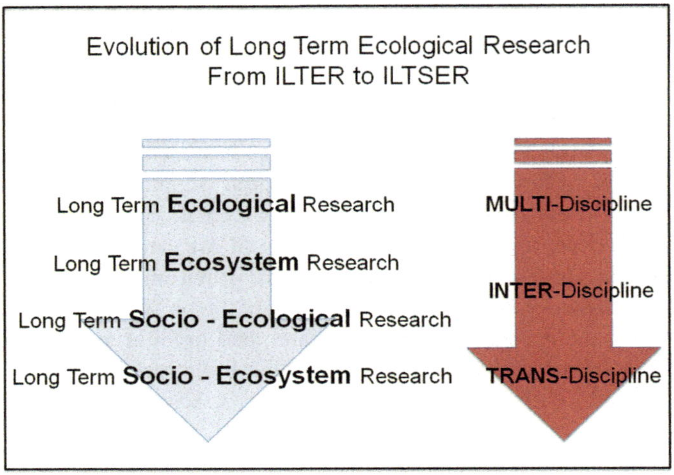

Fig. 14.1 Evolution, during the last 30 years, of the focus and epistemic tools of long-term research, within the international scientific community interested in global environmental problems

funding system which promoted either small scale studies (e.g. a few species in a few m^2 for a 1–2 years) or large scale studies but conducted in short time (e.g. 2–3 years study of the ice or sediment cores thousands of year old). Initially, most LTER groups were working on natural reserves, looking to extend their understanding of ecological processes for longer periods of time (decades) and at larger scales (hectares and km^2) in a "secure" environment. The ecosystem approach followed naturally when scientists from different disciplines started to work on the same place for many years (site based research concept), accumulating the necessary knowledge to deal with the complexity of ecological systems. The socio-ecological research came later in order to understand the human drivers behind the transformation of natural ecosystems and with an interest in supplying scientific information for proper ecosystem management. However, the need for a transdisciplinary research has emerged with the new socioecosystem paradigm, in which humans are not just another species taking advantage of ecosystem services, but a complex human-biological-physical entity that evolves into a tight integration of biophysical and cultural components, living and non-living, at different scales. *Socioecosystem research* requires a shift from viewing humans as external drivers of natural systems to that of agents acting within socio-ecological systems (Grimm et al. 2000; Redman et al. 2004; Haberl et al. 2006). On these grounds, a new initiative within the LTER community has been launched as a strategic research initiative called "Integrative Science for Society and the Environment" (ISSE), proposed to elevate environmental science to a new level of integration, collaboration, and synthesis necessary for addressing current and emerging environmental research challenges (Collins et al. 2007).

14.4.3 The Importance of Site-Based Research

Most ILTER members are country-level networks of academic groups committed to maintain their research efforts on a particular site during many years. This "site-based research" character of ILTER is one of its most important assets. It not only allows for the accumulation of knowledge through time, as was stated before, but also is the only way to develop the necessary trust between the academic community and the local stakeholders that require a transdisciplinary research approach. Capacity building is another advantage of site-based research since working with complex systems requires the recognition of uncertainty and, therefore, the need for a strong and long-lasting learning process. Teaching and tutoring students in socio-ecosystem research requires identifying a particular aspect to focus during a short time, without loosing the long-term and large-scale context of their thesis research theme. This is crucial and is easier to accomplish within a long-term and site-based research group.

14.4.4 Partnerships Approach

ILTER is not alone in this quest for global change, socio-ecological, and earth stewardship research. Its international scope, its flexible research agenda, and its commitment for long-term and site-based research, makes ILTER a natural partner for many global initiatives (Parr 2013). As stated above, members of ILTER actively participate in local, national, regional, and international initiatives. One of the main objectives at ILTER meetings is to engage in collaborative research activities and to foster partnerships with international players such as the Group on Earth Observations Biodiversity Observation Network (GEO BON); the Global Land Project (GLP); UNESCO International Hydrological Program (UNESCO IHP); etc. We are currently designing a multi-site level project to participate in the new Program for Ecosystem Change and Society (PECS-ICSU) which has a strong socioecosystem and transdisciplinary approach (Carpenter et al. 2012).

14.4.5 ILTER Heterogeneity and North–South Inequalities

ILTER comprises nearly 600 research sites located in a wide array of ecosystems. Of course not all ILTER sites have the same experience and capabilities. Of the ILTER groups that do ecosystem research, many are involved in socio-ecologic studies while a minority conducts socioecosystem and transdisciplinary research. However the interest among the groups to do transdisciplinary research has increased in recent years. As we said before, we are currently building a collaborative strategy for the PECS Program, which certainly will stimulate further socioecosystem and transdisciplinary research in the network. Forest biome dominates ILTER sites (40 % of the sites), but aquatic sites are also well represented: fresh water (25 %), costal (7 %) and marine sites (5 %). There are also mountain sites (10 %), deserts (6 %), and grasslands (6 %). We even have a few urban LTER sites (less than 2 %). On top of this ecosystems diversity, there are also diverse socio-economic conditions, bringing important heterogeneity of socioecosystems types and arrangements, which makes ILTER a very profitable platform for earth stewardship and sustainability research. However, it is also important to recognize that ILTER does not differ from other International Programs in which there is an important unbalance of North/South research effort. Rozzi et al. (2012) have pointed out the Northern hemispheric research dominance, and highlighted a particularly notorious gap of ILTER research sites in the temperate and sub-Antarctic regions of South America (between 40° and 60° south). In fact, only 3 % of the current ILTER's research sites, listed on its webpage, are located on the southern hemisphere and only 8 % of the sites belong to the inter-tropical zone (between 23°N and 23°S). A similar gap has been detected in the Northern Africa and Middle East region, as well as in the North of Asia. It does not necessarily mean that no one is working in these regions, but the fact is that there are very few groups associated with ILTER there. However, ILTER is taking actions to revert this situation, fostering the participation of new partners though a new type of membership called "associated sites". Under this admission category, a research group from a country without a formal ILTER national-level network will be able to join ILTER through an affiliation process with an already accepted member. The associated sites will have the commitment to participate in building their national-level network. With this mechanism, ILTER is expecting to incorporate good research groups, which are already making individual efforts to conduct LTER research in developing countries.

14.4.6 The Bottom Up Approach

In very large organizations like ILTER, with a highly heterogeneous membership, it is difficult to coordinate research activities in which all members participate. Our approach has been to standardize methods (to facilitate data sharing and comparison), foster diversity (to increase collective intelligence), identify common interest (to induce collaboration), facilitate the communication between groups (to generate opportunities), and allow for self-organization of activities through what we call "bottom up" initiatives. By scooping from the bottom, ILTER not only increase the possibility of getting new and exciting ideas, but also is facilitating the integration process within the network. Since "bottom up" initiatives can come from any network group, the member's participation is encouraged, the commitment is self-imposed, the sense of community is amplified, and the effort of running the network is distributed. "Bottom up" initiatives are very good for dealing with local limitations and identifying good opportunities for collaboration. They may also promote redundancy and bring stability to the network (Csermely 2006; Ahn et al. 2010).

14.5 Some Precisions About Socioecosystem Research

As we have asserted here, socioecosystem research for a sustainable earth stewardship urges significant changes in the way we do science. We should not underestimate its urgency, nor the level of commitment required. Thus, it is important to make some comments about the speed and magnitude of these changes, because there is a tendency to overstate the roll of scientists, increasing the already heavy load on the research community. Firstly, it is important to point out that we recognize that what it is needed to stroll along the sustainability path is a socio-economic *development model* blended in a socioecosystems framework (integral, nested multi-level, non-linear, complex, self-organized, human-biological-physical system). However, what we have been discussing here is just the need for a change to a more transdisciplinary scientific *research model* that will feed into this new approach for earth stewardship. There is an important difference between a *transversal approach* (working with different sectors of society) and a *transdisciplinary approach (*working with different sources of knowledge). The former is a *development tool*; the latter is an *epistemological stance*. We need both. However, scientists do not necessarily need to become producers, policy makers, business people or developers but, in order to conduct research in a truly transdisciplinary fashion, they have to participate in real development situations, as another stakeholder embedded in the collective. Participating in transversal work is the only way to learn about this "other knowledge" requirement in real transdisciplinary research. A good analogy is a university hospital in which scientific research on health is conducted with real patients. However, rather than working as a health service unit for the local community, the university hospital selects particular cases for treatment based on their research interests. Transdisciplinary research is conducted in real case studies, and that is why "site based research" is so important.

Another aspect that requires awareness by scientists interested in socioecosystem research for a sustainable earth stewardship is the recognition of our working under conditions of high uncertainty. We are not only confronted with highly complex systems, but the climate change scenario is increasing even more this uncertainty. Adaptive management is a conceptual tool developed to deal with this uncertainty, provided that suitable management options are at hand and reversibility of very dangerous environmental impacts is possible (Holling 1978). We no longer expect to have a complete understanding of the process for making management decisions. Rather, managers decide based on the best available knowledge, but keep a monitoring program to feed back into their decision-making process. If the system is performing as expected, the decision is maintained; if on the contrary it is not, the decision is tuned or changed accordingly. Scientists are not managers, but they should also recognize their limitations as knowledge providers under these highly uncertain conditions. An *adaptive learning* approach has to be followed, but the only way to do it, is working on real situations where *adaptive management* is conducted.

The multi-level character of the socioecosystems is another heavy load for scientists interested in sustainability research for earth stewardship. We not only need to incorporate the social, biological, and physical aspects in our research, but also, to consider the multiple spatial-temporal strata in which socioecosystems operates. How to tackle such a complex system? The environmentalist slogan "think globally and act locally", conveying some systems thinking perspective, may help. However, between the global and local tier, there are plenty of other levels to consider (municipal, state, national, regional, continental, hemispheric, etc.). In order to deal with such complexity, it is recommended to choose one particular tier to focus our research, and concentrate on the interactions between that particular level of interest with the immediate upper (or supra) and lower (or sub) ranking. One can be aware of further upper and lower layers (beyond the immediate supra and sub ranks), but only as observers, reducing the level of observations as the scales get farther away from the focus of interest. In this way, one will be able to understand the immediate context's factors, which are inducing the behavior of our socioecosystem (at the focus of interest) and also the local and particular conditions that our chosen scale of focus is directly influencing, without loosing the whole perspective. Sometimes, it is also necessary consider a particular levels of the hierarchy that most strongly influence your level of interest. For example, the critical level above the national level might be global (rather than regional) because of globalization of trade and climate.

Finally, we cannot leave this discussion without talking about the role of technology. Although it is true that in many respects technology brought us into an environmental conflict of global proportions, there is no way we can deal with the problem and walk a sustainable earth stewardship course without the aid of technology. However, technological development should also be aligned with this socioecosystem paradigm. Human nature is technological because it is the blend of knowledge and conscious intent prompted by environmental interactions, and thus it is the way humans live with their surroundings. Our environmental awareness should encourage a technology design shift conscious that humans do need their ecosystems, not only because they depend on them, but truly because with them we constitute socioecosystems. The idea of Jordan (1998) encouraging "working with nature" suggests that an understanding of the many interactions and processes that occur in nature, should enlighten us to embed them in our technological design. We need to align our technological quest with our socioecosystem character.

Rozzi (2012) has pointed out that a particular *habitat* induces in living things *habits* that eventually match to astonishing perfection that particular *habitat*. It is a fact of life. Species appear, adapt, and extinguish following this interactive rational, and if something changes, ecosystems self-organize following through this dynamic systemic imperative. With the help of technology we have created artificial habitats, giving us the impression that we do not need our original environment any more. And we have developed habits that obviously do not match with our original habitat. With the advancement of technology we have come to think that we are separated from the rest of the species, and we dream of traveling in an aseptic spacecraft (just humans and machines) conquering other worlds: it is a false impression (Margulis 1998). Even when we have visited the moon several times and even set technological foot in

Mars, currently we are not contemplating the idea of establishing a colony there, because there are no ecosystems on the moon nor apparently on Mars. Everybody can have an artificial climate in cars, but only a few can afford to have it in houses. We can imagine an artificial climate in a small city, but it is highly impractical. It is certainly beyond our current skills and knowledge at regional or global scales. The current environmental crisis is not only evidencing the always incomplete adaptive nature of technology, but also is disclosing our socioecosystem nature and demonstrating our dependence on ecosystems to maintain us. The challenge is to learn how to fulfill human needs through coevolution with nature, rather than aiming to subdue it (Jordan 1998).

14.6 Final Remarks

We have to recognize that the changes we describe in the way we do science in order to align with the socioecosystem nature of human enterprise, have already been taking place very slowly (for decades), and some of them in a serial fashion (one after the other). Some of the changes are now very well established in the scientific community and many others still need to gain recognition by it (see for example Carmel et al. 2013). In any event these changes do not mean a substitution of one type of research for another. Rather, a complementarity of approaches for better understanding our world is what is emerging. However, we are convinced that the epistemological paradigm we have described, is a reaction to some of the limitations the current scientific paradigm has in identifying and dealing with the severe global-scale environmental crisis that we are facing. The level of implementation of these necessary changes varies highly between countries and academic communities, as is also the level of opposition from them to explore alternative approaches. However, the process is gaining momentum and it is a matter of time before we see this new approach fully flourishing. The sooner the better, since time it is not precisely our ally in the face of the currently high-speed planet's degradation process.

Acknowledgements The authors thank Ricardo Rozzi for his invitation to participate in this volume. Also, commentaries on the text by Antonio Zirión, Terry Chapin and Ricardo Rozzi, as well as the English revision by Roy May were very useful and highly appreciated. Lyliana Rentería, Raúl Ahedo and Atzimba López helped with the preparation of the manuscript. The document was prepared during a visit by Manuel Maass to the Finnish Environment Institute (SYKE) with a DGAPA-UNAM sabbatical scholarship and within CONACYT support for ILTER.

References

Ackoff RL (1999) Ackoff's best: his classic writings on management. Wiley, New York

Ahn YY, Bagrow JP, Lehmann S (2010) Link communities reveal multiscale complexity in networks. Nature 466(7307):761–764

Bertalanffy L (1950) An outline of general system theory. Br J Philos Sci 1:134–165

Burgos A, Páez R, Carmona E et al (2013) A systems approach to modeling community-based environmental monitoring: a case of participatory water quality monitoring in rural Mexico. Environ Monit Assess 185(12):10297–10316

Callicott JB (2007) Lamarck Redux: temporal scale as the key to the boundary between the human and natural worlds. In: Brown CS, Toadvine T (eds) Nature's edge: boundary explorations in ecological theory and practice. State University of New York Press, Albany, pp 19–34

Carmel Y, Kent R, Bar-Massada A et al (2013) Trends in ecological research during the last three decades – a systematic review. PLoS ONE 8(4):1–8

Carpenter SR, Folke C, Norstrom A et al (2012) Program on ecosystem change and society: an international research strategy for integrated social–ecological systems. Curr Opin Environ Sustain 4:1–5

Chapin FS III, Power ME, Pickett STA et al (2011) Earth stewardship: Science for action to sustain the human–Earth system. Ecosphere 2(8):Art. 89. doi:10.1890/ES11-00166.1

Collins SL, Swinton SM, Anderson CW et al (2007) Integrative science for society and environment: a strategic research initiative. National Science Foundation, p 35. http://www.lternet.edu/decadalplan/

Crutzen PJ, Stoermer EF (2000) The Anthropocene. Glob Chang Newsl 41:12–13

Csermely P (2006) Weak links: the universal key to the stability of networks and complex systems. Science 126:2827. doi:10.1007/978-3-540-31157-7

Daily GC, Alexander S, Ehrlich P et al (1997) Ecosystem services: benefits supplied to human societies by natural ecosystems. Issues Ecol 2:1–16

Ehrlich PR (2002) Human natures: genes, cultures and the human prospect. Penguin Books, Rutherford

Ehrlich PR, Ehrlich AH (1991) Healing the planet: strategies for resolving the environmental crisis. Addison Wesley, Reading

García R (1994) Interdisciplinariedad y Sistemas Complejos. In: Leff E (ed) Ciencias Sociales y Formación Ambiental. Ed. Gedisa/UNAM, Barcelona, pp 45–71

Golley FB (1993) A history of the ecosystem concept in ecology. Yale University Press, London

Gosz JR (1996) International long-term ecological research: priorities and opportunities. Trends Ecol Evol 11(10):444

Gosz JR, Waide R, Magnuson J (2010) Twenty-eight years of the US-LTER program: experience, results, and research questions. In: Muller F, Schubert H, Klotz S (eds) Long term ecological research: between theory and application. Springer, Dordrecht, pp 54–75

Grimm NB, Grove JM, Pickett STA et al (2000) Integrated approaches to long-term studies of urban ecological systems. Bioscience 50(7):571–584

Haberl H, Winiwarter V, Andersson K et al (2006) From LTER to LTSER: conceptualizing the socioeconomic dimension of long-term socioecological research. Ecol Soc 11(2):13. http://www.ecologyandsociety.org/vol11/iss2/art13/

Holling CS (ed) (1978) Adaptive environmental assessment and management. Wiley, London

Hopkins BC (2010) The philosophy of Husserl. McGill-Queen's University Press, Montreal/Kingston, pp 90–132

Husserl E (1913) Ideen zu einer reinen Phänomenologie und phänomenologischen Philosophie. Erstes Buch: Allgemeine Einführung in die reine Phänomenologie. Max Niemeyer Verlag, Halle

ICSU (International Council for Science) (2010) Grand challenges in global sustainability research: a systems approach to research priorities for the decade. International Council for Science, Paris

Jordan CF (1998) Working with nature: resource management for sustainability. Taylor & Francis, Amsterdam

Kates R, Parris TM (2003) Long-term trends and a sustainability transition. PNAS 100(14):8062–8067

Kaya JJ, Regierb HA, Boylec M et al (1999) An ecosystem approach for sustainability: addressing the challenge of complexity. Futures 31:721–742

Maass JM (2012) El manejo sustentable de socio-ecosistemas. In: Calva JL (ed) Cambio climático y políticas de desarrollo sustentable, Tomo 14 de la colección Análisis Estratégico para el Desarrollo. Juan Pablos Editor-Consejo Nacional de Universitarios, México, pp 89–99
Maass JM, Equihua M (2014) La Red Internacional de Investigación Ecológica a Largo Plazo a 20 años de su creación: sus avances y retos. Bosque 35(3):415–419
Maass JM, Martínez-Yrízar A (1990) Los ecosistemas: definición, origen e importancia del concepto. Ciencias Núm Especial 4:10–20
Margulis L (1998) Symbiotic planet: a new look at evolution. Basic Books, New York
Norgaard RB (2008) Finding hope in the millennium ecosystem assessment: an essay. Conserv Biol 22(4):862–869
O'Neill RV (2001) Is it time to bury the ecosystem concept? Ecology 82(12):3275–3284
Odum EP (1953) Fundamentals of ecology. W.B. Saunders, Philadelphia
Odum EP (1969) The strategy of ecosystem development. Science 164:262–270
Parr T (2013) The international long-term ecological research network and its role in global research and policy. In: Tabarelli M, Duarte da Rocha CF, Romanowski HP et al (eds) PELD – CNPq: dez anos do Programa de Pesquisas Ecológicas de Longa Duração do Brasil : achados, lições e perspectivas. Ed. Universitária da UFPE, Recife, p 446
Power ME, Chapin FS III (2009) Planetary stewardship. Front Ecol Environ 7(8):399–399
Redman CL, Grove JM, Kuby LH (2004) Integrating social science into the Long-Term Ecological Research (LTER) network: social dimensions of ecological change and ecological dimensions of social change. Ecosystems 7(2):161–171
Rockström J, Steffen W, Noone K et al (2010) A safe operating space for humanity. Nature 46:472–475
Rozzi R (2010) Field environmental philosophy. Dialogue Universalism 20(11/12):85–109
Rozzi R (2012) Biocultural ethics: recovering the vital links between the inhabitants, their habits, and habitats. Environ Ethics 34(1):27–50
Rozzi R, Arango X, Massardo F et al (2008) Field environmental philosophy and biocultural conservation: The Omora Ethnobotanical Park Educational Program. Environ Ethics 30(3):325–336
Rozzi R, Armesto JJ, Gutiérrez JR et al (2012) Integrating ecology and environmental ethics: earth stewardship in the southern end of the Americas. Bioscience 62:226–236
Simonsen SH, Biggs R, Schlüter M et al (2014) Applying resilience thinking: seven principles for building resilience in social-ecological systems. Stockholm University, Stockholm
Sokolowski R (2012) Introducción a la Fenomenología. (trans: Marín Ávila E), 1st edition in Spanish. Círculo Latinoamericano de Fenomenología (Coord. Zirión A et al) Jitanjáfora Morelia Editoral, México
Spangenberg JH (2011) Sustainability science: a review, an analysis and some empirical lessons. Environ Conserv 38(3):275–287
Steffen W, Sanderson A, Tyson PD et al (2004) Global change and the earth system: a planet under pressure. Springer, Berlin/Heidelberg/New York
Swank WT, Crossley DA Jr (eds) (1988) Forest hydrology and ecology at Coweeta. Springer, New York
Toledo VM (1995) Peasantry, agroindustriality, sustainibility: the ecological and historical grounds of rural development. Interamerican Council on Sustainable Agriculture. Working Papers 3:1–27
Toledo VM (2001) Biodiversity and indigenous peoples. Encycl Biodivers 3:451–463
Vaughan H, Waide RB, Maass JM et al (2007) Developing and delivering scientific information in response to emerging needs. Front Ecol Environ 5(4):W8–W11
Vitousek PM (1992) Global environmental change: an introduction. Annu Rev Ecol Syst 23:1–14

Chapter 15
Hermeneutics and Field Environmental Philosophy: Integrating Ecological Sciences and Ethics into Earth Stewardship

Jorge F. Aguirre Sala

Abstract Integrating Earth Stewardship with Field Environmental Philosophy (FEP) addresses two major challenges. The first is to ensure that the economic view of land administration is self-limiting to make it compatible with Earth stewardship. The second is to link Latin American conceptual and methodological approaches to international initiatives. Toward these aims, this essay shows the contribution that hermeneutics provides to FEP and its integration of ecological sciences and ethics. The focus is the theoretical framework of the FEP methodological approach developed by Ricardo Rozzi and his students at the Chilean Network of Long Term Socio-Ecological Research (LTSER-Chile), and the integration of FEP's methodology to the International Long Term Ecological Research (ILTER) network. The contributions of hermeneutics to FEP are shown by discussing the translations of the meanings of key concepts such as "Earth" or "soil" and the holistic concepts of "environment" or "biosphere." The biophysical and symbolic-linguistic domains of these concepts are linked through these translations. The route to achieve this is to: (1) establish the need for a methodology that links contrasting economic, ecological, and ethical views of the Earth; (2) identify the role that FEP plays in the theoretical framework and the development of a methodological approach to integrate ecology and ethics; (3) introduce the hermeneutical steps supportive of the FEP methodology; and (4) illustrate the FEP and other Latin American transdisciplinary initiatives that can contribute to the integration of ethics and ecology in the ILTER network and the Earth Stewardship initiative.

Keywords Earth stewardship • Field environmental philosophy • Fusion of horizons • Hermeneutic • Latin America

---

J.F. Aguirre Sala (✉)
Department of Humanities, University of Monterrey, Garza García, NL, Mexico
e-mail: jorge.aguirre@udem.edu

R. Rozzi et al. (eds.), *Earth Stewardship*, Ecology and Ethics 2,
DOI 10.1007/978-3-319-12133-8_15

15.1 The Need for a Methodology that Integrates Ecological Sciences and Environmental Ethics

Earth stewardship may have several intentions that define ways to value stewardship and the Earth itself. Some of these modes of administration have caused ecological crisis but could be avoided by using a methodological approach that integrates Earth stewardship, ecological sciences, and ethics, and provides a deep meaning to the concept of Earth that also can be translated metaphorically. This integrative methodology would enhance an understanding of the Earth as a living element, as a biosphere.

A holistic methodology that integrates sciences and ethics could enhance the Earth Stewardship initiative (*sensu* Chapin et al. 2011) by achieving a metaphorical translation of ecological concepts and ethical values. Without this methodological approach Earth stewardship could be constrained by the prevailing conceptual and practical framework of utilitarian economy. However, to overcome the current global environmental crisis and achieve sustainable life, the contributions of ecology and philosophy, including aesthetic, spiritual, and ethical values, are necessary in order to broaden the spectrum of purely economic values (Rozzi et al. 2012, p. 233). A major constraint to the integration of environmental science and ethics is, as mentioned above, the lack of methodologies guiding interdisciplinary work between ecologists and philosophers (Poole et al. 2013). To address this methodological limitation, based on a Long Term Socio-Ecological Research (LTSER) site located in southern Chile, Ricardo Rozzi and collaborators have proposed the Field Environmental Philosophy (FEP) methodological approach (Rozzi et al. 2006, 2008a, 2010, 2012).

In this chapter, a hermeneutical or interpretive framework is adopted to analyze the FEP methodological approach. This adds a level of analysis that contributes to the consolidation of this and other methodologies for integrating ecological sciences and environmental ethics, and in this way contributes to a solid foundation of the Earth Stewardship initiative. FEP requires the input of hermeneutics to achieve a better translation of concepts, as well as incorporating positive attitudes and self-restraint in the context of Earth stewardship initiatives.

15.2 Hermeneutics: Narrow Economic Versus Ecological Views of the Land

How can hermeneutics help link socio-ecological research and FEP with Earth stewardship? To answer this central question, it is necessary to specify some subsidiary but fundamental questions: Why should Earth stewardship not ignore axiological or ethical aspects? Why should the biosphere concept not be examined at the margin of productive development framework? To achieve sustainable life should we focus on the utilitarian management of resources or on the intrinsic value of biological species? And, especially: What theoretical framework does FEP provide for tracing a hermeneutical path towards an ecological Earth stewardship?

One archaic reason for conceiving Earth as a purely economic resource for exploitation derives from a biased interpretation of the biblical Genesis. The words "...fill the Earth and subdue it" (1:28) have been interpreted from an anthropocentric and utilitarian perspective, with a reductionist interpretation that lacks a historical and hermeneutic context (see White 1967). Considering that humans are created in the image and likeness of God, this text has been interpreted as teaching that humans are privileged beings with the right and divine endorsement to use nature without restriction. The "land" or "Earth" (not the creation) is an object to be dominated and conquered. In addition, a monotheistic theological justification prevents conceiving nature as a sacred reality with its own subjectivity –if its subjectivity were recognized, then nature would demand respect rather than be subjected to the desires of exploitation of the human will.

This reifying theological conception of the Earth and the land ruptures the link between humans and nature. Similarly, the natural sciences also have established a reifying conception based on the epistemological dichotomy: "subject-object" (see Klaver 2014). Such separation has its sources in the epistemology of early modern sciences. Descartes (1596–1650) believed that non-human living organisms were equivalent to machines and that all material reality was *res extensa* –i.e., a mere measurable thing. This is the sphere of the object. In contrast, human beings are conscious of themselves, they are *res cogitans* –i.e., a "thinking thing," and therefore a subject that controls everything else.

This reifying attitude was instilled in land use planning and management through the foundational ideas of economic liberalism expressed by John Stuart Mill. In his influential book *Principles of Political Economy* (Stuart Mill 1978), the "Earth" or "land" is considered to be "soil" –i.e., as a factor of production and capital. Mill introduced these ideas specifically in Chapter XII, Book I of his work entitled "Of the Law of the Increase of Production from Land," whose name expresses the focus on production. Later in Book III, Chapter VI "Summary of the Theory of Value," Mill adopted the capacity of production as the only value of land. Following this utilitarian view, today "natural resources" commonly refer to nature.

With archaic biblical slogans and the emerging epistemology and political economy of modernity, the objectification and exploitation of the land and the Earth have been endorsed. Mill's idea continues even today, stimulating attempts to evaluate the biosphere in monetary terms. In the framework of market economy, in 1997 the value of ecosystem goods and services delivered by the biosphere was estimated in an annual amount of 33 trillion dollars (Costanza et al. 1997). This example illustrates how the prevailing ways in which the Earth is conceived and understood today, and how this understanding influences the prevailing ways in which humans assess, inhabit, and manage the Earth today (Rozzi 1999).

Contemporary ecological sciences have disclosed, however, that the notion of Earth or land as a mere factor of production and capital should be expanded into the notion of biosphere (Naeem 2013). Moreover, the concept of sustainability itself goes beyond a purely economic condition. Sustainability is more than the perpetual availability of inputs for the market economy, as critically stated by ecologist Shahid Naeem (2013) in his chapter "Ecosystem Services: Is a Species Servicing One Planet Likely to Function?." In a pragmatic sense, the sustainability and the survival of living things are a necessary, but are not a sufficient, condition to make sense of life and the relationships among living beings. The "Earth" (understood by Mill as an object of exploitation) or the "biosphere" (understood by Naeem as a subject that must be defended by an environmental ethic) require much more than a simple physical well-being or an absence of diseases, because it also requires an axiological sense of existence. Mark Sagoff emphasizes that a simple life can be more valuable than an opulent life, and that economic growth could be morally undesirable, even if it would be ecologically sustainable (Sagoff 1995). A view of the biosphere that corresponds to the biocultural ethic of a FEP will include this axiological sense. Environmental philosophy is not a mere theoretical reconciliation. It has ethical interests that Rozzi (2012, p. 341) explains synthetically:

> (a) it proposes limits of action to the prevailing neo-liberal policy (the essence of which has been to free itself from restrictions for entrepreneurship and economic growth) and (b) it extends the moral community beyond those who govern and benefit from the market (to include the majority –marginalized and oppressed human populations), and beyond the human species (to include all beings with whom we co-inhabit in the biosphere).

The transformation and industrialization that enable economic development and human well-being are legitimate, but in a regulated manner. So then, FEP has the axiological mission to establish a new narrative to limit the deterioration of the biosphere and to expand the benefits of human development to all co-inhabitants. It has also the methodological mission of integrating environmental science and ethics in order to achieve a genuine Earth Stewardship.

15.2.1 The Theoretical Framework and Methodology of Field Environmental Philosophy (FEP)

FEP emerged as a proposal from the work of a Latin American Long Term Socio-Ecological Research (LTSER) network. Based on the conceptual framework of a biocultural ethic (Rozzi 2013), FEP addresses biophysical and symbolic-linguistic levels of existence. It seeks to integrate these levels into an ethics, under a socio-ecological approach of co-habitation within the biosphere. It provides an alternative perspective to the one stated by Mill and economic liberalism of land use and exploitation, which is the perspective that prevails today.

If we posed the mission of FEP in a few words, we would ask ourselves: How can we persuade neoliberals to consider the "Earth" not as a mere natural resource but as a "biosphere," and to have respect for it? How can we make them see that the biosphere should be respected in its processes of interdependence and its existential sense? How do we convince them to abandon a view of nature that reduces it to industrial resources? In summary, FEP addresses the question of how to establish a new narrative that encompasses the many minority cultural traditions that oppose the reifying meaning of nature promoted by liberal economy, early modern science, and a distorted interpretation of the Bible? The Australian philosopher Arran Gare (1998) has proposed to erect narratives that go beyond a narrow economic sense into a new tradition, but this enterprise requires a theoretical framework and method, as proposed by FEP.

15.2.2 Theoretical Foundation of FEP

FEP's theoretical foundation has two premises: (1) the link among *Habitats, Habits* and *co-in-Habitants* (the "3Hs" of the biocultural ethic), and (2) the integration of social components into an ecosocial justice, and biocultural conservation and education programs (Rozzi 2013).

15.2.2.1 The Links Among Habitats, Habits and co-in-Habitants

In the beginning, the Greek word *ethos* did not mean ethics, but a den: the place where an animal lives (Scott and Liddell 1996). This idea broadened to include human practices and it came to mean the abodes of humans. With an ecological hermeneutic, Rozzi (2008a, p. 116) describes how ethos can be understood as a habitat, and how ethos was used later as a verb: to inhabit. When a form of inhabiting becomes recurrent it forms a habit. With the framework of biocultural ethics, Rozzi (2013, pp. 20–22) emphasizes that to address current socio-environmental problems it is essential to better understand the relationships among the human habits, the communities of co-inhabitants, and the habitats, where they inhabit. For this endeavor, interdisciplinary teamwork among ecologists and philosophers who integrate research on the habits and the habitats of specific communities of co-inhabitants, enable a recovery of the archaic meaning of *ethos* and add novel insights from ecological-evolutionary sciences to the understanding of contemporary ethics.

It is important to note that any habitat influences and, in turn, is influenced by the ways in which it is inhabited. When those ways of inhabiting are established regularly, then they produce habits (Rozzi et al. 2008a, p. 116). The habits become customs, thereby shaping the *ethos* of behavior. In this way, *habitats* and *habits* are the original basis of the ethics which guide behavior on a regular basis; i.e., they regulate the character of the beings that inhabit them (the co-inhabitants).

In addition, we should not forget that habits influence habitats and that this action should be called intentional behavior. On the other hand, the influence of habitats on habits could be called conditioned conduct. Consequently, the personality of the beings that live and interact in habitats arises as a hybrid of intentional behavior and conditioned conduct. It is important to note that unlike the

Western modern traditions of thought, especially within the tradition of Christianity, contemporary ecological sciences suggest that personality is not a unique attribute of humans (van Dongen et al. 2010). Consequently, animals, plants, and other entities may be viewed as having a personality, as being non-human persons (Rozzi 2013). In other words, the soil, the air, the rivers, and the sea are not simply "natural resources" in a narrow economic sense, but are a biosphere with dignity. However, we ought to suspect that the etymologies of the words *ethos* or *ecology* and *economy* (science and laws of the "house or home;" i.e., the habitat) are not sufficient to confront the interests of the market for the exploitation of "natural resources." The conceptual framework still requires integrating social and axiological components.

15.2.2.2 Integrating Social Components and Biocultural Education

Rozzi et al. (2010, p. 50) foster an inclusive process to conserve biodiversity within an intercultural dialogue. In hermeneutics, such integration technically is called a *fusion of horizons of meaning* (Gadamer 2000). When the interdisciplinary method of integrating ecological sciences and environmental ethics (ecology studies the habitats and ethics examines the habits) is understood, then the next step is to merge these horizons of meaning in order to understand the Earth as more than a mere resource. Obviously, this concept is foreign to the prevailing idea of economic exploitation. However, the hermeneutical task is precisely to *take ownership of that which is foreign* (Gadamer 2000); i.e., to appropriate a symbol of high significance that is *foreign* or *strange*. In this case, to translocate the aforementioned meaning of *land-resource* into the symbolism of *biosphere*. To appropriate the new meaning will imply a recognition of *the other*, simultaneously implying a transformation of self-understanding.

15.2.3 The FEP Methodology

The FEP methodology has defined four steps (Rozzi et al. 2012, p. 234): (1) Interdisciplinary ecological, ethno-ecological, and philosophical research. This was the first point expressed in this essay to highlight the need for linking ecological sciences and ethics, in order to achieve a deeper concept of Earth stewardship. Such diversity of disciplines, cultures, perceptions, and values generate the need to move to the next step: (2) Composition of metaphors and narratives through communication. Hermeneutics guides the creation of comprehensive metaphors that can help to reassess the dignity of the Earth. If this is fully understood, then indiscriminate exploitation will limit itself and will give way to other forms of relationship with the biosphere. FEP's cycle is completed with two further practical steps: (3) Field activities guided with an ecological and ethical orientation, and (4) Implementation of areas for in situ biocultural conservation.

15.3 Hermeneutical Steps that Support the FEP Methodology

We should not forget a central purpose of the FEP hermeneutical itinerary for Earth Stewardship: to persuade the neoliberal mindset to consider the *land* as *biosphere*, in order *to foster respect for the planet Earth*. This requires transposing the sign "land" into the symbol "biosphere," through the new meanings generated by the metaphor. Metaphor, as Ricoeur has shown, has an instructive value: "to instruct by means of an unexpected relationship between things that seem totally unrelated at first" (Ricoeur 2001, p. 52).

15.3.1 Embracing One's Own Perspective

The hermeneutical journey begins with embracing our own perspective. That is, to recognize our own horizons of meaning. To recognize that every subject belongs to a historical context and to a tradition, which is conserved and transmitted through one's own language. Along with the language, one has a set of pre-conceptions. These pre-conceptions should not be condemned, but understood as anticipated ways of seeing, feeling, thinking, and acting in the world. Hermeneutics does not wish to change preconceptions, including prejudices, in spite of the negative evaluation that was attributed to them by Bacon and the Enlightenment. On the contrary, it values them as cores that help to understand the worldview of the one who looks at and interprets the world, because hermeneutics considers them as " ... [the] judgment that is formed before the final validation" (Gadamer 2000, p. 337). In other words, pre-conceptions are not judgments that lack foundations and are therefore false; instead, they are judgments whose content is useful to evaluate.

Preconceptions, including prejudices, are the lenses through which we view the world, and through which a tradition is constituted. They are the bone that has formed the mode of being of a person. Consequently, they are the key to discovering what and why something is significant in an ethical attitude. The preconceptions of the individuals are the *historical reality of their being* (Gadamer 2000, p. 334).

Preconceptions or prejudices, as prior data to all experiences of the world, are the pre-structure of our understanding. They become evident only through self-reflection or when confronting *otherness*, which obligates subjects to exit their own conceptual parameters. Confronting an anomaly, one discovers something in herself or himself: the possession of prejudices and the awareness that the world is interpreted through the lenses of one's own horizon of meaning. For example, an anomaly in the flora of the Magellanic sub-Antarctic ecoregion in southwestern South America, which presents the singularity of having a greater diversity of non- vascular than vascular plant species (which is the global standard), stimulated an ecologist, such as Ricardo Rozzi and his research team at Omora Park, to "change the lenses to assess biodiversity richness" and focus on the little non-vascular plants instead of the large vascular plants for defining conservation priorities (Rozzi et al. 2008b).

If the horizon is "the field of view that covers and encloses all that is visible from a given point" (Gadamer 2000, p. 372), then the horizon of meaning defines the world. It even defines the selfhood of the person. In this way, the cultural habitat also determines the habits of the co-inhabitant; in turn, this habit influences the natural and cultural habitat. Therefore, "the horizon of meaning is not fixed but it is in perpetual motion" (Gadamer 2000, p. 337). That is, the interests determine what is to be seen and, in turn, what is seen recreates the interests. For example, the prevailing interest within the sciences of biological conservation has been the vertebrate fauna and the vascular flora. However, by undertaking their own perspective grounded in the Magellanic sub-Antarctic ecoregion, Rozzi and his colleagues needed a change of lens with regard to the traditional perspective of conservation sciences.

In broader terms, every being is exposed within a habitat formed by other co-inhabitants (Rozzi 2013). Part of this *biophysical* and *symbolic -linguistic* habitat is tradition. In the case of extractive economy, the change of perspective has been toward a tradition that changed the vision from a habitat (or community of co-inhabitants) toward an object to be exploited. For this reason, Lynn White (1967) insisted on the need for an ethical shift from anthropocentrism to ecocentrism. That is, to focus the attention on the habits that co-inhabitants have rooted as a tradition in their habitats. And toward this end, with hermeneutics they will liberate their own horizons of meaning from the pretension of exclusive truth, by testing their own prejudices.

To recognize preconceptions and prejudices, and to see beyond them, enables a valuation of the mode of conceiving things. Thus, by giving attention to the way of viewing what is outside ourselves, we can perceive our own horizon of meaning. As Gadamer has concluded (2000, p. 337), this is the context of significant historical understanding. That is, the attention should be directed toward the *subject that interprets* reality. Humans project their prejudices when they face any foreign affair, and they show themselves in these prejudices. The aim is not to discredit their prejudices or the tradition to which they belong. Instead, the aim is to gain awareness about their own prejudices in order to stop considering them as "the only truth." In this way, it is possible to keep them restrained and to be aware of them. This contributes to avoiding reductionist positions.

15.3.2 Assuming a Historical Perspective

The second hermeneutical step is to *assume a historical perspective*: the subjects that interpret the text must understand themselves from traditions of thought and from the interests of the historical period in which they are immersed.

Now, what would be an understanding of the biosphere that does not reduce it to mere "soil" and something to be traded in the market? To answer this question it is necessary to take into account two dimensions: the habitat (including the biophysical dimension as well as traditions, history and language) and the interpreter who is a co-inhabitant (cfr. Rozzi 2013).

The habitat guides the way we see, think, and act in the biosphere, and generates the habits that establish and unite with tradition. The interpreters or co-inhabitants, maintain their habits while remaining immersed in their traditions. If these two dimensions are not explicitly taken into account, then the interpreting subjects will be blind to the perception from which they have the opportunity to recognize themselves while perceiving.

By assuming a historical perspective, different historical contexts can be distinguished and understood (Gadamer 2000, p. 267). This is a necessary hermeneutical distance to separate ourselves from prejudices, without pretending to delink the interlocutor from their own particular historical contexts. Without assuming the historical perspective, the speakers fall into a monologue; that is, into a trap that prevents their paying attention to their own way of viewing reality.

The true meaning of things can only be found with hindsight and historical perspective. Thanks to this hindsight and historical perspective it is possible to "solve the real critical issue of hermeneutics, to distinguish the *true* prejudices [or preconceptions] under which we *understand*, from the *false* prejudices that cause misunderstandings" (Gadamer 2000, p. 369). That is, true prejudices are nothing but an anticipated way in which we intend to understand the world. It is thanks to this categorization that we know in truth. In contrast, false prejudices are anticipations that distort the data that we get from reality. Awareness of these anticipations, not the prejudices themselves, allows the evolution from pre-understanding toward understanding. In this way, tradition is highlighted as a worldview and not as an ultimate criterion of truth. That is, the habit of understanding the world is visualized as if it were the habit of another self –another co-inhabitant, in the terms of biocultural ethics (cfr. Rozzi 2013). As a consequence, the dialoguing partners of various worldviews will discover that when they talk among themselves or confront each other, they are not facing a foreign situation, but are all immersed *in* it. Only then will the conditions for the fusion of horizons exist.

15.3.3 The Fusion of Horizons of Meaning

The fusion of horizons of meaning *assumes (reorders and limits) identity from difference*. This last step implies that humans who exploit the "Earth" in the utilitarian sense now accept that they are part of the biosphere in the ecological sense, and that they and the biosphere are equals. From the perspective of FEP and the biocultural ethic the biosphere is understood as a community of co-inhabitants. From this understanding the implications for respect, restraint and conservation are derived.

Persons are more responsible for others than for themselves when they take historical distance from themselves and perceive their way of understanding the biosphere and the links to the other. And that greater responsibility for the other mandates self-restrains before imposing bans on others. That is, if there is no recognition of your own and the identities of others, then it will not be possible to value differences. It is from this point of view that equality (not the identification or the equivalence) of dignity among diverse beings should be applied.

This hermeneutical contribution is not exclusive to FEP, but it has been approached in various schools of environmental thought. For example, in the systemic proposal of "deep ecology," Arne Naess (1995) has essayed a synthesis inter-relating modes of knowledge and ecological life. From another perspective, the environmental philosopher Baird Callicott (1994) has analyzed critically the reductionist and economic interpretation of *Genesis*, and has proposed a comparative analysis of multiple ecological worldviews and ways of relating to nature to overcome the economic model that imposes its presumed objectivity.

Confronted with the reification of the biosphere as an *Earth to "fill and subdue"* we need to proceed with precision: recognize that equality does not imply being identical because there are beings with different levels of biological organization and diverse biotic processes. Thus, as co-inhabitants in the same *biospheric habitat,* we are equal but not identical. For this reason, hermeneutics is dialogic; "We define our identity always in dialogue with, sometimes in struggle against, the things our significant others want to see in us" (Gadamer 2000, p. 53). The need for dialogue is because equality is not assimilation or identification with the other, but it is "moving" in the *logos*.

"Moving" towards new horizons of meaning begins with linguistic translocations: new views are adopted by adopting new words, forms of inhabitation, and habits of others who share the same habitat. That experience leads to recognizing ourselves as co-inhabitants and to recognize the other in the same condition or habitat, but not as an identical inhabitant. This leads to understanding others as more than objects of inquiry, and therefore avoids reifying them. Equal status, preserving the dignity of each being, implies a privilege and a responsibility that grants rights and obligations, and preserves identity by posing a "new citizenship" (Aguirre 2012). It is based on a citizenship that is responsible with ecological justice, especially with "inter-specific justice" grounded on the principle of biospheric hospitality for other living beings (Lecaros 2013).

"Moving" toward inter-specific justice and biospheric hospitality is not simply engaging in a conversation to understand the dialoging partner or to overcome differences through agreements; but rather to follow the others in their projects. Not only to allow their projects, but also to foster them. To put it in a few words with the well known Heideggerian words: the other is not in an indifferent state of being thrown around ("Geworfenheit"), merely being there, and no more. Instead, the other is a *pro-ject*, living with a purpose that adds appreciation to the moral value already present in itself in its own existence. In ecological terms this implies recognizing any "other" as part of the processes and levels of biological organization in which all living and non-living beings are embedded.

The co-inhabitant "other," with equal rights to co-inhabit, although not identical in habits, requires a recognition that shifts boundaries and horizons. For this reason, we never should close the paths to what the "other" has to say. In this vein, Aldo

Leopold (1949), a central figure in US environmental ethics, with his well-known metaphor invites us to "think like a mountain."

In the fusion of horizons we aim to not only understand *an object*, but we want to understand *the other*, precisely *as a non-object*, as a dialogue partner and interpreter –with the aim of not reifying the other. The expectations about the *other* do change, and these changes open ways to discoveries and encounters, not to reductionism.

15.4 A Latin American Approach to Integrate Ethics and Ecology into Earth Stewardship

The International Long Term Ecological Research (ILTER) network is fostering an understanding of "humanity as complex, self-organized, multi-level, and highly integrated socio-bio-physical entities" (Maass and Equihua in this volume [Chap. 14]). As Manuel Maass and Miguel Equihua show, this requires a new ontological paradigm, new epistemological tools, and transdisciplinary research approaches. In this ILTER context, the notion of "natural resources" should shift toward the notion of "biosphere," a shift that might be catalyzed the FEP methodology. It may seem contradictory to assign intrinsic value to non-human persons that provide services to humans, or to appreciate the value of "in-itself" of realities that are not alive but are interrelated with living beings. However, it is fully possible. For instance, in the transition from LTER to LTSER (Long Term Socio- Ecological Research) networks, cross-cultural analysis show that many cultures assign intrinsic value to living and anon-living beings. It is important to note that many cultures value the bodies of deceased ancestors and animals, due to their relationship with humans. It is clear that this value assignment is not of biotic dependency. Similarly, other non-living and intangible realities--uch as climate, social and cultural factor--are recognized as having a significant impact on matters which have a direct or indirect link with living organisms. Thus, research at the Chilean LTSER has disclosed how these realities are deemed to be "living" entities (Rozzi et al. 2012, p. 234). Similar bio-cultural links between living and non-living realities are also present in Andean ecological worldviews (see Sarmiento and Mamani-Bernabé in this volume [Chaps. 5 and 6]). Hermeneutically and analogously it can be said that climates, cliffs, soils, societies, landscapes, and other entities are "alive".

Analogously, in medical sciences a whole universe of inorganic realities axiologically receives the adjective of "healthy or healthier." For example, the adjective healthy is applied to a job as much as to a resting period, to an exercise as much as to a moment of bed rest. Further, for purposes of diagnosis and prescription the adjective healthy is applied to food as well as to excrements. Is it the case that inanimate and intangible things are actually alive and healthy, or that these things can die or get sick? A suitably hermeneutic-analogical perspective of the biosphere can include this concept through the methodological approach of FEP, and thereby provide a contribution to the Earth Stewardship initiative.

From a Latin American perspective we can cite more examples of metaphorical uses that enable a new understanding of the biosphere as a reality in which we are not only immersed, but a reality that we should also respect. An example can be found at the Bioethics Research Institute of Monterrey (*Instituto de Investigaciones en Bioética de Monterrey*), Mexico, which is supporting a research program about the La Silla River, metaphorically called the "Last Living River" (Canales 2014). It is called the "Living River" because it is the only river in the metropolitan area that still bears a continuous water flow, and whose ecosystem barely "survives" due to changes in the course of the river, non-planned settlements, and industrial discharges. The research aims to identify and describe the violation of moral values and principles, based on the premise of equal dignity among all beings. The aim of this transdisciplinary approach is to fuse the horizons of meaning of bioethics with the horizons of meaning of sustainable development. Another example of a transdisciplinary approach can be found in southern Chile at the Omora Ethnobotanical Park, which using the FEP methodology has proposed the metaphor of the "miniature forests of Cape Horn" and the activity of "ecotourism with a hand-lens." Both the metaphor and the field activity provide visitors with a language and a guided field experience that allow them to appreciate the ecological, aesthetic, economic, and ethical values of the luxuriant diversity of mosses and lichens in the Magellanic sub-Antarctic ecoregion (Rozzi et al. 2008b).

These Latin American experiences offer a transdisciplinary dialogue and partnership, which contributes to an intercultural Earth stewardship (Rozzi et al. 2012). The best service of hermeneutics is the recognition of horizons of belonging among members of different cultures, to achieve their fusion. According to the philosophy of Bryan Norton (1991), the best argument for protecting biodiversity is to underscore the value biological species and ecosystem processes. However, this cannot be achieved without the deeper understanding given by metaphors, thanks to which it is possible to understand that non-biologically alive realities are actually alive in an axiological sense, or that non-human persons can have dignity as much as human persons in an ethical sense. Toward this end, FEP and other Latin American initiatives highlight the need to proceed with persuasive metaphorical translations of meanings, and with poetic symbolization. In this way, those with reductionist positions not only will realize that different stakeholders are in the same boat, but also that the boat and the waters where we sail must be in the minds and the hearts of all participants.

References

Aguirre J (2012) Hermeneutic contributions to the citizen project. Archiv für Rechts- und Socialphilosophie 98(4):535–543

Callicott B (1994) Earth's insights: a survey of ecological ethics from the Mediterranean basin to the Australian outback. University of California Press, Berkley/Los Angeles

Canales E (2014) Río La Silla: implicaciones bioéticas de los impactos ambientales sobre el último río vivo de la zona metropolitana de Monterrey. Investigación doctoral del IIB, Monterrey

Chapin FS III, Pickett STA, Power ME et al (2011) Earth stewardship: a strategy for social-ecological transformation to reverse planetary degradation. J Environ Stud Sci 1:44–53
Costanza R, d'Arge R, de Groot R et al (1997) The value of the world's ecosystem services and natural capital. Nature 387:253–260
Gadamer H (2000) Verdad y Método I. Sígueme, Salamanca
Gare A (1998) McIntyre, narratives and environmental ethics. Environ Ethics 20:3–21
Klaver I (2014) Environment imagination situation. In: Rozzi R, Pickett STA, Palmer C et al (eds) Linking ecology and ethics for a changing world: values, philosophy, and action. Springer, Dordrecht/Heidelberg/New York/London, pp 85–107
Lecaros J (2013) La ética medio ambiental: principios y valores para una ciudadanía responsable en la sociedad global. Acta Bioethica 19(2):177–188
Leopold A (1949) La Ética de la Tierra. Ambiente y Desarrollo (2007) 4:29–40
Naeem S (2013) Ecosystem services: is a planet servicing one species likely to function? In: Rozzi R, Pickett STA, Palmer C et al (eds) Linking ecology and ethics for a changing world: values, philosophy, and action. Springer, Dordrecht/Heidelberg/New York/London, pp 303–321
Naess A (1995) The shallow and the deep long-range ecology movements: a summary. In: Deep ecology for the 21st century. Shambahala Publications, Boston
Norton B (1991) Environmentalists toward unity. Oxford University Press, New York
Poole AK, Hargrove EC, Day EP et al (2013) A call for ethics literacy in environmental education. In: Rozzi R, Pickett STA, Palmer C et al (eds) Linking ecology and ethics for a changing world: values, philosophy, and action. Springer, Dordrecht/Heidelberg/New York/London, pp 349–371
Ricoeur P (2001) La metáfora viva. Ediciones Cristiandad y Trotta, Madrid
Rozzi R (1999) The reciprocal links between evolutionary-ecological sciences and environmental ethics. BioScience 49:911–921
Rozzi R (2012) Catalyzing an interregional planetary dialogue on environmental philosophy. Environ Ethics 34:341–342
Rozzi R (2013) Biocultural ethics: from biocultural homogenization toward biocultural conservation. In: Rozzi R, Pickett STA, Palmer C et al (eds) Linking ecology and ethics for a changing world: values, philosophy, and action. Springer, Dordrecht/Heidelberg/New York/London, pp 9–32
Rozzi R, Massardo F, Anderson C, Heidinger K, Silander J (2006) Ten principles for biocultural conservation at the southern tip of the Americas: the approach of the Omora Ethnobotanical Park. Ecol Soc 11(1):43. http://www.ecologyandsociety.org/vol11/iss1/art43/
Rozzi R, Arango X, Massardo F, Anderson C, Heidinger K, Moses K (2008a) Field environmental philosophy and biocultural conservation: the Omora Ethnobotanical Park educational program. Environ Ethics 30(3):325–336, [en español: "Filosofía ambiental de campo y conservación biocultural: el programa educativo del Parque Etnobotánico Omora." Environ Ethics 30(S3):5–128]
Rozzi R, Armesto J, Goffinet B et al (2008b) Changing lenses to assess biodiversity: patterns of species richness in sub-Antarctic plants and implications for global conservation. Front Ecol Environ 6:131–137
Rozzi R, Anderson CB, Pizarro JC et al (2010) Filosofía ambiental de campo y conservación biocultural en el Parque Etnobotánico Omora. Revista Chilena de Historia Natural 83:27–68 [supplementary materials]
Rozzi R, Armesto JJ, Gutiérrez J et al (2012) Integrating ecology and environmental ethics: earth stewardship in the southern end of the Americas. BioScience 62(3):226–236
Sagoff M (1995) Carrying capacity and ecological economics. BioScience 45(9):610–620
Scott R, Liddell HG (1996) A Greek-English Lexicon. Oxford Press, New York
Stuart Mill J (1978) Principios de Economía Política. F.C.E, México
Van Dongen WFD, Maldonado K, Sabat P et al (2010) Geographic variation in the repeatability of a personality trait. Behav Ecol 21:1243–1250
White L (1967) The historical roots of our ecological crisis. Science 155:1203–1207

Chapter 16
Arts and Humanities Efforts in the US Long-Term Ecological Research (LTER) Network: Understanding Perceived Values and Challenges

Lissy Goralnik, Michael Paul Nelson, Leslie Ryan, and Hannah Gosnell

Abstract Calls for interdisciplinary approaches to environmental problem-solving are common across the biophysical and social sciences. Recently, some of these collaborations have incorporated the creative arts and humanities, including projects across the 24 sites of the US Long-term Ecological Research (LTER) network. A substantial body of artistic and written work has been produced by LTER-affiliated sites. However, there has been no systematic analysis of this work. We used a cross-site, social scientific analysis to understand the extent and nature of arts and humanities inquiry in the LTER network and to assess perceptions about the values and challenges associated with it. We found that 19 of the 24 LTER sites agree or strongly agree that arts and humanities inquiry is important and relevant for the sites. Perceived values of this work include its goodness in and of itself, as well as its ability to foster outreach and public involvement and to inspire creative thinking. Contrarily, participants identified funding, available labor, and available expertise as limiting factors in the growth of arts and humanities inquiry in the LTER network. Respondents highlighted themes relevant to the relationship between ecological science and ethics, including participants' willingness to accept fostering empathy, an identified value of arts and humanities inquiry, as pertinent to LTER network goals and research on some level. This ethical potential of arts and humanities inquiry in the LTER network provides an opportunity to bridge ecological research with arts and humanities inquiry in ways that are meaningful for Earth stewardship.

Keywords Empathy • Ethics • Ecology • Interdisciplinary • Intrinsic value • Place-based

---

L. Goralnik (✉) • M.P. Nelson • L. Ryan
Department of Forest Ecosystems and Society, Oregon State University, Corvallis, OR, USA
e-mail: Lissy.goralnik@oregonstate.edu; mpnelson@oregonstate.edu; leslie.ryan@oregonstate.edu

H. Gosnell
Geography, Environmental Sciences, and Marine Resource Management, College of Earth, Ocean and Atmospheric Sciences, Oregon State University, Corvallis, OR, USA
e-mail: gosnellh@geo.oregonstate.edu

R. Rozzi et al. (eds.), *Earth Stewardship*, Ecology and Ethics 2,
DOI 10.1007/978-3-319-12133-8_16

16.1 Background

Calls for interdisciplinary approaches to environmental problem-solving are common across the biophysical and social sciences (Lubchenco 1998; Klein 2004; Nisbet et al. 2010; Sörlin 2012). Recently, some of these collaborations have included the creative arts and humanities. The US National Science Foundation (NSF) sponsored an extended art-science workshop at San Francisco's Exploratorium and a joint workshop with the National Endowment for the Arts to develop a national agenda for art-science collaboration (Malina 2011; Harrell and Harrell n.d.). Agencies and institutions as varied as the United States Geological Survey, the Joint Fire Sciences Program, the European Organization for Nuclear Research (CERN), the European Science Foundation, and the NSF-funded Long-Term Ecological Research (LTER) network are inviting artists and humanities scholars to participate at their sites and, on occasion, in their research.

16.2 Arts and Humanities in the US LTER Network

The LTER network has made a commitment to a "culture of collaboration" (Collins et al. 2007; Carpenter et al. 2007). While this initiative specifically targets social science inquiry, the interdisciplinary focus creates space for other disciplines as well. The twenty-four sites of the US Long Term Ecological Research (LTER)[1] network represent an array of biomes, from conifer forests to grasslands, tundra to coral reefs. Goals of the network include:

> the study of phenomena over long periods of time [and] significant integrative, cross-site, network-wide research….[to] provide the scientific community, policy makers, and society with the knowledge and predictive understanding necessary to conserve, protect, and manage the nation's ecosystems, their biodiversity, and the services they provide. (http://www.lternet.edu/network/)

The LTER network conducts ecological research at broad spatial and temporal scales that contributes to understanding, conservation, protection, and management across ecosystems.

The first documented arts and humanities interactions in the LTER network were writer's residencies in the H.J. Andrews Experimental Forest in 2002, hosted by the Spring Creek Project for Ideas, Nature, and the Written Word at Oregon State University. Several sites have since developed arts and humanities programs, including Harvard Forest in Massachusetts, Bonanza Creek in Alaska, and North Temperate Lakes in Wisconsin. In 2010 these sites and others joined to form Ecological Reflections, an informal collection of venues that host science and art interactions (http://www.ecologicalreflections.com/). A substantial body of artistic

[1] In this chapter, the Long Term Ecological Research (LTER) network refers to the network of 24 sites funded by the US National Science Foundation (NSF) in the United States of America.

and written work has been produced by affiliated LTER sites, examples of which have been displayed at: NSF headquarters in Washington DC in 2013; the 2012 Ecological Society of America meeting in Portland, Oregon; the 2012 LTER All-Scientists Meeting in Estes Park, Colorado; and in galleries across the country, as well as published in *Orion* and *Terrain.org*.

16.3 Survey

To date, however, there has been no systematic analysis of the arts and humanities work emerging across the LTER network. Therefore we have employed a cross-site, social scientific analysis to understand the extent and nature of this work and to assess perceptions about the values and challenges associated with it. In May 2013 we received a grant from the LTER Network Office to explore three guiding questions:

1. What kind of arts and humanities inquiry exists across the Network and where is it taking place?
2. What is the perceived value of this work?
3. What are the perceived challenges to maintaining or further developing arts and humanities inquiry across the LTER Network?

In August 2013 we sent all 24 LTER Principal Investigators a Qualtrics online survey (http://www.qualtrics.com/), and encouraged them to use the personnel at their site to respond. The instrument consisted of 14 Likert-scale, draggable bar, and optional short answer questions. It took the respondents between 5 and 25 min to complete. Our response rate was 100 %.

16.3.1 What Kind of Arts and Humanities Work Exists Across the Network and Where Is It Taking Place?

Through anecdotal evidence, we assumed that perhaps 50 % of the 24 LTER sites had hosted some kind of arts and humanities inquiry. When we asked participants how their site engaged with arts and humanities inquiry—*Not at all, Sporadically, Consistently*—only three sites answered *Not at all.* Already the survey was revealing (Fig. 16.1).

Twenty-one of 24 sites have engaged with arts and humanities inquiry in some way. Six sites reported hosting this type of interdisciplinary inquiry consistently, including: (i) a long-running writers-in-residence program at the H.J. Andrews Experimental Forest in Oregon, (ii) an ongoing Arts and Ecology research experience for undergraduates (REU) program at Sevilleta LTER in New Mexico, (iii) Art and Ecology workshops for public school art teachers and an artist-in-residence

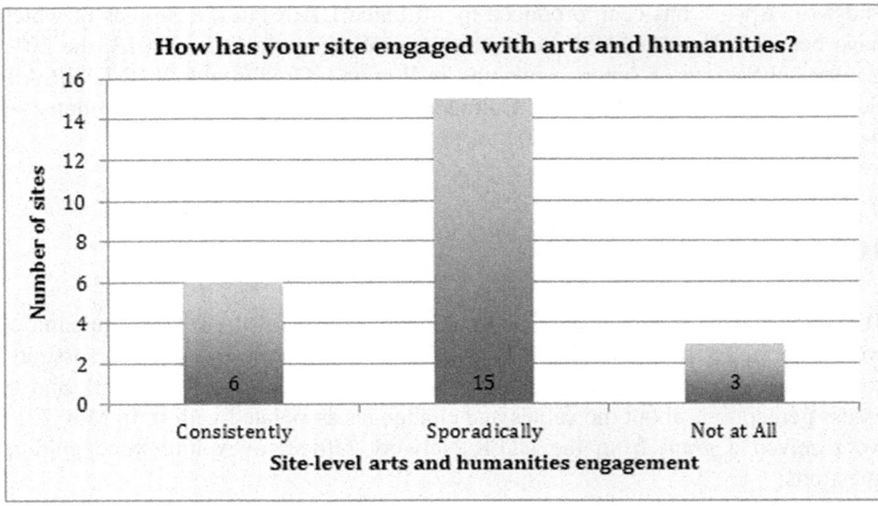

Fig. 16.1 Level of site participation with arts and humanities efforts

program at Virginia Coastal LTER, and (iv) a yearly visual and performing arts exhibit connected to Bonanza Creek LTER in Fairbanks, Alaska. Though they did not describe their programs in the survey, the other two sites that identified themselves as hosting consistent arts and humanities inquiry were (v) Baltimore Ecosystem Study LTER, where they host an artist-in-residence program and visiting artist field trips through their BES Art and Science Integration Program (BES-ASIP) and (vi) Harvard Forest LTER, where they also host an artist-in-residence program and ongoing historical research, as well as house the Fischer Museum.

We also asked respondents about the types of work their sites have hosted (Fig. 16.2). We provided lists of visual, literary, and performing art genres and offered *Other* for categories we might have missed. The most prevalent genres were: painting (15 sites), photography (10 sites), and literary prose (8 sites). Respondents also wrote-in: observational drawing, ephemeral art/meditation, electronic visual arts, participatory art/digital art, and legend/myth.

The general nature of the survey precluded some nuance in the data and this flaw suggests a potential challenge in cross-network approaches to fostering arts and humanities inquiry in the future. In an open comments section at the end of the survey one respondent wrote: "One problem with this survey … is that it mixes too many different types of scholarship, art and humanities. There is no way to broadly articulate answers to the questions above when painting, poetry, photography, film and history are all merged." The function, intent, impact, and audience of arts and humanities work varies across genres and individual participants. As sites nurture specific projects, there will be opportunities to ask how these diverse approaches can come together to tell a connected story across landscapes, similar to the way cross-site science aims to do.

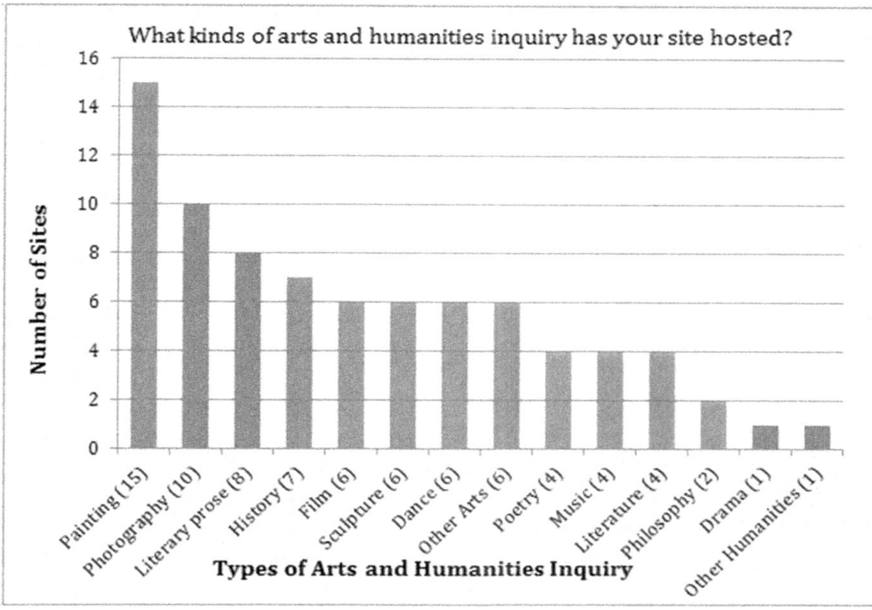

Fig. 16.2 Description of site engagement with arts and humanities inquiry by genre

In our survey, we were also interested in how participants perceive the relevance of arts and humanities inquiry within the LTER network. To answer this question, we asked participants to respond to the following statement: Arts and humanities inquiry is relevant to and important for LTER sites (Fig. 16.3).

Nineteen of the 24 sites agree or strongly agree that arts and humanities inquiry is important and relevant for the sites. No site disagrees.

16.3.2 What Is the Perceived Value of This Work?

We used a draggable bar question, which allows respondents to rank statements between 0 and 100, to understand the relative importance of a series of potential values for arts and humanities inquiry (Fig. 16.4). This question type is useful because it is interactive and allows for relatively easy ranking of multiple items. While research on the response consistency between this and other question types is somewhat mixed (Downes-Le Guin et al. 2012), we found it a worthwhile tool to observe trends across the field of responses. To facilitate analysis, we grouped results in a 7-category Likert-style format (Figs. 16.5, 16.6, 16.7 and 16.8).[2]

[2] We implemented the Likert conversion to facilitate our analysis. Therefore the scale was not available for participants during their survey experience.

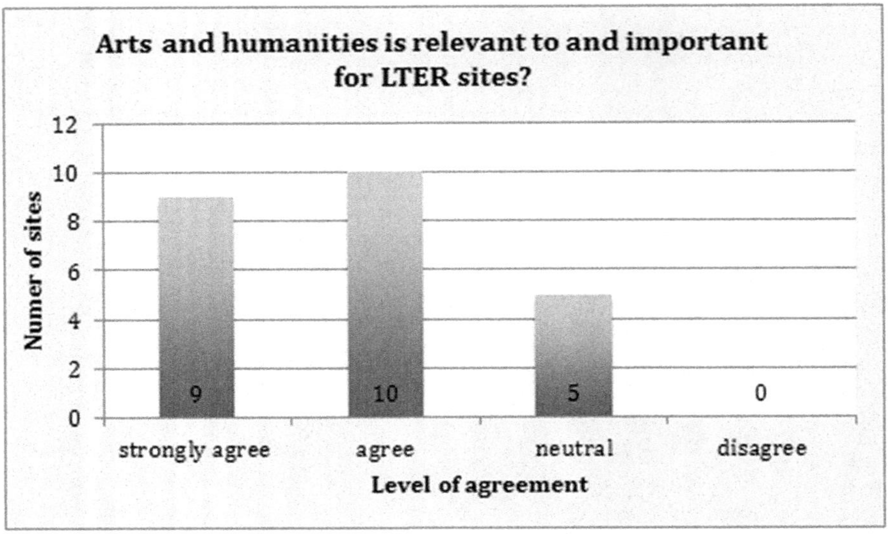

Fig. 16.3 Perceived value of arts and humanities inquiry for LTER sites

As illustrated in Fig. 16.7, respondents perceive arts and humanities inquiry to be valuable. Among the 13 values defined in the questionnaire,[3] three proposed values were ranked ≥80 % by 14 of the 24 respondents. These values included: (1) *Arts and humanities inquiry fosters outreach and public involvement*, (2) *Is good in and of itself*, and (3) *Inspires creative thinking*. Arts and humanities inquiry is also valued because it: (4) *Provides opportunities for education* and (5) *Broadens our understanding of the natural world*. Half the respondents rated these five responses ≥80 %. There were twice as many responses in the top tier of ≥80 % relative value than in the lowest tier of responses <20 % relative value.

The three least valuable perceived attributes of arts and humanities inquiry are its ability to: (1) *Play a role on grants*, (2) *Stimulate collaboration*, and (3) *Enhance site science in important ways*. Eight or more respondents rated these responses ≤20 % importance. A number of other values were close behind, including: (4) *Contributes to environmental problem-solving*, (5) *Enables interdisciplinary scholarship*, and (6) *Stimulates empathy*. Six or more respondents ranked all six of these proposed values ≤20 % relative importance.

To demonstrate whether arts and humanities inquiry actually facilitates these outcomes would require additional research. However, the perception is that the value of arts and humanities inquiry lies more in fostering education, under-

[3] Proposed values provided on the survey were determined by the researchers with consultation from several colleagues and then refined during survey development and pilot testing.

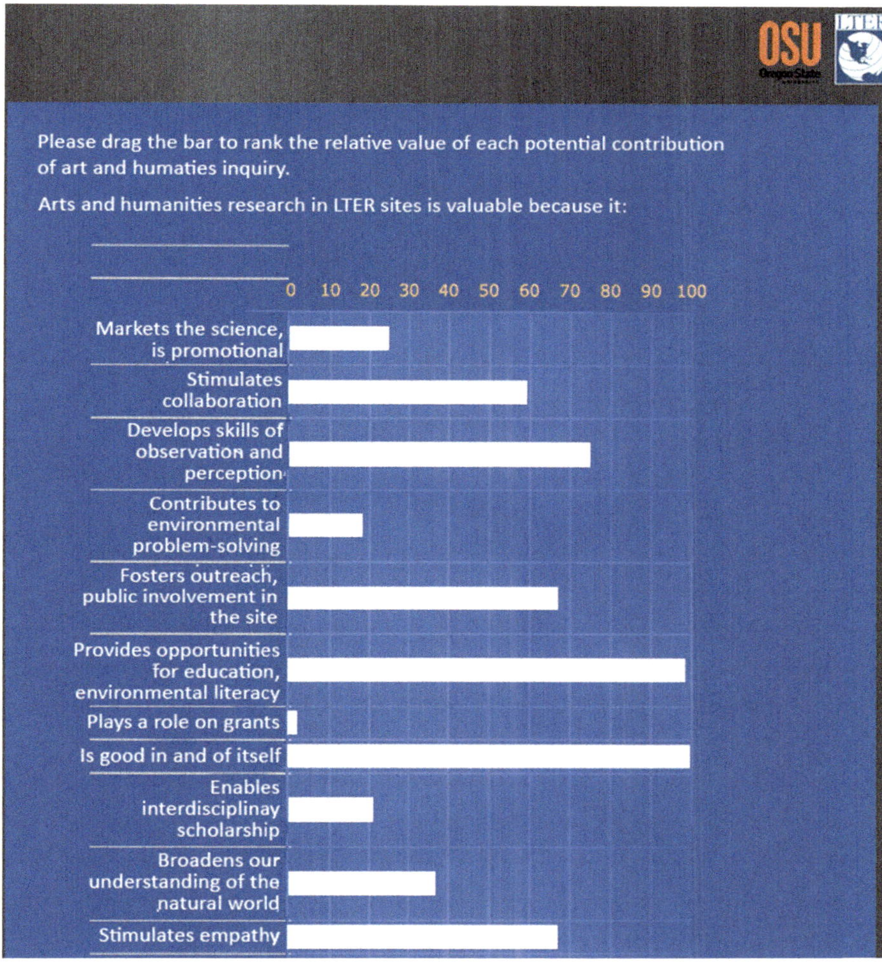

Fig. 16.4 Example of a draggable bar question

standing, and outreach than it does in enabling more traditional scientific metrics of grant-funding, collaboration, and problem-solving. Stimulating empathy stands outside these broad categories. Not only does it fall in the middle of all received responses, but its relevance to ethics—especially within a question about values—is particularly interesting, as is the relationship between participant response to empathy here and to a related question later in the survey, discussed below.

0%	Definitely Disagree
1-19%	Strongly Disagree
20-39%	Disagree
40-59%	Neutral
60-79%	Agree
80-99%	Strongly agree
100%	Definitely agree

Fig. 16.5 Likert conversion of draggable bar response values

Arts and humanities research in LTER sites is valuable because it:	
1	Markets the science, is promotional
2	Stimulates collaboration
3	Develops skills of observation and perception
4	Contributes to environmental problem-solving
5	Fosters outreach, public involvement in the site
6	Provides opportunities for education, environmental literacy
7	Plays a role on grants
8	Is good in and of itself
9	Enables interdisciplinary scholarship
10	Broadens our understanding of the natural world
11	Stimulates empathy
12	Enhances our science in important ways
13	Inspires creative thinking

Fig. 16.6 Provided responses for perceived value of arts and humanities research in LTER sites

16.3.3 What Are the Perceived Challenges to Maintaining or Further Developing Arts and Humanities Inquiry Across the LTER Network?

In addition to understanding the perceived value of arts and humanities inquiry, we also wanted to know about perceived challenges. Toward this end, using a draggable bar question we asked respondents to rank eleven provided challenges that may impact the integration of arts and humanities inquiry in LTER sites[4]; we also offered an *Other* category for challenges we did not anticipate (Fig. 16.9).

[4] The list of proposed challenges were determined by the researchers and refined during survey development and pilot testing.

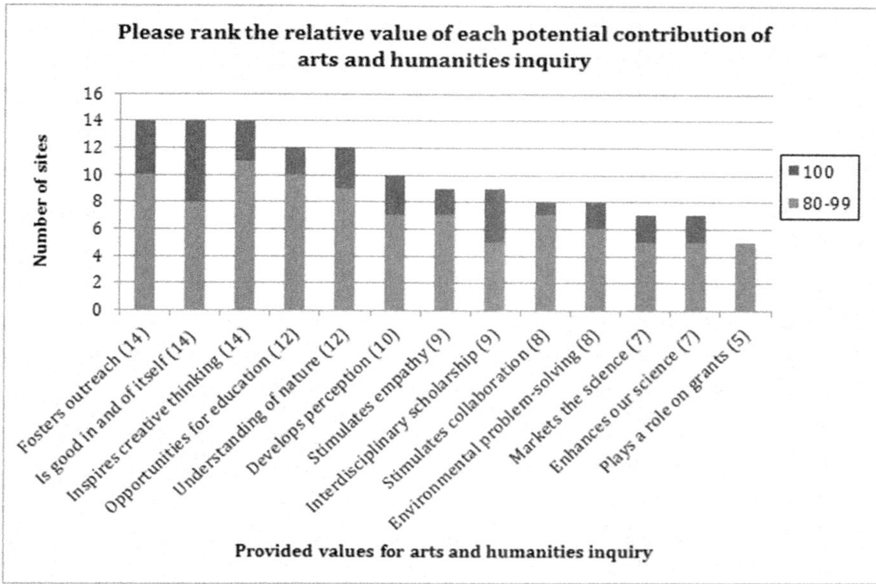

Fig. 16.7 Perceived value responses ranked >80 % relative value

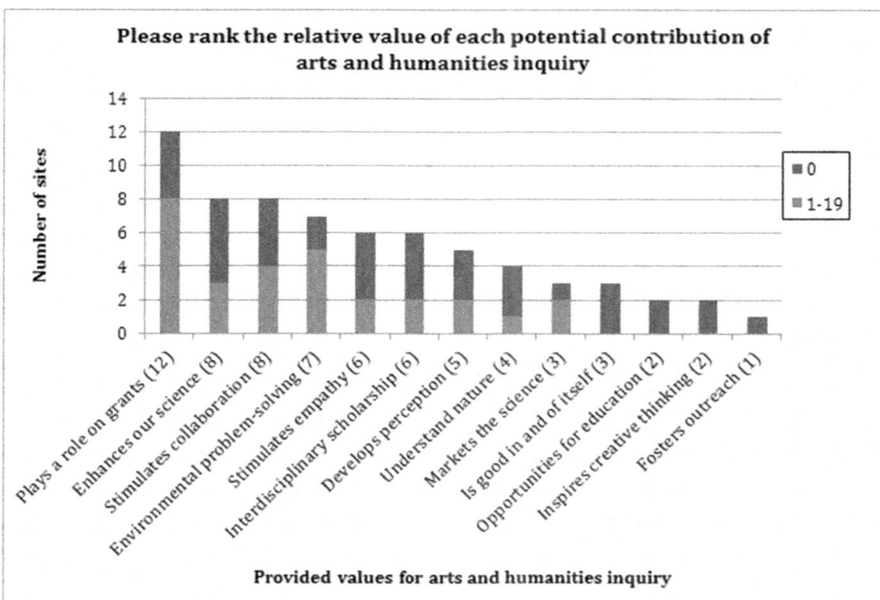

Fig. 16.8 Perceived value responses ranked <20 % relative value

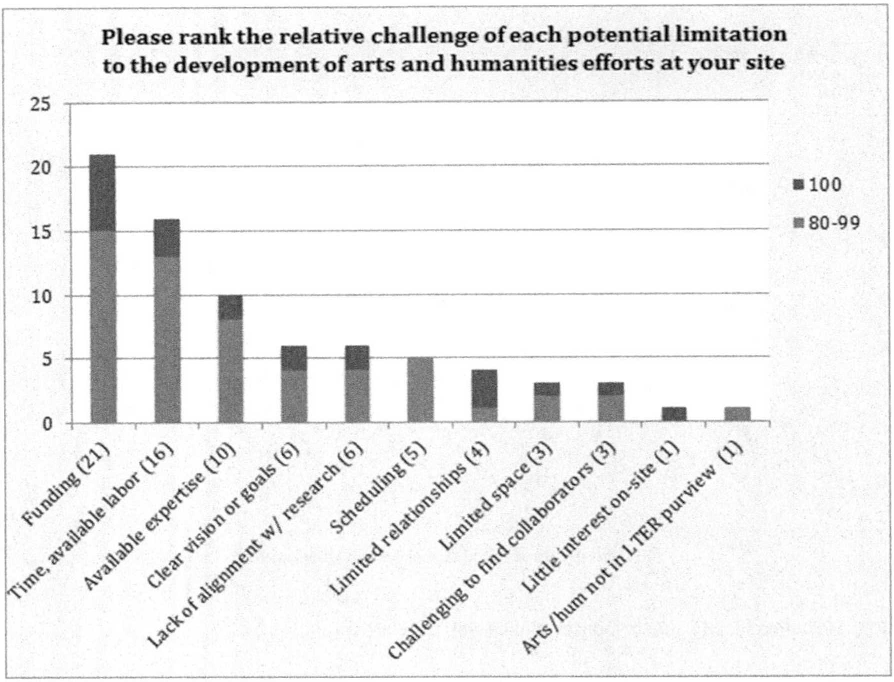

Fig. 16.9 Perceived challenge responses ranked >80 % relative value

Participants consistently ranked three challenges—*Funding, Time or Available Labor*, and *Available Expertise*—much higher than the other responses. We expected these responses to top the list, both because LTER sites are grant-funded and research requires input, including money and labor, and also because these are standard responses to limitations in academic settings. This result, though, is important, because with the right resources, these perceived hurdles could also be addressable.

Challenges seem to fall into three general categories: practical, logistical, and ideological. Practical challenges include funding, available labor, and expertise; these kinds of challenges primarily involve external resources. Logistical challenges include internal resources and procedure, like site space and scheduling. Both these categories of challenges present obstacles that are potentially surmountable. Grants exist, experts are available, schedules and appointments can be restructured. Ideological challenges, however, are more difficult to address. If sites are just not interested in this work, then the work has little future. Only two sites ranked the ideological challenge "Do not view arts and humanities within the purview of LTER sites" with a relative value ≥80 %. This particular challenge ranked lowest of all 11 responses. Therefore, the majority of LTER sites perceive the challenges presented by arts and humanities inquiry to be primarily practical and logistical.

Only one participant answered *Other*, but ranked it at 91 %, therefore it deserves mention. The respondent wrote that arts and humanities inquiry does not: "Match with mission as defined by university administration," suggesting that universities themselves might stand in the way of these kinds of broad interdisciplinary collaborations. This statement raises a crucial question: If this work is not done at universities, then where might it be done? The amount of freedom sites have to nurture projects may be limited by constraints outside their control, including university administration, LTER grant protocol, network research agendas, or the research foci of their site. Welcoming arts and humanities scholars to implement independent projects is a different proposition than diverting resources to support this inquiry, and some of the sites do not believe they have the freedom to do the latter.

These challenges were echoed in an open comments section at the end of the survey. We received written responses from 17 of the 24 participants, nearly all of whom think that arts and humanities inquiry in the LTER Network is a good idea. But a number of recurring themes illuminate why the future development of arts and humanities inquiry in the LTER Network will require more than just the belief in a good idea.

The most consistent refrain in the open comments pertained to funding: "If funds were available, we could strengthen collaborations with visiting and local artists," wrote one site. Another explained: "We would like to curate and present this work at … the national level but funds are non existent." A third emphasized that they would be interested in this work "only if it pays for itself." These responses, coupled with an equally consistent refrain about a lack of labor, parallel responses from the survey. A remote site explained: their limited arts and humanities engagement as due to "extreme constraints on space, time, and effort that we can devote to ANY activity." The demands of current workloads lead to "sporadic rather than sustained" efforts. One site offered that they would be interested in this work if it "does not lead to further diffusion of our already too-diffuse efforts."

Participants also discussed the absence of a "clear vision or goals," which was tied as the fourth highest ranking challenge on the survey. Some sites have broad ideas about program expansion or project development but lack vision or expertise. One site has "interest in taking that next step and doing work that is truly synthetic, though it's not clear to any of us what that means, looks like, or what would facilitate that kind of work." Another repeats this sentiment: "At the site level there is interest, but we could strongly benefit from guidance/leadership from those with greater experience and a clearer vision of what role the humanities can actually play in research beyond appeal to the senses." Several sites are just initiating engagement, while others are "in the process of thinking these important issues through right now" or "have started a working group." There is forward motion, but the common direction is undeveloped.

Another hurdle identified in the open comments section is a lack of relationships with artists or humanists, which did not rank highly on the survey, tying with "limited space" for third to last of the provided challenges. "We have reached out to artists/humanists with varied results," one site shares. Another explains that they are "grappling a bit with … finding the right relationships. There are a lot of artists working on urban and environmental issues in our ecosystem. The trick for us is to how to engage them in our work—what is in it for them when they appear to be doing quite well on their own." This feedback loop between the art and the science, and a clear picture of how they might inform each other, is a persistent theme that would benefit from further thought. "We have sporadically tried to connect with artists/humanists with variable success" one site explains. "Distance is a problem, but also conceptual barriers on both sides as to what the inherent value and or intent is of the effort is." Both physical and ideological distance can present a barrier. But sites appear to realize the potential benefit of these relationships for their understanding of their particular site. One site shares that there is a "rich history of art involving nature which we would like to connect to."

Despite these challenges, participants were enthusiastic about the development of arts and humanities inquiry. Several sites expressed energy for a network-wide initiative to help with momentum and direction, and a number of sites described current project- and relationship-building. Our "program is developing rapidly with tremendous interest and participation from the community," writes one site. Others are "establishing an artist-in-residence program," collaborating with regional colleges to do work "involving ecology, music and visual arts," "Plan[ning] to expand current Art and Ecology workshops.....[, and] adding a Nature Writing class this winter." A number of sites plan to nurture current projects, while others intend to develop new work by seeking research opportunities, adding genres, or developing programming. This momentum creates opportunities for research on the impact and effectiveness of this work, potential collaborations between artists and scientists, and participation with the LTER network by wider and more diverse audiences.

16.4 Relevance to LTER Goals and Mission

If arts and humanities inquiry is consistent with established LTER network goals, then demonstrating its relevance and value gets easier, and so might addressing some of the logistical and practical challenges. In a draggable bar question, we provided participants a list of 12 responses related to the goals and mission of the LTER network: six (*Understanding, Synthesis, Information, Legacies, Education, Outreach*) taken directly from the "LTER Goals" on the LTER network website (LTER Goals n.d.), three (*Conservation, Communication, Environmental Impact*) using language from the "LTER Network Vision and Mission Statements" (LTER Network Vision and Mission Statements n.d.), and three (*Relationship Building, Human Dimensions, Long-term Ecological Research*) written to reflect the LTER network's commitment to place-based, long-term research (Wattchow and Brown 2011; Billick and Price 2010; Farnum et al. 2005; Kurdryavtsev et al. 2012; Cross 2001) (Fig. 16.10).

Similar to the earlier question about perceived value, respondents associated arts and humanities inquiry most closely with (1) *Outreach* and (2) *Communication*, followed by (3) *Relationship Building*, (4) *Human Dimensions*, and (5) *Education*

Please rank the relative value of each potential contribution of arts and humanities inquiry to LTER goals. Arts and humanities research contributes to or enables:	
Understanding	To understand a diverse array of ecosystems at multiple spatial and temporal scales
Synthesis	To create general knowledge through long-term interdisciplinary research, synthesis of information, and development of theory
Information	To inform the LTER and broader scientific community by creating well-designed and well-documented databases
Legacies	To create a legacy of well-designed and documented long-term observation, experiments, and archives of samples and specimens for future generations
Education	To promote training, teaching, and learning about long-term ecological research and the Earth's ecosystems, and to educate a new generation of scientists
Outreach	To reach out to the broader scientific community, natural resource managers, policymakers, and the general public by providing decision support, information, recommendations and the knowledge and capability to address complex environmental challenges
Conservation	To protect or manage ecosystems, biodiversity, and environmental services
Communication	To foster dialogue between the scientific community, policy makers, and society
Environmental Impact	To contribute toward the advancement of the health, productivity, and welfare of the global environment
Relationship Building	To develop empathetic relationships with the natural world and stimulate inspiration, awe, and wonder
Long-Term Ecological Research	To participate in studies of ecological processes that play out at time scales spanning decades to centuries, provide a context to evaluate the nature and pace of ecological change, interpret its effects, and forecast the range of future biological responses to change
Human Dimensions	To understand human drivers on natural systems, investigate the impacts of ecosystems on humans, and explore human perceptions of and attitudes about the natural world

Fig. 16.10 Provided list of LTER goals, mission, and intellectual commitments

(Fig. 16.11). While we expected *Outreach*, *Communication*, and *Education* to rank highly, we were surprised by the high ranking of both (3) *Relationship Building* ("To develop empathetic relationships with the natural world and stimulate inspiration, awe, and wonder") and (4) *Human Dimensions* ("To understand human drivers on natural systems, investigate the impacts of ecosystems on humans, and explore human perceptions of and attitudes about the natural world"). Both responses relate to human-nature relationships and they represent either new or implicit commitments of the LTER Network. However, eleven sites, nearly half, ranked these responses $\geq$80 % relative value.

Alternatively, the LTER goals and mission statements least relevant to arts and humanities inquiry, as identified by respondents, are (1) *Information* and (2) *Long-Term Ecological Research*. These are followed closely by (3) *Legacies*, (4) *Synthesis*, and (5) *Understanding* (Fig. 16.12). It is not clear if respondents see these responses as unrelated to arts and humanities inquiry, or whether instead they think LTER science already does these well, and therefore arts and humanities can (and should) contribute in different ways. It is clear that respondents identify a strong relationship between arts and humanities inquiry and several stated LTER goals, specifically *Outreach*, *Communication*, and *Education*, as well as to *Human Dimensions*

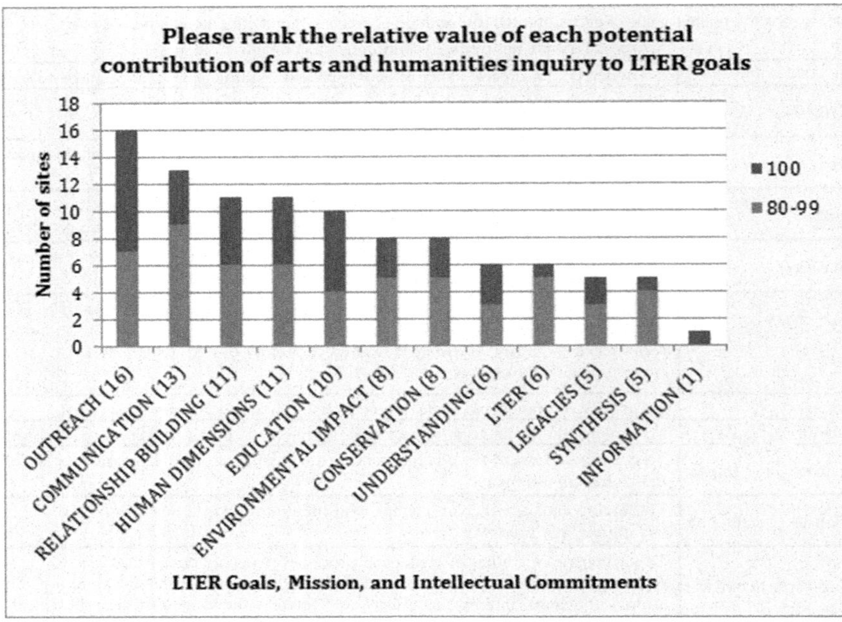

Fig. 16.11 Responses ranked >80 % relative value

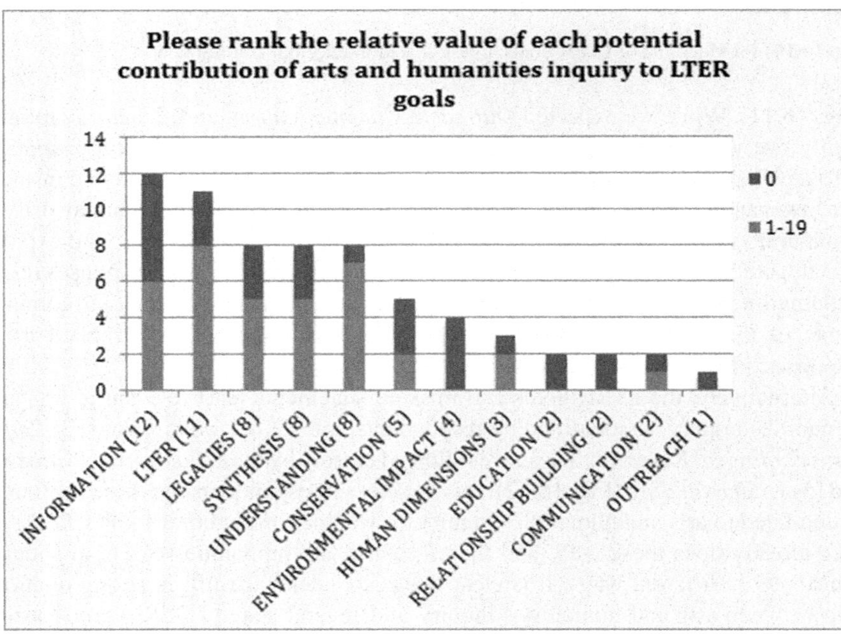

Fig. 16.12 Responses ranked <20 % relative value

and *Relationship Building*. All five of these responses were ranked ≥60 % relative value by 17 of 24 sites. This level of response seems to confirm that the arts and humanities are perceived as having a meaningful role in contributing to LTER goals and mission.

16.5 Discussion

The survey illuminated several themes relevant to the relationship between ecological science and ethics that will inform the future steps of this research. Two of the most interesting themes relate to ethically-relevant perceived values of arts and humanities inquiry.

More than half the participants, 14 of 24, ranked arts and humanities inquiry ≥80 % relative value because it *Is good in and of itself*, placing this response alongside *Fosters Outreach* and *Inspires Creative Thinking* as the three highest ranked perceived values of this work. This is a nod toward the intrinsic value of arts and humanities inquiry, distinct from its contributions to science, outreach, or environmental- or conservation problem-solving. Two of these three highest ranked values—*Is good in and of itself* and *Inspires creative thinking*—are intangible outcomes, thus not easily observed products that directly serve the science or the sites. As the LTER network guides future arts and humanities inquiry, there will be opportunities to discuss how these kinds of outcomes might be assessed or honored, so arts and humanities projects can best be nurtured in these ways and as integral elements of the LTER program.

Another ethically-relevant theme is the participants' willingness to accept fostering empathy, on some level, as relevant to LTER network goals and research. Common definitions of empathy (*Oxford*, *Merriam-Webster*) include an imaginative quality, emotion or feeling, understanding, awareness, sensitivity, experience, and an *other*, whose feelings or experience are shared. Thus empathy is described as an imaginative affective awareness of another's experience. This perspective, scholars argue (Hoffman 2000; de Waal 2006, 2009; Slote 2007) provides entrance to moral judgments and agency. When we inhabit another's point of view we are better able to understand the *other's* needs and act wisely on their behalf (see Aguirre Sala 2015 in this volume [Chap. 15]).

It is worth considering whether arts and humanities inquiry in the LTER network can (or should) be doing the work of ethical reflection and development, e.g. stimulating qualities like empathy or relationship-building with the natural world, or if this is perhaps a more desirable outcome than the consequentialist contributions of outreach or education, which relegate arts and humanities inquiry to serving as a means to an end in support of the science. These kinds of consequentialist roles might even be at odds with the strong support for the value of art and humanities inquiry as "good in and of itself."

If arts and humanities inquiry is perceived as valuable for its ability to stimulate empathy for the natural world, then it still functions as a means to an end, in this case empathetic relationships. There is nothing inherently wrong with serving as a means to an end, and this kind of indirect moral valuation does not necessarily contradict direct, or intrinsic, moral valuation, in this case valuing arts and humanities efforts in and of themselves. For example, one can value a family dog as a being worthy of direct moral consideration and also value it for the joy it brings to one's life as a pet. But the two kinds of valuation can conflict when decision-making requires prioritization. For example, if promoting arts and humanities inquiry as an educative tool requires restrictions on this work that preclude it from manifesting the characteristics we value as good in and of itself, then we must choose either to facilitate arts and humanities inquiry as a means to particular end, e.g. education, or to allow it to flourish in its own way.

Empathy as a means to an end might, however, enable a kind of middle ground. Empathetic awareness of the natural world is not an outcome that serves sites or the science directly, so the stakes are a little different than considering arts and humanities as valuable for other instrumental contributions, like education or outreach. The 'end' is a new ethical relationship with the natural world in general, not just with a specific place. This is quite different than an 'end' in the form of a product or a service. Therefore arts and humanities inquiry facilitated for the development of empathetic awareness would serve as a catalyst for sensitivity, imaginative understanding, and emotional engagement with the natural world, all of which might lead us to act wisely on its behalf. This approach seems compatible with a valuation of the work as good in and of itself.

Because empathy figures meaningfully in ecological sciences, arts, and both ecology and arts education literature, it might also provide a bridge to connect interdisciplinary approaches to long-term inquiry about place. In ecological literature empathy is often associated with natural history learning and knowledge. Scholars argue that good ecological research depends on a sensitivity to natural patterns and processes, an ability to listen to the natural world, careful description, and highly developed skills of observation. All of these qualities reflect an empathetic awareness of the natural world and are cultivated through natural history learning about place (Cooper 2000; Dayton and Sala 2001; Fleischner 2011).

The "ecology of place" (Billick and Price 2010), which is place-based long-term ecological research, much like that across the LTER network, relies on a similar appreciation of natural history. Esteemed ecologists (Pulliam and Waser 2010; Pecharsky et al. 2010; Paine et al. 2010; Krebs 2010; Louda and Higley 2010) are re-placing the value of natural history in contemporary ecological scientific progress by describing its integral role in the practical and theoretical success of their work. Understood alongside the relationship between natural history and empathy (Cooper 2000; Fleischner 2011; Dayton and Sala 2001), these ecologist-authors are affirming the importance of empathetic relationships in the facilitation of deep understanding of the natural world and in conservation practice. In essence, they are arguing for an emotional, as well as an intellectual, engagement with the natural world.

This element of emotional affiliation with the natural world connects scholarship in ecology with scholarship in the environmental arts and humanities. Emotional connections to the natural world can take many forms, but often they manifest as inspiration, awe, and wonder (Carson 1965; Moore 2005; Vucetich and Nelson 2013; Dayton and Sala 2001; Vucetich 2010), which is the language we included in our survey. This kind of relationship is accessible to all people, not just scholars, artists, or environmentalists (Dayton and Sala 2001), and it can be enabled by art (Curtis 2009). But art is capable of inspiring more than just warm and fuzzy feelings about nature.

Often when we think about emotional relationships with the natural world, we think of romantic vistas and childhood exploration. But right action on behalf of the natural world ought to be driven by the way the world is, not the way we wish it to be; the natural world is more complex than romantic notions of grandeur. Art can communicate this complexity. In addition to stimulating feelings of love and awe, "Art can also increase an emotional indignation about insufficient nature protection or can increase a cognitive interest in nature" (Reid et al. 2005 qtd. in Curtis 2009, p. 182). This cognitive interest alongside emotional investment is a catalyst for moral engagement with the natural world (Gruen 2009). Therefore art can prompt us to care about the natural world in ways that have the potential to inform action.

Based on these interconnections between awe and wonder, empathy, ecology, and the arts in the literature we included two different references to empathy on the survey. One, which we proposed as a potential value of arts and humanities inquiry, did not include a definition of empathy. We simply asked if respondents felt arts and humanities research in LTER sites is valuable because it: *Stimulates Empathy*. The second included a definition of empathy and was proposed as an LTER goal. We asked respondents if arts and humanities research contributes to or enables *Relationship Building: To develop empathetic relationships with the natural world and stimulate inspiration, awe, and wonder*. The first reference ranked in the middle tier of responses; the second reference ranked in the top tier of responses.

We were encouraged that participants did not reject notions of empathy outright. In fact, in the first question, 9 of 24 participants ranked the reference to empathy $\geq$80 % relative value, and in the second question 11 of 24 participants ranked the reference to empathy $\geq$80 % relative value. Of course, we are not sure what empathy meant to the participants in this context or how they considered it in relation to their work as ecologists. Does their acceptance of the concept suggest they see their work as related to empathy? Or do their answers suggest they see empathy within the realm of LTER goals and research? If they do consider empathy as an acceptable, even operational, element of the LTER program, do they think arts and humanities might facilitate this kind of relationship with the natural world in ways LTER science is not yet doing?

Vucetich and Nelson (2013, p. 19) describe empathy as, "A vivid knowledge-based imagination of another's circumstance, situation, or perspective." This is "a capacity that depends on objective, empirical knowledge…about the conditions and capacities of others." In many ways this definition describes the domain of ecology. The connection between ecology and empathy, filtered through the relationship between empathy and ethics (Gruen 2009; Moore and Nelson 2010), clarifies the bond between ecological research and ethics. If arts and humanities inquiry can enhance the empathetic quality of ecological work for scientists and also inspire empathetic awareness for audiences, then arts and humanities inquiry is both contributing to the work of ecology and doing work ecologists deem important. Teasing out these connections and perhaps even demonstrating them empirically could further illuminate the role of arts and humanities within the LTER Network. The open reception the respondents gave empathy on the survey, likely an unfamiliar metric for environmental inquiry, invites further work.

16.6 Implications for an Earth Stewardship Initiative

The relationship between empathy, ethics, and ecology—facilitated by long-term observation of and commitment to place—underlies the goals of the Earth Stewardship Initiative. For Earth stewardship is the effort to "respectfully cohabitate with" the planet with the goal "to maintain not only human welfare but the welfare of the whole community of life" (Rozzi et al. 2012, p. 234). This notion of community building and maintenance is central to contemporary environmental ethics (Leopold 1949; Moore 2004; Goralnik and Nelson 2011) and lies at the heart of the kind of empathetic relationship-building we discuss here.

As well, the goal of Earth stewardship is "to enhance ecosystem resilience and human well-being" (Earth Stewardship) and "to rapidly reduce anthropogenic damage to the biosphere" (Power and Chapin 2009). Certainly, such a stewardship effort demands a great deal of ecological information about the world, and across multiple scales. Ecologists and ecological networks can contribute to Earth stewardship by learning how ecosystems work and how the resilience of those ecosystems is likely to be altered in the near future. But information alone cannot deliver Earth stewardship. Stewardship is "bigger than ecology" (Power and Chapin 2009). It is as much an ethic as it is about science– a decision about how we *ought* to live in relationship to the world around us.

In order to "profoundly reorient our endeavors" we must "radically redefine our relationship with the planet" (Power and Chapin 2009, p. 399). In short, "Earth stewardship requires a new ethic of environmental citizenship" (Earth Stewardship). This kind of commitment to relationship demands work, for relationships are reciprocal, contextual, and require virtues like humility, empathy, and patience. The pursuit of Earth stewardship, therefore, logically requires a fusion of the biophysical and social sciences with the humanities (most notably with ethics). The history of ecological science is populated with leaders who opened the door to ethics, who recognized "the choices faced by human society are ethical ones, for which the ecological sciences provide essential knowledge to inform responsible societal decisions" (Rozzi et al. 2012, p. 233). As noted above, empathy is a moral framework amendable to ecology. As well, the LTER network appears amenable to the empathy framework, and therefore to at least some kind of ethical exploration of our relationship with and obligations to the natural world. Findings from our survey indicate that the important and "inevitable fusion" (to quote Aldo Leopold) of ecology and ethics – a pillar of Earth stewardship – might be realized within the LTER network.

Acknowledgements This research was supported by Long Term Ecological Research Network subaward grant 976021-874U-2 from the University of New Mexico (NSF Prime award # 0936498). We obtained approval from Oregon State University Institutional Review Board (project #5827). Omar Barroso helped with Fig. 16.4. The authors would like to acknowledge support from the HJ Andrews LTER program, funded by the National Science Foundation's Long-Term Ecological Research Program (DEB 0823380), as well as LTER colleagues Fred Swanson and Mary Beth Leigh.

References

Aguirre Sala J (2015) Hermeneutics and field environmental philosophy: integrating ecological sciences and ethics into Earth stewardship. In: Rozzi R, Chapin FS III, Callicott JB et al (eds) Earth stewardship: linking ecology and ethics in theory and practice. Springer, Dordrecht, pp 235–247

Billick I, Price MV (2010) The ecology of place. University of Chicago Press, Chicago

Carpenter S, Benson BJ, Biggs R et al (2007) Understanding regional change: a comparison of two lake districts. Bioscience 57(4):323–335

Carson RL (1965) The sense of wonder. Harper and Row, New York

Collins S, Swinton SM, Anderson CW et al (2007) ISSE: integrative science for society and the environment, a strategic research initiative. LTER Network Publication. Retrieved Apr 2014 from: http://www.rff.org/rff/Events/upload/25844_1.pdf

Cooper NS (2000) Listening to nature: ethics within ecology. Biodivers Conserv 9:1009–1027

Cross J (2001) What is sense of place? 12th headwaters conference, Western State University

Curtis DJ (2009) Creating inspiration: the role of the arts in creating empathy for ecological restoration. Ecol Manage Restor 10(3):174–184

Dayton PK, Sala E (2001) Natural history: the sense of wonder, creativity and progress in ecology. Sci Mar 65(Suppl 2):199–206

De Waal F (2006) Primates and philosophers: how morality evolved. Princeton University Press, Princeton

De Waal F (2009) The age of empathy: nature's lessons for a kinder society. Random House, New York

Downes-Le Guin T, Baker R, Mechling J et al (2012) Myths and realities of respondent engagement in online surveys. Int J Mark Res 54(5):1–21

Ecological Society of America (ESA). Earth stewardship. Retrieved June 2014 from: http://www.esa.org/esa/?page_id=2157

Farnum J, Hall T, Kruger LE (2005) Sense of place in natural resource recreation and tourism. USDA, USFS, PNWRS. General technical report: PNW-GTR-660

Fleischner TL (2011) Why natural history matters. J Nat Hist Educ Exp 5:21–24

Goralnik L, Nelson MP (2011) Framing a philosophy of environmental action Aldo Leopold, John Muir, and the importance of community. J Environ Educ 42(3):181–192

Gruen L (2009) Attending to nature: empathetic engagement with the more than human world. Ethic Environ 14(2):23–38

Harrell DF, Harrell SV (n.d.) Strategies for arts + science + technology research. Executive report on a joint meeting of the National Science Foundation and the National Endowment for the Arts. Retrieved Oct 2013 from: http://cms.mit.edu/news/Harrell-NSF-NEA-WorkshopExecutiveReportFinalDraft.pdf

Hoffman M (2000) Empathy and moral development. Cambridge University Press, Cambridge

Klein JT (2004) Interdisciplinarity and complexity: an evolving relationship. Emerg Complex Org 6(1, 2):2–10

Krebs CJ (2010) Case studies and ecological understanding. In: Billick I, Price M (eds) The ecology of place. University of Chicago Press, Chicago, pp 283–302

Kurdryavtsev A, Stedman RC, Krasny ME (2012) Sense of place in environmental education. Environ Educ Res 18(2):229–250

Leopold A (1949) A sand county almanac and sketches here and there. Oxford University Press, New York

Louda SM, Higley LG (2010) Responsive science: the interplay of theory, observation, and experiment in long-term, place-based research. In: Billick I, Price M (eds) The ecology of place. University of Chicago Press, Chicago, pp 303–326

LTER Goals (n.d.) The long term ecological research network. Retrieved June 2014 from: http://www.lternet.edu/node/22

LTER Network Vision and Mission Statements (n.d.) The long term ecological research network. Retrieved June 2014 from: http://www.lternet.edu/node/20

Lubchenco J (1998) Entering the century of the environment: a new social contract for science. Science 279:491–497

Malina RF (2011) Alt.Art-Sci: we need new ways of linking arts and sciences. Leonardo 44(1):2

Moore KD (2004) Pine island paradox. Milkweed Editions, Minneapolis

Moore KD (2005) The truth of the barnacles: Rachel Carson and the moral significance of wonder. Environ Ethics 27:265–277

Moore KD, Nelson MP (eds) (2010) Moral ground. Trinity University Press, San Antonio

Nisbet MC, Hixon MA, Moore KD et al (2010) Four cultures: new synergies for engaging society on climate change. Front Ecol Environ 6:329–331

Paine RT, Wootton T, Pfister CA (2010) A sense of place: Tatoosh. In: Billick I, Price M (eds) The ecology of place. University of Chicago Press, Chicago, pp 229–250

Pecharsky BL, Allan DJ, McIntosh AR et al (2010) Understanding the role of predation in open systems. In: Billick I, Price M (eds) The ecology of place. University of Chicago Press, Chicago, pp 185–206

Power ME, Chapin FS (2009) Planetary stewardship. Front Ecol Environ 7:399

Pulliam RH, Waser NM (2010) Ecological invariance and the search for generality in ecology. In: Billick I, Price M (eds) The ecology of place. University of Chicago Press, Chicago, pp 69–92

Reid N, Reeve I, Curtis DJ (2005) Creating inspiration: how visual and performing arts shape environmental behaviours. Report for Land and Water Australia, Canberra Project

Rozzi R, Armesto JJ, Gutiérrez J et al (2012) Integrating ecology and environmental ethics: earth stewardship in the southern end of the Americas. Bioscience 62(3):226–236

Slote M (2007) The ethics of care and empathy. Routledge, London

Sörlin S (2012) Environmental humanities: why should biologists interested in the environment take the humanities seriously? BioScience 62(9):788–789

Vucetich JA (2010) Wolves, ravens, and a new purpose for science. In: Moore KD, Nelson MP (eds) Moral ground. Trinity Press, San Antonio, pp 337–343

Vucetich JA, Nelson MP (2013) The infirm ethical foundations of conservation. In: Beckoff M (ed) Ignoring nature no more: the case for compassionate conservation. University of Chicago Press, Chicago, pp 9–25

Wattchow B, Brown M (2011) A pedagogy of place: outdoor education for a changing world. Monash University Press, Victoria

Chapter 17
The Technosphere and Earth Stewardship

Charles L. Redman and Thaddeus R. Miller

Abstract Scientists develop conceptual frameworks in an effort to better understand and manage the world around them. The dominant framework for most authors in this book and others concerned with Earth Stewardship is a coupled human-natural systems framework. This framework continues to provide new insights and promising management strategies. However, we argue that the addition of a third major domain, infrastructure/technology would more accurately reflect the key dynamics in today's world and allow more sustainable outcomes. Further we argue that scientists associated with each of these domains adhere to overlapping, but distinct sets of rules and fundamental assumptions that inhibit successful interdisciplinary collaboration. Rectifying this misalignment should be a cornerstone of future Earth Stewardship.

Keywords Conceptual frameworks • Interdisciplinary • Sustainability • Technology

The past 10,000 years have witnessed enormous advances in practical and theoretical knowledge, as demonstrated by revolutionary achievements such as agricultural technologies, urban life, treatment of disease, and information and communication technologies, as well as new social institutions and awe inspiring humanistic creations. These accomplishments have brought us many rewards, not the least of which have been our ability to extend life expectancies of individuals and resource systems that underpin the extraordinary growth of our population as a species. However, at the same moment that we appear to be masters of the world around us, there is a growing disillusionment that we are unable to solve some of the most basic and enduring challenges faced by humanity. As our knowledge advances in specific domains we also come to recognize that taken as a whole, the world is

C.L. Redman (✉)
School of Sustainability, Arizona State University, Tempe, AZ, USA
e-mail: Charles.redman@asu.edu

T.R. Miller
Nohad A. Toulan, School of Urban Studies and Planning, College of Urban and Public Affairs, Portland State University, Portland, OR, USA
e-mail: trm2@pdx.edu

R. Rozzi et al. (eds.), *Earth Stewardship*, Ecology and Ethics 2,
DOI 10.1007/978-3-319-12133-8_17

extraordinarily complex and that our current conceptual frameworks for perceiving, analyzing, and managing the world around us have proven inadequate to address many of the most important and urgent problems we face. Further, our ethical frameworks are ill equipped to deal with choices that may require geographical, interspecies and intergenerational trade-offs. The emerging call for earth stewardship in this volume and elsewhere (Chapin et al. 2011; Rozzi et al. 2012) recognizes the need for new ethical frameworks that account for the socio-ecological complexity and interdependencies. While this development is promising, in this chapter we argue that there are fundamental discontinuities in the way we come to understand and manage social, ecological and technological issues. More specifically, we contend that those concerned with sustainability and earth stewardship must more robustly account for the centrality of technology in human-environment interactions, adjusting our conceptual frameworks to explore socio-eco-technological systems (SETS).

17.1 Interdisciplinary Collaboration and Conceptual Frameworks

Conceptual frameworks are necessary tools for dealing with the world around us (i.e., Earth Stewardship). They provide us with a series of assumptions about how the world works, they encode operating principles and drivers, and they suggest a language of communication and a set of priorities for investigation. Yet one must keep in mind that conceptual frameworks are simplifications that are structured by how we, as individuals, think the world operates, what we think is important to know about the world, and what constitutes knowledge. Much of this comes to us from our academic disciplinary training and for the most part disciplinary frameworks are effective at solving problems as they define them. For at least 50 years it has been recognized that, although effective by some definitions, discipline-based conceptual frameworks have limits and might only resolve a part of the problem and reflect only a portion of reality. This has stimulated scientists from different disciplines to work together in increasingly popular interdisciplinary teams.

Many scientists have viewed the conjunction of social sciences and ecological approaches to be the most effective in understanding and managing the world around us. The two authors of this chapter and most of the authors in this volume have spent much of their careers utilizing frameworks that attempt to merge these two perspectives. In the literature, variants of this combination are often referred to as *social ecological systems* (SES) as in Holling and Gunderson (2002) or *coupled human and natural systems* (CHANS) as in Lui et al. (2007). This combination of the many activities of humans with the resources and dynamics of the biophysical environment, has become a compelling framework for most ecological scientists who are also concerned with the impact of humans, as well as for many individuals in social science and other disciplines, and forms the basis of countless academic articles, environmental impact statements, and policy documents. The thrust of this chapter is that despite the success of the social and ecological system framework, holistic, sustainability thinking would be better advanced by adopting a three part framework of a social, ecological, and technical/infrastructural system (SETS).

Before advancing the case for a new, expanded framework, it necessary to outline some of our frustration with the current dual framework that led us to seek a new integration. Notwithstanding the ubiquity and apparent success of the social ecological framework, we suggest that its employment has two major shortcomings that lead us to propose an alternate framework and approach. First, although interdisciplinary approaches encompassing social and ecological perspective have enriched scientific knowledge and transformed natural resource management, they often fall short of actualizing their goal of integrating key elements of their respective perspectives. Second, although community decision-makers and design professionals acknowledge in theory the relevance of a CHANS (or SES) framework, the patterns and dynamics investigated by those employing these approaches are not of primary importance to the decision-maker's work. Hence, elected officials, government administrators, planners, engineers, public health professionals, and corporate decision-makers may utilize some of our CHANS insights or discoveries in specific projects, yet when looking at the majority of their work, what we do and discuss has little impact on their decisions.

We are led to this conclusion based on a variety of observations. Two experiences convinced us of the need to transform the conceptual framework we employ if we are to significantly contribute to a sustainable future. One of us, Redman, has had the good fortune for the past 17 years to co-direct the Central Arizona-Phoenix Long Term Ecological Research project, among the most successful coupled human and natural system based research projects operating today (Grimm and Redman 2004). Earth, life, and social scientists have been working effectively together on this and related projects to make many significant observations and discoveries as demonstrated by over 300 published scientific articles. Although local planners and administrators are interested in what we are doing, it is very difficult to identify specific instances where results of our project have influenced public policy or decisions. Recognition of this situation has led project leaders in recent years to focus on bridging this gap, and those early accomplishments have strongly influenced the recommendations we make in this article. The second situation that both of this chapter's authors experienced is in the planning and initial years of Arizona State University's School of Sustainability (2004–2010; see Miller et al. 2011; Redman 2013). In the planning for what was to be a unique academic unit, faculty and staff were assembled from the widest range of disciplines possible. A variety of engineers (and architects) were very prominent in the planning, first years of teaching, and the initial graduate student cohort. This reflected the inclusive approach we were taking to sustainability in designing the School, and the importance of transforming the way we designed and built infrastructure in the broader world if we were to achieve sustainability. Yet, by the end of the third year of the school's existence, of the three full-time engineering-trained faculty, two had left the university, one was reassigned within the university, and new student cohorts included almost no students with an engineering background. In contrast, the early hiring of new faculty was based on very broad job descriptions, yet resulted in all tenure-track hires being in economics and other social sciences. Neither the departures nor hires were explicitly related to the dominance of the SES framework in the thinking of most faculty and students, but we, the authors, believe that indirectly it was a significant contributor. More than anything else, recognition of this pattern has led the two of us and many others to think seriously about the cascading influence of the conceptual framework we employ, and whether one focused on an SES approach is an adequate reflection of real world dynamics and an effective tool to generate sustainable management strategies.

We propose that *infrastructure* should be considered as one of three equally important domains in a new conceptual framework: social, ecological, and technological/infrastructural system (*SETS*; see Ramaswami et al. 2012 for a similar suggestion for curriculum development). Infrastructure currently is inadequately dealt with in the SES or CHANS frameworks or when it is richly considered, as by engineers or urban planners, it is often done in relative isolation from the social and ecological systems. Infrastructure is defined as the basic physical (hard) and organizational (soft) structures needed for the operation of a society or enterprise. The term typically refers to the technical structures that support a society, such as roads, bridges, water supply, sewers, electrical grids, telecommunications, and so forth, and can be defined as the physical components of interrelated systems providing commodities and services essential to enable, sustain, or enhance societal living conditions.. Here we are also including the technological systems associated with the services provided and the institutions that manage the hard infrastructure and associated technologies (communication systems such as the internet or cell phone systems being clear examples). It might be preferable to use the term 'technosphere' to refer to this combination of infrastructure and technological systems.

Technosphere is defined as the part of the physical environment affected through building or modification by humans (McGraw-Hill Science Dictionary). Infrastructure (or more broadly the technosphere) is clearly recognized as important in any conceptual formulation, but most often considered as a secondary part of the social system or in the case of green infrastructure as a component of the environmental system. This modest level of interest in infrastructure by SES scientists works against their influencing major decision-makers in society. Infrastructure is where investment focuses. When a community plans for the future and for a more sustainable world, new infrastructure and technological systems are at the center of the image. Moreover, infrastructure serves many purposes, often mediating the interactions within and between the social and environmental domains. Whether or not infrastructure is within the social domain or a domain in itself as in our SETS conceptual framework would not be important if all of these systems were thought of in the same terms and as having the same dynamics and underlying principles of operation. Our argument here is that they do not, and that there is a significant *intellectual misalignment* between those of us who deal, respectively, with the social, environmental, and infrastructural subsystems that acts to inhibit our understanding, management, and planning for the future.

Intellectual misalignment is the primary hurdle to effective interdisciplinary and transdisciplinary collaboration, and although significant strides have been taken in actualizing effective collaboration, there are still barriers to be broached. We argue that among the most serious of those divides for moving toward a more sustainable world, is the misalignment between those who focus on social-ecological systems and those who focus on infrastructural systems (within a SETS or other comprehensive framework), and that secondarily there remain barriers between social and ecological integration within a SES framework. A variety of factors contribute to the differences in approaches common to each of these domains. These reflect the fact that each of us is trained in particular disciplines with their own set of rules and priorities, but who sometime go on to gain additional perspectives from training or collaborative experiences. Difficulty in communication is probably the most fundamental barrier, and it may be based on different definitions for the same word or concept, as well as familiarity with different literatures. A more complex barrier derives from disciplinary differences in what constitutes acceptable knowledge. That is, what are the acceptable methods for acquiring new knowledge, what assumptions are the basis of inquiry, what dynamics are important for understanding the operation of a system? All of these questions are implicitly (and sometimes explicitly) answered by disciplinary training which defines appropriate domains of study and objectives of research. We are not suggesting that all disciplinary training can be discarded; disciplines exist for good reasons, and those that are most enduring often have evolved to be quite effective at what they do!

Several approaches have been taken to address the barriers presented by interdisciplinary collaboration (Miller et al. 2008). The most common recommendation is to spend time together, and to learn as much as possible about the other's approach. We agree that this advice should be taken and that it underlies much successful collaboration today. In a way, this chapter accepts this as a first step, and attempts to lay out what is needed to address the intellectual misalignment that continues to inhibit those who do work together. In working together, one may find elements common to other approaches that can be jointly pursued and mutually valued. Tracing land use histories for landscapes of mutual interest (and other place based approaches), are very useful for both ecologists and social scientists (Redman and Foster 2008). A common form of collaboration is when collaborators "compromise" their methods and objectives so that both approaches can be applied. However we do not find this form to be productive. Although it may lead to interesting results, the approach is so "watered down" that it does not satisfy either party. Instead of compromise, we suggest that each of the collaborating parties explore the logic and concepts of their partners, as an effort to find value in another approach. This effort to overcome intellectual misalignment is not easy, yet we believe that it pays significant dividends. Our effort to layout this misalignment of logic and of concepts in each of the three major domains, is not meant as a definitive analysis of each perspective, or even of specific instances of their application, but rather, to stimulate discussion on reducing barriers and moving forward on an integrated, comprehensive pathway toward sustainability.

17.2 Social-Ecological-Technological Systems

Although we are focusing on the elements of these three perspectives that are misaligned, we affirm at the outset that elements of logic and central concepts are shared by all three perspectives. Most scientists view their respective domain as systems consistent with the logic and rules of general systems theory governing the operation of each of the three perspectives. Hence, a focus on boundary definitions, system membership, linkages, feedback, regulation, scale of operation, cross-scale interactions, and complexity are integral to each of the approaches. Other concepts, such as resilience and sustainability, commonly are used in all three domains, but their usage represents very distinct ontologies. The potential for convergence in their meanings only recently has been explored by some scientists (Fiksel 2006; Redman 2014). In what follows, we suggest some of the distinct elements of the logic, or rules, that govern each of the three perspectives, as well as some of the basic concepts that are common to each perspective (see Table 17.1 for summary). These are the building blocks of conceptual frameworks that scientists employ. In so doing, these frameworks not only emanate from their training and background, they also reflect their ethical beliefs about the operation of the world around them. Once again we emphasize the caveat, that what follows is not a rigorous attempt to deconstruct specific conceptual frameworks, but rather to propose simple suggestions for further discussion, that largely are based on the experience of the authors as they have interacted with advocates of each of these perspectives in collaborative ventures.

Table 17.1 Misalignment of logic and central concepts of three perspectives in conceptual frameworks

Ecological perspective	Social perspective	Technological/infrastructural perspective
System tends toward balance	System is asymmetrical, flawed	System can be optimized
All members have a function	Member contribute disproportionately, if at all	Design component form as their function
Disturbance shapes population	Disturbance may lead to regime shift	Disturbance is to be controlled, minimized
Change via evolution, succession, adaptive cycles	Change via cycles, planning, reflexivity, externalities	Change via design for service delivery, path dependent due to sunk costs
Habitat, niche, fragmentation, species/area relationship	Property, rent, land use	Return on investment (ROI), connectivity
Food web, carrying capacity	Limitless, innovation, taste	Limitless due to substitution, technological fixes
Resilience of system to adapt to stress	Resilience of community to cope with stress	Resilience to return to former condition after stress

Although the logic of an ecosystems approach has many rules and principles, we will focus on a limited number that appear to contrast with social scientific thinking, as we have observed through our participation in SES investigations. In an ecosystem, all members of the system are connected and contribute in some manner to overall ecosystem functioning. If undisturbed, ecosystems tend toward balance and stability. All members of a single population (species) react/behave in a similar fashion, leading to the expectation of similar results for repeated field observations, and to the expectation that laboratory experiments may mimic field observations. Disturbance is a primary determinant of population, community, and ecosystem characteristics. Population dynamics are self-limiting and time constrained. For many analyses, structure and function are related closely and may be viewed as the same (Smith and Stirling 2010).

A huge variety of concepts drive ecosystems thinking, many of which have been modified for use in relation to social systems or SES. Some involve patterns of change over time, such as evolution with its associated concepts of adaptation and selection. On an organismic or community level, the life cycle has widespread application; the alternative models of community succession and the adaptive cycle, also have widespread application. Geographic distributions also are important with the niche concept, fragmentation of landscapes, and the edge effects of these patterns. Resource related concepts are also central, such as the centrality of the food web, the species/area relationship, and the guiding rule of carrying capacity. Virtually all of these concepts have been borrowed, with or without modification, by social scientists. Often this has led to new insights, and sometimes forms the basis of analyses of SES. However, we believe that this sharing has unintended consequences since perceived similarities are fewer than first thought.

Generalizing about the logic and rules of a social scientific approach is difficult, because the number of disciplines is quite large, and because there are significant differences among them. Once again, attempting to keep this discussion manageable, we focus on elements of logic and basic concepts that are fundamental to the perspective and that are sometimes borrowed by the other two perspectives, or which contrast sharply with elements in those perspectives.

Members of a social system are seen as connected in the system, but interestingly, different from ecosystems, being a member does not mean that one contributes to overall social system functioning; for many social scientists, human systems are out of balance and their structure is flawed. Compounding these differences is the view that individuals within a population or community, are likely to behave/react differently, that is, individuals have agency. Related to this is the fact that power and access to resources are assumed to be unevenly distributed in social systems, and that these asymmetries are basic drivers of system operation and change. Another key difference with ecosystems, is that humans are reflexive; they attempt to anticipate the future and act to modify it to their perceived benefit. How one perceives the world around them, what beliefs one holds about how systems operate, and what values one prioritizes, all affect one's knowledge about the world (observation is not a physical act alone). Hence, knowledge is contingent; contingent upon the observer, upon the context within which the observation/decision is being made, and contingent upon one's values and beliefs. Following this reasoning, there is no single, scientific reality, and problems to be investigated and acted upon should be framed by a variety of interested parties.

Many concepts underlie the social sciences. We highlight the following because, for the most part, they distinguish human activities and organizations from the rest of the biological world and from the inanimate objects of infrastructure. Of enormous importance is the human ability to learn, encode knowledge, and to transmit it to others over time. This has given rise to culture, which can be understood as patterns of learned behavior that facilitate the operation of society, help situate people within their groups, and distinguish groups from each other. These patterns are enhanced by the human ability to learn from the past. To anticipate and plan for the future, humans self-organize themselves into enduring family and kinship relations, as well as by assigning tasks, abilities, and beliefs. The success of societies often is related to individual human ability or proclivity to exhibit leadership, and in other cases to self-organize for taking collective action. In addition to the various biological drivers that we share to some extent with other species, people also act on values that are passed on to them through culture. Most people believe in basic 'rights' that they are entitled to, and a system of right from wrong that guides their own behavior. Individual behavior often is related to perceived individual advantages, but also includes altruistic behavior benefiting others, such as one's offspring or members of one's own group. Other distinguishing concepts organize group behavior; among the most important is the ability to organize interactions in order to receive an advantage or profit from activities such as the exchange of goods or provision of services. Humans have further enhanced this ability by endowing these temporary advantages with lasting qualities by transforming them into monetary instruments, differential future access to productive resources, or social status.

17.3 Technology and the Challenge to Earth Stewardship

The core argument of this chapter is that infrastructural-technological systems operate with a distinct logic, or set of rules and concepts, that only can be understood and managed by considering them independently in SETs conceptual frameworks. Those working primarily in the infrastructural domain rely on a variety of physical/chemical/mechanical 'laws' to envision and design their products, but also have adopted ideas originating in the other two domains. Infrastructure is highly valued in society since it usually provides services such as transportation, housing, energy, water, etc. that might not be possible without it, or at least at a more reasonable price than would be possible without it. Other infrastructure buffers individuals and communities against the effects of disamenities, such as seasonal variation in temperature, waste water disposal, or extreme natural hazards. Because of these services and protections, infrastructure is used as a primary tool for managing risk, and therefore for allowing the expansion of communities into geographic locations that are susceptible to hazards. A central characteristic of infrastructure (or technological systems) that should guide our conceptualizing, is that once put in place, it creates a powerful path dependency, meaning that it has enormous influence over subsequent decisions. Part of that derives from the fact that it is usually long lasting, and has a large upfront cost that diminishes the subsequent cost of utilizing it. Because of this infrastructure (or technological systems), once in place it is a major barrier to alternative solutions (high cost of entry), thereby encouraging future investments for extending or enhancing the original infrastructure. The existence of infrastructure almost always gives rise to institutions that are oriented toward building, managing, and advancing this infrastructure and the services it provides (or disamenities it buffers against). Similar to the path dependency the physical infrastructure creates, associated institutions and their members create a social path dependency, further encouraging the continuation and expansion of this particular infrastructural/technological solution.

A wide variety of concepts drive infrastructure/technology system thinking. Some of them have to do with the design of the infrastructure and how it will provide the desired services. Fundamental laws and principles of the physical world are the building blocks of the infrastructural world, such as the second law of thermodynamics, laws of gravitation, momentum, aerodynamics, and general properties of materials. Given these physical characteristics and the parameters of the services desired and the cost requirements, designers of infrastructure believe that there is an optimal solution, or at least a small series of optimal solutions, to be selected among. Moreover, given desired outcomes, it is believed that there may be more than one way or set of materials that could provide similar services. This principle of substitutability is central to the design sciences, but is controversial when the substitution replaces a natural or ecologically provided service, with one that is human constructed or implies an equivalence between two alternate social outcomes.

Designers of infrastructure increasingly have attempted to incorporate into infrastructural/technological thinking, issues that are important to those in the social or environmental domains. In an effort to go beyond the relatively restricted view that "form equals function," practitioners have considered indirect impacts of the fabrication, use, and disposal of their products. Industrial ecology and life cycle analysis take this broader view and reflect ideas borrowed from ecological sciences (Allenby 2005). Similarly, measurement and efforts to redesign urban metabolism, are efforts to think of formerly separate infrastructures/technologies as interdependent wholes. With other concepts such as resilience, whose traditional interpretation is that of being an object's ability to return to its original form or condition after being stressed, some practitioners have borrowed elements of the SES use of the concept to describe a broader phenomenon in risk management for designing systems able to cope successfully with external stresses, rather than resisting them (that is, safe-to-fail rather than fail-safe; see Park et al. 2011). Other innovative areas include the actual merging of these domains, such as green infrastructure that attempts to maximize the use of natural features and processes to provide services previously provided by built (hard) infrastructure, or efforts to maximize the use of real time data and elements of artificial intelligence to solve the problems arising from the movement of goods, people, and services in smart cities of the future.

Infrastructure and technology have been central to much of what we consider progress in the human endeavor. In fact, many would consider them to have been pivotal in setting us aside from other species and enabling our dominance of the world around us. Virtually every one considers that refined and improved infrastructures and technological systems will be central elements in any sustainable future we can envision. Acknowledging this pivotal role infrastructure and technology has played in getting us to where we are and where we are going, we can also question whether the largely, separate sectorial way we have been conceptualizing, managing, and designing the social, environmental, and infrastructural domains of our world has resulted in all that we hope for. It is our contention that barriers raised by misalignment in our thinking about each of these domains separately, has prevented us from attaining our potential for designing and managing a more desirable future in which ecological integrity is ensured, human well-being is enhanced, and social equity advanced (Leach et al. 2010).

The solution, we think, is not to search for a singular, overarching, holistic framework (Sarewitz 2010). Rather, the answer is in developing a more pluralistic understanding of each of these domains and how they interact, propagating resilience or vulnerabilities. Only then can we begin to appreciate and understand the thorny ethical and practical choices we are faced with. It is our hope that a richer understanding and subsequent realignment of the SETS perspectives will allow transform what had been a serious barrier to change into a pathway to change and sustainability. Experiments such as field environmental philosophy methodology developed in the Chilean International Long-Term Ecology Research site (Rozzi et al. 2012) are promising as they acknowledge the need to engage in practical and contextual ethical analyses and understanding. In order for earth stewardship to act as a viable guide, scholars and practitioners alike must first work to understand the central place of technology and the limitations of understanding within and between social, ecological and technological domains.

References

Allenby BR (2005) Reconstructing Earth: technology and environment in the age of humans. Island Press, Washington, DC

Chapin FS III, Pickett STA, Power ME et al (2011) Earth stewardship: a strategy for social–ecological transformation to reverse planetary degradation. J Environ Stud Sci 1:44–53

Fiksel J (2006) Sustainability and resilience: toward a systems approach. Sustain Sci Pract Policy 2(2):14–21

Grimm NB, Redman CL (2004) Approaches to the study of urban ecosystems: the case of Central Arizona-Phoenix. Urban Ecosyst 7:199–213

Holling CS, Gunderson LH (2002) Resilience and adaptive cycles. In: Gunderson LH, Holling CS (eds) Panarchy: understanding transformations in human and natural systems. Island Press, Washington, DC, pp 25–62

Leach M, Scoones I, Stirling A (2010) Dynamic Sustainabilities. Earthscan, London

Lui J, Dietz T, Carpenter SR et al (2007) Complexity of coupled human and natural systems. Science 317:1513–1516

Miller TR, Baird TD, Littlefield CM et al (2008) Epistemological pluralism: reorganizing interdisciplinary research. Ecol Soc 13(2):46 [online]. http://www.ecologyandsociety.org/vol13/iss2/art46/

Miller TR, Munoz-Erickson T, Redman CL (2011) Transforming knowledge for sustainability: towards adaptive academic institutions. Int J Sustain High Educ 12(2):177–192

Park J, Seager TP, Rao PSC (2011) Lessons in risk-versus resilience-based design and management. Integr Environ Assess Manag 7(3):396–399

Redman CL (2013) Transforming the silos: Arizona State University's School of Sustainability. In: Chase G, Barlett P (eds) Sustainability in higher education: stories and strategies for transformation. MIT Press, Cambridge, MA, pp 229–239

Redman CL (2014) Should sustainability and resilience be combined or remain distinct pursuits? Ecol Soc 19(2):37

Redman CL, Foster DR (eds) (2008) Agrarian landscapes in transition: comparisons of long-term ecological and cultural change. Oxford University Press, New York, p 296

Ramaswami A, Weible C, Main D, Heikkila T, Siddiki S, Duvall A, Pattison A, Bernard M (2012) A social-ecological-infrastructural systems framework for interdisciplinary study of sustainable city systems. J Ind Ecol 16(6):801–813

Rozzi R, Armesto JJ, Gutiérrez JR et al (2012) Integrating ecology and environmental ethics: earth stewardship in the southern end of the Americas. Bioscience 62(3):226–236

Sarewitz D (2010) Against holism. In: Frodeman R, Klein JT, Mitcham C (eds) Oxford handbook of interdisciplinarity. Oxford University Press, Oxford

Smith A, Stirling A (2010) The politics of social-ecological resilience and sustainable socio-technical transitions. Ecol Soc 15(1):11

Chapter 18
Using the Ecosystem Services Framework in a Long-Term Socio-Ecological Research (LTSER) Platform: Lessons from the Wadi Araba Desert, Israel and Jordan

Daniel E. Orenstein and Elli Groner

Abstract The establishment of Long-Term Socio-Ecological Research (LTSER) platforms is part of a paradigmatic shift in the way ecosystems are studied and managed: from a narrow species-level focus to a holistic socio-ecological systems approach. The need for this shift is based on increasingly urgent global environmental challenges and the realization that traditional ecological research methods and foci have been insufficient for meeting these challenges. While the theoretical foundation for this shift and guidelines for implementing it are increasingly well defined, there is little actual experience in implementation on the ground. We recount our experiences in establishing an LTSER platform in Wadi Araba, a hyper-arid desert in southern Israel and Jordan, focusing on the challenges in establishing a cooperative agenda between the two countries. We discuss the use of the ecosystem service (ES) conceptual framework for guiding our research program and our efforts to create a dialogue between research scientists and community members, and identify some of the ethical issues inherent in trans-border research and in the application of the ES framework.

Keywords Arava • Ecosystem services • Socio-ecological research • Transboundary

D.E. Orenstein (✉)
Faculty of Architecture and Town Planning, Technion – Israel Institute of Technology, Haifa, Israel
e-mail: DanielO@ar.technion.ac.il

E. Groner
Ramon, The Dead-Sea and Arava Science Center, Mitzpe Ramon, Israel
e-mail: elli@adssc.org

R. Rozzi et al. (eds.), *Earth Stewardship*, Ecology and Ethics 2,
DOI 10.1007/978-3-319-12133-8_18

18.1 Introduction

Wadi Araba of Israel and Jordan is a hyper-arid desert in the demographic and geographic periphery of the two countries. Although the climate is extreme and characterized by low primary productivity, the region hosts unique biodiversity – including rare hyper-arid species of conservation importance (Dolev and Perevolotsky 2004), one of the earth's northern-most tropical coral reefs, and a major flyway for migratory birds between northern Asia and Europe and central and southern Africa. The area is the first stopover for millions of birds migrating from Africa to Europe in the spring (Frumkin et al. 1995). Concurrently, both countries are interested in distributing their rapidly growing populations away from their geographic core regions and are making great efforts to entice populations to move to these regions. The region is characterized by rapid economic development in the form of transportation, trade (Eilat and Aqaba serve as important port cities for each country) and tourism development. These activities are claiming growing amounts of land, not only for urban development, but for agriculture, solar power generation, and large-scale infrastructures including airports, rail networks, and the proposed Red Sea – Dead Sea conduit.

Into this contentious setting, the Israel Long-term Ecological Research (LTER) Network made a strategic decision to establish a Long-Term Socio-Ecological (LTSER) Platform. The LTSER concept is new to the 17-year old Israel LTER network. Israel was accepted into the International LTER network in 1997 with the recognition of two research sites in the Negev Desert, which had a history of ecological research and monitoring. The Israeli sites were grouped with the (then East) Europe LTER network (Groner and Shachak 2011). Since then, 11 more sites have been added to the Israel network, and more recently, two platforms were added. Until 2009, almost all the research conducted within the Israel LTER network focused on ecology, biology, hydrology, climatology and other natural science disciplines. Accordingly, the data base resulting from LTER monitoring reflected this disciplinary bias (Dick et al. 2014).

In 2009, the Israel LTER management committee decided to expand the disciplinary foundation of the network to include the social sciences and humanities. This decision was based on trends in Europe towards inter- and trans-disciplinary research (Haberl et al. 2006), and because local scientists adopted the view that the incorporation of social science into ecological research was crucial for addressing contemporary conservation challenges. It was decided to establish two LTSER platforms: one based on the five existing LTER sites in the Northern Negev (Orenstein et al. 2012), and a second, completely new, platform in Wadi Araba. Concurrent to the establishment of the Israeli Wadi Araba platform, the new Jordanian LTER network established a Jordanian platform across the international border.

The LTSER concept evolved within the European LTER network as part of a broader trend among scientists responding to the realization that global ecological challenges such as climate change, biodiversity loss and resource depletion were rooted in social and economic dis-function and that the problems must be analyzed through a multi-disciplinary lens (Haberl et al. 2006; Singh et al. 2013a). LTSER platforms go beyond the LTER site concept in that they link biophysical processes to social processes and governance and include science communication in their mission (Rozzi et al. 2012). The LTSER platform facilitates a place-based, stakeholder-driven, trans-disciplinary[1] research program designed to define the ecological and socio-ecological challenges of a region (from the perspective of both experts and local communities) and research them in order to produce policy-relevant results (Mirtl et al. 2013). As the LTSER concept proliferated throughout Europe and beyond, Israeli LTER scientists realized that the concept was ideal for a country where no areas were devoid of intensive human activity, and the drivers of ecological change (population growth, consumption, political conflict, and land use policy) were (and are) deeply embedded within the values and psyche of the population. Like the LTSER concept, the ecosystem service (ES) approach leads to trans-disciplinary research for society. For this reason and others, many LTSER platforms have adopted the ES framework (Mirtl et al. 2013).

In this chapter, we share some insights from our research on ES of the past 5 years in the Wadi Araba LTSER platform. We hold a mirror to our own research endeavors and interweave history, research results, and anecdotes to highlight the organizational challenges and ethical dilemmas in establishing the platform and applying the ES conceptual framework. Much has been written on the establishment and effective management of LTSER platforms (Singh et al. 2013b); our approach here is admittedly more eclectic, characterized by anecdotes, discussion protocol, and open questions. Such details are seldom presented in the scientific literature (although see Mirtl et al. 2013, which documents the challenges in establishing an LTSER Platform). It is our hope that a candid and honest narrative will resonate with field researchers who might identify with our challenges. We further hope that our experiences will provide inspiration for overcoming those challenges, toward the goal of fostering Earth Stewardship.

18.2 Ecosystem Services (ES) as a Research Framework

Concurrent to the process of establishing the LTSER platform, and similar to established LTSER platforms in Europe (Mirtl et al. 2013) and urban LTER sites in the U.S. (Grimm et al. 2013), the Wadi Araba LTSER team adopted the ES conceptual framework to study socio-ecological interactions between local communities and their environment. The decision to adopt the ES framework was self-evident for socio-ecological research. We began with the assumption that an ecosystem does not a-priori provide services - that is, an ecological survey of an area cannot yield an

[1]Trans-disciplinary research combines traditional research-based knowledge from multiple disciplines with local knowledge derived from local residents and other stakeholders.

ES inventory in the absence of studying the recipients of those services. What would be necessary to develop our understanding of the region was to simultaneously:

(a) survey the local residents with regard to their values, behaviors and opinions regarding their ecosystem,
(b) interview local experts of all disciplines to procure expert knowledge regarding human interaction with the local ecosystem and its services, and
(c) research the status of biodiversity and ecosystem integrity in the region.

We thereby combined social and natural science methods to provide a holistic understanding of the socio-ecological system via the currency of ecosystem services. Our final step will be to turn these data in to policy relevant information. By including policy makers and land use planners in the basic research (as interviewees, for example) and bringing them close to the research in an advisory capacity, we have cultured a conduit for uptake of our research results throughout the process.

The ES framework has been criticized for its over-reliance on economic valuation and its lack of social valuation methods. The act of social-based research is a response to the ethical dilemmas inherent in the ES concept and its over-reliance on monetization (Kosoy and Corbera 2010; Luck et al. 2012). Social-based ES research within the LTSER platform provides a powerful framework for integrating local communities, their values and opinions, into the local research and policy agenda. The social-based approach to ES assessment has proven itself as a catalyst for constructive, community-level engagement in multiple venues (e.g. Bryan et al. 2010; Maynard et al. 2010; Chan et al. 2012).

Through our research in Wadi Araba, including multiple studies employing questionnaire-based surveys and in-depth interviews with local residents, we have established a solid foundation for the claim that the local population is strongly connected to the local landscape (Sagie et al. 2013; Orenstein and Groner 2014) and that despite the perception that deserts are low in ecosystem service provision due to low primary productivity, they are extremely rich in cultural services. The population exhibits high affinity to the desert landscape, its mountains and open spaces. This has led to a persistent debate about whether landscapes devoid of biological life (or whose beauty is attributed primarily to geodiversity rather than biological features) are providing an ecosystem service. The United Kingdom National Ecosystem Assessment (UK-NEA) proposes that both biological and geological features combine to provide ecosystem services. The inclusion of geology as a provider of ES may resolve our conundrum (Gray 2011). Others suggest that cultural value of landscapes is a unique phenomenon that should not be considered within the rubric of ecosystem services, but rather should be considered services unto themselves (Brown 2013). Other research findings include:

1. Ecological knowledge of respondents was generally low; the population lacked a holistic understanding of the ecosystem, the importance of biodiversity, and the implications of their economic activities on ecosystem processes.

2. Cultural value of landscapes, view, open spaces and mountains is very high, and they are considered the most beloved assets of the region. Respondents in Israel were most motivated to political activism when faced with threats to their aesthetic environment (rather than ecological environment).
3. There are both similarities and differences regarding what aspects of the ecosystem are appreciated and used by different populations – along national, urban-rural, and gender lines.
4. Differences along socio-demographic lines may become less significant, however, when core values (derived through factor analysis of survey data) are considered (Diamantopoulos et al. 2003; de Groot and Steg 2008; Orenstein and Katz-Gerro in review). Analyzing core values may liberate us from dividing groups along traditional socio-demographic lines, and allow us to look more at the individual as a product of values rather than as an affiliate of a socio-economic or demographic group.
5. Local residents in both countries express pro-environmental and [selective] pro-development opinions, and reconcile the two through support of "sustainable" economic activities. However, the development trajectories in Israel and Jordan are at very different stages and opinions and behaviors may be affected by economic factors (see below).

While our research has shown similarities between Jordanians and Israelis with regard to their perceptions of their natural environment, we also detect a socio-economic fault line that passes down the valley, alongside the geological fault (the Syrian-African Rift). The two populations share virtually the same ecosystem, and yet – by global standards – one community is impoverished and one is relatively wealthy, one Muslim and the other Jewish, one has relatively low formal educational achievement and one high (socio demographic data available in Sagie et al. 2013; Orenstein and Groner 2014).

The degree of collaboration between Jordanian and Israeli researchers in the LTSER platforms varies with time and individuals, as it is strongly affected by social and political currents. However, despite the perennially difficult political climate, trust has been established on a personal level, enabling fairly stable coordination and continuity of research. In August 2010, the ILTER annual meeting was held in the Israel's Negev desert and Jordan sent a delegation. As a step to lessen political pressures, the collaborative research results are sometimes published separately, although one paper was written together within the context of a Pan-European research team (Dick et al. 2014).

The cross-border differences continue in the communities themselves. While both Jordanians and Israelis ranked characteristics of the extreme environment rather low (e.g. heat, aridity, brightness, sand storms), Jordanians expressed lower affinity for these characteristics than Israelis. Some environmental characteristics, such as sand dunes and open space, appealed to Israelis but not to Jordanians. And shrub vegetation, which provides fodder for grazing animals, was the only environmental characteristic that Jordanians ranked higher than Israelis. Outdoor recreational activities were notably different, with Jordanians more often engaging in campfires and off-road vehicle driving, while Israelis reported swimming in the Gulf of Aqaba/Eilat more often (Orenstein and Groner 2014).

These differences may be due to the host of socio-economic differences between the populations on either side of the border. We suggest that affluence contributes to the relative resilience of Israelis to environmental extremities. Further, with economic resources, Israelis are able to turn these extremities into economic opportunity via specialty crops, algae farming, solar power production, and eco-tourism. In Israel, locals have capitalized on the desert in order to market their products. A senior manager of a local dairy conveyed, "the taste of our chocolate milk is not connected to chocolate, but to the Kibbutz and its pleasant people, with the background of the desert with palm trees… we're not selling chocolate milk, we're selling an idea, an image." Both poverty and economic structure (the agricultural resources in the rural sector of the Jordanian Araba are in private hands) may prevent the Jordanians from perceiving their environment as a potential economic opportunity.

Respondents were also asked for their opinions regarding various environmental and development issues. In general, both Israelis and Jordanians expressed environmental concern, and for many of the questions, results were similar (Table 18.1). However, for each question where economic and environmental issues were presented in conflict with one another, the Jordanians tended towards a development preference, while the Israelis tended more towards environmental protection. The exception was the demographic question. In Israel, the issue of population growth in the demographic peripheries has long been one of national importance. The local population has internalized the belief that their long-term economic sustainability is dependent on local population growth, as reflected in the survey results. The Jordanians, on the other hand, are evenly split on the issue.

The actions of one country have direct impact on the ecosystem services of the other in the narrow landscape of 10 km width. Though separated by a border, the region is linked climatically, hydrologically and ecologically. In Jordan, hunting takes place and the population of large animals is low, while on the Israeli side of the border, hunting has been outlawed since the 1950s, when hunting led to the decline and local extinction of several species. Large herbivores and carnivores sometimes move from Israel to Jordan to feed, before returning to Israel. Land use also differs across the border. In Israel, the percentage of agricultural land cover is much larger, and most of the rare Wadi Araba sand dune ecosystems in the country have been cultivated for agriculture (Yom Tov and Mendelsohn 1988). The sand dunes on the Jordanian side are more prevalent and protected along the border zone, though their proximity to the border also makes them inaccessible to the local population (Sagie et al. 2013). While the geomorphology and abiotic conditions are similar on both sides of the border, the pressures and drivers are different and this results in different densities of large animals and different species assemblages of smaller animals (Shanas et al. 2006, 2011).

Table 18.1 Results from an opinion survey of residents in the southern Wadi Araba of Jordan and Israel. Results are based on 407 respondents in an opportunity sample (details in Orenstein and Groner 2014). 38 % of the respondents were from urban localities (Eilat and Aqaba) and 62 % from the rural areas; 37 % were Jordanian and 63 % Israeli; 43 % were female and 57 % male. Educational achievement, income levels, religion and occupation were assessed to be representative of the general population of the region

		Strongly disagree (%)	Disagree (%)	Neutral (%)	Agree (%)	Strongly agree (%)	Don't know/no opinion (%)	Average response (1 = strongly disagree; 5 = strongly agree)
There are not enough people living in the southern Arava/Eilat	Jordan	5	**25**	15	**36**	13	7	3.29
	Israel	9	19	**20**	**30**	**20**	2	3.35
Most tourists come to the region because of the natural environment (geology, ecology, aesthetics)	Jordan	5	**25**	15	**36**	13	7	3.29
	Israel	5	**24**	13	**29**	23	6	3.44
The economic benefits of building a hotel outweigh the environmental costs[a]	Jordan	7	16	**19**	**25**	18	15	3.35
	Israel	**33**	**19**	16	9	11	11	2.39
It is important to construct an international airport in the Arava (in Israel)	Jordan	11	**19**	17	**22**	17	13	3.18
	Israel	**21**	19	15	20	**22**	2	3.04
"Environmentalists" are too extreme in their desire to prevent development in the Arava	Jordan	2	17	**33**	17	12	19	3.26
	Israel	15	19	**30**	20	13	4	2.97
I am very environmental in my behaviors	Jordan	2	5	14	**45**	**33**	1	4.04
	Israel	1	8	**33**	**37**	20	2	3.69
My community is very environmental in its behaviors	Jordan	5	22	**23**	**31**	13	5	3.27
	Israel	4	18	**29**	**33**	12	4	3.33
Economic development should always take precedent over environmental protection	Jordan	11	**25**	**23**	**23**	14	4	3.03
	Israel	**25**	**44**	17	7	6	2	2.24
Economic development and environmental protection can occur together	Jordan	2	2	6	**48**	**39**	3	4.24
	Israel	1	8	8	**47**	**35**	2	4.08
Developing tourism infrastructure in the Arava is important for the future of the region	Jordan	1	2	9	**41**	**45**	3	4.29
	Israel	2	5	9	**44**	**39**	1	4.15

(continued)

Table 18.1 (continued)

		Strongly disagree (%)	Disagree (%)	Neutral (%)	Agree (%)	Strongly agree (%)	Don't know/no opinion (%)	Average response (1 = strongly disagree; 5 = strongly agree)
My economic wellbeing depends on a clean, healthy environment	Jordan	1	3	14	**41**	**38**	3	4.15
	Israel	2	9	20	**34**	**33**	2	3.88
It is important to protect Arava sand dunes from development[b]	Jordan	1	6	**25**	**42**	21	5	3.80
	Israel	1	6	13	**31**	**46**	3	4.19
It is important to protect biodiversity in the southern Arava and the Gulf of Eilat, even if it means foregoing economic opportunities	Jordan	7	20	**25**	**25**	20	3	3.32
	Israel	0	6	19	**32**	**39**	4	4.07
I enjoy spending time in nature	Jordan	3	2	11	**47**	**35**	1	4.11
	Israel	2	2	8	**37**	**50**	0	4.31

[a]Israeli question was place-specific, referring to the Sasgon Valley example described in the text
[b]Israeli question was place-specific, referring to the Samar Dunes example described in the text

18.3 Where Has the Ecosystem Services Conceptual Framework Led Us?

The objective driving our ES-centered research agenda is the desire to produce policy-relevant research and, like most ES researchers, to protect biodiversity and ecosystem integrity (Cowling et al. 2008). The next stage in our research program is to present our research findings to policy makers and facilitate a community dialogue regarding future regional development. Our assumption was that the ES framework would facilitate a three-way dialogue among local communities, ecologists, and land use managers / policy makers, but a more implicit assumption was that this process would lead to better ecological outcomes than would occur in the absence of the process. Current paradigms in planning and resource management all point to the essential need for collaborative planning with local communities and suggest that top-down conservation planning has not proven successful in many cases (Fraser et al. 2006; Adams and Hutton 2007; Cowling et al. 2008; Clark 2011).

These lofty goals are rife with challenges. Below we draw upon quotes and anecdotes from our various research venues to highlight some of the recurring dilemmas we face.

18.3.1 Ecosystem Services or Biodiversity?

"If I go to a community and tell them that their development is going to cause a species extinction and they respond that they don't "value" that species or that the species doesn't provide them with a service, how does that help me in my job to protect biodiversity?"

— Ecologist, Israel Nature and Parks Authority

"This is a tree and that is a tree – what does it matter which tree, so long as it's nice to look at."

— Local Israeli government planner in response to seeing several landscape options based on different species assemblages

While ES literature is explicit regarding the dependence of ES on biodiversity, we often face decisions that pit ES against biodiversity, or a particular set of ES (e.g. cultural or provisioning) against another (regulating or supporting). Environmental NGO officials and Nature and Parks Authority representatives express great skepticism regarding the ES path, suggesting that it leaves little room for preserving biodiversity on the basis of its intrinsic value and always places the burden of proof on conservation agencies to prove the utility of a given species. Other critics of the ecosystem services concept have foreseen such scenarios (Turnhout et al. 2013). Our response, based on international experience and modern conservation planning theory (Dietz et al. 2003; Clark 2011), is that community-integration is crucial and that top-down policies to preserve biodiversity also carry multiple risks to biodiversity (Adams and Hutton 2007). Unfortunately, this claim is difficult to prove on a case-by-case basis because once policy is implemented it is impossible to do a counterfactual study of what would have happened in the absence of the policy. Nature advocates therefore often dismiss the ES framework as hopelessly anthropocentric.

18.3.2 Do the Benefits Arising from Cultural Services for the Many Outweigh the Benefits for the Few?

"Right now this place [a secluded natural spring] is precious to us – a little heavenly corner that we can enjoy; if you open it up, [expletive] will come with their loud music and garbage and destroy it. Why would I want to open the site up for more people?"

— Local resident in the southern Dead Sea Basin in Israel

In a comparison of ecosystem services provided at five LTER sites in Israel's northern Negev, it was found that some management authorities purposely develop sites based on ecological features to attract tourists (Orenstein et al. 2012). Thus, when compared to other sites, sites developed for tourists receive higher values for cultural services. We defined "potential ecosystem services," as ES that are currently not exploited for human use, but have potential to be used depending on changes in cultural, ecological or other circumstances. Some agencies, prioritizing cultural services, realize these potential services by preparing infrastructure and access to natural settings. Survey and interview respondents in Wadi Araba seemed to support such management strategies, by expressing enthusiasm for developing eco-tourism based on cultural ES of the region. This led us to suggest to a group of local residents in the southern Dead Sea that they develop an infrastructure to allow access to several fresh water springs in the region. Aside from capitalizing on potential cultural services, we reasoned that a greater number of visitors would translate into a greater amount of public support should the springs face threats from development. Our suggestion led to the response quoted above. This leads us to a number of open questions: What are we maximizing? How do we use the framework to balance the relative value of cultural services provided by a site? Do more users necessarily translate into a higher value of cultural service benefits?

18.3.3 Biodiversity or Cultural Landscapes?

"I had to think hard for an ecological [biodiversity-related] reason why building the hotel would be bad"

— Local Israeli activist in Wadi Araba protesting the planned construction of a new hotel

During the period of our research, there were two prominent environmental campaigns taking place in the Israeli Wadi Araba. The first was a proposed hotel to be built in the Sasgon Valley at the entrance to the Timna Valley tourist site. Local residents expressed, both in our research and in public protest, that the hotel was a threat to the pristine nature of the valley. The second campaign occurred across the road from the Sasgon Valley, in a small sand dune that was a relic of the once larger Samar Dunes (Fig. 18.1). A tender had been issued by national governmental agencies to mine the remaining sand. While both sites share cultural and landscape importance,

Fig. 18.1 Ecological research conducted at the remaining Samar Sand Dunes in Israel, which have high biodiversity value as well as high value for cultural ecosystem services (Photograph by Elli Groner)

there was a crucial difference between the sites regarding their ecological value. The Samar Dune, once 11 km^2, is home (habitat) for a unique assemblage of species. In the Samar example, ES and biodiversity were both provided as reasons for avoiding mining. The Sasgon Valley is not unique in biodiversity (Nissim 2012). As the quote above exemplifies, the motivation to stop construction had little to do with ecological consequences, but was due to its landscape and wilderness value. In the Sasgon example, opponents to development were limited in their approach when arguing biodiversity, but could harness the terminology of ES to argue against the project (even though, in this case, ES provision was not based on biodiversity). This is an example that biodiversity and ecosystem services may provide contrary conclusions regarding management of a habitat.

Another example arose regarding the ES-biodiversity tradeoff in a debate regarding constructing high tension electric wires through the center of the otherwise open Ramon Crater nature reserve. A public campaign attempted to convince the electricity company to bury the wires instead of erecting lines above ground. Ecologists assessed that this option (burying) would cause extensive damage to biodiversity. Managers were left deciding between biodiversity (underlying all ES) and landscape beauty (a cultural ES). The public was vocally in favor of landscape.

Further north, in the northern Negev Desert, land management agencies and environmental NGOs have been disputing the relative impact of dryland forestry. The managers argue that foresting the area provides crucial ES including carbon sequestration, recreational areas, and prevention of runoff and soil erosion. Others argue that forestry threatens local biodiversity. The public discourse is framed (mistakenly) as a conflict between managing for ecosystem services or for biodiversity. Different land management agencies with different development priorities exacerbate this conflict by choosing one conceptual framework over the other.

Another side to the ES-biodiversity dilemma is that biodiversity and ES can reflect culture. Culture may not always value diversity, as witnessed in the planner's quote above. However, as this chapter shows, cultural diversity can have a direct impact on biodiversity. Acacia trees in the Israeli Wadi Araba depend on large wild herbivores for seed dispersal because local shepherds do not exist anymore on that side of the border. On the Jordanian side, however, large herbivores are hunted and seed dispersal depends on the local domestic herbivores. So not only is herbivore diversity dependent on cultural diversity, ecosystem functions also depend on cultural diversity (e.g. seed dispersal). Cultural diversity can prevent the homogenization of biodiversity or biocultural diversity (*sensu* Rozzi 2013).

18.3.4 Who Should Make the Decisions?

Various experts often suspect that using the ES framework will lead to over-reliance on public opinions to make decisions. Our ecologist colleague quoted above suggests that while the public expresses particular affinities for the outdoors, it has little understanding of the importance of ecosystem integrity or of human reliance on natural systems for their regulating or supporting services. Our results from surveys and interviews indeed confirm a public knowledge gap.

As ecologists, we hold the importance of conservation of biodiversity to be indisputable, and yet, as socio-ecologists, we understand that (1) community participation in natural resource and land-use decision making is crucial, and, (2) accepting community participation means accepting that ecologists themselves become stakeholders, and not the agents of truth and last word in decision making. In assessing our role within the new paradigm of socio-ecology, we must consider the balance between "expert" knowledge and "local" or "stakeholder" knowledge, each providing a unique and complementary knowledge base. This complementarity is represented graphically in Fig. 18.2. In the figure, ecosystem services are assigned high and low value by local stakeholders and experts, respectively (though the two aren't mutually exclusive groups).

High priority ES, as defined by both experts and locals, are considered high-priority, easy targets for management, as both groups agree about their importance (e.g. a rare, charismatic species). Where locals do not consider a given ES important (usually they do not mention it at all), but experts do consider it important (e.g. nitrogen cycling), there is a *knowledge gap*. Where locals give high importance

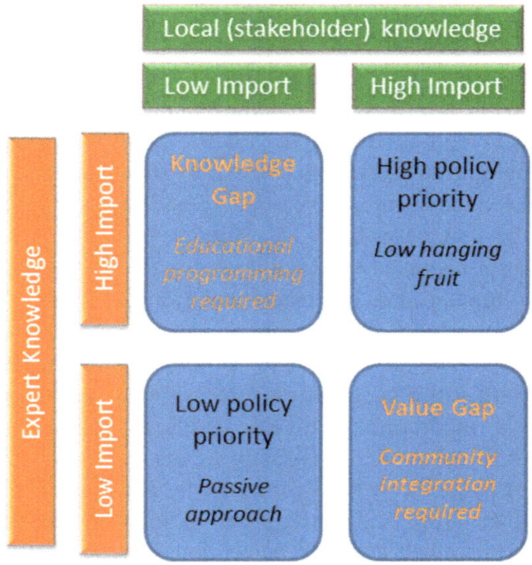

Fig. 18.2 Complementary knowledge sources for policy-relevant ecosystem service research

to ES, but experts do not, then there is a *values gap*. In our region, the knowledge gap exists where the public has little understanding of the broad value of biodiversity and the presence of supporting ES. On the other hand, as in the Sasgon example, ecologists found little ecological importance in the valley, yet the local population emphatically demonstrated that its landscape value was of utmost importance (the values gap).

18.4 Conclusions

ES tradeoffs is a recurring theme in our research and experience. Maximization of one or more services may lead to the degradation of others; maximization of some may yield a decline in biodiversity. This issue has been covered in the literature, both theoretically (Foley et al. 2005; de Groot et al. 2010) and also practically, within the context of community stakeholder engagement (Fraser et al. 2006; Cowling et al. 2008; Chan et al. 2012). We find the model proposed by de Groot and colleagues (2010) useful in explaining this dynamic. Their model suggests that it is possible, like in our case studies, to raise the value of cultural-recreational services through a small amount of development, but this may come at the expense of regulating services, cultural-informational services, and biodiversity. In general, as their model and our experience suggests, land use decisions will often be a function of trade-offs in the composition of a bundle of ES; the optimal bundle of services is a socio-ecological decision that should be made through community discussion.

We are reminded that biodiversity was the predominant concept for conveying the importance of ecological conservation to the public during the past two or more decades. We suggest that the lack of success of that concept, which has been argued to have been largely a paradigm development to strengthen the role of ecologists in development discourse (Takaks 1996), in preventing global species extinctions and habitat destruction gave rise to the new, current concept, of ecosystem services. Like the term biodiversity, we find that the ES concept also is difficult to convey to the general public. The Hebrew translation is particularly difficult, as interview respondents confused the term with a variety of other phenomena, from the private companies that come and haul away garbage and sewage, or those that sell pesticides to farmers. Some mistook the term to mean composting toilets.

Yet we are cautiously optimistic regarding the application of the ES framework within the burgeoning research program of the Wadi Araba LTSER platform. An important added value of ES research is that applying social research methods to ES assessment has catalyzed direct contact between researchers and communities. This contact allows for two-way learning in which respondents to questionnaires, interviews, and participants in focus group discussions have the opportunity to express their opinions, perceptions, and needs with regard to their natural environment. The researcher, through the act of inquiry, provokes the respondent to think about issues that they may have not considered, particularly about the role of ES in their well-being. Both researcher and subject become more sensitive to social and ecological needs.

We concur with Maass and Equihua (2015, this volume [Chap. 14]) regarding the need to make their suggested paradigm shifts, which corroborate well the transition from LTER to LTSER (Haberl et al. 2006). Our experiences provide insights regarding the on-the-ground challenges for implementing those changes in a small corner of the Middle East. Within Israel, the gap between theory and implementation of LTSER remains wide. Between Israel and Jordan, the collaboration is an excellent first step to bridging the gaps in a common ecosystem. It is clear that without the political motivation to work together in order to build bridges between the two countries, such research would not have been possible due to the many obstacles that exist. In both countries the ecologists work together with social scientists. The combination of transboundary, transdisciplinary research is a complex challenge that brings new insights and angles to the study of the Wadi Araba landscape. If mutual benefits (for Jordanians and Israelis) could be derived from the platform in the form of improved ecosystem management, and local people perceive these benefits, this would be an excellent tool for peace making and, as a result, it would strengthen the collaboration and the two LTSER platforms could flourish.

Acknowledgements We gratefully acknowledge the work of our Jordanian colleagues. We thank the editors of this volume for their improvements to the manuscript and our research assistants, Hila Sagie and Inna Kaplan. Roy Zaidenberg developed the model displayed in Fig. 18.2. Portions of this research were supported by a trans-border cooperation grant from the Israel Ministry of Regional Cooperation to the Dead Sea – Arava Science Center.

References

Adams W, Hutton J (2007) People, parks and poverty: political ecology and biodiversity conservation. Conserv Soc 5:147–183

Brown G (2013) The relationship between social values for ecosystem services and global land cover: an empirical analysis. Ecosyst Serv 5:58–68. 10.1016/j.ecoser.2013.06.004

Bryan BA, Raymond CM, Crossman DM et al (2010) Targeting the management of ecosystem services based on social values: where, what, and how? Landsc Urban Plan 97(2):111–122. 10.1016/j.landurbplan.2010.05.002

Chan KMA, Guerry AD, Balvanera P et al (2012) Where are cultural and social in ecosystem services? A framework for constructive engagement. BioScience 62(8):744–756. doi:10.1525/bio.2012.62.8.7

Clark SG (2011) The policy process: a practical guide for natural resource professionals. Yale University Press, New Haven

Cowling RM, Egoh B, Knight AT et al (2008) An operational model for mainstreaming ecosystem services for implementation. Proc Natl Acad Sci 105(28):9483–9488. doi:10.1073/pnas.0706559105

de Groot JIM, Steg L (2008) Value orientations to explain beliefs related to environmental significant behavior: how to measure egoistic, altruistic, and biospheric value orientations. Environ Behav 40(3):330–354. doi:10.1177/0013916506297831

de Groot RS, Alkemade R, Braat L et al (2010) Challenges in integrating the concept of ecosystem services and values in landscape planning, management and decision making. Ecol Complex 7(3):260–272. 10.1016/j.ecocom.2009.10.006

Diamantopoulos A, Schlegelmilch BB, Sinkovics RR et al (2003) Can socio-demographics still play a role in profiling green consumers? A review of the evidence and an empirical investigation. J Bus Res 56(6):465–480. 10.1016/S0148-2963(01)00241-7

Dick J, Al-Assaf A, Andrews C et al (2014) Ecosystem services: a rapid assessment method tested at 35 sites of the LTER-Europe network. Ekologia 33(3):217–231. doi:10.2478/eko-2014-0021

Dietz T, Ostrom E, Stern PC (2003) The struggle to govern the commons. Science 302(5652): 1907–1912. doi:10.1126/science.1091015

Dolev A, Perevolotsky A (2004) The Red Book: vertebrates in Israel. The Israel Nature and Parks Authority and The Society for the Protection of Nature in Israel, Jerusalem

Foley JA, DeFries R, Asner GP et al (2005) Global consequences of land use. Science 309(5734):570–574. doi:10.1126/science.1111772

Fraser EDG, Dougill AJ, Mabee WE et al (2006) Bottom up and top down: analysis of participatory processes for sustainability indicator identification as a pathway to community empowerment and sustainable environmental management. J Environ Manage 78(2):114–127

Frumkin R, Pinshow B, Kleinhaus S (1995) A review of bird migration over Israel. J Ornithol 136(2):127–147. doi:10.1007/BF01651235

Gray M (2011) Other nature: geodiversity and geosystem services. Environ Conserv 38(03):271–274. doi:10.1017/S0376892911000117

Grimm NB, Redman CL, Boone CG et al (2013) Viewing the urban soc-ecological system through a sustainability lens: lessons and prospects from the Central Arizona-Phoenix LTER Programme. In: Singh SJ, Haberl H, Chertow M et al (eds) Long term socio-ecological research. Springer, Dordrecht

Groner E, Shachak M (2011) The history of establishing the LTER network in Israel and its status. Hebrew, Jerusalem, Ma'arag, Jerusalem

Haberl H, Winiwarter V, Andersson K et al (2006) From LTER to LTSER: conceptualizing the socioeconomic dimension of long-term socioecological research. Ecol Soc. 11(2):13 [online]. http://www.ecologyandsociety.org/vol11/iss2/art13/

Kosoy N, Corbera E (2010) Payments for ecosystem services as commodity fetishism. Ecol Econ 69(6):1228–1236. 10.1016/j.ecolecon.2009.11.002

Luck GW, Chan KMA, Eser U et al (2012) Ethical considerations in on-ground applications of the ecosystem services concept. BioScience 62(12):1020–1029. doi:10.1525/bio.2012.62.12.4

Maass M, Equihua M (2015) Earth stewardship, socioecosystems, the need for a transdisciplinary approach and the role of the International Long Term Ecological Research Network (ILTER). In: Rozzi R, Chapin FS III, Callicott JB et al (eds) Earth stewardship: linking ecology and ethics in theory and practice. Springer, Dordrecht, pp 217–233

Maynard S, James D, Davidson A (2010) The development of an ecosystem services framework for South East Queensland. Environ Manage 45(5):881–895. doi:10.1007/s00267-010-9428-z

Mirtl M, Orenstein DW, Wildenberg M et al (2013) Development of LTSER platforms in LTER-Europe: challenges and experiences in implementing place-based long-term socio-ecological research in selected regions. In: Singh SJ, Haberl H, Chertow M et al (eds) Long term socio-ecological research. Springer, Dordrecht

Nissim H (2012) Landscape units in the Arava. M. Sc. thesis. Blaustein Institute for Desert Research, Ben Gurion University of the Negev, Beer Sheva

Orenstein DE, Groner E (2014) In the eye of the stakeholder: changes in perceptions of ecosystem services across an international border. Ecosyst Serv 8:185–196. 10.1016/j.ecoser.2014.04.004

Orenstein DE, Katz-Gerro T (in review) Environmental tastes, opinions and behaviors: social theory in the service of ecosystem service assessment.

Orenstein DE, Groner E, Argaman E et al (2012) An ecosystem services inventory: lessons from the northern Negev Long-Term Social Ecological Research (LTSER) Platform. Geogr Res Forum 32(2012):96–118

Rozzi R (2013) Biocultural ethics: from biocultural to homogenization toward biocultural conservation. In: Rozzi R, Pickett STA, Palmer C et al (eds) Linking ecology and ethics for a changing world: values, philosophy, and action. Springer, Dordrecht/Heidelberg/New York/London, pp 9–32

Rozzi R, Armesto J, Gutiérrez JR et al (2012) Integrating ecology and environmental ethics: earth stewardship in the southern end of the Americas. BioScience 62(3):226–236. doi:10.1525/bio.2012.62.3.4

Sagie H, Morris A, Rofè Y et al (2013) Cross-cultural perceptions of ecosystem services: a social inquiry on both sides of the Israeli–Jordanian border of the Southern Arava Valley Desert. J Arid Environ 97:38–48. 10.1016/j.jaridenv.2013.05.007

Shanas U, Abu Galyun Y, Alshamlih M et al (2006) Reptile diversity and rodent community structure across a political border. Biol Conserv 132(3):292–299. 10.1016/j.biocon.2006.04.021

Shanas U, Galyun YA, Alshamlih M et al (2011) Landscape and a political border determine desert arthropods distribution. J Arid Environ 75(3):284–289. 10.1016/j.jaridenv.2010.10.008

Singh SJ, Haberl H, Chertow M et al (2013a) Introduction. In: Singh SJ, Haberl H, Chertow H et al (eds) Long term socio-ecological research. Springer, Dordrecht

Singh SJ, Haberl H, Chertow H et al (eds) (2013b) Long term socio-ecological research. Springer, Dordrecht

Takaks D (1996) The idea of biodiversity: philosophies of paradise. The Johns Hopkins University Press, Baltimore

Turnhout E, Waterton C, Neves K et al (2013) Technocratic and Economic ideals in the ecosystem services discourse. Conserv Lett. doi:10.1111/conl.12069

Yom Tov Y, Mendelsohn H (1988) Changes in the distribution and abundance of vertebrates in Israel. In: Yom Tov Y, Tchernov E (eds) The zoogeography of Israel: the distribution and abundance of a zoogeographical crossroad. DRW Junk Publishers, Dordrecht

Chapter 19
Socio-Ecological Studies in Urban and Rural Ecosystems in Chile

Olga Barbosa and Paula Villagra

Abstract To embrace a global "Earth Stewardship", researchers associated with the Chilean Long Term Socio-Ecological Research (LTSER-Chile) network, highlighted the urgent need to integrate the variety of ecosystems and cultures, and overcome the bias of information centered in Northern Hemisphere. However this initiative doesn't include the Chilean territory under strongest anthropogenic influence, and where most of the population lives and relies economically: agricultural lands and cities. Here we present two innovative projects contributing to the international Earth Stewardship Initiative: urban planning in the city of Valdivia, and the development of sustainable winery in Mediterranean Chile. Urban concentration amplifies strong social inequities that not only impact individual's economic opportunities, but also the environmental quality of the surrounding landscapes and the ecosystem services they provide. In Chile, a highly urbanized country and prone to natural disasters, access to ecosystem services, provided for example by wetlands and urban forest remnants are essential to mitigate the effects of such catastrophes. Similarly, a conservation initiative with the wine industry was recently developed to preserve the threatened Mediterranean ecosystem, and improve vineyard management practices to minimize impact on native biodiversity and ecosystem services that sustain the wine industry and local communities. For both study cases a complementary approach through capacity building activities with the local community (bottom-up) coupled with building relationships with government institutions and corporations (top-down), have increased the effectiveness of the decision making process, highlighting that an holistic approach to Earth Stewardship should consider a variety of values, and undertake a socio-ecological approach.

O. Barbosa (✉)
Instituto de Ciencias Ambientales y Evolutivas, Universidad Austral de Chile, Valdivia, Chile

Institute of Ecology and Biodiversity, Santiago, Chile

Centro de Desarrollo Urbano Sustentable (CEDEUS), Valdivia, Chile
e-mail: olga.barbosa@uach.cl

P. Villagra
Instituto de Ciencias Ambientales y Evolutivas, Universidad Austral de Chile, Valdivia, Chile
e-mail: paula.villagra@uach.cl

R. Rozzi et al. (eds.), *Earth Stewardship*, Ecology and Ethics 2,
DOI 10.1007/978-3-319-12133-8_19

Keywords Capacity building • Conservation in farmlands • Earth stewardship • Mediterranean ecosystem • Social inequity • Sustainable urban planning

A novel framework for Earth Stewardship as a means of engaging science and society to reduce the rates of anthropogenic damage to the biosphere has been proposed by the Ecological Society of America (ESA) (Power and Chapin 2009; Chapin et al. 2011). In our era of the Anthropocene (Crutzen 2002), the integration of social and ecological dimensions of ecosystems is essential to advance towards Earth stewardship. However, in order to truly embrace "Earth Stewardship" as a planetary phenomenon, researchers associated with the Chilean Long Term Socio-Ecological Research (LTSER-Chile) network have highlighted the urgent need to integrate a wide variety of heterogeneous ecosystems and cultures, and overcome the bias of information centered in Northern Hemisphere sites (Rozzi et al. 2012). To address these needs, the recently created LTSER–Chile has added a new biome (South American Temperate Forests) to the International LTER network, and also has introduced the methodological approach of field environmental philosophy (FEP) (Rozzi et al. 2012). FEP's methodological approach which integrates ecological sciences and environmental ethics into biocultural conservation, thereby contributing to the implementation of Earth stewardship (Aguirre Sala, in this volume [Chap. 15]).

LTSER-Chile is a valuable initiative, however it does not include the Chilean territory under the strongest anthropogenic influence, and where most of the population lives and relies economically: agricultural lands and the cities. The Chilean population is markedly urban (> 87 %, MINVU 2014), concentrated in cities that occupy only 0.23 % Chile's continental land. The Chilean Economy relies on natural resources: mining, fishing and agricultural industries account for approximate 25.8 % GDP (Banco Central 2012). This context raises two major challenges for the Chilean LTSER: to reach the urban population, and to engage the economic sector, both private and public. Private sectors have been incorporated into conservation partnerships worldwide, through approaches such as corporate social responsibility strategies (Tang and Li 2009).[1]

In this chapter we present two innovative projects that can contribute to both the LTSER-Chile network and the international Earth stewardship initiative: urban planning in the city of Valdivia in southern Chile, and the development of sustainable winery in central Chile. Both projects are funded by national scientific research funding agencies, demonstrating an emergent recognition by the Chilean government of the importance of addressing urban ecosystems, and engaging in partnerships with the private sector. We discuss how a socio-ecological approach might essential to advance towards ecosystem stewardship.

[1]The concept of integrating corporate social responsibility into conservation projects, and some specific cases, have raised a strong controversy (see MacDonald 2010).

19.1 The Challenge of Incorporating Stewardship in Chilean Urban Planning

Latin America has experienced one of the fastest growth in urban population since mid-twentieth century (WUP 2011; Rozzi 2013), and cities have sprawled fast into areas of high biodiversity value (Pauchard and Barbosa 2013). According to the ONU-Habitat (2012) report the rates of rural – urban migration are now expected to decrease. In addition, urban concentration amplifies strong social inequities that not only impact individual's economic opportunities, but also the environmental quality of the surrounding landscapes and therefore, the potential ecosystem services that can be beneficial (Pauchard and Barbosa 2013; Rozzi 2013).

Chilean cities rely on ecosystems service provisions in a variety of ways. However, these services are available in a differentiated manner to different socio-economic groups. Urban poor communities are more dependent but have less access to them than wealthy communities, which are also able to substitute these benefits. For example, wealthy neighborhoods may lack public green spaces for recreational use, but they can afford a garden, which acts as partial substitutes for these spaces (Barbosa et al. 2007). Inequality is marked in Chile, and poor communities are more vulnerable to the negative effects of natural disasters. The country is prone to natural disasters such as flooding (e.g. El Niño events), volcanic eruptions and earthquakes, and human-made disasters, such as fires. Natural areas often offer ecosystem services such as food, fuel, and water, which are of particular importance in the event of a catastrophe, and others, such as recreation and beauty, which are relevant in daily life.

Moreover, some approaches to urban planning have resulted in uncontrolled urban sprawl, informal settlements, and negligent location of housing in non-secure areas, causing large infrastructure, economic, and human losses. Recently, during April 2014, more than 2,900 people lost their homes, and 15 lost their lives in the biggest fire ever recorded in Chilean main port city, Valparaiso. The fire, was fueled by nearby extensive tree plantations of exotic fast-growing species (*Eucalyptus sp and Acacia dealbata*), and spread easily due to unplanned territorial development, illegal trash dumps, and the steep geography with fast winds and extremely difficult accessibility. This and other examples have revealed the need to encourage and strengthen socially and -ecologically integrated approaches, and to study risks to human populations and environmental systems from natural disturbances to improve urban planning in South America (LA RED 2013). Complementarily, some sociological studies exploring the key factors that increase the chances of survival of groups under high-risk conditions, have identified community practices that maintain traditional relationships with ecosystems, which are important for mitigating the consequences of natural disasters (e.g. García et al. 2012). However in Chile, this type of integration in policy and research is still missing. While in 1931 the first building code required taking precautions against tsunamis and earthquakes (Art. 178), these indications no longer appear in the 1949 building code, which is still in effect (Herrmann 2014). It was not until, 1974 that the National Emergency Agency

(ONEMI) was created, but with limited powers. Even though the ONEMI includes the perception of risk as important for preventing or mitigating disasters, the agency only is allowed to make recommendation to the local municipalities. Finally it is the municipality that decides the type of intervention applied to the landscape. Today, these models are outdated and differ from international disaster recovery plans and models that highlight the valuable role of ecosystems services (Vale and Campanella 2005; ISDR 2005; Resilience Alliance 2010).

In southern Chile, the fluvial city Valdiva, for example, has a past and recent history of natural disasters where certain ecologically important structures have provided ecosystem services to the population. However, today these places (e.g. urban wetlands and urban forest remnants) hardly are valued by local authorities and thus continually disappear, often being replaced by housing developments. In 1552 when Valdivia was founded as a Spanish fort the city was surrounded by the Valdivia River that had clean and navigable waters, as well as lakes with an abundance of birds, other wildlife, and luxuriant flora as described in historical records (Guarda 2009). Today, several of these Valdivian natural attributes could be considered as cultural ecosystem services due to their beauty, recreational opportunities, and river transport advantages (Guarda 2009). However, this natural urban landscape underwent several changes over time due to virtually continuous processes of reconstruction following natural disasters, housing and infrastructure densification, and constant urban sprawl. By 1885 the lakes mostly had been filled in for housing developments. The need for homes caused by the devastating effects of a 1909 fire led to even more pressure to fill remaining wetland areas. The most dramatic of urban changes in Valdivia occurred after the severe 1960 earthquake, when the South American Plate lurched upward as much as 20 m relative to the subducting Nazca Plate (Barrientos and Ward 2007), adding new wetlands to the already existing ones. Wooden tents used as temporary housing facilities, the *rucos*, were located around these new wetlands (Fig. 19.1a). Over the years, *rucos* were replaced by permanent housing, creating new neighborhoods next to these swampy areas (e.g. Skewes et al. 2012). These changes have formed an urban landscape in which society and nature interact. However, these interactions have not respected these places as sources of water since the 1960s earthquake, nor as sources of city beautification and recreation valued as it was in 1552, when the city was founded. The way that these places have been valued have changed over time, risking their current existence.

A recent study funded by the National Commission for Technology and Scientific Research (CONICYT) revealed some of the social values associated with the wetlands during the aftermath of an earthquake in Valdivia (Villagra et al. 2014). Wetlands were found to be amongst the seven most used urban spaces for earthquake recovery. Nonetheless, their utility varies depending on the presence or absence of a set of biophysical aspects that modify their appearance, and hence, their utilitarian values. In case of an earthquake scenario nowadays, the study shows that the presence of biophysical attributes such as water, vegetation, street infrastructure, and iconic architecture in the urban landscape, cause them to be perceived as useful for shelter, evacuation, and temporary housing. Studies in two other coastal

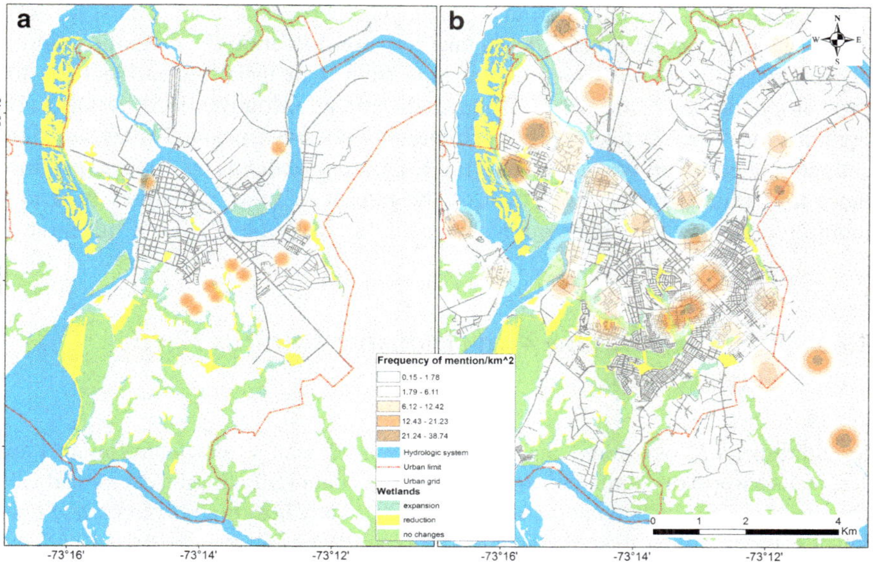

Fig. 19.1 The figure shows land use change in the city of Valdivia close to, and in wetland areas. Panel (**a**) shows Valdivia main city configuration in 1961, 1 year after the 9.5 earthquake, where areas close to wetlands were used as temporal housing (the 'rucos') which evolved later into permanent housing. Panel (**b**), shows that similar areas close to wetland would be used for the allocation of trash and debris in a hypothetical post-earthquake scenario today and the evolution of main urban wetlands

cities, San Francisco and Kobe, have proved that in the presence of water and vegetation found in urban parks and streams, were fundamental for earthquake recovery after the 1909 and 1995 earthquakes respectively (Allan and Bryant 2011; Hayashi 2010). However, in Valdivia, most wetland areas are perceived as places to dispose debris and garbage, regardless of the water and vegetation these places contain, a situation that can be witnessed every day (Fig. 19.1b).

Landscapes perceived as dirty, unmanaged, and dangerous, or even just empty, tend to be valued negatively by planners and urban dwellers (Van den Berg et al. 2007). This is the case for Valdivia where wetlands were once a 'beautiful' landscape useful for sailing and recreation (Guarda 2009), but now are waste lands with overgrown vegetation, lack of public urban infrastructure, and accessibility. Interestingly, a similar study in Concepcion, Chile, proved the opposite situation than in Valdivia. This study found urban wetlands useful for water extraction, temporary refuge, as places for public gathering, rather than for debris and waste disposal (Villagra and Rojas 2013). In Concepcion, the wetlands provide daily recreation and have been taken into account for urban design, intervention, and management, as they were perceived as positive assets by the community following an earthquake scenario. The little value assigned to urban wetlands in Valdivia today, as providers of ecosystems services after a disaster, is probably due to their negative appearance resulting from mismanagement over the last 50 years. However, urban planners and designers should take into consideration the dynamic nature of urban landscape values, which can change from their usual recreational use to a depository for waste and debris. Today, educational initiatives lead by local civic organizations, such as taking school children to 'discover' wetland areas, are aimed at changing community values and attitudes toward wetlands by direct exposure to these important habitats. Nevertheless, the impact of these initiatives need to be monitored by long-term perception studies, in line with the LTSER- Chile strategy.

19.2 Working with the Wine Industry Towards Earth Stewardship

Mediterranean Ecosystems, characterized by mild winters that concentrate rainy-season and warm dry summers, are areas of great, but highly fragile, ecological value (Myers et al. 2000; Olson and Dinerstein 2008). They host high diversity and endemism rates of floras that exceed the combined rates of tropical Africa and Asia (Arroyo and Cavieres 1991; Cowling et al. 1996). Geographically they include portions of USA, Mexico, Chile, Australia, and South Africa, as well as the Mediterranean Basin, and tend to be densely populated regions. The Chilean Mediterranean Region extends between 23 and 39.5° South, and although it represents only 16 % of the country's territory, it harbors almost 50 % of Chilean vascular plants, including 50 % of endemic species (Arroyo et al. 1995; Armesto et al. 2007a, b).

Mediterranean areas around the globe historically have exhibited an intense anthropogenic pressure due to agricultural and urban land uses, and globally have less than 4.5 % under any conservation protection category (Cox and Underwood 2011). Chile is not an exception, with 77 % of the population (INE 2011) and 76 % of the country's GDP (Banco Central 2012), concentrated in its Mediterranean region. Despite its ecological value and the threats to its biological integrity, less than 1 % of the regional surface is under any official protection (Underwood et al. 2009). This has proven to be insufficient for biodiversity conservation (Simonetti 1999; Tognelli et al. 2008; Durán et al. 2013). Two centuries of intensive deforestation, together with intensive grazing by cattle and extremely variable rainfall, have had long-lasting effects on forest cover in south-central Chile, whose effects are persistent even today (Armesto et al. 2010). In addition, economic incentives to forestry in the last two decades, together with economic globalization and free trade, promoted the expansion of new crops, leading to the further decline of woodlands. One of these new crops are grapes for wine production.

The wine industry has expanded persistently in the Chilean Mediterranean region with 63,550 hectares (ha) in 1997 to 128,367 in 2012 (202 % growth), and recently expanding beyond the Mediterranean limits (MINAGRI 2013, Fig. 19.2a). The image of Chile is very much associated with the wine industry. Most vineyards own

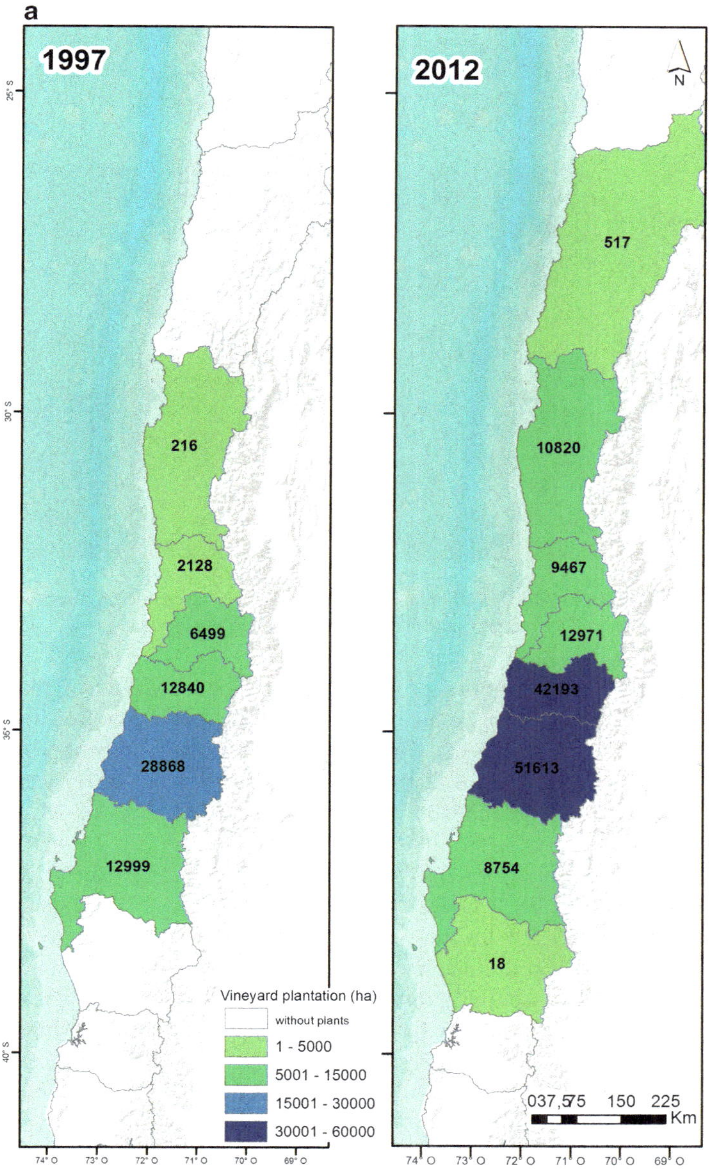

Fig. 19.2 (**a**) Evolution of vineyard expansion by administrative region between 1997 and 2012, in the Mediterranean zone of Chile

Fig. 19.2 (**b**) Photo of Viña Veramonte, with native forest remnants in the background in Casablanca valley, a main wine production valley (Photo Olga Barbosa)

between 2 and 4,000 ha, or more, of natural adjacent land. Moreover, the quality of the wine is defined by its geographic origin, a meaning that is captured in the French term "*terroir*" (Viers et al. 2013). This term highlight the connection between the resulting product (e.g. wine) and the environment (Wilson 2001; Renouf et al. 2006), thus inclining the industry to favor the conservation of Mediterranean ecosystems (Fig. 19.2b).

With the opportunity afforded by innovative new funding by the government of Chile (Fondos BASALES, CONICYT), a group of Chilean scientists embraced a 5-year program working with the wine industry to develop ways to balance the growth of the industry and the conservation of this highly fragile ecosystem. This program was planned essentially through research on the links between the provision of ecosystem services to vineyards, and their susceptibility under climate change and land use change, leading to the establishment of the Wine Biodiversity and Climate Change Program (WBCC).[2] The overall goal is to improve vineyard design with management practices that minimize their impact on native biodiversity and ecosystem services that sustain the wine industry and local communities.

Since the beginning of this program, the number of wineries interested in partnering with the WBCC consistently has increased, from one in 2008 to 14 in the first 5 years of the program, together accounting for nearly 70 % of wine Chilean global exports by volume (Wines of Chile 2013). However, after 2 years of developing

[2]The Wine, Climate Change and Biodiversity Program is a scientific initiative of the Institute of Ecology and Biodiversity and Universidad Austral de Chile, which aims to conciliate biodiversity conservation with the development of the Chilean wine industry. www.vccb.cl

basic research, it was realized that deliveries in both time and spatial scale were not really coupled with the producers' immediate needs. This was especially evident when the world financial crisis hit at the beginning of 2008, when many producers expressed the willingness to continue the ongoing collaboration, but needed to postpone any investment on restoration or conservation programs until the economy stabilized. It is important to realize that WBCC's first approach was essentially focused on the ecological component of the system. However, these systems are susceptible to external variables such as global markets that determine prices and therefore sales, thus proving the vulnerability of any system where components are treated in isolation (Chapin et al. 2009). This unexpected but nonetheless learning process led to the creation of an education and knowledge transfer program for delivering best management practices to producers through the recognition and valuation of native local biodiversity and coupled ecosystem services (Fig. 19.3a).

These WBCC workshops highlighted two important things. The first was to realize how some inherited agricultural practices from the northern hemisphere were so ingrained that Chilean biodiversity had suffered dire consequences over time. The second was the keen interest and participation of farm workers to link biodiversity with their local ecological knowledge and cultural identity. As a result of these workshops and long-term engagement with wine producers, many of the vineyards associated with the program have already set aside areas that contain native forest and shrubland, approaching nearly 20,600 ha in a region where the National System of Protected Areas includes only 64,930 ha in National Parks (CONAF 2013). However, the WBCC program is still voluntary and has not been adopted by the Chilean wine industry as a whole, a situation that contrasts with South Africa (Von Hase et al. 2010).

A survey conducted with an open sample of Chilean wineries has shown that land set aside for conservation proposes is not a rare practice. Out of 45 surveyed wineries, 86 % claimed to have conservation areas on their farms. However, when asked about specific management practices to preserve biodiversity, we found out that only 35 % took steps to ensure that cattle were not allowed into these areas. This common practice in farmlands is based in the erroneous belief that cattle will remove grass biomass and therefore decrease the risk of fires. However, this practice not only severely limits the recruitment of native flora (Henriquez and Simonetti 2001), but it increases propagation of invasive grasses that are highly flammable (Pauchard et al. 2008, 2011). The introduction of nonnative species is of particular concern in Chile, given that 15 % of plant species present in the country are nonnative (Fuentes et al. 2013). Chile has a strict quarantine control over pests, and there is a clear normative to avoid the introduction of exotic species without a permit. However, a national strategy for control and prevention of exotic invasive species based on the damage that can be caused to the native biodiversity, has not been implemented (Pauchard et al. 2011).

Implementation of cover cropping between roads that might involve exotic species can be nevertheless a beneficial practice that prevents soil erosion, improves soil conditions, and provide habitat for natural enemies (Altieri 1999; Nicholls et al. 2001; Patrick-King and Berry 2005). When choosing species for this purpose, commercial mixes, such as those that are used in California, USA, are the only available

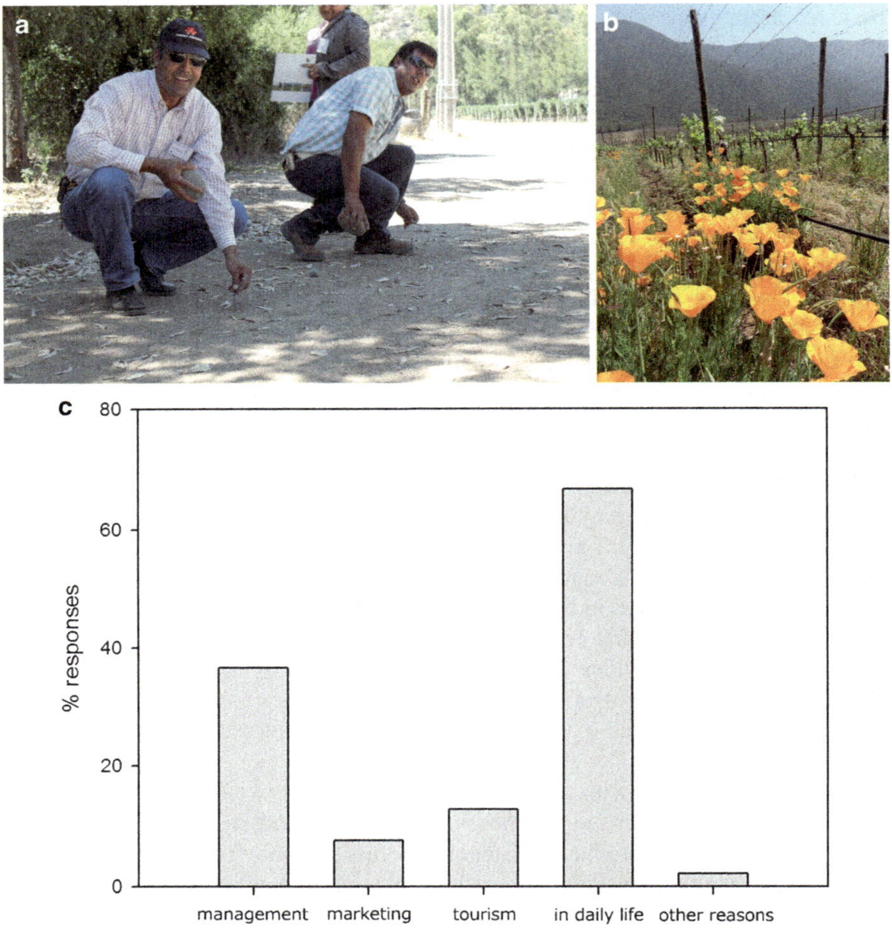

Fig. 19.3 (**a**) Workshop activities using the inquiry methodology (Feinsinger 1987), evaluating soil compaction between vineyards, road and native forest. (**b**) The use of *Eschscholzia californica* as a cover crop in farms that sustain high native species richness in surrounding areas, risking valuable ecosystems. Flower remains up to 4 years after initial seed plantation. (**c**) Interviews post workshops showing responses on "in which areas they would like to apply what they learned during this" (n = 98)

commercial options. One of the most popular mixes of cover crops includes *Eschscholzia californica,* even though it is a highly invasive species (Peña-Gómez and Bustamante 2012). The use of *E. californica* is common and has an ability to grow everywhere. For this reason, it has been propagated by some wine producers who have collected local seeds from roadways and trainlines because they believed erroneously that this would contribute to preserving local genetic diversity (Fig. 19.3b). Therefore they were applying correct restoration techniques but with the wrong species. After wine producers enrolled in the program, they demanded seedmixes without this invasive species. Currently, the development of cover crop mixes with native species, which do not actually exist, is taking place.

In summary, the way in which stewardship is understood and put into practice requires knowledge of local ecosystems, a knowledge that may or may not be available. As some (or probably most) historically used practices that are not good for biodiversity conservation in the Chilean Mediterranean ecosystem are inherited from the global north, local research is extremely important (Rozzi et al. 2012). On the other hand, it has been demonstrated that the main motivations for companies to "go green" are competiveness, legitimation, and ecological responsibility. These motivations are influenced, among other reasons, by individual concern (Bansal and Roth 2000). For these reason, the approach used in workshops has been very valuable in part because it is based on the inquiry methodology of Schoolyard Ecology, in which participants experience the process of learning through their own practices (Feinsinger 1987; Feinsinger et al. 1997). This also has made it possible to work horizontally during workshops, with all employees of the winery and vineyard, and enabling local workers to rediscover and validate their local ecological knowledge without regard to any corporate rank order. In addition these workshops have contributed to corporate social responsibility strategies (CSR) and provide a potential avenue for extending these educational programs to the local community. Indeed, interviews conducted with participants after the workshops indicate that 88 % of respondents believe that protecting native biodiversity inside their farm contributes to conservation goals. Moreover, when asked "in which areas they would like to apply what you learned during this" workshop, 67 % responded "in daily life" (Fig. 19.3c). This shows that enhancing stewardship in vineyard practices can have an extended effect into broader areas.

After this experience, it is strongly believed that diversification of knowledge transfer, in which the variety of individual motivations for conservation are acknowledged (e.g. from CSR to cultural ancestral value) has been one of the reasons why these workshops have been valued greatly by wineries in the program. Through this type of strategies almost all participants find some cultural connection to the environment, which has also been regarded as a powerful social force fostering stewardship and social-ecological sustainability (Berkes et al. 2000; Chapin 2009b). Despite the success of this initiative, the lack of institutional incentives for conservation of private lands remains a real threat to this voluntary strategy.

19.3 Awakening into Biodiversity Leads to Urban and Rural Earth Stewardship

Both study cases show that through capacity building activities with local community, we could rely on a bottom-up approach to improve knowledge about wetland values in Valdivia and natural landscapes around vineyards. However, the combined effect, with a top-down approach (government institutions and corporations), increases the effectiveness of the decision making process. People's knowledge and past experiences influence people's perception (Kaplan et al. 1998), and people's perception influence people's values, attitudes and actions towards caring for the environment (Nassauer 1995a, b; Rozzi 2013). This complementary bottom-up and top-down approach can lead to earth stewardship through increasing community awareness of the diverse values derived from ecosystems.

The socio ecological approach in both study cases has also generated an aesthetic experience, which can be scenic (awareness through landscape beauty) or ecological (awareness through enhanced knowledge of ecosystems) aimed at engaging the community in earth stewardship. Other studies have demonstrated that, for example, an ecological aesthetic approach has been widely used as an educational tool for improving attitudes toward fire management practices in environments where fire is an aid to biodiversity (Gobster 1994). A growing ecological aesthetic (*sensu* Gobster et al. 2007) has also grown in wine producers, and has contributed to improving the interrelationships between ecology and wine production, by incorporating values that lack an evident relationship to economic incentives (e.g. native flora in case of the wineries). Other urban studies also suggest that a scenic aesthetic, can be an effective catalyst for earth stewardship (Felson et al. 2013).

Biophysical landscape elements play a part in the valuation of urban ecosystems, regardless of their 'ugly' and 'dirty' appearance. They are known as 'cues to care' (Nassauer 1995b), or familiar landscape elements which can first trigger people's attraction for landscape and then develop over time a deep concern for it. In the case of Valdivia, the cues to care in urban wetlands are missing, but are much needed to improve the interrelationships between people and wetland systems for earthquake recovery.

In the same way, the intentionality to sustain animal and plant production for human use, underscores the social-ecological connection that has been going for at least 10,000 years, when domestication began (Smith 1998; Naylor 2009). An holistic approach to Earth stewardship should consider a variety of values, and undertake socio-ecological a socio-ecological approach.

Acknowledgements This work was funded by CONICYT PFB-23, Millennium Scientific Initiative grant# PO5-002 through IEB-Chile, FONDAP grant# 1510020, and FONDECYT#11110183 to OB and # 11110297 to PV. We are very grateful to K. Godoy, C. García, K. Gomez, N. Pohl, J. Diaz and M. Marquez for their contribution to this chapter.

References

Allan P, Bryant M (2011) Resilience as a framework for urbanism and recovery. J Landsc Arch 6:34–45

Altieri MA (1999) The ecological role of biodiversity in agroecosystems. Agr Ecosyst Environ 74:19–31

Armesto JJ, Arroyo MTK, Hinojosa LF (2007a) The Mediterranean environment of central Chile. In: Veblen TT, Young KR, Orme AR (eds) The Physical Geography of South America. Oxford University Press, New York, pp 184–199

Armesto JJ, Díaz I, Celis-Diez JL et al (2007b) Análisis y evaluación técnico-científica de Decreto Supremo No. 78. Technical report, p 37
Armesto JJ, Manuschevich D, Mora A et al (2010) From the Holocene to the Anthropocene: A historical framework for land cover change in southwestern South America in the past 15,000 years. Land Use Policy 27:148–160
Arroyo MTK, Cavieres L (1991) The Mediterranean type climate flora of central Chile – what do we know and how can we assure its protection? Noticiero de Biologia 5:48–56
Arroyo MTK, Cavieres L, Marticorena C et al (1995) Convergence in the Mediterranean floras in Central Chile and California: insights from comparative biogeography. In: Arroyo MTK, Zedler PH, Fox MD (eds) Ecology and biogeography of Mediterranean Ecosystems in Chile, California, and Australia. Springer, New York
Banco Central (2012) Series de Indicadores, Actividad Económica y Gasto. http://xurl.es/e4yw2
Bansal P, Roth K (2000) Why companies go green: a model of ecological responsiveness. Acad Manage J 43:717–736
Barbosa O, Tratalos JA, Armsworth PR et al (2007) Who benefits from access to green space? A case study from Sheffield, UK. Landsc Urban Plan 83:187–195
Barrientos SE, Ward SN (2007) The 1960 Chile earthquake: inversion for slip distribution from surface deformation. Geophys J Int 103:589–598
Berkes F, Colding J, Folke C (2000) Rediscovery of traditional ecological knowledge as adaptive management. Ecol Appl 10:1251–1262
Chapin FS III (2009) Managing ecosystems sustainability: the key role of resilience. Chapter 2. In principles of ecosystem Stewardship. In: Chapin FS III, Kofinas GP, Folke C (eds) Resilience-based natural resource management in a changing World. Springer, New York, pp 29–54
Chapin FS III, Folke C, Kofinas GP (2009) A framework for understanding change, chapter 1. In principles of ecosystem Stewardship. In: Chapin FS III, Kofinas GP, Folke C (eds) Resilience-based natural resource management in a changing World. Springer, New York, pp 3–28
Chapin FS III, Carpenter SR, Kofinas GP et al (2011) Ecosystem Stewardship: sustainability strategies for a rapidly changing planet. Trends Ecol Evol 25:241–249
CONAF (2013) Catastro de los Recursos Vegetacionales Nativos de Chile, Monitoreo de Cambios y Actualizaciones, Período 1997–2011. Corporación Nacional Forestal, Julio 2011, Santiago de Chile
Cowling RM, Rundel PW, Lamont BB et al (1996) Plant diversity in Mediterranean-climate regions. Trends Ecol Evol 11:362–366
Cox RL, Underwood EC (2011) The importance of conserving biodiversity outside of protected areas in Mediterranean ecosystems. Plos One 6:e14508
Crutzen PJ (2002) Geology of mankind. Nature 415:23
Durán AP, Casalegno S, Marquet PA et al (2013) Representation of ecosystem services by terrestrial protected areas: Chile as a case study. Plos One 8:e82643
Feinsinger P (1987) Professional ecologists and the education of young children (commentary). Trends Ecol Evol 2:51–52
Feinsinger P, Margutti L, Oviedo RD (1997) School yards and nature trails: ecology education outside the university. Trends Ecol Evol 12:115–120
Felson AJ, Bradford MA, Terway TM (2013) Promoting earth stewardship through urban design experiments. Front Ecol Environ 11(7):362–367
Fuentes N, Pauchard A, Sanchez P et al (2013) A new comprehensive database of alien plant species in Chile based on herbarium records. Biol Invasions 15:847–858
García V, Audefroy J, Briones F (2012) Estrategias sociales de prevención y adaptación. Centro de Investigaciones y Estudios Superiores en Antropología Social/La Red, México
Gobster P (1994) The urban savanna. Reuniting ecological preferences and function. Restor Manag Notes 12:64–71
Gobster P, Nassauer J, Daniel T (2007) The shared landscape: what does aesthetics have to do with ecology? Landsc Ecol 22:959–972
Guarda G (2009) Cuatro Siglos de Evolución Urbana Valdivia 1552–1910. Universidad Austral de Chile, Valdivia

Hayashi M (2010) Water revives Kobe communities after the Great Hanshin Awaji earthquake. University of Hyogo, Awaji City

Henriquez CA, Simonetti JA (2001) The effect of introduced herbivores upon an endangered tree (Beilschmiedia miersii, Lauraceae). Biol Conserv 98:69–76

Herrmann G (2014) Regulation of coastal zones and natural disasters: mitigating the impact of tsunamis in Chile through urban and regional planning. The Berkeley Law Journal of Issues in Legal Scholarship (in press)

INE (2011) Instituto Nacional de Estadísticas. Estadísticas Demográficas, Compendio Estadístico 2011. www.ine.cl

ISDR (2005) Hyogo framework for action 2005–2015: building the resilience of nations and communities to disasters. International Strategy for Disaster Reduction (ISDR), Hyogo. www.unisdr.org/wcdr. Accessed 30 May 2013

Kaplan R, Kaplan S, Ryan RL (1998) With people in mind: design and management of everyday nature. Island Press, Washington, DC

LA RED (2013) Red de Estudios Sociales en Prevención de Desastres Abril. http://www.desenredando.org/. Accessed 20 Jan 2014

MacDonald KI (2010) The devil is in the (Bio)diversity: private sector "engagement" and the restructuring of biodiversity conservation. Antipode 42(3):513–550. ISSN 0066-4812, doi:10.1111/j.1467-8330.2010.00762.x

MINAGRI: Ministerio de Agricultura, división de protección agrícola y forestal, sub departamento de viñas y vinos, inocuidad y biotecnología sección viñas y vinos año publicación (2013) Catastro Vitivinícola Nacional 2012

MINVU (Ministerio de Vivienda y Urbanismo) (2014) Política Nacional de Desarrollo Urbano. MINVU/PNUD, Santiago. http://politicaurbana.minvu.cl/wp-content/uploads/2014/10/L4-Politica-Nacional-Urbana.pdf. Accessed 24 Dec 2014

Myers N, Mittermeier RA, Mittermeier CG et al (2000) Biodiversity hotspots for conservation priorities. Nature 403:853–858

Nassauer JI (1995a) Culture and changing landscape structure. Landsc Ecol 10:229–237

Nassauer JI (1995b) Messy ecosystems, orderly frames. Landsc J 14:161–170

Naylor RL (2009) Managing food production systems for resilience. Chapter 12. In: Chapin FS III, Kofinas GP, Folke C (eds) Resilience-based natural resource management in a changing World. Springer, New York, pp 259–280

Nicholls CI, Parrella M, Altieri MA (2001) The effects of a vegetational corridor on the abundance and dispersal of insect biodiversity within a northern California organic vineyard. Landsc Ecol 16:133–146

Olson DM, Dinerstein E (2008) The Global 200: a representation approach to conserving the Earth's most biologically valuable ecoregions. Conserv Biol 12:502–515

ONU-Habitat (2012) Estado de las ciudades de América Latina y el Caribe, 2012. Rumbo a una nueva transición urbana. www.onuhabitat.org

Patrick-King A, Berry AM (2005) Vineyard delta 15-N, nitrogen and water status in perennial clover and bunch grass cover crop systems of California's Central Valley. Agr Ecosyst Environ 109:262–272

Pauchard A, García RA, Peña E et al (2008) Positive feedbacks between plant invasions and fire regimes: Teline monspesulana (L) K.Kock (Fabaceae) in central Chile. Biol Invasions 10:547–553

Pauchard A, García R, Langdon B et al (2011) The invasion of non-native plants in Chile and their impacts on biodiversity: history, current status, and challenges for management. In: Biodiversity conservation in the Americas: lessons and policy recommendations. Figueroa E. Editorial FEN-Universidad de Chile, Santiago, pp 133–165

Pauchard A, Barbosa O et al (2013) Latin America: rapid urban development and social economic inequity threatens biodiversity hotspots, chapter 28. In: Elmqvist T, Fragkias M, Goodness J (eds) Urbanization, biodiversity and ecosystem services: challenges and opportunities – a global assessment. Springer, Dordrecht

Peña-Gómez FT, Bustamante RO (2012) Life history variation and demography of the invasive plant Eschscholzia Californica Cham. (Papaveraceae), in two altitudinal extremes, Central Chile. Gayana Bot 69:113–122

Power ME, Chapin FS III (2009) Planetary Stewardship. Front Ecol Environ 7:399

Renouf V, Miot-Sertier C, Strehaiano P et al (2006) The wine microbial consortium: A real terroir characteristic. Int J Vine Wine Sci 40:209–216

Resilience Alliance (2010) Assessing resilience in social-ecological systems: workbook for practitioners version 2.0. http://www.resalliance.org/3871.php. Accessed 30 May 2013

Rozzi R, Armesto JJ, Gutiérrez JR et al (2012) Integrating ecology and environmental ethics: earth stewardship in the southern end of the Americas. BioScience 62(3):226–236

Rozzi R et al (2013) Biocultural ethics: from biocultural homogenization toward biocultural conservation, chapter 2. In: Rozzi R, Pickett STA, Palmer C (eds) Linking ecology and ethics for a changing world: values, philosophy, and action. Springer, Dordrecht, pp 9–32

Simonetti JA (1999) Diversity and conservation of terrestrial vertebrates in Mediterranean Chile. Rev Chil Hist Nat 72:493–500

Skewes JC, Rehbein R, Mancilla C (2012) Ciudadanía y sustentabilidad ambiental en la ciudad: la recuperación del humedal Angachilla y la organización local en la Villa Claro de Luna, Valdivia, Chile. EURE 38:127–145

Smith BD (1998) The emergence of agriculture. Scientific American Library, New York, 232pp

Tang L, Li H (2009) Corporate social responsibility communication of Chinese and global corporations in China. Public Relat Rev 35:199–212

Tognelli MF, de Arellano PIR, Marquet PA (2008) How well do the existing and proposed reserve networks represent vertebrate species in Chile? Divers Distrib 14:148–158

Underwood EC, Viers JH, Klausmeyer KR et al (2009) Threats and biodiversity in the Mediterranean biome. Divers Distrib 15:188–197

Vale LJ, Campanella TJ (eds) (2005) The resilient city, how modern cities recover from disaster. Oxford University Press, New York

Van den Berg A, Hartig T, Staats H (2007) Preference for nature in urbanized societies: stress, restoration, and the pursuit of sustainability. J Soc Issues 63(1):79–96

Viers JH, Williams JN, Nicholas KA et al (2013) Vinecology: pairing wine with nature. Conserv Lett 6(5):287–299

Villagra P, Rojas C (2013) Dimensiones física y cultural de la resiliencia post-desastre: ¿Son compatibles en ciudades Chilenas? Revista Geográfica del Sur IV(6):85–102

Villagra P, Rojas C, Ohno R et al (2014) A GIS-base exploration of the relationships between open space systems and urban form for the adaptive capacity of cities after an earthquake: the cases of two Chilean cities. Appl Geogr 48:64–78

Von Hase A, Rouget M, Cowling RM (2010) Evaluating private land conservation in the cape lowlands, South Africa. Conserv Biol 24(5):1182–1189

Wilson JE (2001) Geology and wine 4. The origin and odyssey of terroir. Geosci Can 28:139–141

Wines of Chile, Studies and Reports. http://www.winesofchile.org/studies-reports/exports/. Accessed 20 Oct 2014

WUP (2011) United Nations Department of Economic and Social Affairs/Population Division. 2012. World urbanization prospects: The 2011 revision United Nations. United Nations, New York

Part III
Integrating Ecology and Ethics as a Foundation for Earth Stewardship Action

Chapter 20
Stewardship Versus Citizenship

Eugene C. Hargrove

Abstract Although "stewardship" may be an environmentally useful term in some contexts, it is also limiting in many ways because it is tied narrowly to three religious traditions, Judaism, Islam, and Christianity. Thus, while it may be extremely helpful in reaching people who are members of one of these three traditions, people who are not followers of these religions may have difficulty with environmental material that is presented to them under the banner of stewardship. They may feel that the message they are receiving is colonizing, imperialistic, and/or totalizing. An alternative term that is religiously neutral is "environmental citizenship," a term first used extensively by Environment Canada, but also used to some degree elsewhere, including the United Nations. It is possible that in some countries both stewardship and citizenship could be pursued in parallel. However, because of the Culture War, which began in the first decade of the nineteenth century, when Catholics first began coming to the United States in large numbers and were displeased with the Protestant religion and ethics they found in the public schools, ethics was largely removed from them by 1860. Since then watch groups in most major religions have formed to watch the public schools to prevent its reintroduction. Given that *stewardship* is a recognizable religious term, a program of Earth or environmental stewardship is more likely to be opposed than a program on Earth or environmental citizenship. Thus, it might be best to focus on citizenship in the public schools and stewardship among Judeo-Christian-Islamic religious audiences.

Keywords Christianity • Citizenship • Islam • Judaism • Stewardship

20.1 Introduction

In 2011 the Ecological Society of America (ESA) 96th Annual Meeting had as theme "Earth Stewardship: Preserving and enhancing the earth's life-support systems." The ESA's stewardship initiative aims to find simultaneous solutions to a

E.C. Hargrove (✉)
Center for Environmental Philosophy, University of North Texas, Denton, TX, USA
e-mail: Eugene.Hargrove@unt.edu

R. Rozzi et al. (eds.), *Earth Stewardship*, Ecology and Ethics 2,
DOI 10.1007/978-3-319-12133-8_20

suite of interconnected problems that threaten the ability of the Earth to provide the services and resources on which we depend (Chapin et al. 2011). If the ESA's Earth Stewardship initiative is to address a global scale, then it needs to better represent the biocultural heterogeneity of the contrasting regions of planet (Rozzi et al. 2012). In this chapter I argue that to effectively address global scale socio-ecological challenges and convene participation at a planetary scale, the concept of environmental citizenship might better suited than stewardship.

20.2 Stewardship and Dominion

In the beginning, the term *dominion* was similar to what stewardship has come to mean today. When the term was translated into European languages, however, it became erroneously associated with another term, *domination.* To capture something of the original sense of dominion, people began using *stewardship,* a term that appears only five times in the Bible and mostly in the context of stories about bad managers. Citizenship has much to commend it environmentally. First, it can be associated with Aldo Leopold's remark that humans are "plain citizens" of the biotic community, opening a door into Leopold's writings and thought. Second, it can be used to distinguish values related to being a consumer from values related to being a good citizen. This distinction is important because economists typically argue for policy in terms of consumer preferences, erroneously substituting them for citizen preferences which can be quite different. Finally, citizenship can help tie ethics and politics together in a sense promoted long ago by Aristotle. Ethics and politics are basically the same, that is, formed in terms of the same elements of moral character, but with the first focused on the good of the individual and the second on the good of the group. Although stewardship need not be abandoned, since spreading the word about the need to protect the environment to Jews, Muslims, and Christians is certainly worthwhile, a focus on citizenship can spread that word farther without danger of religious and cultural backlash.

Environmental stewardship under the label "land stewardship" has been promoted by such Christian environmental thinkers and practitioners as Wendell Berry and Wes Jackson. Berry (1981, p. 81) writes: "To see and respect what is there is the first duty of stewardship … That is an ecological principle and a religious one." Essentially for him, the purpose of stewardship is to protect the Earth: "… in losing stewardship, we lose fellowship; we become outcasts from the great neighborhood of Creation. It is possible—as our experience in *this* good land shows—to exile ourselves from Creation, and ally ourselves with the principle of destruction…" (Berry 1981, p. 281). Jackson, likewise, strikes a religious tone, noting that Mennonite "famers, like their close religious relatives, the Amish, believe that the highest calling of God is to farm and be good stewards of the soil. Within an agricultural context, they are usually regarded as the most ecologically correct farmers in America. The strong ethic of land stewardship is, without doubt, largely responsible" (Jackson 2011, p. 10). Jackson is a founder of the Land Institute in Kansas and has worked closely with the Land Stewardship Project in Minnesota.

In addition to such nonprofit efforts in stewardship, there are also some governmental efforts. The Environmental Protection Agency of the United States supports environmental stewardship which it defines as "the responsibility for environmental quality shared by all those whose actions affect the environment."[1] The United Kingdom has an advanced funding program for farmers and land managers in England to promote environmental stewardship through its Department for Environment, Food and Rural Affairs (Defra).[2] Introduced in Great Britain in 2005, this stewardship agri-environment scheme includes a wide variety of activities: wildlife conservation, landscape enhancement, historical environment protection, promotion of public access and understanding of the countryside, natural resources protection, the prevention of soil erosion and water pollution, and support of environmental management of upland areas. Its secondary objectives include genetic conservation of cattle breeds and fruit trees. It even aims to help the environment adapt to climate change.[3]

Stewardship also found its way into early environmental philosophy literature in opposition to dominion. In his book, *Man's Responsibility for Nature: Ecological Problems and Western Traditions,* originally published in 1974, John Passmore tried to head off the creation of the field of environmental philosophy, arguing that Western civilization depends on dominion defined as the domination of nature and that stewardship, though mentioned in the Bible, is a weak alterative. Passmore asserted that "an ethic … is not the sort of thing that one can simply decide to have; 'needing an ethic' is not in the least like 'needing a new coat.' A 'new ethic' will arise out of existing attitudes, or not at all" (Passmore 1980, p. 56). Passmore's point was that dominion was a much better coat in the context of Western traditions than stewardship. Later, Robin Attfield (1983) in *The Ethics of Environmental Concern* argued that, to the contrary, stewardship was compatible with Western traditions and the appropriate model for nature conservation and preservation, siding with environmentalists such as Berry and Jackson.

Although dominion defined as the domination of nature is clearly not a good model for environmentalism, there is a great deal of confusion within the debate between dominion and stewardship. Stewardship is a curious replacement for dominion because originally in the Hebrew language *dominion (radah)* had the same meaning that *stewardship* has today. It was Adam and Eve's obligation to take care of the Garden of Eden. However, when *dominion* was translated out of Latin into European languages, it was mistranslated as *domination*. It then became associated with God's commandment to "subdue" [*kabash*] the Earth.[4] Environmental historian J. Donald Hughes (1975, p. 124) has argued further that *dominion* did not become environmentally harmful until it was linked with Artistotle's views in the

[1] Environmental Protection Agency at http://www.epa.gov/stewardship. Established in 1994, the Pesticide Environmental Stewardship Program (PESP) is an EPA partnership program that works with the nation's pesticide-user community to promote Integrated Pest Management (IPM) practices.

[2] See Department for Environment, Food and Rural Affairs at https://www.gov.uk/environmental-stewardship.

[3] Ibid. at https://www.gov.uk/environmental-stewardship#what-is-environmental-stewardship.

[4] Genesis 1:28.

late Middle Ages. Aristotle wrote in the *Politics* that plants exist for the sake of animals, tame animals for use and food, and wild animals, if not all, for the greater part, for food and for the provision of clothing and various instruments for human beings.[5] Viewed in this way, nature is simply a resource for humans to use for their personal benefit.

Lloyd H. Steffen (1992) in "In Defense of Dominion" has argued that turning to *stewardship* from *dominion* was not a particularly good choice, given the role that stewardship actually plays in the Bible. It is associated mostly with stories of bad managers: "The Hebrew word *radah* [*dominion*] means 'govern, rule, have dominion.' The verb was employed to refer to the rule of kings over territory, masters over servants, and the rule of God either over land or in the midst of God's enemies."[6] *Dominion* in terms of its original meaning "identifies and promotes action of a limited kind that expresses an attitude of non-maleficence, even benevolence, toward that which God created and designated good. Dominion respects the integrity of creaturely existence and assigns human responsibility and accountability for any dominion activities undertaken."[7] In contrast, "Biblically, stewardship refers to a notion of 'one who is over a house'—a household manager … Stewardship is put to work in Scripture to illustrate injustice and abusive power relations. The parable of the dishonest steward (Luke 16:1–13) points out that stewards can, as stewards, abuse their position and misuse their power."[8] Stewardship has come to some prominence in environmental circles only because the mistranslation *dominion* as domination has irreparably made it unusable. However, as Lloyd notes, *stewardship* is far from a perfect replacement for *dominion:* "When translated into environmental terms, the stewardship concept entails a view of the Earth as property (to which rights of ownership inevitably attach) and upholds the values of anthropocentrism: that the steward serves the master by managing or mastering the household." He adds that it "might open Christians to a renewed broadside from Christianity's critics,"[9] referring to the 10-year Lynn White (1967) debate throughout the 1970s over Christian responsibility for causing the environmental crisis.

20.3 Stewardship as a Colonizing Term

A major additional problem for stewardship is that it, like dominion, is limited in its application because it is closely connected with three religious and cultural traditions: Christianity, Judaism, and Islam. It is not associated with any other major Asian religious traditions, for example, Hinduism and Buddhism, or the various

[5] Aristotle, *Politics,* 1256b7-22.

[6] Steffen (1992), pp. 64–65.

[7] Ibid., p. 78.

[8] Ibid.

[9] Ibid.

indigenous cultural and religious traditions in North America, South America, Africa, and the Pacific.

Although cultural borrowing does sometimes occur, an idea or concept from one culture inserted to another can often be counterproductive. It can be called colonizing, imperialistic, and totalizing. Such has been the fate of the national park idea and the related wilderness concept in Africa and southern Asia. Because both call for the exclusion of humans from natural areas, there has been considerable social disruption and resentment as local peoples were prevented from their traditional, centuries-old relationships with the land. In addition, teaching Western environmental values has proved to be very difficult because children usually form their values tacitly[10] long before entering elementary school. G. W. Burnett and Kamuyu wa Kang'ethe (1994, p. 159) in "Wilderness and the Bantu Mind" point out that "efforts to instruct Kenyan school children in Western wildlife and wildland values are predicated, at least in part on an assumption of an indigenous attitude to wilderness that is unacceptable to the West." They continue:

> Efforts to instruct Africans in Western wilderness values have proceeded with little, or no, articulation of how Africans might already understand wilderness, and consequently, what ideas the interventionist seeks to change. It would be far easier if the Bantu concept of wilderness could be articulated and developed as an indigenous philosophy of wilderness (Burnett and wa Kang'ethe 1994, p. 159).

The fate of stewardship internationally could be much like that of the national park idea and the wilderness idea in non-Western countries. Burnett and wa Kang'ethe speculate that if the West had left Bantuland alone, the Bantu might have come to love wilderness just as the Puritans did in New England, but the imposition of national parks has made that impossible.[11]

20.4 Stewardship Versus Citizenship

An alternative way to promote something like the original meaning of *dominion* probably without crosscultural problems is to focus on the notion of ecological or environmental citizenship, a notion that is religiously neutral and not associated specifically with Western culture. Environmental citizenship began with Environment Canada,[12] which developed programs for children on the web on this theme (although more recently it seems to have abandoned this project). The United Nations Environmental Programme has also embraced the term for a time, although references are also currently mostly missing from its website. When UNEP was more active in developing global environmental citizenship, Alicia Bárcena, Senior Advisor on Global Environmental Citizenship wrote: "... new relationships and

[10] See Polanyi (1967), p. 4. This book is an extension of Polanyi's basic position developed more extensively in Polanyi (1974).

[11] Ibid., p. 160.

[12] See http://www.ec.gc.ca/default.asp?lang=En&n=FD9B0E51-1.

interactions are creating the basis for a global environmental citizenship, with rights and responsibilities to the planet as its capacity to support human life is pushed to the limits. Society must collectively manage its own future." She added: "Global environmental citizenship is about asserting the ethical responsibilities of individuals, organizations, countries and corporations to create new forms of solidarity to protect all life on Earth."[13]

Although the Aldo Leopold Foundation emphasizes land stewardship,[14] Leopold himself spoke of citizenship instead of stewardship. He wrote in *The Land Ethic*, "In short, a land ethic changes the role of *Homo sapiens* from conqueror to plan member and citizen of it. It implies respect for his fellow-members, and also respect for the community as such."[15] Leopold's use of the term *citizen* provides a lot of alternatives in educational contexts. There is the possibility, of course, of tying the passage to stewardship for Judeo-Christian-Islamic audiences.

A second approach, following the work of Mark Sagoff, is to contrast *citizen* with *consumer*.[16] As Sagoff points out, we humans have preferences both as citizens and as consumers and these can be in conflict. Economists currently document our consumer preferences through surveys and declare these consumer preferences to be our citizen preferences, which is actually a category mistake,[17] since what we prefer as consumers may be different than what we prefer as citizens. For example, although we love to drive our cars and hate to take the bus, we may nevertheless as citizens support gasoline taxes to pay for public transportation.[18] Ricardo Rozzi agrees with Sagoff in criticizing the narrowness of the prevailing economic language. He affirms that "Long-term socio-ecological research programs have mostly emphasized economic values while the broader dimensions of ethics have been overlooked" (Rozzi et al. 2012, p. 226). A focus on citizen preferences as opposed to consumer preferences is also compatible with Leopold's concern that farmers only pay attention to economic considerations and have not developed an ethical relation to the land.[19]

A third possibility is to tie Leopold's view to the philosophy of Aristotle. According to Aristotle in his *Nicomachean Ethics*, the character traits of the moral agent and the citizen are the same except for the fact that ethics is from the standpoint of the individual and politics (or citizenship) is from the standpoint of the group.[20] Leopold was concerned that political actions on behalf of the environment could not take place without support of the general public. His recognition that

[13] Alicia Bárcena, "Global Environmental Citizenship," Our Planet 8.5, January 1997, UNEP 15, http://www.ourplanet.com/imgversn/85/barcena.html.

[14] See http://www.aldoleopold.org/Programs/stewardship.shtml.

[15] Leopold (1949), p. 204.

[16] Sagoff (2008), pp. 47–52.

[17] Ibid., p. 51.

[18] Ibid., p. 48.

[19] Leopold, "The Land Ethic," p. 214.

[20] Aristotle, *Nicomachean Ethics,* Book 1, Chap. 2.

ethics and politics were based on the same moral character caused him to emphasize moral education and write "The Land Ethic."

There is also considerable technical, theoretical material that can be used in developing environmental citizenship within a given society. The two major traditions are liberal and civic republican citizenship. In addition, Andrew Dobson has argued that a new tradition is needed, which he calls post-cosmopolitan citizenship.[21] It is also possible to tap into a wide variety of positions and perspectives on environmental citizenship.[22] While it is possible to become as academic and theoretical as needed, the success of environmental citizenship, like that of environmental stewardship, will depend on its acceptance at a grass-roots level, where a more general conception may work better.

If Passmore is correct that "A 'new ethic' will arise out of existing attitudes, or not at all," then trying to promote a view with as few drawbacks and handicaps as possible is probably the best approach. Within Christianized Western society, environmental stewardship probably has few handicaps, especially among Christians, and perhaps among Jews and Muslims as well, who also have a conception of stewardship within their religious traditions. Nevertheless, an alternative approach for minorities not sharing in the main religious traditions would be appropriate. In moving to non-Western societies, the handicaps of Judeo-Christian-Islamic stewardship may be more problematic if stewardship in those countries comes to be viewed as colonizing, imperialistic, or totalizing, as noted earlier in the paper. It is true that crosscultural borrowing does occur, but it is unpredictable whether a notion clearly attached to another culture will catch on, and successful borrowing is usually not the norm.

The advantage of citizenship over stewardship is that it is religiously and culturally neutral. As a result, environmental citizenship can more likely be formed out local conceptions of what citizenship is. There is an interest in developing an international environmental ethic. Whether such an ethic is possible and could ever be created is debatable. All of the elements of such an ethic have to come from somewhere and when some of the elements are regarded as foreign intrusions, efforts will be more difficult. Promoting environmental citizenship will probably not experience such problems since it can be presented as an extension of local conceptions of citizenship.

20.5 Stewardship and Citizenship

Although so far I have been presenting stewardship and citizenship as conflicting and competing approaches, it is important to note that there is a place for both: instead of stewardship versus citizenship, we could talk in terms of stewardship *and* citizenship. In countries where stewardship is a strongly established concept

[21] See Dobson (2003).

[22] See, for example, Dobson and Bell (2006).

religiously, stewardship may be a good if not the best approach especially for people within appropriate religious traditions. At the same time, citizenship could also be promoted. People who are willing to embrace environmental stewardship religiously can no doubt also embrace environmental citizenship. Citizens in such countries who find stewardship culturally offensive can be approached in terms of citizenship alone. Likewise, in countries where Judeo-Christian-Islamic religions are not dominant, a similar approach can be followed but with a focus on the more neutral notion of citizenship with perhaps a minority focus on stewardship.

The environmental or Earth citizenship approach will not directly lead to an international environmental ethic. However, it may in the long term. In the United States, there is a tendency to encourage states to experiment with solutions to problems independently. This approach prevents the entire country from suffering from a policy that works in some states but not in others. Furthermore, trying out competing solutions at the state level improves the chances that a policy may eventually be found that works everywhere for everyone. The development of culturally diverse environmental ethics throughout the world will likely improve the environment on the short term more effectively and perhaps through cultural borrowing eventually lead to an internationally inoffensive environmental ethic, should it actually be possible. Citizenship is compatible with this approach in all countries, since citizenship is unlikely to be considered a foreign intrusion in any of these countries and stewardship, though more limited, will likely be beneficial in those countries religiously compatible with it. Environmentalists and environmental groups who wish to promote stewardship alone should probably restrict their efforts to places where Christianity, Judaism, and Islam can strongly support it.

However, even in the United States an education program based on Earth or environmental stewardship may be difficult because of the two-century-old Culture War,[23] which began in the first decade of the nineteenth century, when Catholics entering the United States were displeased to find Protestant religion and ethics in the public schools. To attempt to resolve this problem and get the Catholics to send their children to the public schools, ethics and religion was gradually removed from them. By 1860 it had been largely eliminated (Jorgenson 1987, p. 110), and thereafter watch groups formed in most of the major religions to ensure that ethics would not be reintroduced. Today, when teachers try to teach ethics they are usually accused of trying to indoctrinate the children into their personal ethical values. When they respond that they are providing alternative perspectives and are therefore not indoctrinating, they are accused of teaching relativism: that ethics is just a matter of how individuals feel emotively and it has no non-arbitrary meaning or standards. Such criticism began to be applied to higher education when Ronald Reagan became President of the United States, though less successfully, especially through the National Endowment for the Humanities. Because *stewardship* is a recognizable religious term, a program of Earth or environmental stewardship is more likely to encounter strong opposition, especially at the primary and secondary levels, than

[23] See Hunter (1991) for an overview of the Culture War, and Hunter (2000), for a detailed discussion of the impact on ethics education.

one called Earth or environmental citizenship. Thus, even in the United States, it might be better to focus on citizenship in the public schools and reserve stewardship for educational efforts among Judeo-Christian-Islamic religious groups.

References

Attfield R (1983) The ethics of environmental concern. Columbia University Press, New York

Berry W (1981) The gift of good land. Counterpoint Press, Berkeley

Burnett GW, wa Kang'ethe K (1994) Wilderness and the Bantu mind. Environ Ethics 16:145–160

Chapin FS III, Pickett STA, Power ME et al (2011) Earth stewardship: a strategy for social-ecological transformation to reverse planetary degradation. J Environ Stud Sci 1:44–53

Dobson A (2003) Citizenship and the environment. Oxford University Press, Oxford/New York

Dobson A, Bell D (eds) (2006) Environmental citizenship. MIT Press, Cambridge, MA/London

Hughes JD (1975) Ecology in ancient Greece. Inquiry 18:115–125

Hunter JD (1991) Culture wars: the struggle to define America, making sense of the battles over the family, art, education, law, and politics. Basic Books, New York

Hunter JD (2000) The death of character: moral education in an age without good or evil. Basic Books, New York

Jackson J (2011) Nature as the measure: selected essays of Wes Jackson. Counterpoint Press, Berkeley

Jorgenson LP (1987) The state and the non-public school: 1825–1925. University of Missouri Press, Columbia

Leopold A (1949) The land ethic. In: A sand county almanac: and sketches here and there. Oxford University Press, New York, p 204

Passmore J (1980) Man's responsibility for nature: ecological problems and western traditions, 2nd edn. Duckworth, London

Polanyi M (1967) The tacit dimension. Doubleday and Co/Anchor Books, Garden City

Polanyi M (1974) Personal knowledge: towards a post-critical philosophy. Harper and Row/Harper Torchbooks, New York/Evanston

Rozzi R, Armesto JJ, Gutiérrez J et al (2012) Integrating ecology and environmental ethics: earth stewardship in the southern end of the Americas. BioScience 62(3):226–236

Sagoff M (2008) The economy of the earth: philosophy, law, and the environment, 2nd edn. Cambridge University Press, New York, pp 47–52

Steffen LH (1992) In defense of dominion. Environ Ethics 14:63–80

White L Jr (1967) The historical roots of our ecologic crisis. Science 155:1203–1207

Chapter 21
The Ethics of Participatory Processes: Dynamic Flux, Open Questions

Peter J. Taylor

Abstract Collaboration and participation are widely emphasized in environmental planning and management. This chapter describes a discussion group on the ethics of participatory processes, raises the possibility of translating the non-equilibrium or dynamic flux view of ecological complexity into a view of ethics and social action, and introduces five ideals for a "dynamic flux ethics"—engagement, participation, cultivating collaborators, transversality, and fostering curiosity. These ideals are linked to a schema woven out of the discussion group's contributions. What sense of stewardship might come from participatory processes informed by this initial exploration of dynamic flux ethics is left as an open question.

Keywords Curiosity • Dialogue • Dynamic flux • Engagement • Transversality

> The most important parts of any conversation are those that neither party could have imagined before starting. Isaacs (1999), *Dialogue*

> [T]he challenge [is to] bring... into interaction not only a wider range of researchers, but a wider range of social agents, and to... keep... them working through differences and tensions until plans and practices are developed in which all the participants are invested. Taylor (2005), *Unruly Complexity*

> "As I said at the beginning," [he] shouted, "you'll cut and run." Jon stood holding the door. The edge of the wood was between his fingers. "I told you. I have these questions to ask. Open questions." Williams (1985), *Loyalties*

A key principle of dialogue is to balance advocacy (making a statement) with inquiry (seeking clarifications and understanding). In a chapter that will be read after the author has finished writing, this principle is difficult to follow. Nevertheless, let me try to create for readers some of the experience of participatory processes, starting by putting two questions on the table: Do the ethics of participatory processes lead participants to Earth Stewardship? Given that I am unsure of this,

P.J. Taylor (✉)
Program in Critical and Creative Thinking, Program on Science, Technology and Values, College of Education and Human Development, University of Massachusetts, Boston, MA, USA
e-mail: peter.taylor@umb.edu

R. Rozzi et al. (eds.), *Earth Stewardship*, Ecology and Ethics 2,
DOI 10.1007/978-3-319-12133-8_21

where do my loyalties lie in relation to the purpose of the volume or, more broadly, of sustaining diverse life on earth? To indicate why these are open questions as well as to articulate my sense of ethics, participation, and Earth Stewardship, I start by describing a discussion group on the ethics of collaborative or participatory processes that met during the 2011 Cary conference "Linking Ecology and Ethics for a Changing World." I then raise the possibility of translating the non-equilibrium or dynamic flux view of ecological complexity (Pickett 2013) into a view of ethics and social action. The final section introduces five ideals for a "dynamic flux ethics," referring at points to a schema woven out of the discussion group's contributions. There is advocacy in these three sections but of a form conducive, I hope, of further inquiry by readers. Indeed, you may seek clarifications and understanding of suggestions made and questions raised in this chapter, but try to balance a desire to be convinced or to have compelling examples provided with a sense of conversation with a fellow inquirer into "what exists and what could exist" (Foucault 1996). To enhance the sense of unfolding inquiry, some matters that were settled for me before starting are placed in appendices.

21.1 The Ethics of Collaborative or Participatory Processes: From Discussion Group to a Picture

As Taylor et al. (2011) note, "collaboration has become a dominant concern in environmental planning and management since the 1990s (Margerum 2008) [and] the need to organize collaborative environmental research can be traced back at least as far as the tropical rainforest ecosystem projects led by H.T. Odum in the 1950s and'60s." This emphasis makes sense at two levels (elaborated in Appendix 1): Collaboration produces results that are either a 1. *Sum of the Parts* (combining multiple perspectives, extending over time, and spanning distance); or 2. *Greater than the Sum of the Parts* (generating new perspectives, ensuring durability of outcomes, developing capacities).

With a view to combining multiple perspectives, generating new ones, and developing capacities, it seemed appropriate during a conference aiming to link "Ecology and Ethics for a Changing World," to convene a discussion group on the ethics of collaborative or participatory processes, and, moreover, to run the group using participatory processes. A record of the group's three sessions, including some audio files, was kept and can be accessed at Taylor (2011). Table 21.1 provides an outline of the sessions. Because text cannot capture the experience of engaging in participatory sessions, the outline is included to intrigue readers enough to explore for themselves the processes listed and to convey the source of a schema that is to be built on in Sect. 21.3.

Table 21.1 Outline of sessions of a participatory discussion group on the ethics of participatory processes

Session 1—Autobiographical introductions
At the start, four principles for participatory processes were presented:
P1. Facilitators should not try to lead without arranging assistants and support
P2. Participants always know a lot about the topic at hand, so bring that to the surface and acknowledge it
P3. Respect for other participants and for ourselves is the first objective, on which basis participants are more comfortable taking Risks that lead to Revelations (new insights) and, through the experience of generating those insights, get Re-engaged with our work and lives (Taylor et al. 2011)
P4. Do not leave any session without taking stock of where we have come, individually and collectively
The session consisted of the following activities (with corresponding principles in parentheses):
• Guided Freewriting (Taylor and Szteiter 2012, pp. 89–90) starting from "When I think about the questions I have about participatory processes around environment, science, ethics, action, what comes to mind includes…" (P2, P3)
• Share in pairs our hopes for the discussion group (P2, P3)
• Autobiographical introductions: Each person takes 5 min to convey how you came to be the kind of person who would be invited to this Cary conference and join a discussion group on ethics of participatory process (P3)
• Share in pairs "connections and extensions" seen among the introductions, including things you didn't include that you might have (P2, P3, P4)
• Two assistants arranged to confer with facilitator about next session (P1)
• Closing circle: "Something you're taking away from this session to chew on" (P4). (The audio linked to Taylor 2011 includes mention, among other things, of the diversity of motivations to participate and diverse kinds of participation, the difficulty of communicating and incorporating different values and perspectives, and the challenge of moving to action and making a difference.)
Session 2—Dialogue and Rapid small-group work
Two activities explored whether and how ethics of some kind can inform participatory process in relation to linking environment, science, and action:
• Dialogue process (P2, P3, P4)—90 min of listening and structured turn-taking on the topic (Taylor and Szteiter 2012, pp. 70–75). (One provocative query emerging from the dialogue was whether ethics is possible *without* participatory processes.)
• Rapid small-group work (P2)—20 min to create and report on a "Program for developing an ethical framework for participatory processes, with special attention to interaction among diverse social agents." (The activity served primarily as a warm-up for the homework and third session. Themes from end-of-session reports included sustained engagement in listening, having stories be heard, and the tension between incremental progress and taking on Big Issues.)
Between-session homework: Compose five statements, questions, or reservations that are important to you concerning development of an ethical framework for participatory processes

(continued)

Table 21.1 (continued)

Session 3—Future Ideal Retrospective
• Future Ideal Retrospective activity (P2) to synthesize the between-session homework. (In brief, each participant imagined at some *future* time being part of a project that embodied an ethical framework for participatory processes. Looking back (*retrospective*) to explain to someone what contributed to making that *ideal* situation possible, three to five words answers were printed on Post-Its. After collecting and copying these for participants, the task was to find and name clusters of Post-Its. (Links to a description of the process and to the collated Post-Its are given in Taylor 2011.)
• Closing circle (P4): "one thing you're taking away from these sessions to keep developing." (The audio linked to Taylor 2011 includes, among other things, mention of the value of freewriting, the need for being willing to participate, and the difficulty of being in the process without knowing the intended product.)

The intended follow up to the last session of the discussion group was that each person would complete and share the Post-It syntheses then email exchanges might build on these. Because I had prior experience in—and a disposition for—the clustering and naming exercise, I readily generated clusters and grouped these into successively more inclusive clusters, which I shared and have subsequently depicted as Fig. 21.1 (and discuss in Sect. 21.3). However, I know of no other follow-up from the participants. In short, the group did not get to a place where we had developed "plans and practices… in which all the participants are invested." (For description of a multi-stage workshop process for moving to such a result from an initial "Practical Vision," see Stanfield 2002.)

Nevertheless, at least for me, the sessions affirmed that participatory processes can result in a "project that is richer, deeper, and has more dimensions than what you came in with. The more angles… that are brought out by the process, the more likely you are to create something you did not anticipate" (Taylor and Szteiter 2012, 149) (P3). It was by chewing on the clusters in Fig. 21.1 and tensions among them that I was moved to articulate the *ideals of engagement* outlined in Sect. 21.3. For those ideals to make sense to readers let me first share a reflection from just after the 2011 Conference, in which I asked what might have happened if ecological science rather than ethics had taken the lead.

21.2 From Dynamic Flux Ecology to Dynamic Flux Ethics

The scientists, philosophers, and interdisciplinary scholars gathered at the 2011 Conference shared a concern with environmental degradation. One model for stemming that degradation is that people need to have a different ethic about non-human nature to govern their actions, the assumption here being that a person's ethics governs their actions, not vice versa. This model was evident in the repeated reference by

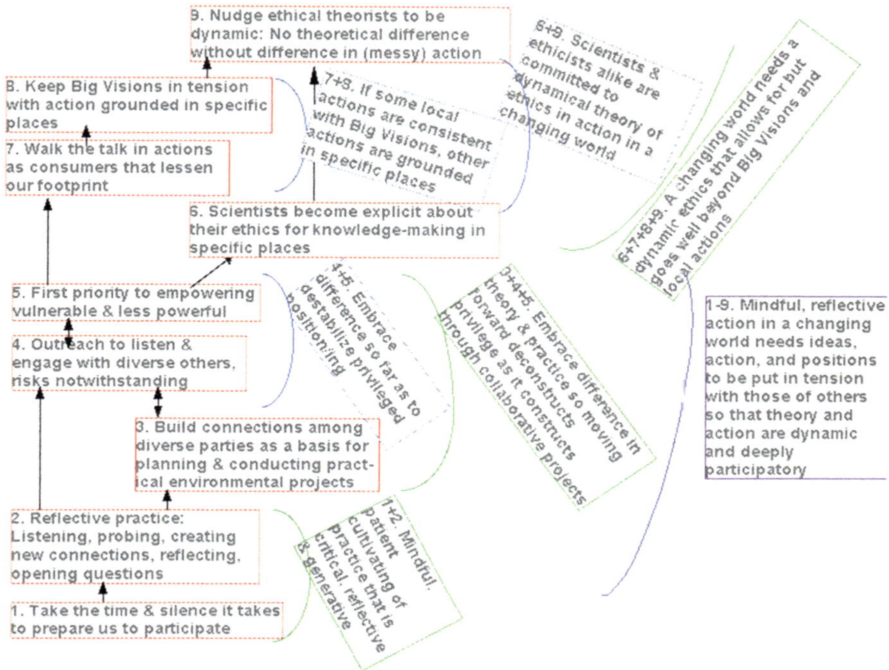

Fig. 21.1 A schema for ethics of participatory processes linking ecology and social action that emerged from a Future Ideal Retrospective activity. The numbered items up the *left* are the names given by the author to clusters of Post-Its, which had been generated by participants responding to the Future Ideal Retrospective prompt (see Table 21.1 for explanation and Taylor 2011 for links to the original Post-Its). These clusters were arranged by the author and linked with *arrows* so as to convey that, if a lower item happens, that makes it more likely that ones above it happen. The clusters are then grouped, as indicated by the *curves* and *numbers*, into successively more inclusive clusters to the *right*. Interpretation of some of the clusters is given in due course in Sect. 21.3; the other clusters are open for readers to give their own meanings to

Conference speakers to Aldo Leopold's land ethic, but also in the reference to animal rights. A variant of this model is to pay attention to religious views about nature, highlighting the ones that seem to be pro conservation or stewardship and downplaying the views that favor exploitation of resources. It is expected of religions that they promote some ethical framework; moreover, they have the authority and numbers to make their views count—to mobilize people into action.

A second model is that economics governs people's actions, collectively as well as individually, so push for an economics based on a different set of values. Factor in especially the benefits of "services" provided by non-human nature—by ecosystems—rather than take them for granted, placing costs to the environment outside economic calculations. A hybrid of the two models, but in a form that provides a counterweight to economics, is *biocultural conservation*, which centers on valuing the conservation jointly of habitats, cultural forms, and peoples threatened by environmental degradation (Rozzi 2013).

The two models and their variants do not stem from efforts to build theory about ecological complexity. (Observing this is not to discount the ecological research needed to measure ecosystem services or to characterize the habitat and co-inhabitants to be conserved.) The ecologist Steward Pickett spoke at the Conference about paradigms in ecology leading up to the present, concluding that values other than economic ones have to come into play to apply knowledge about ecology's dynamic fluxes (Pickett 2013). His conclusion brings us back, however, to the hybrid model above and to a focus on values, not ecological theory. *What might we see if we translated what is entailed in the non-equilibrium view into the realm of human actions?*

The non-equilibrium, dynamic flux view of ecology, as I would summarize it (Taylor and Haila 2001), is as follows: Since the 1980s ecologists became increasingly aware that situations may vary according to historical trajectories that have led to them; that particularities of place and connections among places matter; that time and place is a matter of scales that differ among co-occurring species; that variation among individuals can qualitatively alter the ecological process; that this variation is a result of ongoing differentiation occurring within populations—which are specifically located and inter-connected—and that apparent interactions among the species under study can be the indirect effects of other "hidden" species (i.e., having dynamics not explicitly considered in the study or models).

There is surely an analogous dynamism to the ways that people, in their contingent, changing social organizations, are able to direct and redirect their actions. We could, therefore, pay attention to the ways that situations—social organizations—may vary according to historical trajectories that have led to them; that particularities of place and connections among places matter; that time and place is a matter of scales that differ among co-occurring social groups and institutions; that variation among individuals can qualitatively alter the social and environmental process; that this variation is a result of ongoing differentiation occurring within populations—which are specifically located and inter-connected—and that interactions among the groups and institutions under study can be artifacts of the indirect effects of groups and institutions with dynamics not explicitly considered.

This picture of human action turns the values-centered models of ethics inside out. Values become a contingent snapshot of themes that *appear* to be directing an individual or group—themes that people may or may not make explicit, discuss, debate, and use to negotiate their actions. As an analogy, in ecology and environmental science, we get some guidance, but not very much, by pointing to the evolutionary imperative for organisms to survive and reproduce. Similarly, we should expect to learn a little, but not too much, from focusing on the ethical basis that is, or could be, *inside* the heads or hearts of people. Instead, we might replace values-centered ethics with a *dynamic flux ethics*. Yet what would that look like? And what could one do with it? The answer to the second question remains to be seen. The final section provides my answer to the first question.

21.3 From Engagement to Curiosity: Ideals to Inform an Ethics of Participatory Processes

There is, as yet, no dynamic flux ethics linking ecology and social action. In the space available it is possible, however, to identify five broad ideals that could inform such an ethics. The first ideal follows as a matter of necessity from the picture of dynamic flux; the other ideals flow each from the one before it.

On the presumption that the dynamic flux of ecological and social complexities cannot be well understood from an *outside* view (in which complexities are, say, reduced to a unifying metric such as energy, ascendancy, or ecosystem services), positions of *engagement* must be taken *within* the complexity (Taylor 2005, p. 203ff). Engagement denotes deliberate involvement in a situation in ways that presume that other people will also take an active role. As suggested by cluster 9 in Fig. 21.1, ethical propositions need to make a difference in the social complexities sketched in the Sect. 21.2. Moreover, if a values-centered ethic is advocated—say, animal rights, biocultural, or land ethic—it should be kept in tension with "action grounded in specific places" (cluster 8). Engagement has, in a sense, long been emphasized in Adaptive Environmental Management (Gunderson et al. 1995): research or knowledge production needs to be linked with planning for action and action itself in an ongoing process so that knowledge, plans, and action can be continually reassessed in response to developments—predicted and surprising alike.

On-going re-assessment means that engagement invites *participation* or collaboration. As mentioned above (see also Appendix 1), collaboration in environmental research allows multiple perspectives to be combined, and, in view of the problematic boundaries of ecological situations, for study to extend over time and span distance. It can also generate new perspectives, ensure durability of outcomes, and develop people's capacities—including their capacity to collaborate (Taylor et al. 2011). We should note that the call for participation is sometimes a smokescreen for the powerful to maintain control (Peters 1996). What makes for *ethical* participation is indicated by the initial quote about "bringing into interaction… a wide range of social agents" (reflected in cluster 3 in Fig. 21.1). Moreover, as clusters 4 and 5 suggest: Give "first priority to empowering the vulnerable and less powerful" and undertake "outreach to listen and engage with diverse others, risks notwithstanding."

Generating knowledge about dynamic fluxes and about the effects of people's actions within those fluxes is only part of the rationale for engagement and participation. The objective of developing people's capacities invites attention as well to the process, with a view, whatever the content or outcome, to *cultivating collaborators*. In what I have previously called flexible engagement: "researchers in any knowledge-making situation [should take up the challenge] of connecting quickly with others who are almost ready to foster—formally or otherwise—participatory processes and, through the experience such processes provide their participants, contribute to enhancing the capacity of others to do likewise" (Taylor 2005, p. 210). In this spirit, the placement of clusters 1 and 2 at the bottom left in Fig. 21.1 is meant to suggest that all the other aspects of a dynamic flux ethics are enhanced by "Tak[ing] the time and silence it takes to prepare us to participate" through "listening, probing, creating new connections, reflecting [and] opening questions."

The qualities of engagement, participation, and cultivating collaborators are illustrated by the case of community planning in a district in Northern Ontario included as Appendix 2. However, as the postscript to that case shows, the community's capacities were stretched and its plans undermined by decisions made at a distance by a multinational employer. This experience points to the need for an additional quality to engagement, namely, that it cuts across and connects different strands, processes, and social realms. Such *transversality* of engagement means not only taking seriously the creativity and capacity-building that arises from well-facilitated participation among people who share a place or livelihood, but also incorporating knowledge-making of non-local or trans-local researchers—including knowledge about the dynamics that produce adverse trans-local decisions and about ways to try to mitigate their effects.

A corollary of transversality is that cultures or cultural forms are not foundational entities for understanding the history of a place or situation and its prospect for the future. Granted, it may sometimes be effective as a tactic to focus on biocultural conservation—just as invoking the Endangered Species Act in the United States provides a way to check environmentally unsound economic development (but see Sellers 1999 for some interesting history behind that tactic). Yet, as emphasized by the anthropologist Eric Wolf (1982), the cultural form to be conserved may be the contingent and perhaps transient outcome of connections among places and distant peoples. For example, as rubber began to be used in nineteenth century Europe, the Mundurucú deep in the Amazon changed from villages centered around male-headed, manioc-growing and hunting units, to numerous small female-centered households, "each linked separately to the trading post in a web of exchanges [of latex for commodities]" (Wolf 1982, pp. 17–18, 326ff). Such economically mediated changes may be just what a biocultural ethic seeks to resist—after all, the cultural shift for the Mundurucú was tied up with their growing indebtedness. Yet, given the long reach of commodity chains, such resistance cannot be focused on one social location. The ideal of transversality means finding ways in the Global North to be accountable for the effects that our consumption—as well as the economic production and other actions (e.g., military interventions) that support our consumption—have on people distant from us geographically, culturally, socioeconomically. This challenge of "walk[ing] the talk in actions as consumers that lessen our footprint" (cluster 7 in Fig. 21.1) increases even further if we add distant from us *in time*—in the future—to this list.

Additional corollaries of transversality stem from recognizing that when, as researchers or activists, we are faced with *complex* connectedness and dynamic flux, our sense of how to change and sustain a new orientation is often crystallized by *simple* themes, such as "Reduce CO_2 below 350 ppm," "Maintain biodiversity as essential for human survival," "Promote Earth Stewardship," or "Facilitate participatory approaches." A challenge, then, for a dynamic flux ethics is to acknowledge the impact of simple themes without discounting the additional, more idiosyncratic knowledge researchers have about the complexity of their social context (Taylor 2004, 2005, p. 198). A complementary challenge is to "Keep Big Visions in tension with action grounded in specific places" (cluster 8 in Fig. 21.1). For example, when conservation biologists deeply value the species threatened by the clearing of a tropical rainforest, transversality of engagement would mean that they learn about the social and economic dynamics that embed the people who are clearing the forest as well as those that embed anyone—local or trans-local—who seeks to resist that destruction.

Neither Fig. 21.1 nor this chapter as a whole provides a concrete framework for or illustrations of the addition of transversality to the ideals of engagement, participation, and cultivating collaborators. Whether a dynamic flux ethics would lead to Earth Stewardship remains, therefore, an open question. My last ideal, then, concerns a sense of stewardship characterized not by firm positions or readily identified loyalties, but by mutual recognition among inquirers—among people trying to make sense of their own circumstances as they seek ways to change what has been given to them by dint of history, place, and the unfolding actions of others. In Raymond Williams's novel *Loyalties*, the ending of which is quoted at the start of this chapter, an elderly character who was once a partisan fighting against Franco's overthrow of the Spanish Republic but is now tending a forest plot for conservation, argues with a relative from the next generation, noting that the scientific career of the younger man has taken him away from the community of his birthplace. Political involvement, the older man contends, cannot be a simple matter of staying loyal to one's roots. Given the "powerful forces" that shape social and environmental change, we can "in intelligence" grapple with them "by such means as we can find" and take a deliberate path of action, but "none of us, at any time, can know enough, can understand enough, to avoid getting much of it wrong" (Williams 1985, pp. 357–8). The final ideal, then, that I would associate with a dynamic flux ethics is *fostering curiosity*—embracing the questions opened up once we set out to put engagement, participation, cultivation of collaborators, and transversality into practice.

> The word [curiosity] pleases me..: it evokes "concern"; it evokes the care one takes for what exists and could exist; a readiness to find strange and singular what surrounds us; a certain relentlessness to break up our familiarities and to regard otherwise the same things; a fervor to grasp what is happening and what passes; a casualness in regard to the traditional hierarchies of the important and the essential… I dream of a new age of curiosity. (Foucault, *The Masked Philosopher*, 1996)

Acknowledgements Helpful comments on a draft were given by Marien González Hidalgo, Kurt Jax, Steward Pickett, and an anonymous editor. This chapter would not have been possible without those who took the risk to join the discussion group on the ethics of collaborative or participatory processes at the 2011 Cary conference and, through their respectful interactions, generated insights that stimulated me to further inquiry. The line of thinking conveyed in this chapter is, however, one I pursued on my own, aware of the limitations this implies.

Appendix 1: Why Emphasize Collaboration in Environmental Research?

(Drawn from Taylor et al. 2011)

A. Sum of the Parts

Combining multiple perspectives

- When research is tied up with planning and management that involves meetings and networks of representatives of established and emerging stakeholder groups, research projects also need to integrate knowledge and questions from the different groups and kinds of research (Margerum 2008; Wondolleck and Yaffee 2000).
- When researchers are concerned about social justice, they can shape their inquiries through on-going work with and empowerment of people whose lives stand to be most affected by some change in social policy or technological development, such as digging of deep wells for irrigation (Greenwood and Levin 1998).
- When the knowledge and research skills of more than one person/speciality are needed, multi-disciplinary research teams are established.
- When the labor of research, especially in data collection, is beyond any research group, amateurs—"citizen scientists"—can be sought as collaborators (Wikipedia n.d.; Barrow 2000).
- Workshops and other organized multi-person collaborative processes in environmental research constitute a self-conscious example of what sociologists of science and technology have called "heterogeneous engineering" (Law 1987, i.e., the mobilization of heterogeneous resources by diverse agents spanning different realms of social action) (Taylor 2005, p. 93ff).

Extending over time

- The nature of environmental complexity means that ongoing assessment (as against a one-time analysis) is needed, so an ongoing organization or group is formed to conduct the assessment, as recognized in the field of Adaptive Environmental Assessment and Management (Resilience Alliance n.d.; Gunderson et al. 1995).

Spanning distance

- Researchers in separate projects and disparate locations use the tools of eco-informatics to link their data into a larger picture (Halpern et al. 2008).

B. Greater than the Sum of the Parts (i.e., outcomes over and above A.)

Generating new perspectives

- Knowledge and further research questions can be generated that the collaborators (individually or in sum) did not have when they came in (Olson and Eoyang 2001).

Durable

- Guided by skillful facilitators, collaborators can become invested in the plans, policy, and ongoing collaborations that emerge from the research (Stanfield 2002, p. 17ff).

Developing capacities

- Collaborators develop skills and dispositions for collaboration in various settings, as warranted by the rise of citizen participation and of new institutions of "civil society" (Burbidge 1997; Taylor 2005, p. 204ff).

Appendix 2: A Case of Participatory Community Planning in Northern Ontario

(Drawn from Taylor 2005)

As described in Stanfield (2002), the workshops of the Institute of Cultural Affairs (ICA) elicit participation in a way that brings insights to the surface and ensures the full range of participants are invested in collaborating to bring the resulting plans or actions to fruition. Such investment was evident, for example, after a community-wide planning process in the West Nipissing region of Ontario, 300 km north of Toronto. In 1992, when the regional Economic Development Corporation (EDC) enlisted ICA to facilitate this process, industry closings had increased the traditionally high unemployment to crisis levels. The EDC wanted specific plans, but it also sought significant involvement from community residents. Twenty meetings with over 400 participants moved through the first three phases—vision, obstacles, and directions. The results were synthesized by a steering committee into common statements of the vision, challenges, and strategic directions. A day-long workshop attended by 150 community residents was then held to identify specific projects and action plans, and to engage various groups in carrying out projects relevant to them. A follow-up evaluation five years later found that it was not possible simply to check off plans that had been realized because the initial projects had spawned many others. Indeed, the EDC had been able to shift from the role of initiating projects to that of supporting them. It made more sense, therefore, to assemble the accomplishments under the headings listed in the original vision and strategy documents. Over 150 specific developments were cited, which demonstrated a stronger and more diversified economic base, and a diminished dependence on provincial and national government social welfare programs. What is especially noteworthy about this example is that the community came to see itself as responsible for these initiatives and developments, eclipsing the initial catalytic role of the EDC-ICA planning process. The EDC saw beyond their catalytic role and came to appreciate the importance of the emergent process and initiated a new round of facilitated community planning in 1999 (West Nipissing Economic Development Corporation 1993, 1999).

Postscript: In late 2002, a major employer in the West Nipissing region, Weyerhaeuser, closed its containerboard plant. A local newspaper article (Haddow 2003) quoted a Weyerhaeuser spokesperson: "[T]he decision to close the facility is not a reflection on the employees of Sturgeon Falls and their abilities and efforts… It was made for economic reasons beyond their control." The spokesperson went on to explain that "the company's preference would have been to keep all facilities running, but the market changes and current economic conditions forced their hand." "If we as a company do not adapt, then we will not survive and none of our employees will have jobs." The community sprang into action and threatened lawsuits, but the plant closure was not reversed.

References

Barrow MV (2000) A passion for birds: American ornithology after Audubon. Princeton University Press, Princeton

Burbidge J (ed) (1997) Beyond prince and merchant: citizen participation and the rise of civil society. Pact Publications, New York

Foucault M (1996 [1980]) The masked philosopher. In: Lotringer S (ed) Foucault live. Semiotext(e), New York, pp 302–307

Greenwood DJ, Levin M (1998) Introduction to action research: social research for social change. Sage, Thousand Oaks

Gunderson LH, Holling CS, Light SS (eds) (1995) Barriers and bridges to the renewal of ecosystems and institutions. Columbia University Press, New York

Haddow SH (2003) Gripped by uncertainty: Sturgeon falls springs into action following the closure of its primary employer. Northern Ontario Business, 1 Jan

Halpern BS, Walbridge S, Selkoe KA et al (2008) A global map of human impact on marine ecosystems. Science 319:948–952

Law J (1987) Technology and heterogeneous engineering: the case of Portuguese expansion. In: Bijker WE, Hughes TP, Pinch TJ (eds) The social construction of technological systems: new directions in the sociology and history of technology. MIT Press, Cambridge, MA, pp 111–134

Margerum RD (2008) A typology of collaboration efforts in environmental management. Environ Manage 41:487–500

Olson EE, Eoyang GH (2001) Facilitating organization change: lessons from complexity science. Jossey-Bass, San Francisco

Peters P (1996) "Who's local here?" The politics of participation in development. Cult Surv Q 20:22–60

Pickett STA et al (2013) The flux of nature: changing worldviews and inclusive concepts. In: Rozzi R, Pickett STA, Palmer C (eds) Linking ecology and ethics for a changing world: values, philosophy, and action. Springer, Dordrecht/Heidelberg/New York/London, pp 265–279

Resilience Alliance (n.d.) Retrieved 28 Apr 2014, from http://www.resalliance.org/index.php/about_ra

Rozzi R et al (2013) Biocultural ethics: from biocultural homogenization toward biocultural conservation. In: Rozzi R, Pickett STA, Palmer C (eds) Linking ecology and ethics for a changing world: values, philosophy, and action. Springer, Dordrecht/Heidelberg/New York/London, pp 9–32

Sellers C (1999) Body, place and the state: the makings of an "environmentalist" imaginary in the post-World War II U.S. Radic Hist Rev 74:31–64

Stanfield RB (2002) The workshop book: from individual creativity to group action. Canadian Institute of Cultural Affairs, Toronto

Taylor PJ (2004) "Whose trees/interpretations are these?" Bridging the divide between subjects and outsider-researchers. In: Eglash R, Crossiant J, Di Chiro G et al (eds) Appropriating technology: vernacular science and social power. University of Minnesota Press, Minneapolis, pp 305–312

Taylor PJ (2005) Unruly complexity: ecology, interpretation, Engagement University of Chicago Press, Chicago

Taylor PJ (2011) Ethics of participatory process, in relation to linking environment, science, and action. Retrieved 6 May 2014, from http://sicw.wikispaces.umb.edu/cary

Taylor PJ, Fifield SJ, Young CC (2011) Cultivating collaborators: concepts and questions emerging interactively from an evolving, interdisciplinary workshop. Sci Cult 20:89–105

Taylor PJ, Haila Y (2001) Situatedness and problematic boundaries: conceptualizing life's complex ecological context. Biol Philos 16:521–532

Taylor PJ, Szteiter J (2012) Taking yourself seriously: processes of research and engagement. The Pumping Station, Arlington

West Nipissing Economic Development Corporation (1993) Vision 20/20: shaping our futures together, Executive summary. Apr

West Nipissing Economic Development Corporation (1999) Vision 2000 plus, Executive summary, June

Wikipedia (n.d.) Citizen science. Retrieved 29 July 2008, from http://en.wikipedia.org/wiki/Citizen_science

Williams R (1985) Loyalties. Chatto & Windus, London

Wolf E (1982) Europe and the people without history. University of California Press, Berkeley

Wondolleck JM, Yaffee SL (2000) Making collaboration work: lessons from innovation in natural resource management. Island Press, Washington, DC

Chapter 22
IUCN/WCPA Protected Areas Program: Making Space for People and Biodiversity in the Anthropocene

Ernesto C. Enkerlin-Hoeflich, Trevor Sandwith, Kathy MacKinnon, Diana Allen, Angela Andrade, Tim Badman, Paula Bueno, Kathryn Campbell, Jamison Ervin, Dan Laffoley, Terence Hay-Edie, Marc Hockings, Stig Johansson, Karen Keenleyside, Penny Langhammer, Eduard Mueller, Marjo Vierros, Leigh Welling, Stephen Woodley, and Nigel Dudley

Abstract Protected areas have emerged as a cultural feature and perhaps the largest land resource allocation decision in human history. Yet they are not without controversy on their adequacy for conservation and social justice. We argue that protected areas not only are necessary for conservation, they also contribute to human well-being and social justice in the Anthropocene. The World Parks Congresses have been a major forum for advancing global protected area policy and practice.

E.C. Enkerlin-Hoeflich (✉)
World Commission on Protected Areas, IUCN (International Union for Conservation of Nature), Ave. Eugenio Garza Sada 2501 Sur, Monterrey, México
e-mail: ernesto.enkerlin@iucn.org; http://www.iucn.org/

T. Sandwith
Global Protected Areas Program, International Union for Conservation of Nature (IUCN), Gland, Switzerland
e-mail: Trevor.sandwith@iucn.org

K. MacKinnon
World Commission on Protected Areas (WCPA), International Union for Conservation of Nature (IUCN), Cambridge, UK
e-mail: kathy.s.mackinnon@gmail.com

D. Allen
Healthy Parks Healthy People US, US National Parks Service, Washington, DC, USA
e-mail: diana_allen@nps.gov

A. Andrade
Conservation International, Bogotá, Colombia
e-mail: aandrade@conservation.org

R. Rozzi et al. (eds.), *Earth Stewardship*, Ecology and Ethics 2,
DOI 10.1007/978-3-319-12133-8_22

Recently the IUCN-World Commission on Protected Areas and the IUCN-Global Protected Areas Program has been moving toward a vision parallel and complementary to the proposed Earth Stewardship initiative of the Ecological Society of America. This novel view of IUCN is also called "The Promise of Sydney" because it will be the focus of the 2014 World Parks Congress in Australia. ICUN's novel view suggests that protected areas are an effective way to put Earth stewardship into action.

Keywords Capacity building • Climate change • Governance • Health • Indigenous peoples • Protected areas • Marine issues • World Parks Congress • Youth

---

T. Badman
World Heritage Programme, International Union for Conservation of Nature (IUCN), Gland, Switzerland
e-mail: tim.badman@iucn.org

P. Bueno
Parques Nacionales Naturales de Colombia, Bogotá, Colombia
e-mail: paula.bueno@parquesnacionales.gov.co

K. Campbell
Healthy Parks Healthy People, Parks Victoria, Melbourne, Australia
e-mail: kathryn.campbell@parks.vic.gov.au

J. Ervin
The Nature Conservancy, Burlington, VT, USA
e-mail: jervin@tnc.org

D. Laffoley
World Commission on Protected Areas (WCPA), International Union for Conservation of Nature (IUCN), Peterborough, UK
e-mail: danlaffoley@btinternet.com

T. Hay-Edie
United Nations Development Programme (UNDP), New York, NY, USA
e-mail: terence.hay-edie@undp.org

M. Hockings
World Commission on Protected Areas (WCPA), International Union for Conservation of Nature (IUCN), University of Queensland, Brisbane St Lucia, QLD, Australia
e-mail: m.hockings@uq.edu.au

S. Johansson
Agricultural and Environmental Services Department, World Bank, Washington, DC, USA
e-mail: stig.johansson@metsa.fi

K. Keenleyside
Parks Canada's National Parks Directorate, Gatineau, QC, Canada
e-mail: karen.keenleyside@pc.gc.ca

P. Langhammer
School of Life Sciences, Arizona State University, Tempe, AZ, USA
e-mail: penny.langhammer@asu.edu

22.1 Introduction

Protected areas[1] have emerged as a cultural feature and perhaps the largest land resource allocation decision in human history. They are considered necessary, but not sufficient, tools to avert or reduce the rate of biodiversity loss. There is mounting evidence that they are effective in maintaining biodiversity, but biodiversity continues to be lost in spite of the rapidly growing number of protected areas. Today protected areas are, in reality, a suite of land/sea-based mechanisms to achieve nature conservation, and more properly should be known as "conservation areas" because protection alone has never been sufficient to achieve their intended objectives. In the face of global climate change, and more broadly global environmental change, they are not protected from large scale processes such as rising CO_2 concentration, ocean acidification, accumulation in other biogeochemical cycles, increased severity and variability of hydro meteorological events, and rising temperatures, among many other problems.

Over the last decade the term "Anthropocene" (Crutzen and Stoermer 2000), has gained acceptance and simultaneously generated debate around biodiversity conservation. It even has been proposed that we concentrate on processes and leave biodiversity as such as a casualty of triage in a world that does not give intrinsic

[1] For the purpose of this paper protected areas will mean those fulfilling the International Union for Conservation of Nature definition of: a clearly defined geographical space, recognized, dedicated and managed, through legal or other effective means, to achieve the long term conservation of nature with associated ecosystem services and cultural values (http://www.iucn.org/about/work/programmes/gpap_home/pas_gpap/)

E. Mueller
World Commission on Protected Areas (WCPA), International Union for Conservation of Nature (IUCN), San José, Costa Rica
e-mail: emuller@uci.ac.cr

M. Vierros
International Organizations Center, United Nations University Institute of Advanced Studies (UNU-IAS), Yokohama, Japan
e-mail: vierros@ias.unu.edu

L. Welling
Climate Change Response, U.S. National Parks Service, Fort Collins, CO, USA
e-mail: Leigh_Welling@nps.gov

S. Woodley
World Commission on Protected Areas (WCPA), International Union for Conservation of Nature (IUCN), Ottawa, Canada
e-mail: stephen.woodley@iucn.org

N. Dudley
World Commission on Protected Areas (WCPA), International Union for Conservation of Nature (IUCN), Natural Solutions, Bristol, UK
e-mail: nigel@equilibriumresearch.com

value to its conservation. Instead conservation is conducted with an anthropocentric focus on "resource" scarcity that drives decision making processes.

To confront global environmental change, the Ecological Society of America (ESA) has launched the "Earth Stewardship" initiative (Chapin et al. 2011). This initiative provides a synergistic approach with the Global Protected Areas Program of the International Union for Conservation of Nature (IUCN) and its World Commission on Protected Areas. IUCN's program supports countries and communities to designate and manage systems of protected areas on land and in the oceans. However, both the IUCN Global Protected Areas Program and the ESA Earth Stewardship initiative confront serious limitations of geographic biases. Not only geographic regions, but also cultural diversity needs to be better represented in both initiatives (Rozzi et al. 2012; Li et al. 2015, in this volume [Chap. 13]). In this chapter we present a concise overview of the current status of preparations for the IUCN VI World Parks Congress (WPC), which offer a timely option to orient protected areas toward novel modes of stewardship for the well-being of humans and biodiversity as a whole.

22.1.1 IUCN World Parks Congresses

A driving force highlighting the importance of protected areas and proposing policies regarding them, have been the various WPC organized by the IUCN with the leadership of the World Commission on Protected Areas and the IUCN Global Protected Areas Program, which have been recognized as harbingers of change: a unique, once-in-a-decade meeting in which protected area professionals come together to share their practices, discuss policy, and meet people from very different parts of the world, who are working towards a common goal and often face similar professional challenges. Each WPC also has created a groundswell of change by introducing new ideas, launching new commitments, and signaling important developments in policy. These Congresses stand out as a series of milestones in the development of the world's protected area system (Phillips 2003; Dudley et al. 2005).

In 2003, the fifth WPC in Durban, South Africa, opened with a moving speech by Nelson Mandela and his call for more involvement of youth, and created the bulk of the text of the Convention on Biological Diversity's (CBD) Program of Work on Protected Areas (POWPA) (CBD 2004), which remains a basic reference and key strategy statement for protected area development (Fig. 22.1). But many essential aspects did not get much attention in Durban. By their nature, global policies quickly become dated, as we learn more and as conditions change: yesterday's preoccupations quickly fade away and new issues emerge into the mainstream.

A broader range of issues is reflected in the Strategic Plan for Biodiversity 2011–2020 that was agreed at the tenth Conference of the Parties to the CBD in Nagoya, Japan in 2010 (CBD 2010). A new target for increasing protected areas to 17 % on landscapes and 10 % on seascapes is juxtaposed with objectives for many other

Fig. 22.1 Nelson Mandela delivered an inspiring speech calling for more involvement of communities and youth at the V World Parks Congress, Durban, South Africa 2003

critical issues for biodiversity conservation and sustainable development. CBD's Aichi Target 11 positions protected areas firmly within the broader goals of sustainability and community well-being through the next generation and beyond. This approach borrows from new insights from ecological sciences (Callicott 1997). To facilitate as broad a discussion as possible on a range of issues, we outline some core themes for IUCN, and discuss its implications for policy and conservation. These will be integrated into guiding principles, a plan for action and vision statement under development called The Promise of Sydney launched at the VI World Parks Congress in November 2014 (www.worldparkscongress.org).

22.2 Reaching Conservation Goals

In the decade since the last WPC, the science of conservation has advanced rapidly, but so too have the pressures on protected areas and the requirements for scaling-up responses. Critics have claimed that protected areas are not the most effective tool for conservation, citing their limited size and relative isolation, proposing, instead, less well-defined approaches for ecosystem management and regulations. We clearly need to look at the future. If the Aichi Targets are meant to be interim goals for 2020, what should IUCN ultimate goals for nature conservation look like?

What does a truly sustainable protected planet look like? What science is available to inform this question? To address these pressing questions and about a future beyond the Aichi Targets, we need to catalyze inter-institutional collaboration between IUCN and professional societies of ecologists to achieve a more effective integration of conservation science and protected area management is a priority for determining conservation goals and communicating them in appropriate terms to decision makers.

22.3 Responding to Climate Change

Protected areas are now viewed as potential instruments for mitigating climate change by securing carbon-rich habitats in new or enhanced protected areas, and by facilitating adaptation through the provision of ecosystem services and cultural benefits that enable society to cope with the consequences of climate change. But at the same time, climate change is being viewed increasingly as a major threat to protected areas and resources for biodiversity conservation are being selfishly diverted into climate change adaptation. Plant and animal ranges may shift outside the borders of these areas set aside for their survival, and the specter of ocean acidification hangs over many coastal and marine protected areas. There is an urgent need for understanding the critical role that protected area systems can play in climate change response strategies. People and societies throughout history have adapted with different levels of success, and the promotion of culturally diverse approaches enhances adaptive capacity for facing climate change impacts.

22.4 Healthy Parks Healthy People

Previous links between health benefits of parks and protected areas, tended to focus on ecosystem services such as providing medicines and fresh water. The 2010 International *Healthy Parks Healthy People* Congress in Melbourne, Australia, launched a movement that has spread around the world. The recent advent of the Healthy Parks Healthy People approach has established broader understanding of the diverse health benefits of nature. These include regulating disease, mitigating climate events such as floods, and providing natural pollination controls. They also include the bio-cultural benefits of nature for physical, mental, and spiritual health, through respecting cultural heritage and diversity, supporting livelihoods, and fostering social well-being to sustain life. Healthy Parks Healthy People addresses the interconnection of people and parks (ecosystems) for health co-benefits.

22.5 Supporting Human Life

Beyond health benefits, the last 10 years has seen an explosion of interest in other benefits of protected areas, from links with faith groups and sacred natural sites, to the role of parks in stabilizing soils and protecting coastlines (see Kerber 2015 in this volume [Chap. 25]). IUCN has identified three critical benefits:

- disaster risk reduction,
- provision of freshwater, and
- maintenance of food security.

Each of these benefits has multiple facets. Natural ecosystems in protected areas can mitigate natural disasters by stabilizing soils, protecting coastlines, providing spillover areas for floods, and preventing avalanches and landslip. Forests and wetlands supply downstream communities with pure water. Marine protected areas maintain fish stocks, and terrestrial reserves preserve wild crop varieties critical for agricultural breeding programs. The Economics of Ecosystems and Biodiversity (TEEB) studies have provided a baseline of information, and a variety of tools for measuring. Getting proper recognition for these wider values also is still a challenge amongst state governments and other beneficiaries of these ecosystem services. Most governments gain more benefits from protected areas than they invest, yet even the limited funding available continues to decline in many countries.

22.6 Reconciling Development Challenges and Meeting Human Aspirations

Sustainable development[2] is about increasing human well-being without compromising nature or future development prospects. While governments struggle to maintain food and water security, and ensure jobs and sustainable livelihoods, they often are faced with hard choices and trade-offs. Research is needed on the intersection between protected areas, and the many development goals and challenges facing national governments. The mission of the United Nations Development Program (UNDP) and the World Bank, is to support countries to achieve sustainable development, while maintaining key ecosystem services and promoting climate resilient natural and human communities. The UNDP, World Bank, Conservation International, and the IUCN's Business and Biodiversity Program, offer solutions

[2] Sustainable development is used as it is the formal wording in UN and other multilateral agreements. The broader context should be sustainability as development is a means towards reaching human aspirations without compromising the future.

and tools for protecting areas that can be integrated in development planning and economic decision-making, and provide sector-specific experience and guidance in managing the intersection between protected areas and development. Protected areas need to be viewed as part of the national economy, and to be incorporated into national development strategies and frameworks.

22.7 Enhancing Diversity and Quality of Governance

Two trends emerged directly from the WPC of 2003: the increasing recognition of indigenous peoples' and community-conserved territories and areas (ICCAs) by governments, and a rapid increase in self-declared protected areas by indigenous peoples or local communities, most notably in Australia where over 20 million hectares have been declared as Indigenous Protected Areas in little more than a decade. The movement is gaining momentum and the ICCA Consortium, recently established, is providing global policy guidance. However, wider issues of governance still remain under-developed. The governance element of the CBD POWPA remains poorly implemented compared with other parts of the Program, with many governments lagging behind in applying good governance principles to existing or new protected areas, or in recognizing ICCAs, rights of communities, or privately protected areas (PPAs). The global policy focus on ICCAs needs to be complemented by a focus on shared governance and PPAs. Aichi Target 11 can only be achieved realistically with the contribution of all the different governance types and other effective area-based conservation measures.

22.8 Respecting Indigenous and Traditional Knowledge and Culture

Representatives of indigenous people came to the WPC 2003 with the specific aim of eliminating protected areas from their countries: two groups who frequently want the same result, protection of natural ecosystems, had drifted dangerously apart. People wanting to eliminate any remaining blocks on unrestrained development have been happy to encourage such divisions. In the years since Durban, important steps have been taken towards healing the rifts between some indigenous peoples' groups and protected area authorities. This is demonstrated by an increased number of collaboratively managed protected areas, indigenous protected areas (Australia), self-declared protected areas, officially recognized ICCAs, and other partnerships between local communities and protected areas. Adoption of the UN Declaration on the Rights of Indigenous Peoples; development of agreements such as the *Akwe Kon* guidelines, facilitated by the CBD; better understanding of issues of governance quality within protected areas; and the wider application of Free, Prior and Informed Consent, together helped to build safeguards and new attitudes. But there is still a long way to go: governments who treat minorities badly are unlikely to make an exception within their protected area management. More examples of successful collaborations are needed to build skills and confidence, and attitudes need to change within many government departments and NGOs.

22.9 Inspiring a New Generation

Young people represent a living and breathing force of great potential whose voices must be heard, stories told, and experiences shared. This new generation must be inspired to connect with nature (see Berchez et al. 2015, in this volume [Chap. 23]). Three strategies are necessary:

- Connecting a new generation to nature by focusing on exciting and inclusive ways of inviting people, who have not had outdoor opportunities, to connect with nature in safe but transformative ways.
- Investing in children by addressing the challenges of connecting school age children with nature in a world where nature is increasingly scarce, exploring the benefits of, and examining innovative ways, in which they can experience nature through exposure to parks.
- Empowering inspired young people by developing forums in which they can engage in collective actions, networking, co-learning, experience-sharing, and capacity-building/raising to inspire people across all generations to connect and engage together for Parks, People and Planet.

22.10 Marine Protected Areas

Oceans and coasts face a wide range of threats, some of which are similar to threats facing land ecosystems (e.g. invasive alien species, pollution, habitat loss, exploration for mineral resources), while some others are specific to marine habitats (e.g., ocean acidification and warming, land-based run-off, unsustainable and/or illegal fishing, and dredging/sea dumping). Although the ocean is a critical source of food and livelihoods for millions in coastal communities, many fish stocks have collapsed, or are collapsing. Cooperation with the fisheries sector to ensure sustainability needs improvement and overfishing and illegal fishing still remain major threats in many marine areas. Because the sea is traditionally and legally viewed as a commons, privately protected areas are much less relevant than in terrestrial environments. This means that policy priorities must continue to focus on persuading those with decision-making power – communities, nations, and international organizations – of the need for urgent and increasingly ambitious action, and providing the tools and advice to manage marine protected areas effectively under rapidly changing conditions. The recent trend of establishing very large marine protected areas (MPAs) that encompass whole ecosystems, and community-based MPAs that support local livelihoods, are two approaches that will help us meet our marine conservation goals (see Berchez et al. 2015, in this volume [Chap. 23]; Nevill 2009).

22.11 Capacity Development

The pace at which new protected areas have been established often has outstripped the ability to manage them effectively; there simply is not enough well-trained staff available, particularly as management needs and priorities change rapidly. IUCN's WCPA has a history of providing technical advice through its Best Practice Protected Area Guidelines series, provision of experts, and individualized training sessions. However, it is no longer sufficient. Field rangers often miss out on training, through lack of basic educational opportunities, inability to read English, French, or Spanish, and lack of access to materials. IUCN works to fill this gap through the development of online training materials based around minimum competency standards, an accreditation system for courses offered on protected areas in tertiary educational establishments, and through focused teaching. There remain many gaps and priorities ensuring that the curriculum is comprehensive and is adopted by the premier educational and training institutions, for preparing a new generation of qualified and competent professionals.

22.12 World Heritage

World Heritage represents in many ways the best of the best and particularly the sovereign decisions of countries to make special efforts in their conservation. The task by IUCN World Heritage Program keeps growing as more sites are added to the list, and as the World Heritage Committee grows in political importance (and as a result becomes increasingly politicized itself). World Heritage Sites, which cover more than 10 % of all protected areas globally, also need to change their role to provide leadership to global efforts on protected areas. Another major aim is to bring natural World Heritage designation closer, philosophically and in practice, to the conservation of the greater number of cultural sites on the World Heritage list. Both face similar challenges in terms of development, the need to maintain naturalness or authenticity, and their role in educating present and future generations about our common heritage.

22.13 A New Social Compact for Effective and Just Conservation

Protected areas will only work, and continue to work in the future, if they are supported by a broad range of people; the pressures against conservation are too great for protected areas to survive in the hands of a few enthusiasts. A New Social Compact is required to bring together people from very different backgrounds, to work together from a common understanding about values, challenges, and opportunities. An inspirational platform must be created so that diverse rights holders, stakeholders, and interest groups can dialogue and commit to building solidarity in human networks and shared understandings of the intrinsic and functional value of nature.

22.14 Conclusion

Protected Areas can contribute much to Earth Stewardship. As such, "The Promise of Sydney" formalized at the 2014 World Parks Congress in Australia, should guide transformative change over the next generation. Under IUCN's new orientation, protected areas teach that value is far more than economic. They themselves can be understood as Earth stewardship in action (Abecasis et al. 2013; see Berchez et al. 2015; Chapin et al. 2015, in this volume [Chaps. 12 and 23]). Protected areas are related integrally to human well-being, not only physical but also cultural. They contribute to justice by protecting traditional and cultural knowledge of indigenous people and are constant reminders that nature and culture are World Heritages. Additionally, a biocultural ethic introduces an ecosocial justice perspective that affirms that "unsustainable practices that are detrimental to the life of other human and other-than-human beings need to be sanctioned and/or remedied". These perspectives shall be incorporated in a "new social compact" emerging from the WPC. Complementarily, the worldviews, forms of knowledge, values, and ecological practices of cultures that are sustainable should be respected, and eventually adapted through intercultural exchanges (Rozzi 2013, p. 10).

IUCN's new orientation aims to better integrate cultural diversity and, at the same time, to achieve ecojustice. The discourse and action in protected areas for the next 25 years should be fully integrated into the broader aim of sustainability at all levels, making this an enhanced cultural feature. Perhaps more importantly, it should be prioritized as an integral part of planning and resource allocation in the international community, such as the Sustainable Development Goals. Less than that would certainly not allow societies around the world to meet their legitimate aspirations in the IUCN vision of a "just world that values and conserves nature."

Acknowledgements Many thanks to all the IUCN staff supporting the VI World Parks Congress and the Australian hosts particularly the Government of Australia through the National Parks Agency and the New South Wales Department of the Environment. Important contributions have been made by many members of WCPA and staff of the IUCN and other collaborating individuals and agencies. Special thanks to all stream and theme leaders at the World Parks Congress; WCPA leadership and to Tom Brooks. E. Enkerlin would like to make a special recognition to fruitful discussions with Brendan Mackey and Nelly Correa.

References

Abecasis RC, Schmidt L, Longnecker N et al (2013) Ocean & coastal management implications of community and stakeholder perceptions of the marine environment and its conservation for MPA management in a small Azorean island. Ocean Coast Manag 84:208–219

Berchez F, Mansilla A, Ghilardi-Lopes N et al (2015) Ecology and education in marine protected areas: insights from Brazil and South America. In: Rozzi R, Pickett STA, Callicott JB et al (eds) Earth stewardship: linking ecology and ethics in theory and practice. Springer, Dordrecht, pp 351–366

CBD [Convention on Biological Diversity] (2004) Conference of the parties decision VII/28. Paragraph 8. http://www.cbd.int/convention/results/?id=7765&l0=PA

CBD [Convention on Biological Diversity] (2010) Conference of the parties decision X/2. Strategic plan for biodiversity 2011–2020. http://www.cbd.int/decision/cop/?id=12268

Callicott JB (1997) Conservation values and ethics. In: Meffe GK, Carroll CR (eds) Principles of conservation biology. Sinauer Associates, Sunderland, MA, pp 24–49

Chapin FS, Pickett STA, Power ME et al (2011) Earth stewardship: a strategy for social-ecological transformation to reverse planetary degradation. J Environ Stud Sci 1:44–53

Chapin FS, Pickett STA, Power ME et al (2015) Earth stewardship: an initiative by the Ecological Society of America to foster engagement to sustain planet earth. In: Rozzi R, Pickett STA, Callicott JB et al (eds) Earth stewardship: linking ecology and ethics in theory and practice. Springer, Dordrecht, pp 173–194

Crutzen PJ, Stoermer EF (2000) The "Anthropocene". Glob Chang Newslett 41:17–18

Dudley N, Higgins-Zogib L, Mansourian S (eds) (2005) Beyond belief: linking faiths and protected areas to support biodiversity conservation. WWF International, Gland

Kerber G (2015) Stewardship, integrity of creation and climate justice – religious ethics insights. In: Rozzi R, Pickett STA, Callicott JB et al (eds) Earth stewardship: linking ecology and ethics in theory and practice. Springer, Dordrecht, pp 383–393

Li B, Parr S, Rozzi R (2015) Geographical and thematic distribution of publications generated at the International Long-Term Ecological Research Network (ILTER) sites. In: Rozzi R, Pickett STA, Callicott JB et al (eds) Earth stewardship: linking ecology and ethics in theory and practice. Springer, Dordrecht, pp 195–216

Nevill JC (2009) Overfishing, uncertainty, and ocean governance: Lord Perry's question revisited. Dissertation, University of Tasmania, p 406

Phillips A (2003) Turning ideas on their heads: a new paradigm for protected areas. George Wright Forum 20:8–32

Rozzi R (2013) Biocultural ethics: from biocultural homogenization toward biocultural conservation. In: Rozzi R, Pickett STA, Palmer C et al (eds) Linking ecology and ethics for a changing world: values, philosophy, and action. Springer, Dordrecht/Heidelberg/New York/London, pp 9–32

Rozzi R, Armesto JJ, Gutiérrez J et al (2012) Integrating ecology and environmental ethics: Earth stewardship in the southern end of the Americas. BioScience 62(3):226–236

Chapter 23
Ecology and Education in Marine Protected Areas: Insights from Brazil and South America

Flávio Berchez, Andrés Mansilla, Natália P. Ghilardi-Lopes, Evangelina Schwindt, Kelen Leite, and Ricardo Rozzi

Abstract South American coastal habitats include a wide range of benthic ecosystems, many of which are unique and constitute hotspots of biodiversity. Marine protected areas (MPAs), instituted mostly during the second half of the twentieth Century, are considered a key management tool to conserve regional biodiversity, prevent overexploitation, and generate economic benefits. Educational actions to promote changes in basic values, principles, and attitudes – although considered also as a main objective – frequently have a poor conceptual basis. In conjunction

F. Berchez (✉)
Botany Department, Instituto de Biociências, Universidade de São Paulo, São Paulo, SP, Brazil
e-mail: fberchez@ib.usp.br

A. Mansilla
Laboratorio de macroalgas Antarticas y Subantarticas, Universidad de Magallanes, Punta Arenas, Chile

Institute of Ecology and Biodiversity, Santiago, Chile
e-mail: andres.mansilla@umag.com

N.P. Ghilardi-Lopes
Centro de Ciências Naturais e Humanas, Universidade Federal do ABC, São Bernardo do Campo, SP, Brazil
e-mail: natalia.lopes@ufabc.edu.br

E. Schwindt
Grupo de Ecología em Ambientes Costeros, Centro Nacional Patagónico (CENPAT-CONICET), Puerto Madryn, Argentina
e-mail: schwindt@cenpat.edu.ar

K. Leite
Estação Ecológica Tupinambás, ICMBio, São Sebastião, SP, Brazil
e-mail: kelenluciana@gmail.com

R. Rozzi
Department of Philosophy and Religion Studies, University of North Texas, Denton, TX, USA

Institute of Ecology and Biodiversity, Santiago, Chile

Universidad de Magallanes, Punta Arenas, Chile
e-mail: rozzi@unt.edu

R. Rozzi et al. (eds.), *Earth Stewardship*, Ecology and Ethics 2,
DOI 10.1007/978-3-319-12133-8_23

with the evaluation of their effectiveness by long-term, site-based ecological and socio-economic research, in Brazil MPAs are aiming to implement a holistic approach. This will allow the development and testing of environmental practices that integrate ecology, economy, ethics, and conflict resolution in the different uses of marine space. However, ecological long-term studies, socio-economic long-term evaluation, and the integration of education and ethics are still incipient. With the recent creation of some independent networks in different South American countries related to the assessment of biological communities, marine biologists of this continent are now focusing more on: (1) sharing methodologies and data to allow comparative and integrated continental analyses, and (2) integrating social components, including not only economic but also ethical values and participatory approaches. Toward this aim, the Chilean Long Term Socio- Ecological Research network (LTSER-Chile) has developed a Field Environmental Philosophy program that could be adapted to MPAs educational programs, and also contribute to the integration of ecology and ethics in theory and praxis for an Earth Stewardship initiative.

Keywords Benthic ecosystems • Environmental education • Long term research • Marine protected areas • Monitoring

23.1 Marine Protected Areas (MPAs) in South America

South American coastal habitats include a wide range of benthic ecosystems, many of which are unique and constitute hotspots of biodiversity (Miloslavich et al. 2011), such as the kelp forests on the Cape Horn Biosphere Reserve and the coral reefs of the Tropical Atlantic. These ecosystems can occupy extensive areas, such as the rhodolith beds and mangrove forests along the Tropical Southwestern Atlantic. In addition to global threats imposed by Global Environmental Change (GEC), these ecosystems are endangered by local and regional stressors, thereby risking a series of ecosystem services provided by them (Turra et al. 2013; see Orenstein and Groner 2015 in this volume [Chap. 18]).

The time for mitigating GEC is over, and the application of adaptation measures has come to the forefront (Heffernan 2012). The past decade presented an exceptional number of unprecedented extreme weather events (Coumou and Rahmstorf 2012); some of them, such as hurricanes and heat waves, directly impacting marine communities. *Marine protected areas (MPAs) are considered a key management* tool to buffer GEC by conserving biodiversity, preventing overexploitation of marine communities, and presenting potential economic benefits such as enhancement of local fisheries, sustainable tourism opportunities, and maintenance of other ecosystem services (Sala et al. 2013; Lubchenco et al. 2003; Kearney et al. 2013; Huntington et al. 2010). MPAs strengthen ecological resilience to climate variability, by offering habitat for range-shifting species, a key element for biotic community responses to long-term climate change (Bates et al. 2013). However, to effectively confront GEC, MPAs should undertake a more holistic role, including the development and assessment of environmental management and education practices that integrate ecology and economy, as well as ethics and conflict resolution in the uses of marine space.

Figure 23.1 shows the current distribution of MPAs in Latin America, and the world context. It is notorious the scarcity of MPAs in most of the South American Pacific coast. In addition, South American MPAs are very recent, most of them created after the 1980s (Schiavetti et al. 2013). Today there are 404 coastal and marine protected areas: 336 in Brazil (Schiavetti et al. 2013), 8 in Chile, 14 in Uruguay (Gambarotta 2006) and 46 in Argentina. Outside these MPAs, there are some vast marine-terrestrial protected areas, such as the Cape Horn Biosphere Reserve or the Namuncurá – Burdwood Bank, in southern Chile and Argentina, respectively. The Burdwood Bank was created in July 2013, and represents the first entirely oceanic MPA of Argentina. It has a total surface of 17,000 km^2 with 1,800 km^2 permanently closed to fishing. With this new MPA, the total surface of marine and coastal protected areas in Argentina is still less than 5 %. So, in this and other South American countries, the goal of protection of 10 % of the ocean established by the Convention of Biological Diversity (http://www.cbd.int/sp), is far from being reached.

In addition to the limitations in area, South American MPAs present other basic deficiencies, such as the absence and/or low efficiency of management plans, monitoring programs, the lack of adequate infrastructure, personnel and enforcement (Gerhardinger et al. 2011). Furthermore, the integration between federal, state, municipal and, private protected areas is very low. This leads to superposition, conflicts, and a poor understanding of the missions and responsibilities of each instance. In this context, although environmental education formally is included within the Conservation Units Systems of most countries, educational actions are frequently absent, deficient, or inadequate to the context of each protected area (Berchez et al. 2007).

MPAs are regulated by national laws or other means and should be created in order to provide ecological, social, and economic benefits to the reserved areas and its borders. They were initially proposed as a means to preserve marine biodiversity and unique habitats, and as an opportunity for recreation, education, and research (Sala et al. 2013). However, the focus frequently has been placed on the implementation of rules and restrictions. For example, the Brazilian legislation (Brasil 2000) considers as core goals of these areas "to discipline the occupation processes, protect the biological diversity, and to secure the sustainability of natural resources, observing the natural attribute quality." In Argentina, the main goal for the creation of MPAs is to protect the reproductive sites of marine birds and mammals (Campagna et al. 2007). During the last decade this restrictive vision has moved toward a more holistic ecosystem–conservation approach (SAyDS 2007).

Despite the limitations and lack of clear and specific guiding elements to planning, management, and monitoring of South American MPAs, positive outcomes regarding biodiversity conservation and fisheries management have been achieved (Floeter et al. 2006; Francini-Filho and Moura 2008a, b; Edgar et al. 2014; Sala et al. 2013). These achievements gradually have been incorporated into the

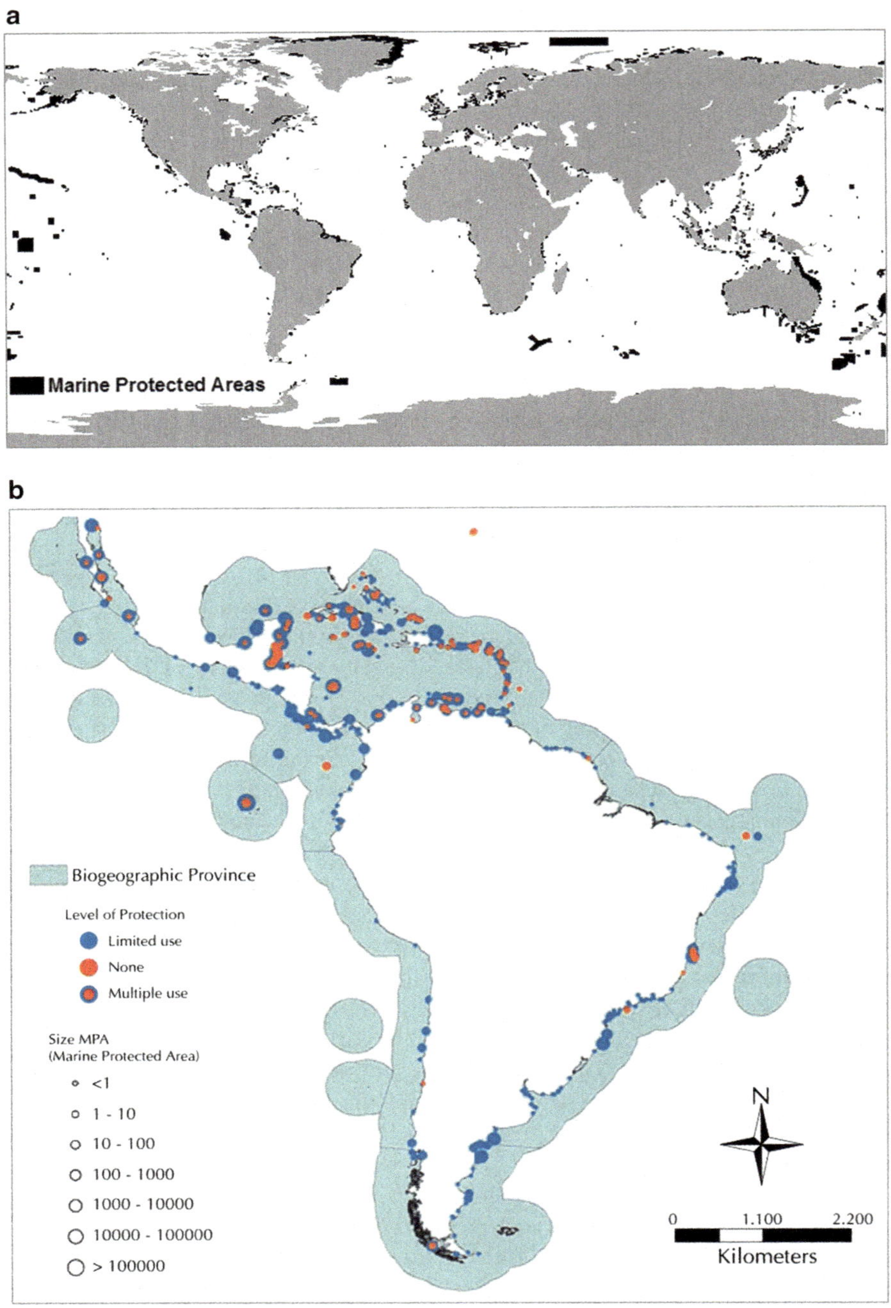

Fig. 23.1 (**a**) Marine protected areas in the world (IUCN and UNEP-WCMC 2010) and (**b**) Latin America (Modified from Guarderas 2007). Distribution of marine protected areas in Latin America and the Caribbean classified by degree of protection against extractive activity

toolbox for marine and coastal management. Yet, socio-economic and ecological benefits generated by MPAs still remain difficult to predict and are debated around the world (Edgar et al. 2014).

MPAs streamlining requires a precise diagnosis of its current status, to better identify outcomes and deficiencies, allowing for due corrections. Actions should be continuously followed by long-term assessments programs of both environmental and social dimensions. For South American MPAs, these evaluations are rare, and most of the available data includes only traditional taxonomic surveys, which are insufficient even for appropriate biodiversity assessments.

In the following sections we present an overview of the ongoing long-term research at South American MPAs, with focus on Brazil and Chile. We provide some successful examples of holistic programs that could lead towards the implementation of an Earth stewardship.

23.2 Some Ecological Long-Term Studies at Brazilian and Chilean MPAs

The establishment of continuous long term monitoring ecological sites is essential to facilitate the early detection of ecological changes, and to apply correct management measures. Populations and communities of marine species often respond quite differently to human pressures within well-designed MPAs (Edgar et al. 2014; Fraschetti et al. 2012). Continuous evaluations in MPAs are especially important in order to assess the impacts related to GECs (Turra et al. 2013). However, global monitoring efforts still are constrained by major geographical gaps. Ecological studies and environmental observatories have overlooked some regions of the Earth that have ecological attributes that are essential to the functioning of the biosphere as a whole (Lawler et al. 2006; Rozzi et al. 2012).

Even basic ecological information regarding the marine community structure in South America countries is extremely scarce, as documented for Brazil (Ghilardi et al. 2008) and Chile (Navarrete et al. 2010). For example, the structure of a vast ecosystem such as the Rhodolith Beds of the Tropical South-western Atlantic Realm remain almost unknown (Spalding et al. 2007). They were described only during the last decade by a few studies restricted to the Eastern Brazilian Ecoregion (Amado-Filho et al. 2010; Pereira-Filho et al. 2011; Berchez et al. 2009).

In relation to long-term marine monitoring and research programs, Chile has relied on the work of a few study sites and research teams. Two paradigmatic sites in Central Chile (33°S, 71°W) are: (i) the marine protected area of Las Cruces Biological Station created in 1982 by the Pontifical Catholic University of Chile (Navarrete et al. 2010), and (ii) the Montemar Institute of Marine Biology built in 1941 as a field station of the University of Chile, to work in partnership with local fishermen. When the University of Valparaíso was formed in the 1980s, its Faculty of Ocean Sciences undertook its administration. With similar goals, in southern Chile the marine stations of Dichato and Mehuin (39°S, 73°W) were inaugurated by the University of Concepcion and the Austral University of Chile, respectively.

Unfortunately, the first of this stations was damaged severely by a Tsunami in 2010 (Harris 2010), and the second one experienced many administrative difficulties. At the end of the 1990s, in the sub-Antarctic region of Western Patagonia, the coastal stations of Huinay (39°S, 73°W) in Aysen, and the Omora Ethnobotanical Park (55°S, 67°W) in Cape Horn, were created in association with the Pontifical Catholic University of Valparaiso and the University of Magallanes, respectively.

In Brazil the programs running for more than a decade are confined to two stations on the Coral Reef Ecosystem of the northeast Brazilian Coast: one developed 14 years ago, based on the AGRRA Protocol (Leão et al. 2010), and the other 17 years ago, based on the Reef Check Program (Hodgson 1999). Fortunately, during the last 3 years new large initiatives have been developed, including the South American Research Group on Coastal Ecosystems (SARCE)[1] and ReBentos networks of continental and regional (Brazil) scopes, respectively. Discussion is also under way in the federal agency that controls the Brazilian Federal Reserves or protected areas (Chico Mendes Institute - ICMBio), as to the development of a protocol to be employed by stations in all MPAs. However the same is not true for most state or municipal protected areas. This scenario highlights the need for developing integrated monitoring programs.

Most methodologies employed in these projects are targeted towards detecting specific responses, and thus only give a limited comprehension of the community structure. This is a limiting factor not only for detecting variations, but also for interpreting their causes. These insufficiencies are the same that have been consistently criticized by most marine reserves assessments, such as limited sample replication, non-random reserve placement, or inadequate controls for temporal and spatial variability (Huntington et al. 2010; Huntington and Lirman 2012). With the development of these recent assessments, Brazilian and South American scientists and decision makers are concerned with the diversity of protocols and efforts. This challenges the possibilities of comparing and integrating data across sites. SARCE is the only program with an integrated continental scope, with sites in Chile, Peru, Venezuela, Colombia, Brazil, Uruguay, and Argentina. There is, therefore, the need for integration also from a regional political view.

Although programs directed towards the monitoring of physical components are increasing, the integration between biological and physical data, essential to the understanding of community dynamics and defining its main drivers, is still low. To improve the integration of biophysical studies, proposals such as the integration of the ReBentos and Coastal Zone Climate networks (http://www.rebentos.org; http://redeclima.ccst.inpe.br) are being developed.

Linking the planktonic, nektonic, and terrestrial compartments is essential. Furthermore, marine ecosystem functions are largely determined by matter and energy transformations mediated by microbial community interaction networks. It has been found recently that viruses are also a crucial components of marine ecosystems, and their abundance exceeds bacteria and phytoplankton by at least an

[1] The SARCE network was established in 2010 . Today, it includes more than 30 researchers from 9 South American coastal countries and has sampled with a standardized protocol in more than 50 sites around the continent.

order of magnitude (Bidle and Falkowski 2004; Hurwitz et al. 2013). None of the current programs consider the influence of these components, which strongly compromises the interpretation of the causes of changes.

23.3 MPAs and Socio-economic Long-Term Evaluation

The integration of socio-economic research into Ecological Long-Term Studies, followed by the integration of noneconomic ethical components to evaluations and consequent educational actions, represents a significant step forward due to the inclusion of the human component in these processes (Maass and Equihua 2015; Redman and Miller 2015 in this volume [Chaps. 14 and 17]). They complement ecological assessments facilitating both the comprehension of causes and consequences, and the undertaking of correct mitigation or adaptation measures when necessary. Furthermore, they provide a basis for "a dialogue about how humans value nature" (Rozzi et al. 2012; Pimm 1994) from the understanding of individual and social perceptions about ecological phenomena. The lack of success of many initiatives could be attributed to the absence of this information to support MPAs management.

Definition of sectors and services of MPAs usually is related to a formal, or informal, management plan. The definition of areas for conservation, economic use, or educational purposes, is frequently a controversial task. Economic benefits, such as traditional exploitation, fisheries, or ecotourism, are frequently the more important concern of the general public. Stakeholders' perceptions based in their cultural basis and beliefs, plays a fundamental role in decisions, and afterwards in the effectiveness of implementation. Thus, the evaluation of stakeholders' perception is essential and should precede educational action and correction measures. A significant increase of the knowledge on perception about marine conservation and MPAs arouse from studies related to GEC.

Perceptions about the environment are influenced not only by its physical nature (Matos 2009), but also by cultural, social, and cognitive aspects (Saheb and Asinelli-Luz 2006). Both individual and social dimensions lead to behavior choices or political decisions, respectively (Whyte 1977). People with different values and interests draw different inferences from the same evidence about global climate changes (Kahan 2012). Experiences related to extreme climate events or local weather conditions, such as cold (Wallace et al. 2014) or hot weather (Zaval et al. 2014), wildfires (Moritz 2012), or hurricanes (Tollefson 2012), may influence global warming beliefs and are very important in changing perceptions about GEC (Howe et al. 2012; Myers et al. 2012). Frequently people overestimate their own knowledge and are unlikely to change their view (Leviston et al. 2013).

The media are fundamental instigators for shaping public opinion, even more importantly than scientific knowledge itself (Zaller 1992). Hence basic scientific consensus to communicate through the mass media (Lewandowsky et al. 2013; Kahan 2012), together with a basic political consensus (Brechin 2012; Brulle et al. 2012), is essential to changing public perceptions on GEC (Gore 2006). However, in spite of its importance, there is no comprehensive program with this aim specifically directed towards regional South American MPAs, where programs are restricted basically to ecological evaluations.

Confronted with pressing near-future challenges related to GECs, measures to implement long term socio-ecological research in South American MPAs are urgent. The recent establishment of new Long Term Socio- Ecological Research (LTSER) programs, such as LTSER-Chile, offers a holistic approach that offers an alternative to integrate into and complement with, the several MPA's efforts already in development.

23.4 Integrated Educational Experiences in South American MPAs

For most conservation unit categories, the development of educational actions is mandatory. For example, Brazilian legislation "SNUC" (Brasil 2000) defines as a main objective "to promote education, educational conditions, environmental interpretation and the recreation in contact with nature" (http://www.mma.gov.br/areas-protegidas/sistema-nacional-de-ucs-snuc). However, environmental education (EE) in Brazilian MPAs is still incipient. Activities for visitors are frequent at marine parks. However, most cases lack instructive activities. Hence, they offer only a superficial tourism, not a genuine ecotourism nor an educational experience. Structured EE activities are rare (Ghilardi and Berchez 2010), and most of them have a poor conceptual basis (Berchez et al. 2007).

Since the 1977 Tbilisi Conference Statement, evolving environmental educational concepts were synthesized during the Rio 92 Conference (Pedrini and Brito 2006). There is growing consensus that, for the betterment of living conditions as well as of those related to the environment, changes in basic values and attitudes are fundamental (La Trobe and Acott 2000). Toward that aim, the Field Environmental Philosophy (FEP) program developed at Omora Park in the Cape Horn Biosphere Reserve in Chile provides a valuable methodological approach that can be adapted to other MPAs (see Aguirre Sala 2015 in this volume [Chap. 15]). FEP's methodology integrates ecology and philosophy, and research results are communicated through metaphors and field activities guided by an ecological and ethical orientation and implemented through special trails or areas. FEP is based on a biocultural ethic that addresses not only human welfare but also the welfare of the whole community of life (Rozzi and Massardo 2011; Rozzi 2013).

EE requires holistic approaches that integrate the teaching of biology, ecology or other disciplines that focus on cognitive objectives with activities that target both emotional experiences and skills acquisition, to achieve enduring transformative processes (Ghilardi-Lopes 2014). Furthermore, EE actions should develop the capacity to contextualize these gains by applying new beliefs and behaviors to everyday life. Objectives related to social transformation, including the capacity to stimulate collective mobilizations and the evolution of social groups and forms of culture, or to promote political changes, are essential EE features (Berchez et al. 2007).

For Brazilian MPAs, only a few examples of attempts for more integrative work can be highlighted. One of them is the Reef Check Program which, for over a decade in the northeast Brazilian Region, in association with the Chico Mendes Institute for Biodiversity Conservation (ICMBio),[2] has developed long term monitoring of benthic communities, and also has included an educational component. Following a Citizen Science approach (Silvertown 2009; Osborn et al. 2005), local people are recruited to perform, after due training, periodic assessments on MPAs, using shared protocol and tables. Besides recruiting labor, the program provides practical comprehension and knowledge of the environment, and allows evaluation by the citizens themselves of environmental health conditions, stimulating the sense of belonging and thus leading to action and facilitating their action as disseminators of scientific knowledge within their groups of influence. Other positive points of the program include integrating the participation of non-local people, and continuing education. Outstanding results have been obtained, such as the professionalization of young collaborators who serve as MPA instructors for undergraduate education. Another example is the Underwater Trail Project that was begun in 2002 at the Anchieta Island State Park (Berchez et al. 2005, 2007). It initiated a training program for EE instructors, who promote the emergence of new well-based educational actions (Box 23.1). Among its distinctive characteristics, one is the association of the project actions with continuous education research, evaluating their success and deficiencies.

Box 23.1. Underwater Trail Project

The Underwater Trail project is a long-term experiment in marine environmental education developed by the University of São Paulo, beginning in 2002 (Berchez et al. 2005, 2007). Its objectives are to develop, apply, and test scientific research and education actions at the Anchieta Island State Park (23°S; 45°W – Fig. 23.2). This is an unpopulated, state insular protected area, located near (7 km) the coast. Due to its proximity to the city of São Paulo, this protected area receives up to 2,000 people a day during the summer periods. People arrive in boats, and stay confined basically along a beach 350 m long, and a trail with the same extension.

Educational models have been developed to stimulate transformative processes in the participants. These models integrate cognitive and emotional dimensions with skills acquisition. The visits include interpretation, direct contact with biodiversity, and the use of diving and trekking equipment. The sequence of activities consider the following 11 steps:

(continued)

[2] ICMBio is the Brazilian Ministry of the Environment's administrative arm.

Box 23.1. (continued)

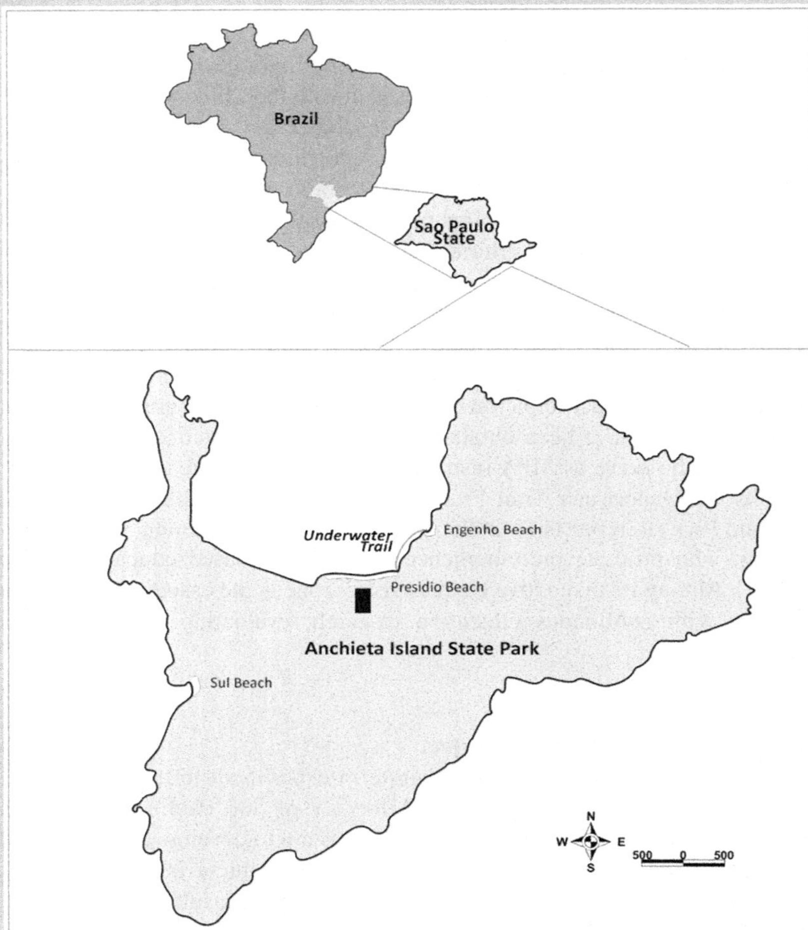

Fig. 23.2 Brazil, São Paulo State and Anchieta Island State Park (Robim et al. 2008). The *blue line* shows the area in which the models "Underwater Free Diving Trail," "Underwater Scuba Diving Trail," and "Ecosystems Trail" of the underwater trail project take place

1. Welcome and introduction,
2. stretching, relaxing, and sensitization,
3. MPA objectives, historical cultural and social background, nowadays conflicts and challenges,
4. nearby ecosystems and their sensitivities,
5. geological origin,
6. equipment characteristics, concepts and use,

(continued)

Box 23.1. (continued)

7. ecosystem functional interpretation,
8. organism functional interpretation,
9. economic relationships,
10. global environmental changes, and
11. closing.

The sequence is planned to have a transdisciplinary character and to stimulate participants to later apply learning to their everyday life when they return home. Following this scheme, seven activities models have been developed to address the expectations of different public, and to expose visitors to a variety of ecosystems: (i) Underwater Free Diving Trail (Fig. 23.3), (ii) Underwater Scuba Diving Trail, (iii) Natural Aquarium Trail, (iv) Canoeing Trail, (v) Outside Water "Panels" Diving Trail (Fig. 23.4), (vi) Ecosystems Trail, and (vii) Vertical Trail. The activities at each of the interpretive trails last for approximately one hour. The activities, initially targeted to the general public visiting the MPA, were expanded to elementary and high school local students. Up to now, a total of 20,351 people have attended the activities, 1,405 of them belonging to public schools.

Fig. 23.3 People participating in Underwater Free Diving Trail

(continued)

Box 23.1. (continued)

Fig. 23.4 Outside Water "Panels" Diving Trail

Regarding the training of educators, most of them are undergraduate students at variety of disciplines at the University of São Paulo. These students serve educators who have a multiplying effect.

These students are very interested in environmental education (EE). Up to 300 students apply each year for the 50 annual vacancies in the teacher training course. Teachers from the public schools selected for the activities also are trained in the same way, and later are engaged as field monitors of their own students, under supervision of more experienced members of the group. Students receive credits for an elective course, and also receive a formal certificate documenting their training as EE monitor. A system of certification, with both horizontal and vertical steps, up to the category of examiner, is being tested. The training system is based in written protocols, explaining the models, including their educational contents, techniques, operational and safety procedures. Educators are urged to complement this basic content with their own experience. Books, chapters, articles, and web-pages have been developed in order to support and complement this training (e.g. Ghilardi-Lopes et al. 2012).

Up to now 733 educators have been trained, including 75 teachers from public schools. Part of this group, including one person from the first year (2002) and four from the second, are still participating as senior "non-profit" members. Since this work is completely volunteered, the high involvement indicates that they are motivated by ethical rather than purely economic values. At least 15 of these educators has been contracted as staff for Brazilian Protected Areas (BPA), a fact that shows the need that BPA have for people trained with this theory and practice integration in EE, having also multiplier capacity. Since the beginning, the EE actions have been annually evaluated by scientific research (Berchez et al. 2005), with results expressed in 22 papers or chapters. Studies have focused on the achievement of the educational objectives, the ecological impact of the activity, and more recently on the evaluation of visitors' perceptions about global environmental change.

23.5 MPAs Educational Agenda: Implications for Earth Stewardship

Involving decision-makers in the process is essential to transferring successful educational experiences to policy (Abecasis et al. 2013). The exchange of experiences among different stakeholders towards a common goal contributes to enlarging their vision along with that of managers. However, construction of the management plan and afterwards planning of educational activities should rely in science-based investigative data and previous experiences (Lubchenco et al. 2003).

Although the last decades have experienced a considerable increase in the number of MPAs, the improvement of their services still is a huge task. The integration of disperse experiences throughout South America is a primary task in order to improve efficiency, to save resources, and to allow the exchange of data and experiences. The sharing of successful local results identified by research programs, through frequent and intensive meetings, workshops, formal or informal courses, or distinct types of media, could be the way to accelerate development, in order to fulfill Earth Stewardship demands (see Chapin et al. 2015 in this volume [Chap. 12]).

MPAs represent areas of reduced human pressure, are expected to act as repositories of species and as habitat refuges (Costa et al. 2013). They are also marine regions where economic activities are controlled. For these reasons, protected areas are ideal to act as incubators of novel conservation experiences, integrating ecology, economy, and ethics, thereby acting as an educational instrument for spreading these concepts and abilities to society as a whole. Like LTSER sites, MPAs should integrate ecological sciences and ethical theories and values in educational programs (Rozzi et al. 2012). These basic concepts and practices also are required for an Earth Stewardship initiative, which demands an integration of theory and practice, overcoming disciplinary constraints and producing innovative responses to address environmental, economic, and social challenges.

References

Abecasis RC, Schmidt L, Longnecker N et al (2013) Ocean & coastal management implications of community and stakeholder perceptions of the marine environment and its conservation for MPA management in a small Azorean island. Ocean Coast Manag 84:208–219. doi:10.1016/j.ocecoaman.2013.08.009

Aguirre Sala J (2015) Hermeneutic and field environmental philosophy to integrate ecological sciences and ethics into earth stewardship. In: Rozzi R, Chapin FS III, Callicott JB et al (eds) Earth stewardship: linking ecology and ethics in theory and practice. Springer, Dordrecht, pp 235–247

Amado-Filho GM, Maneveldt GW, Pereira Filho GH, Manso RCC, Bahia RG, Barros-Barreto MB et al (2010) Seaweed diversity associated with a Brazilian tropical rhodolith bed. Cienc Mar 36(4):371–391

Bates AE, Barrett NS, Stuart-Smith RD, Holbrook NJ, Thompson PA, Edgar GJ (2013) Resilience and signatures of tropicalization in protected reef fish communities. Nat Clim Chang 4(1):62–67. doi:10.1038/nclimate2062

Berchez F, Carvalhal F, Robim MJ (2005) Underwater interpretative trail: guidance to improve education and decrease ecological damage. Int J Environ Sustain Dev 4(2):128–139

Berchez F, Ghilardi N, Robim MJ et al (2007) Projeto Trilha Subaquática: sugestão de diretrizes para a criação de modelos de educação ambiental em unidades de conservação ligadas a ecossistemas marinhos. OLAM Ciên Tecnologia 7(3):181–209

Berchez F, Tiago CG, Rosso S et al (2009) Structure of a coralline algal bed on southeastern Brazil. Braz J Ecol 13:49–57

Bidle KD, Falkowski PG (2004) Cell death in planktonic, photosynthetic microorganisms. Nature reviews. Microbiology 2(8):643–655. doi:10.1038/nrmicro956

Brasil (2000) Lei 9.985, de 18 de julho de 2000. Regulamenta o art. 225, § 1o, incisos I, II, III e VII da Constituição Federal, institui o Sistema Nacional de Unidades de Conservação da Natureza e dá outras providências. Diário Oficial da União, Brasília

Brechin SR (2012) Shaping US climate opinion. Nat Clim Chang 2:236–237

Brulle RJ, Carmichael J, Jenkins JC (2012) Shifting public opinion on climate change: an empirical assessment of factors influencing concern over climate change in the U.S., 2002–2010. Clim Change 114(2):169–188. doi:10.1007/s10584-012-0403-y

Campagna C, Falabella V, Tagliorette A et al (2007) El mar des-protegido: zonificación y conservación de la biodiversidad. In: Boltovskoy D (ed) Atlas de Sensibilidad Ambiental de la costa y el Mar Argentino. http://atlas.ambiente.gov.ar

Chapin FS III, Pickett STA, Power ME et al (2015) Earth stewardship: an initiative by the Ecological Society of America to foster engagement to sustain planet earth. In: Rozzi R, Chapin FS III, Callicott JB et al (eds) Earth stewardship: linking ecology and ethics in theory and practice. Springer, Dordrecht, pp 173–194

Costa B, Erzini K, Caselle J et al (2013) "Reserve effect" within a temperate marine protected area in the north-eastern Atlantic (Arrábida Marine Park, Portugal). Mar Ecol Prog Ser 481:11–24. doi:10.3354/meps10204

Coumou D, Rahmstorf S (2012) A decade of weather extremes. Nat Clim Chang 2(7):491–496. doi:10.1038/Nclimate1452

Edgar GJ, Stuart-Smith RD, Willis TJ et al (2014) Global conservation outcomes depend on marine protected areas with five key features. Nature 506(7487):216. doi:10.1038/Nature13022

Floeter SR, Halpern BS, Ferreira CEL (2006) Effects of fishing and protection on Brazilian reef fishes. Biol Conserv 128:391–402

Francini-Filho RB, Moura RL (2008a) Dynamics of fish assemblages on coral reefs subjected to different management regimes in the Abrolhos Bank, eastern Brazil. Aquat Conserv Mar Freshwat Ecosyst 18(7):1166–1179

Francini-Filho RB, Moura RL (2008b) Evidence for spillover of reef fishes from a no-take marine reserve: an evaluation using the before-after control-impact (BACI) approach. Fish Res 93(3):346–356

Fraschetti S, Bevilacqua S, Guarnieri G et al (2012) Idiosyncratic effects of protection in a remote marine reserve. Mar Ecol Prog Ser 466:21-+. doi:10.3354/Meps09937

Gambarotta JC (2006) Estado actual, propuestas y perspectivas de manejo de las Áreas Protegidas Costeras. In: Menafra R, Rodríguez-Gallego L, Scarabino F et al (eds) Bases para la conservación y el manejo de la costa uruguaya. Vida Silvestre Uruguay, Montevideo, p 668

Gerhardinger LC, Godoy EAS, Jones PJS et al (2011) Marine protected dramas: the flaws of the Brazilian national system of marine protected areas. Environ Manage 47(4):630–643

Ghilardi NP, Berchez F (2010) Projeto Trilha Subaquática - modelos de educação ambiental marinha para o grande público. In: Pedrini AG (ed) Educação ambiental marinha e costeira no Brasil. EdUERJ, Rio de Janeiro, pp 71–91

Ghilardi NP, Pereira Filho GH, Berchez F (2008) Current knowledge status of the ecology of hard bottom benthic communities in Brazil and the need for new approaches. Oecol Brasiliensis 12(2):197–205

Ghilardi-Lopes NP (2014) Engaging people in the conservation of 'non-charismatic' benthic marine organisms: a Brazilian case study. Eingana 37:11–13

Ghilardi-Lopes NP, Hadel VF, Berchez F (2012) Guia para Educação Ambiental em Costões Rochosos. Artmed, Porto Alegre

Gore A (2006) An inconvenient truth: the planetary emergency of global warming and what we can do about it. Rodale Books, New York

Guarderas PG (2007) Marine conservation in Latin America and the Caribbean: an analysis of marine protected areas (MPAs). Masters Thesis, Oregon State University

Harris RE (2010) University of Concepcion Marine Station, Dichato, Chile. J Plankton Res 32(8):1103–1104

Heffernan O (2012) Adapting to a warmer world: no going back. Nature 491(7426):659–661. doi:10.1038/491659a

Hodgson G (1999) A global assessment of human effects on coral reefs. Mar Pollut Bull 38(5):345–355

Howe PD, Markowitz EM, Lee TM et al (2012) Global perceptions of local temperature change. Nat Clim Chang 3(4):352–356. doi:10.1038/nclimate1768

Huntington BE, Lirman D (2012) Coral species richness estimates are sensitive to differences in reef size and regional diversity. Limnol Oceanogr Methods 10:110–116. doi:10.4319/lom.2012.10.110

Huntington BE, Karnauskas M, Babcock EA et al (2010) Untangling natural seascape variation from marine reserve effects using a landscape approach. PLoS One 5(8):e12327. doi:10.1371/journal.pone.0012327

Hurwitz BL, Hallam SJ, Sullivan MB (2013) Metabolic reprogramming by viruses in the sunlit and dark ocean. Genome Biol 14(11):R123. doi:10.1186/gb-2013-14-11-r123

IUCN, UNEP-WCMC (2010) The World Database on Protected Areas (WDPA) 2010. UNEP-WCMC, Cambridge, UK

Kahan D (2012) Why we are poles apart on climate change. Nature 488:255

Kearney R, Farebrother G, Buxton CD et al (2013) How terrestrial management concepts have led to unrealistic expectations of marine protected areas. Mar Policy 38:304–311. doi:10.1016/j.marpol.2012.06.006

La Trobe HL, Acott TG (2000) A modified NEP/DSP environmental attitudes scale. J Environ Educ 32:12–20

Lawler JJ, White D, Neilson RP et al (2006) Predicting climate-induced range shifts: model differences and model reliability. Glob Chang Biol 12(8):1568–1584. doi:10.1111/j.1365-2486.2006.01191.x

Leão ZMDAN, Kikuchi RKP, Oliveira MDM et al (2010) Status of Eastern Brazilian coral reefs in time of climate changes. Pan Am J Aquat Sci 5:224–235

Leviston Z, Walker I, Morwinski S (2013) Your opinion on climate change might not be as common as you think. Nat Clim Chang 3:334–337. doi:10.1038/nclimate1743

Lewandowsky S, Gignac GE, Vaughan S (2013) The pivotal role of perceived scientific consensus in acceptance of science. Nat Clim Chang 3:399–404. doi:10.1038/nclimate1720

Lubchenco J, Palumbi SR, Gaines SD et al (2003) Plugging a hole in the ocean: the emerging science of marine reserves. Ecol Appl 13(1):S3–S7

Maass M, Equihua M (2015) Earth stewardship, the need for a transdisciplinary approach and the role of the International Long-Term Ecological Research Network (ILTER). In: Rozzi R, Chapin FS III, Callicott JB et al (eds) Earth stewardship: linking ecology and ethics in theory and practice. Springer, Dordrecht, pp 217–233

Matos LFOR (2009) Percepção ambiental de estudantes de uma escola da região central de Cuiabá, MT. Masters Thesis, Universidade Federal de Mato Grosso

Miloslavich P, Klein E, Díaz JM et al (2011) Marine biodiversity in the Atlantic and Pacific coasts of South America: knowledge and gaps. PLoS One 6:e14631

Moritz MA (2012) Wildfires ignite debate on global warming. Nature 487(7407):273

Myers TA, Maibach EW, Roser-Renouf C et al (2012) The relationship between personal experience and belief in the reality of global warming. Nat Clim Chang 3(4):343–347. doi:10.1038/nclimate1754

Navarrete S, Gelcich S, Castilla J (2010) Long-term monitoring of coastal ecosystems at Las Cruces, Chile: defining baselines to build ecological literacy in a world of change. Rev Chil Hist Nat 83:143–157

Orenstein DE, Groner E (2015) Using the ecosystem services framework in a Long-Term Socio-Ecological Research (LTSER) platform: lessons from the Wadi Araba Desert, Israel and Jordan. In: Rozzi R, Chapin FS III, Callicott JB et al (eds) Earth stewardship: linking ecology and ethics in theory and practice. Springer, Dordrecht, pp 281–296

Osborn D, Pearse J, Roe C (2005) Monitoring rocky intertidal shorelines: a role for the public in resource management. In: Magoon OT, Converse H, Baird B, Jines B, Miller-Henson M (eds) California and the world ocean '02: revisiting and revising California's ocean Agenda. American Society of Civil Engineers, California, pp 624–636

Pedrini AG, Brito MIMS (2006) Educação Ambiental para o desenvolvimento ou sociedade sustentável? Uma breve reflexão para a América Latina. Rev Educação Ambient Ação 17:20

Pereira-Filho GH, Amado-Filho GM, Moura RL et al (2011) Extensive Rhodolith beds cover the summits of Southwestern Atlantic Ocean Seamounts. J Coast Res 28(1):261–269

Pimm SL (1994) A history of the ecosystem concept in ecology – more than the sum of the parts – Golley FB. Nature 370(6486):188. doi:10.1038/370188b0

Redman CL, Miller TR (2015) The technosphere and earth stewardship. In: Rozzi R, Chapin FS III, Callicott JB et al (eds) Earth stewardship: linking ecology and ethics in theory and practice. Springer, Dordrecht, 269–279

Robim MJ, Vilar DD, Fonseca FF, Cocco MDA (2008) Caminhos para a conservação do patrimônio cultural em Unidades de Conservação. O caso do Parque Estadual da Ilha Anchieta. OLAM – Ciênc Tecnol 8(2):104–130

Rozzi R (2013) Biocultural ethics: from biocultural homogenization toward biocultural conservation. In: Rozzi R, Pickett STA, Palmer C et al (eds) Linking ecology and ethics for a changing world: values, philosophy, and action. Springer, Dordrecht/Heidelberg/New York/London, pp 9–32

Rozzi R, Massardo F (2011) The road to biocultural ethics. Front Ecol Environ 9(4):246–247

Rozzi R, Armesto JJ, Gutierrez JR et al (2012) Integrating ecology and environmental ethics: earth stewardship in the southern end of the Americas. BioScience 62(3):226–236. doi:10.1525/bio.2012.62.3.4

Saheb D, Asinelli-Luz A (2006) As representações de meio ambiente de professores e alunos e a pedagogia de projetos: um estudo de caso em classes de alfabetização. Rev Eletrônica Mestr Educ Amb 16:163–178

Sala E, Costello C, Dougherty D et al (2013) A general business model for marine reserves. PLoS One 8(4):e58799. doi:10.1371/journal.pone.0058799

SAyDS (2007) Efectividad del manejo de las áreas protegidas marino-costeras de la Argentina. Secretaría de Ambiente y Desarrollo Sustentable de la Nación: Vida Silvestre Argentina, Buenos Aires

Schiavetti A, Manz J, DosSantos CZ et al (2013) Marine protected areas in Brazil: an ecological approach regarding the large marine ecosystems. Ocean Coast Manag 76:96–104. doi:10.1016/j.ocecoaman.2013.02.003

Silvertown J (2009) A new dawn for citizen science. Trends Ecol Evol 24(9):467–471

Spalding MD, Fox HE, Allen GR et al (2007) Marine ecoregions of the World: A bioregionalization of coastal and shelf areas. BioScience 57(7):573–583

Tollefson J (2012) Hurricane sweeps US into climate-adaptation debate. Nature 491(7423):167–168

Turra A, Cróquer A, Carranza A et al (2013) Global environmental changes: setting priorities for Latin American coastal habitats. Glob Chang Biol 19(7):1965–1969

Wallace JM, Held IM, Thompson DWJ et al (2014) Global warming and winter weather. Science 343(6172):729–730

Whyte AVT (1977) Guidelines for field studies in environmental perception. UNESCO, Paris

Zaller JR (1992) The nature and origins of mass opinion. Cambridge University Press, Cambridge, UK

Zaval L, Keenan EA, Johnson EJ et al (2014) How warm days increase belief in global warming. Nat Clim Chang 4(2):143–147. doi:10.1038/nclimate2093

Chapter 24
Earth Stewardship, Climate Change, and Low Carbon Consciousness: Reflections from Brazil and South America

Eduardo Viola and Larissa Basso

Abstract The planet has entered the Anthropocene, a period in which humanity is the main driver of Earth systems changes. The effects of these fast and profound changes have been devastating for several fellow species – extinct or threatened to be extinct – and endanger a great part of the human population. In order to rebalance human interests and the equilibrium of the planet, Earth Stewardship, paradigm in which humanity recognizes itself as a co-inhabitant and steward of the planet, should be implemented. However, deep-rooted cultural traits can be stumbling blocks to the development of this new planetary ethics, and this is the case for South America. This chapter discusses the foundations of a new planetary ethics in South America and the challenges for its implementations, exemplified by one of the most contemporary complex issues, climate change.

Keywords Anthropocene • Biophysical planetary boundaries • Climate change • Environmental policy • Long term decision-making • Planetary consciousness

24.1 Introduction

If humanity is to advance towards a planetary consciousness, Earth Stewardship is the paradigm to be adopted. According to it, human beings are co-inhabitants of the planet and should consider the consequences of their development patterns, not only for current and future human generations, but also for other species. The challenges of implementing this paradigm are great, especially after the beginning of the Anthropocene, period in which humanity became the main driver of the transformations of Earth systems. In the last 150 years, human beings were responsible for fast

E. Viola (✉) • L. Basso
Climate Change and International Relations Research Programme,
Institute of International Relations, University of Brasília, Brasília, DF, Brazil
e-mail: eduviola@gmail.com; larissabasso@gmail.com

R. Rozzi et al. (eds.), *Earth Stewardship*, Ecology and Ethics 2,
DOI 10.1007/978-3-319-12133-8_24

and profound changes in Earth systems: three of the nine identified biophysical planetary boundaries leading to a systemic planetary disruption have already been trespassed. Climate change, the most studied of the nine, requires implementing a low carbon consciousness to be mitigated, so deeply rooted values that have hindered not only the equality of opportunities for different human populations, but also the survival of several other species, need to be substituted with new ones, based on communal bonds among humans and their fellow species.

This transition is not easy, and faces different stumbling blocks, usually connected to culture. Due to common traits of their history and their chosen development paths, South American countries share several cultural obstacles to a planetary consciousness. Even in countries where environmentalism and climate change policies are more advanced, such as Brazil, the traits are present and block a steady environmental development.

The objective of this chapter is to explore the links between Earth stewardship, climate change, and a low carbon consciousness. For this purpose, it is divided in four parts. In the first, the Anthropocene, climate change and the values of the low carbon ethics needed to mitigate it are the focus. In the second, South American characteristics that relate to climate change and the development of low carbon consciousness in the continent will be explained. Then a brief analysis of the environmental movement and the trajectory of climate change and low carbon development in Brazil, and the features it shares with South America, are presented. Finally, the links between an environmental consciousness in general, and a low carbon one in particular, are made with the development of the Earth Stewardship paradigm in South America. The article concludes that South America has important contributions to the new paradigm, but relevant challenges lie ahead in its full implementation in the continent.

24.2 The Anthropocene, Climate Change, and Planetary Consciousness

Approximately 11,000 ago, the last glaciation ended and the Earth entered a geological epoch of warmer and more stable climate called the Holocene. During this geological period, humanity changed from a few groups of hunters-gatherers to a complex population of 7 billion people divided by nationalities, culture, and ways of living. The expansion of human population – both in numbers and in per capita exploitation of Earth's resources (Crutzen and Stoermer 2000) – was such that it became the driver of the transition to a new geological epoch, the Anthropocene. This is unprecedented: from the evidence currently available, all previous changes in Earth systems were caused either by Earth's natural dynamics or the dynamics of the cosmos. Instead, in the Anthropocene human beings are the main drivers of transformations in Earth systems (Crutzen and Stoermer 2000; Viola et al. 2013).

Science has identified nine biophysical planetary boundaries that, when trespassed, can lead to a systemic planetary disruption of Earth systems: climate change, biodiversity loss, nitrogen/phosphorus cycle, stratospheric ozone depletion, ocean acidification, global freshwater use, change in land use, aerosol pollution, and chemical contamination (Rockstrom et al. 2009). Three of the boundaries already have been trespassed: first, the nitrogenous cycle, due to the massive use of nitrogenized fertilizers since the 1950s; second, biodiversity loss, especially since the 1970s, when an expanded human population began impacting natural ecosystems and other species to the point of exhausting them; third, climate stability, due to the accumulation of Greenhouse Gases (GHG) in the atmosphere – currently, 400 part per million (ppm),[1] when the safe boundary is 350 ppm. By trespassing these boundaries, human beings demonstrate that the modes of development adopted by the majority of human groups during the last centuries have not taken into account their effects on Earth's equilibrium: an anthropocentric ethics has dominated the human mindset,[2] and even human survival in the long term is now threatened (North 1990). Humanity is at a crossroads: contemporary issues require a new planetary consciousness in order to be successfully tackled, yet policy making is stuck in an outdated framework incapable of incorporating this new consciousness.

Take, for instance, climate change. From the nine planetary boundaries, climate change is the broadest studied and understood by science, and yet tackling it is still a great challenge, due to the nature of the issue (Rockstrom et al. 2009). It was defined as a truly wicked or diabolical problem because its complexity defies current international problem solving (Prins et al. 2010; Jamieson 2011; Steffen 2011). First, it is intrinsically global, once it is led by changes in the concentration of GHG in the atmosphere. However, there are no institutions for direct global policy making: human policy making is mostly carried out by nation states that only occasionally cede sovereignty to international institutions. Second, climate change is not linear, because vulnerabilities and responsibilities to the problem are skewed: the poorest of every society are usually the ones that contribute the least to the problem, but also the ones that mostly suffer their effects. Third, it operates in a time scale that is beyond human daily experience, so the appeal of passing the burden to future generations is always present.

Tackling climate change requires the adoption of a low carbon development model, a pattern in which GHG emissions are decoupled from economic growth.

[1] The concentration of GHG in the atmosphere was 401.85 ppm in 12 Jun 2014 according to the Carbon Dioxide Information Analysis Center. Data available at http://cdiac.ornl.gov, accessed 19 Jun 2014.

[2] It is important to clarify that, whenever the article does not specify the population, it is referring to the contemporary global society that encompasses most Western and many non-Western societies such as China, India, South Africa, Egypt, Indonesia, Japan and South Korea. However, societies that are very poor, such as many sub-Saharan African countries, so far have not impacted the global equilibrium, neither did small subnational social systems in all societies that are guided by values that consider the ecological equilibrium, local and planetary.

This represents a revolution in current ways of living: changes are required in construction, transportation, appliances, patterns of production and consumption, and waste management, among others. At bottom, it requires cultural changes that redefine success and wellbeing (Socolow and English 2011). Two major obstacles for such shifts to take place are: (i) cultural changes take decades or even centuries to be completed, while climate change is an urgent issue – given that its effects have started to be felt through the increase in frequency and intensity of extreme weather events that produce massive human suffering and material losses (Grimm and Jacobs 2013; US Global Change Research Program 2014); and (ii) climate change requires thinking about the planet in a holistic manner, acknowledging its finite resources and the rights of all beings to share them, which means not only thinking about the consequences of one's decision for all humanity, but also for the other species. Despite ideas developed by Buddhists, Stoics, Kant, and environmental philosophers of the last century, these concerns have only been marginally incorporated into climate change policy-making.[3] Acknowledging this problematic trend and incorporating this critical understanding into the contemporary mindset requires changes in values, replacing deep-rooted values and epistemologies[4] with new ways of thinking.

First, change from materialism towards post-materialism must occur. Currently, success and well-being are coupled with consumerism; one Earth is not sufficient to supply goods for all human beings to consume at the level of consumption of most industrialized societies. Success and well-being must be decoupled from luxury and disposable consumption, and more frugal lifestyles should be valued (Socolow and English 2011). It is not the case of going back to live in caves or to give up important achievements that made human life longer and more comfortable, but to understand that enabling all humans to have access to these comforts without pressing the planet even more requires excesses to be cut off. Less – and multifunctional – items, lasting longer and capable of being repaired, and production methods that reduce harmful impacts on natural resources and other species, should be central concerns.

Second, dramatic change is needed in reproductive ethics and welfare systems (Viola et al. 2013). There are 7.2 billion people in the world, too many; human population must downsize – stabilizing at 9–10 billion is not an option for Earth's equilibrium. A fast transition to fertility rates below the level of replacement (2.1 children per woman) is key. For fertility to decrease, women must have strongly protected reproductive rights, and higher levels of education and personal independence. Very few societies have reached responsible fertility rates – Italy, Spain,

---

[3] The United Nations Framework Convention on Climate Change (UNFCCC) was signed and ratified by all countries; the agreement states that humanity should reduce carbon emissions and enter upon a decarbonization path; however, the Kyoto Protocol is the only implementation tool of UNFCCC objectives, and almost a decade after entering in force, it has not achieved its goals – in fact, carbon emissions have increased since the ratification of the Protocol. International rhetoric in climate change is decoupled from effective implementation tools.

[4] Some of the values that predominate in contemporary global society, against which the necessary chances are defined, are consumerism and excessive aesthetic concerns, national interests defined narrowly (as seen before) and short-term thinking.

Germany, Nordic countries, China, Japan, South Korea –, but they now face problems related to an ageing population (Timonen 2008). Therefore, responsible fertility rates need to be coupled with dramatic changes in lifestyles and health systems, which need to be based in the concept of healthy and responsible longevity. Furthermore, pension schemes must be reformulated to match the increase in life expectation: gradual change from full employment to full retirement, increased age for retirement, and abolition of the earlier retirement for women in countries where this difference exists (Holzmann and Hinz 2005).

Third, change towards post-sovereignty is required. Sovereignty should be adapted to contemporary requirements, in two different levels. First, at the level of the States: sovereignty should no longer be defined as a static defense against foreign influence over domestic issues, but instead as an entitlement that enables participation in a pluralistic dialogue to find solutions to global concerns. Second, at the level of the individual: nationality should be seen as one of many features of a person, and not as a dividing line that determines who is entitled to the resources of the planet.

Finally, change from short-term to long-term to decision-making is imperative. Long-term considerations favor a holistic approach to decisions, once links between different issue areas are understood and measured against each other. This approach enables integration, seeing the whole picture before the decision is made: it is not about exchanging living in the present for living towards the future, but acknowledging the cause-and-consequence links of different decisions, not only to one's own life but to the life of the community. When decisions are better informed, they lead to more long-term and satisfying results.

Climate change is just one issue challenging effective international policy-making; keeping humanity from trespassing other planetary boundaries requires deeper changes, embedded in a new planetary ethics in which every human being becomes an Earth steward. The challenge is great, but can be faced. Analyzing the several stumbling blocks is the first step towards meeting it. In the next two sections, South American and Brazilian struggles towards successfully adopting a low carbon consciousness – one that mitigates climate change – will be analyzed, so that, in the last section, the links to a wider Earth Stewardship concept can be established.

24.3 Climate Change and Low Carbon Consciousness in South America

South American countries share features that are relevant to climate change and the transition to low carbon development. First, the continent is significantly vulnerable to climate change: (i) semi-arid areas, such as the Brazilian north-eastern region, would become more arid; (ii) low costal and densely populated areas, such as the city of Recife and delta regions like the Plate and Orinoco rivers, are extremely vulnerable to changes in sea levels; (iii) glacial retreat in the Andes threaten the supply of water of large cities located in dry regions, such as Lima, Quito, and Mendoza; (iv) due to change in rainfall, agricultural areas all across the continent will lose productivity, and eastern Amazonia could be transformed into a savannah; (v) electricity production also will be affected greatly.

Hydroelectricity is, in fact, one of the continent's low carbon features: no other region of the world produces as much electricity from hydropower as South America. The decision to invest in hydropower was based on energy security concerns. In the 1970s, when the oil crises hit the world, many countries decided to rely on their hydro potential for electricity, and large hydropower plants were built. It is curious that even countries with large oil reserves use hydro for electricity generation – the best example is Venezuela, a member of the Organization of the Petroleum Exporting Countries (OPEC). In the continent, great hydropower potential remains to be explored. However, important parts of this potential are in areas that could affect Amerindian people, biodiversity, or magnificent natural beauties, raising a great controversy in the continent and around the world.

The Amazon region has been at the center of environmental policy discussions in the continent. Due to Brazilian policies to increase the population in the area and exploit it commercially, Amazon deforestation has been a major issue during the last quarter of the twentieth Century. The Amazonian countries, led by Brazil – 70 % of the Amazon forest is in its territory –, have answered sharply to foreigners that have criticized the use of the Amazon. These countries have highlighted their sovereign rights over the area, and argued that social, economic, and environmental sustainability are impossible to be obtained without economic development, especially in a continent in which the latter remains a major issue.

Indeed, coupling economic development and environmental sustainability is the continent's platform in international negotiations. Even if the links between the issues are real and have been acknowledged internationally, they are frequently distorted in South American discourse. Legacy of the colonization process, the discourse of Northern imperialism and foreign exploitation of South American resources is deep-rooted in South American culture. The world is blamed for the continent's lower development standards, and every new issue that rises in international arenas is met with suspicion. The discourse might win votes for populist governments, but it is too simplistic to explain the real picture.

In the end, South America is a victim of itself (Viola et al. 2013; Halperin Donghi 1970; O'Donnel 1999). One of the greatest stumbling blocks to South American development is the low level of republicanism around the continent – using the Kantian definition, which implies understanding the difference between private and public spheres, and the importance of a civic attitude towards public matters.[5] Cronyism is rooted deeply in South American politics. From the time the states

[5] Republican values are in force when there is: (i) a clear distinction between the public and private interests, (ii) equality of opportunities for all, through quality public education, (iii) legal equality of all, (iv) due process of law, (v) state intervention in the economy to balance private and public interests, and (vi) a welfare state with relatively low social inequality.

were Iberian colonies – following the tragic reduction of native South Americans by war, epidemics, and inhuman labor exploitation–, personal relationships have settled the tune of most ventures that takes place in the region. In South America, economic sectors have excessive influence over governmental decision-making and the political system. When hardship hits, instead of investing in productivity and efficiency, great companies look for the national government to solve their problems; in exchange, they fund election campaigns. When the plutocratic links reach such unbearable levels that corruption scandals become common, the resort to populism – another common discourse in the region – does not bring any relief. Only Chile and Uruguay have had significant progress toward effective republicanism; in the last two decades, these countries have gradually increased the quality of representative democracies without falling into the populist trap (Viola and Leis 2008).

Another obstacle, and intrinsically linked to low republican values, is short-term thinking. Long-term planning requires understanding the big picture, not only about oneself, but also about others – understanding that each individual is part of a community and that their behaviors are interdependent. When communal ties are loose, it is difficult for long-term thinking to flourish. Short-term thinking in climate change consideration leads to South American failure to adopt the vanguard position that would be expected from a continent that leads in low carbon assets, energy being the most evident of them (Vergara et al. 2013). South America could lead low carbon development, taking advantage of the technology that has been created to develop its great low carbon energy sources. Instead, the continent is increasing its dependency on fossil fuels, and missing the opportunity to develop local renewable energy industries that could contribute to increase the region's welfare (Vergara et al. 2013).

Still, the picture has been changing in recent years. Following the increase of extreme weather events in the continent and the negative consequences for the populations, the consciousness of climate change is raising. Slowly, societal groups are embracing the cause and trying to push for change: highly educated and cosmopolitan urban activists, and some cities and towns, are changing their development paths.[6] Amerindian groups, especially from the Andean region, are creating links between their pledges and the climate cause, legitimizing their pleas with supposedly climate friendlier ways of living (Ribeiro 2009). Nevertheless, most South American countries' positions in international climate negotiations do not follow the recent progress in national and subnational arenas. Most of them are very defensive, insist on historical responsibilities and ecological debt from the North to the South, and understand international negotiations as a source of financial resources to protect their forests. Last, but not least, there is no South American coalition in the climate regime.

[6] Bogota, Buenos Aires, Caracas, Curitiba, Lima, Rio de Janeiro, Santiago, and Sao Paulo are members of the C-40, a group of cities established during the Earth Summit in 2012, engaged in developing initiatives to tackle shared environmental problems. Website: http://www.c40.org/cities

24.4 Brazil: Environmentalism, Climate Change, and the Roots of a Low Carbon Consciousness

Although some remarkable Brazilian thinkers had environmental concerns early in the XIX Century (José Bonifacio, Joaquim Nabuco, Alberto Torres), they did not impact Brazilian culture beyond very small circles of intellectuals; therefore, environmental concerns in Brazil flourished mostly as the result of European and American influence (Viola 1987; Padua 2004). The myth of abundance of natural resources, rooted in the Brazilian mindset since colonial times, made environmental concerns appear exaggerated in the eyes of Brazilian public opinion until the 1970s. By that time, the scientific discourse against industrial pollution had been heard and an environmental movement had started to develop, mainly in the south and southeast of the country. In the second half of the 1980s, claims over the preservation of the Amazon forest developed simultaneously in Brazilian and Euro-American public opinion (Mainwaring and Viola 1984; Viola 1988, 1997). Still, until 1989, the Brazilian government argued that those claims were related to foreign imperialism and threatened Brazilian national sovereignty.

Environmentalism gained momentum in the late 1980s and in the beginning of the 1990s. One of the greatest marks of the new environmental awareness was the inclusion of a chapter dedicated to the topic in the Brazilian Constitution of 1988, enacted after the re-democratization of the country (Box 24.1). Internationally, Brazil became more active in environmental negotiations; it was chosen as the host country for the 1992 Earth Summit. Nevertheless, the Brazilian negotiating mindset was still coupled with traditional South American narrow claims of sovereignty over natural resources and the priority of economic development. Brazil strongly opposed being legally bound by the Kyoto Protocol to reduce deforestation and to reduce the curve of growth of carbon emissions in the modern sector of the economy. Moreover, it was a member of the conservative *ad hoc* coalition (with USA, China, Canada, Russia, and India) that dramatically weakened the climate treaty in the end of the 1990s and during the 2000s, nullifying the progressive efforts of the European Union and Japan.

For decades, Brazilian economic elites felt Amazon deforestation was beyond tackling. These elites argued deforestation was spread throughout a sparsely inhabited region, and that the country lacked resources to deal with it. The feeling of fatalism was disseminated in the country. It was not until the second half of the 2000s that deforestation was reduced. During the period in which Marina Silva (2003–2008) and Carlos Minc (2008–2010) were Ministers of the Environment, deforestation decreased from 27,000 km^2 (in 2004) to 7,500 km^2 (in 2009) (Viola and Basso 2014).[7] The cutback was due to legal and institutional changes[8]: (a) political

[7] Annual averages. Data from the National Institute of Spatial Research (INPE), http://www.obt.inpe.br/prodes/index.php. In the first two years of Da Silva's tenure (2003–04) there was a dramatic increase in deforestation.

[8] In 1996 the old Forest Code (enacted in 1965) was changed to make compulsory the preservation of 80 % of the vegetation in the Amazon region. In 2006 the Act on the Management of Public Forests created the Brazilian Forest Service in order to manage the forests.

priority was given to the issue; (b) law enforcement and institutional capacity was enhanced; (c) coalitions by multi-stakeholders against the consumption of soy beans and beef produced in deforested areas were formed; (d) the influence of NGOs and the scientific community on the media increased; (e) new and extensive national parks and conservation units were created; and (f) cooperation between state and national governments was boosted (Viola 2013; Viola and Franchini 2013; Viola and Basso 2014).

Climate change effectively entered the Brazilian policy scenario in 2008, when the National Climate Change Plan was enacted. The Plan establishes targets that should be achieved in several areas so as to mitigate climate change.[9] In 2009, the Plan was included in a legal framework, the National Climate Change Law (Law 12187, complemented by Decree 7390/2010). This was a major advancement. Brazil was then among the selected group of countries that had enacted a Climate Change Law constraining carbon emissions, and the first non-OECD nation to do so – even some OECD countries, such as USA, Canada, and Australia, did not have one. Besides incorporating the Plan, the legal framework established that sectorial action plans should target reducing GHG emissions.[10]

Deforestation is no longer the greatest source of Brazilian GHG emissions, even if its amount is still significant. Since 2010, agriculture, cattle ranching, and energy services, answer for 67 % of the emissions (Brazil 2013).[11] The shift is important, because it shows that the fatalist arguments of the 1980s were wrong; but it also represents a challenge. If Brazil wants to reach a low carbon profile, reducing deforestation is no longer enough. Emissions from the energy sector are particularly concerning. The use of fossil fuels in transportation has been very high in Brazil, and since 2006 the federal government has been keeping the prices of oil derivatives artificially low in order to maintain higher rates of economic

---

[9] Increase energy efficiency; substitute charcoal coming from deforestation with charcoal coming from tree plantations; replace old fridges and HCFCs; invest in solar water heating and urban waste recycling; phasing out the use of fire for clearing sugar cane plantations; integrate agriculture and cattle raising systems maintain the share of renewable energy in the Brazilian electricity matrix, encourage the domestic and international use of biofuels, seek further reduction of deforestation, eliminate the net loss of forest coverage, strengthen inter-sector actions to reduce climate change vulnerability, and identify impacts of climate change in the environment and support scientific research about it. Plano Nacional sobre Mudança do Clima, available at http://www.mma.gov.br/estruturas/smcq_climaticas/_arquivos/plano_nacional_mudanca_clima.pdf, accessed 1 May 2014; executive summary in English available at http://www.mma.gov.br/estruturas/208/_arquivos/national_plan_208.pdf, accessed 1 May 2014.

[10] Among the most relevant sectorial plans are the plans to reduce deforestation in the Amazon and Cerrado Savannah, the low carbon agriculture plan, and the plans to reduce emissions from steel industry and energy sectors. There are also several subnational policies, especially in the Southern and South-eastern states, with stricter environmental standards compared to federal ones. Lei Nacional sobre Mudança do Clima, Law nr. 12187/2009, available at http://www.planalto.gov.br/ccivil_03/_ato2007-2010/2009/lei/l12187.htm, and Decree 7390/2010, available at http://www.planalto.gov.br/ccivil_03/_Ato2007-2010/2010/Decreto/D7390.htm, accessed 1 May 2014.

[11] In 2010, agriculture and cattle ranching's share of Brazilian GHG emissions was 35 %; energy services', 32 %; and LULUCF's, 23 %.

growth and to avoid an increase in inflation. This political maneuver increased greatly the consumption of fossil fuels (Viola and Basso 2014). For the same economic reasons, later in 2012 electricity was subsidized. It is important to remember, however, that due to the struggles to prevent the building of new hydropower plants, the share of fossil fuel and thermal power plants in Brazilian electricity generation, has increased since 2001. Adding to that, Brazilian energy efficiency policies are only marginally developed.

In spite of the fact that the presidential term of Dilma Rousseff (2011–2014) has been negative in terms of de-carbonizing the economy, the elaboration of a National Climate Change Adaptation Plan started in 2013 and has advanced. Indeed, it could become a law in the near future.

Brazilian international position in climate negotiations reflects the advances and setbacks in domestic policies. Until 2005, Brazil discourse remained very conservative: (i) it strongly opposed commitments for developing countries to reduce their carbon emissions growth from business-as-usual scenarios; (ii) it promoted an extremely distorted interpretation of the principle of common and differentiated responsibilities as almost everything was differentiated (Viola 2002); (iii) it strongly opposed the introduction of avoided deforestation in the Clean Development Mechanism[12]; and (iv) it formed alliances with highly carbonized energy matrix countries. Brazil's inability to deter deforestation of the Amazon at the time, and the fear of being internationally charged for it, were the main reasons for the conservative position. From 2006 to 2010, Brazil moved gradually towards reformism, accepting measures previously opposed (such as international forest regulation), and pledging a voluntary 36–39 % reduction of its emissions by 2020. It also started to abandon the discourse of Brazil as a poor country that needs international assistance to tackle climate change.

Since 2011, however, Brazil has gone back to the traditional conservative standing, requesting developed countries to commit to a second period of compulsory emissions reduction targets before emerging economies have to accept binding targets. Brazil set the tone of the Earth Summit 2012 (Rio+20) by fighting to exclude climate change from the negotiations and promoting a diffuse definition of green economy, against a more consistent one defended by the European Union. Moreover, in the 19th Conference of the Parties (COP 19) of the UNFCCC, held in Warsaw, Poland, in 2013, the Brazilian government revived the doctrine of historical emissions; in order to define carbon rights, emissions should be measured from 1850. This doctrine has never been accepted by developed countries, and is strongly criticized by most scientist and analysts; it had been abandoned by Brazil in 2009.

Crony capitalism and short-term thinking are features that Brazil shares with most South American countries. Brazilian climate change and low carbon policies have been greatly impacted by vested interests, individualism, and the lack of consideration of the links among environmental, social, and economic policies. Nevertheless, in Brazil public opinion is better informed about climate change than

[12] Due to fears that international financial resources that would be attracted to the Amazon region by CDM projects which would undermine the national sovereignty of the region.

in the rest of the continent, due to the country's prominence in international negotiations, and even one of the coalitions for the presidential elections of 2014 defended advancement toward de-carbonization, nationally and internationally. The following years will tell if this knowledge will be transformed into a low carbon consciousness and demand for low carbon policies, or if the country remains trapped in the illusion that it has a role to play in solving the problem.

Box 24.1. Brazilian Constitution Chapter VI: Environment

Article 225 of the Brazilian Constitution states that:

All have the right to an ecologically balanced environment, which is an asset of common use and essential to a healthy quality of life, and both the Government and the community shall have the duty to defend and preserve it for present and future generations.

Paragraph 1: In order to ensure the effectiveness of this right, it is incumbent upon the Government to: (I) preserve and restore the essential ecological processes and provide for the ecological treatment of species and ecosystems; (II) preserve the diversity and integrity of the genetic patrimony of the country and to control entities engaged in research and manipulation of genetic material; (III) define, in all units of the Federation, territorial spaces and their components which are to receive special protection, any alterations and suppressions being allowed only by means of law, and any use which may harm the integrity of the attributes which justify their protection being forbidden; (IV) demand, in the manner prescribed by law, for the installation of works and activities which may potentially cause significant degradation of the environment, a prior environmental impact study, which shall be made public; (V) control the production, sale and use of techniques, methods or substances which represent a risk to life, the quality of life and the environment; (VI) promote environment education in all school levels and public awareness of the need to preserve the environment; (VII) protect the fauna and the flora, with prohibition, in the manner prescribed by law, of all practices which represent a risk to their ecological function, cause the extinction of species or subject animals to cruelty.

Paragraph 2: Those who exploit mineral resources shall be required to restore the degraded environment, in accordance with the technical solutions demanded by the competent public agency, as provided by law.

Paragraph 3: Procedures and activities considered as harmful to the environment shall subject the infractors, be they individuals or legal entities, to penal and administrative sanctions, without prejudice to the obligation to repair the damages caused.

Paragraph 4: The Brazilian Amazonian Forest, the Atlantic Forest, the Serra do Mar, the Pantanal Mato-Grossense and the coastal zone are part of the national patrimony, and they shall be used, as provided by law, under conditions which ensure the preservation of the environment, therein included the use of mineral resources.

(continued)

Box 24.1. (continued)

Paragraph 5: The unoccupied lands or lands seized by the states through discriminatory actions which are necessary to protect the natural ecosystems are inalienable.

Paragraph 6: Power plants operated by nuclear reactor shall have their location defined in federal law and may not otherwise be installed.

Official translation provided by the Brazilian Supreme Court website: http://www.stf.jus.br/repositorio/cms/portalStfInternacional/portalStfSobre-Corte_en_us/anexo/constituicao_ingles_3ed2010.pdf (accessed June 19, 2014).

24.5 Implications for Earth Stewardship

Sustainability has been on the international agenda since the 1970s, and several principles have guided the discussions that targeted its implementation. Sustainable development has been the widest accepted of them all. Since the 1980s, international commitments reaffirm the importance of reaching a development path that respects the rights of future generations to live in a safe environment and to have opportunities for a prosperous existence. The decades that followed proved not only that humanity has not achieved sustainable development, but also that its prescriptions are not sufficient to harmonize human existence and Earth's equilibrium. Earth Stewardship offers a new paradigm for reaching this balance. It removes human beings from the center and emphasizes their ties with other beings that also have rights to the planet. Humanity is no longer owner of Earth's resources, but rather a part of them, and its existence should neither threaten future generations nor co-existing species.

Adopting this new paradigm in social sciences in general, and in international affairs in particular, would mean a complete reshaping of policy-making (Power and Chapin 2010; Rozzi 2013). First, complexity should be acknowledged; biological systems are complex, so policy responses need to respect it (Ostrom and Cox 2010). One-size-fits-all policies do not work; tailoring is key for effectiveness. Second, tailoring should be guided by considering the complexity of an issue and its connections to the planet. By checking multiple scales and verifying ecological and socioeconomic consequences, policy can align incentives with stewardship behavior and strengthen people's connections to pivotal features of the new paradigm (Chapin et al. 2011).

In international relations, implementing the Earth Stewardship paradigm means reinterpreting power politics. Power has been defined as the capacity to steer international politics according to a country's sole national interests; this understanding produced two devastating world wars and gave rise to totalitarian regimes in the industrial era. It changed slightly between 1948 and 1989. During these years, the United States and the USSR rivalry was the main driver of international affairs, but it was constrained by the nuclear arms race and the real risk of a global nuclear war.[13] Since the 1990s, war between the major powers became very unlikely, as their economic interdependence means both sides lose if involved in escalating conflicts. It is understood that cooperation is needed to tackle global problems. However, due to the narrow definition of national sovereignty that still prevails in most countries, while humankind is capable of avoiding major wars, it cannot act collectively in order to cope with global climate change, the deregulation of the global financial system, extensive poverty, massive loss of biodiversity, failed states, civil wars, and transnational religious conflicts. Understanding power in this confrontational sense is a major constraint for advancing global paradigms; revising it to find effective solutions to contemporary issues is a principal challenge for international politics. Unfortunately, there are no clear signs that the world will be successful in completing this transition in time to deal with urgent environmental matters, so alternatives that balance geopolitical consideration and planetary consciousness must be considered.

Engaging South America further in Earth stewardship means recognizing its demands in international politics and acknowledging its contributions to developing tailored solutions to current global problems. Development has always been a central demand in South America. Because power has been distributed among countries according to the size of their population and their level of development, South American countries resent their past as exploited colonies, and demand opportunities to reach a more egalitarian status with other actors of world politics. Dismissing these concerns is counterproductive because it creates resistance to international policies that are of the interests of South Americans as well, but might be regarded as foreign imperialism by them, thus delaying their implementation. Instead, supporting South American policies that aim at achieving reasonable standards of development[14] would ensure a more proactive and creative participation of South America in solving global problems.

South America has important contributions to an Earth Stewardship paradigm in general, and a low carbon consciousness in particular. First, some of the world's most precious natural endowments are located in the continent and are still well preserved. The Amazon forest is one of them. One of the most biodiverse areas of the world, it is crucial for mitigating climate change, biodiversity loss, water scarcity, land use change, and air pollution; it is home for Amerindian populations whose lifestyles are in harmony with forest dynamics. Patagonia is another example,

---

[13] The geopolitical equilibrium was always on the brink of annihilation based on the doctrine of mutual assured destruction (MAD).

[14] A good standard of development could be defined as a combination of per capita income of at least USD 20,000, GINI index < 0,4, almost 0 % illiteracy, an average of at least 10 years of good quality education, fertility rate below replacement levels, and life expectancy above 80 years (Viola et al. 2013). For other definitions check *World Development Indicators 2014*, by the World Bank, and the *Index of Human Development*, by the UN.

which covers a vast area of complex ecosystems (temperate and sub-Antarctic forests), inhabited by distinct Amerindian populations who respect environmental dynamics in pursuing their existence (Rozzi 2013). Furthermore, South America has advanced considerably in low carbon assets. It produces low carbon energy in greater degree than other continents; it has successfully tackled deforestation; the birth rates of many countries (led by Brazil and Uruguay) have been reduced considerably during the last decades. It has an important role to play by promoting low carbon agriculture, since some of the world's biggest agricultural commodities producers are South American. South American strategies to protect soils and reduce the use of fertilizers would contribute to mitigate the transformations in the nitrogen cycle already under way.

In general, South Americans value their natural endowments, and understand the importance of their preservation, as a result of the engagement of epistemic communities to clarify the links between ecosystems and life in the rest of the continent. It would also be unlikely that a low carbon consciousness would have started to flourish in a continent with considerable fossil fuel reserves and high economic development requirements, if it were not for the engagement of scientists and their resonance in South American societies. If Earth stewardship is to be truly promoted in South America, science needs to engage in understanding the complexity of social dynamics as well, pointing to characteristics that could further the new paradigm, but also to the ones that hinder its development.

Among the first, it is important to acknowledge that several different environmental ethics co-exist in the continent, and some of them combine human development and environment equilibrium. Several Amerindian peoples, for example, have traditionally praised harmony with ecosystems in which they live; their environmental impact has been minimal over the centuries. In some cases, these cultures have influenced the implementation of national legal frameworks that are in the vanguard considering the relations between nature and society, such as Ecuador and Bolivia. Different philosophies are important sources of values and should be regarded as such; however, critical thinking is essential to finding the balance between what could be successfully applied to tackling current issues, and what would worsen other contemporary struggles.

In addition, it is important to note that social issues are embedded in the context in which they develop, and it is very hard to export solutions from one context to another. On the one hand, Amerindian peoples developed philosophies that took global cosmologies into account, but their daily experience was local or regional at most. Some of the contemporary issues such as climate change, on the other hand, are truly global, related to a globalized reality that is not similar to any previous situation in the history of humanity. A successful ethics for dealing with these problems needs to be embedded in contemporary context. It can learn from past experiences, but must develop its own framework to fight the problem according to current characteristics that take into account the complex causes of the issues and the different alternatives to solve them.

Considering this, one of the greatest contributions social sciences engaged in advancing a planetary consciousness in South America could make is clarifying how deep-rooted cultural traits, such as crony capitalism and short-term thinking, are major obstacles to paradigm change in the continent. Earth Stewardship is not a paradigm only from the North, so Southern concerns, such as opportunities for economic development and the respect for diverse lifestyles, should be acknowledge in its matrix. But dismissing critical stumbling blocks to its implementation is not legitimate either. If human beings are really to advance towards a new ethical path in which they understand themselves as co-inhabitants and stewards of the planet, then implementing post-materialism, new reproductive understanding, post-sovereignty, and long-term consideration is a duty owed not only to their fellow human beings, but to other species as well.

References

Brasil (2013) Estimativas anuais de emissões de gases de efeito estufa no Brasil. http://gvces.com.br/arquivos/177/EstimativasClima.pdf. Accessed 12 Jan 2014

Chapin FS III, Power ME, Pickett STA et al (2011) Earth Stewardship: science for action to sustain the human-earth system. Ecosphere 2(8) article 89:01–20

Crutzen PJ, Stoermer EF (2000) The 'Anthropocene'. Glob Chang Newsl 41:17–18

Grimm NB, Jacob KL (2013) Evaluating climate impacts on people and ecosystems. Front Ecol Environ 11:455

Halperin Donghi T (1970) História contemporanea de America Latina. Alianza, Madrid

Holzmann R, Hinz R (2005) Old-age income support in the 21st century: an international perspective on pension systems and reform. World Bank, Washington, DC. http://siteresources.worldbank.org/INTPENSIONS/Resources/Old_Age_Inc_Supp_Full_En.pdf. Accessed 19 June 2014

Jamieson D (2011) The nature of the problem. In: Dryzek JS, Norgaard RB, Schlosberg D (eds) The Oxford handbook of climate change and society. Oxford University Press, Oxford, pp 38–54

Mainwaring S, Viola E (1984) New social movements, political culture and democracy: Brazil and Argentina in the 1980s. Telos 1984(61):17–52

North RC (1990) War, peace, survival: global politics and conceptual synthesis. Westview Press, Boulder

O'Donnel G (1999) Poliarchies and the (un)rule of law in Latin America. In: Mendez J, O'Donnel G, Pinheiro PS (eds) The (un)rule of law and the underprivileged in Latin America. Notre Dame University, Notre Dame

Ostrom E, Cox M (2010) Moving beyond panaceas: a multi-tiered diagnostic approach for social ecological analysis. Environ Conserv 37(4):451–463

Padua JA (2004) Um sopro de destruição: pensamento político e crítica ambiental no Brasil escravista, 1786–1888. Editora Zahar, Rio de Janeiro

Power ME, Chapin FS III (2010) Planetary stewardship, with an introduction from the Editor-in-Chief. Bull Ecol Soc Am 91:143–175

Prins G, Galiana I, Green C et al (2010) The Hartwell paper – a new direction for climate policy after the crash of 2009, London School of Economics and University of Oxford. http://eprints.lse.ac.uk/27939/1/HartwellPaper_English_version.pdf. Accessed 2 May 2014

Ribeiro GL (2009) Non-hegemonic globalizations. Alternative transnational processes and agents. Anthropol Theory 9:1–33

Rockstrom J, Steffen W, Noone K et al (2009) A safe operating space for humanity. Nature 461(24):472–475. http://www.nature.com/nature/journal/v461/n7263/full/461472a.html. Accessed 1 May 2014

Rozzi R et al (2013) Biocultural ethics: from biocultural homogenization toward biocultural conservation. In: Rozzi R, Pickett STA, Palmer C (eds) Linking ecology and ethics for a changing world: values, philosophy, and action. Springer, Dordrecht/Heidelberg/New York/London, pp 9–32

Socolow RH, English MR (2011) Living ethically in a greenhouse. In: Arnold DG (ed) The ethics of global climate change. Cambridge University Press, Cambridge, UK, pp 170–191

Steffen W (2011) A truly complex and diabolical policy problem. In: Dryzek JS, Norgaard RB, Schlosberg D (eds) The Oxford handbook of climate change and society. Oxford University Press, Oxford, pp 21–37

Timonen V (2008) Ageing societies: a comparative introduction. Open University Press, Maidenhead

UNEP (2011) Towards a green economy: pathways to sustainable development and poverty eradication. http://www.unep.org/greeneconomy/Portals/88/documents/ger/ger_final_dec_2011/Green%20EconomyReport_Final_Dec2011.pdf. Accessed 1 May 2014

United Nations Conference On Sustainable Development – UNCSD (2012) The future we want; United Nations General Assembly Resolution A/RES/66/288. http://daccess-dds-ny.un.org/doc/UNDOC/GEN/N11/476/10/PDF/N1147610.pdf?OpenElement. Accessed 1 May 2014

US Global Change Research Program (2014) Climate change impacts in the United States. http://nca2014.globalchange.gov. Accessed 19 June 2014

Vergara W, Alatorre C, Alves L (2013) Rethinking our energy future a white paper on renewable energy for the 3GFLAC regional forum, Inter-American Development Bank, Climate Change and Sustainability Division Energy Division, Discussion paper no. IDB-DP-292, http://publications.iadb.org/bitstream/handle/11319/5744/3gflac%20White%20Paper%20English%20v21%20-%20with%20cover.pdf?sequence=1. Accessed 19 June 2014

Viola E (1987) O movimento ecológico no Brasil: do ambientalismo à ecopolítica. Rev Bras Ciências Sociais 1(3)

Viola E (1988) The ecologist movement in Brazil (1974–86): from environmentalism to ecopolitics. Int J Urban Reg Res 12(2):211–228

Viola E (1997) The environment movement in Brazil: institutionalization, sustainable development and crisis of governance since 1987. In: Macdonald GJ, Neilson D, Stern MA (eds) Latin American environmental policy in international perspective. Westview Press, Boulder

Viola E (2002) O regime internacional de mudança climática e o Brasil. Rev Bras Ciências Soc 17(50): 25–46. http://www.scielo.br/pdf/rbcsoc/v17n50/a03v1750.pdf. Accessed 1 May 2014

Viola E (2013) Brazilian climate policy since 2005: continuity, change and prospective. Centre for European Policy Studies Working document no. 373. http://www.ceps.be/book/brazilian-climate-policy-2005-continuity-change-and-prospective. Accessed 1 May 2014

Viola E, Basso L (2014) Amazonian policy and politics, 2003–13: deforestation, hydropower and biofuels. In: NOREF – Norwegian Peacebuilding Resource Centre Report. http://www.peacebuilding.no/Themes/Emerging-powers/Publications/Amazonian-policy-and-politics-2003-13-deforestation-hydropower-and-biofuels. Accessed 1 May 2014

Viola E, Franchini M (2012) Os limiares planetários, a Rio+20 e o papel do Brasil. Cadernos EBAPE 10(3):470–491

Viola E, Franchini M (2013) Brasil na governança global do clima, 2005–2012: a luta entre conservadores e reformistas. Contexto Internacional 35(1):43–76

Viola E, Leis H (2008) America del Sur en el mundo de las democracias de mercado. Cadal Homo Sapiens, Buenos Aires

Viola E, Franchini M, Ribeiro TL (2012) Climate Governance in an international system under conservative hegemony: the role of major powers. Rev Bras Política Int 55(special edition):9–29

Viola E, Franchini M, Ribeiro TL (2013) Sistema internacional de hegemonia conservadora – governança global e democracia na era da crise climática. Annablume, São Paulo

Chapter 25
Stewardship, Integrity of Creation and Climate Justice: Religious Ethics Insights

Guillermo Kerber

Abstract In religious ethics, Earth stewardship (ES) relates to the care for creation and climate justice. This chapter analyses as a case study the work done by the World Council of Churches over the last decades on environment and climate change and relates it to other faith initiatives. It takes into account the two different stories of creation in the Bible and their consequences in Christian thought. Listening both to the cry of the Earth and the cry of vulnerable people and trying to respond to them, constitute the rationale for developing actions addressed to individuals as well as to institutions at national and international levels. Ethical and spiritual values are presented as countercultural attitudes to confront the present development paradigm which destroys peoples and Earth.

Keywords Climate change • Ethics • Justice • Religion • Spirituality

The concept of Earth stewardship is at the core of religious messages. For Christians, for instance, human beings have been put on Earth to take care of it and to look after it as we read in the first book of the Bible, Genesis, chapter 2 verse 15. This affirmation is shared by Jewish and Muslim traditions. For these three so called Abrahamic religions, God created the Earth (and the whole universe) and put human beings on it to take care of it. Earth stewardship is based on this basic understanding.

But the Bible also has another creation story in chapter 1 of the book of Genesis. In this story, human beings are depicted as the ones who rule, control, and dominate the Earth (Genesis 1, 26–29). This anthropocentric approach has received numerous critiques, also from within the Christian community.

Two different views of humankind are, therefore, presented at the beginning of the Bible. In the first chapter human beings are placed on top of creation to dominate it. In the second one, human beings are placed in creation to take care of it. Earth stewardship, from a Christian perspective, should be understood taking into account this tension. Though dominion, as understood in the Bible, has a different meaning from how we understand this concept today, it needs to be recognized that in some periods of history, the "dominion" perspective prevailed, thereby contributing to the

G. Kerber (✉)
Programme on Ecological Justice, World Council of Churches, Geneva, Switzerland
e-mail: gkm@wcc-coe.org

R. Rozzi et al. (eds.), *Earth Stewardship*, Ecology and Ethics 2,
DOI 10.1007/978-3-319-12133-8_25

depletion of natural resources and ecological destruction. However, although it is controversial (Hargrove 2015), the concept of stewardship has been clearly defined to offer an interpretation which does justice both to the contemporary scientific debate and the message conveyed by the Bible.

How has Earth stewardship been understood by Christian and other religious organizations? I will focus my contribution to this book on the work done by the World Council of Churches (WCC) to be able to deepen some of the contents of this understanding. This means that the contents will reflect a Christian understanding (that is shared by many other Christian organizations) which is closely linked to Jewish and Muslim perspectives on creation. Although other religions differ on basic concepts like God, creation, salvation, etc., they share some of these insights as I will briefly show with some examples later.

The WCC is the broadest and most inclusive among the many organized expressions of the modern ecumenical movement, a movement whose goal is Christian unity (WCC 2014). The WCC brings together churches, denominations, and church fellowships in more than 110 countries and territories throughout the world, representing over 500 million Christians and including most of the world's Orthodox churches, scores of Anglican, Baptist, Lutheran, Methodist, and Reformed churches, as well as many United and Independent churches. While the Roman Catholic Church is not a member of the WCC, there are working relations with it in many domains, including ecology and climate related work. While the bulk of the WCC's founding churches in 1948 were European and North American, today most member churches are in Africa, Asia, the Caribbean, Latin America, the Middle East and the Pacific. Thus, for many years, WCC's perspective has included a strong Southern perspective in its views.

25.1 Earth Stewardship as the Care for Creation and Climate Justice

In 2008 the WCC adopted a Minute entitled "Be stewards of God's creation" (WCC 2008). The Minute builds on previous statements by the WCC, especially a statement adopted a year earlier (WCC 2007a). From a Christian theological perspective, Earth stewardship should be understood as an expression of two biblical imperatives: the integrity of creation and the commitment to justice. The WCC 2007 statement on the "Tenth Anniversary of the Kyoto Protocol" stressed that:

> The Bible teaches the wholeness of creation: Life is created, sustained and made whole by the power of God's Holy Spirit" (Genesis 1; Romans 8). God creates human beings out of the dust of the earth (Genesis 2). Sin breaks relationships among humankind and with the created order (Genesis 3 and 4; Jeremiah 14, Hosea 4,1-3). Bearing the marks of human sin, 'creation waits with eager longing for the revealing of the children of God' (Romans 8:19). God provided all creatures with the conditions to live life as it is meant to be, in a specific relation towards one another. When creation is threatened by climate change we are called to speak out and act as an expression of our commitment to life, justice and love (WCC 2007a, 1).

This wholeness or integrity of creation's perspective implies the need to revisit the way creation is understood by theology (Vischer 2007). From this perspective, creation is not only something that happened "at the beginning," as presented in the book of Genesis' accounts. Creation, for Christian theology, is a permanent divine activity that creates, redeems, and renews. As stated in Psalm 104, 30: "When you send forth your spirit, they are created and you renew the face of the ground." In theological jargon, creation is not only *creatio prima* but also *creatio continua*. In this continuous creation, God as life-giving Spirit brings salvation, reconciles, heals, and liberates the whole creation. This provides the framework for a comprehensive view of creation and salvation (Conradie 2012).

This understanding reacts against a sole understanding of a transcendent God as an absolute entity. God combines absolute transcendence (*superior summo meo*, in Augustine of Hippo's words) with deep immanence (*intimior intimo meo*, again according to Augustine) and transparency (as understood by Teilhard de Chardin). God is seen not only beyond creation but also in creation (Moltmann 1985).

If God is in creation, human beings are also in creation and not above it. There is a close and indissoluble bond of human beings to the Earth. "Man" is "earthling" (*adam* in Hebrew), created out of the soil (*adamah*). Created "in God's image and likeness" (Genesis 1, 27), humans have a special place within creation and responsibility towards it.

For some decades process theologians (Cobb and Griffin 1976; Suchocki 1989) and ecofeminists (Ruether 1993; McFague 1993; Gebara 1999) have contributed much to unpack the contents of a revised theology of creation. They have developed notions such as *panentheism* or the world as the body of God, which have helped to understand the meaning of a stewardship of creation. This approach also tries to respond adequately to the accusation of anthropocentrism in Christianity – as raised by Lynn White (1967), for example, a criticism to which we will come back later.

To have a holistic understanding of earth stewardship, together with a biblical theological reflection on the integrity of creation, the biblical concept of justice constitutes another key component. Justice is at the core of the biblical message. "He [God] secures justice for widows and orphans, and loves the alien who lives among you, giving him food and clothing" (Deuteronomy 10, 18–19). The God of the Bible is a God who does justice, who cares, who loves, and gives security to the poor, the vulnerable, represented by the widow, the orphan, and the stranger in many biblical texts. Accordingly, human beings need to act justly, which means protecting the vulnerable ones: "…cease to do evil and learn to do right; pursue justice and champion the oppressed, give the orphan his rights, plead the widow's case" (Isaiah 1, 16b–17).

Latin American liberation theology has especially looked at the poor, the vulnerable, and the victims. Some authors have related the vulnerability of human beings with the vulnerability of the Earth. Leonardo Boff, a well known Brazilian theologian, says that the cry of the poor is echoed by the cry of the Earth (Boff 1997). Based on the notion of the groaning of creation (Romans 8, 22), Boff calls for widening the meaning of the option for the poor, which has been a key component of liberation theologies in Latin America. According to Boff, this option should include an option for the most threatened beings and species: the poorest human beings, those whose ways of life are threatened with extinction, such as the Kaiapo, the Yanomami, and other indigenous peoples, as well as animal species, such as the golden lion tamarin (*Leontopithecus rosalia*), the musician wren or uirapuro (*Cyphorhinus arada*), and the panda (*Ailuropoda melanoleuca*). Without forgetting these species, Boff recalls that the most threatened being in creation is planet Earth itself as a whole (Boff 1997).

In summary, for Boff, an articulation between creation and salvation is necessary, owing to the new paradigm, the paradigm of the 're-ligation' (connectedness) of all with all. A new paradigm, he declares, calls for a new language, a new imagination, a new politics, a new pedagogy, a new ethics, a new discovery of the sacred, and a new process of self-understanding (spirituality). The cry of the poor should be listened to together with the cry of the Earth, and all humanity should respond effectively to them (Kerber 2012). Earth stewardship should be understood from this perspective.

25.2 Earth Stewardship and Climate Change

For more than two decades, this consideration of the integrity of creation and justice for the poor and the vulnerable especially has been emphasized in the WCC`s work on climate change. This emphasis has a scientific basis. Since the reports of the Intergovernmental Panel on Climate Change (IPCC), the scientific community increasingly recognizes the fact of climate change and that there is a relationship between human action and climate change, particularly regarding CO_2 emissions and, in general terms, the other so-called greenhouse gases. Greenhouse gases trap heat in the earth's atmosphere and tend to heat the planet. The Fourth and Fifth Assessment Reports of the IPCC (2014) recognized the most vulnerable groups and regions are already suffering and will suffer the most the consequences of climate change. These groups include indigenous peoples, the poor, and communities living in low-lying islands. To respond to the challenges of climate change means to look particularly at these communities and act responsibly and audaciously.

Climate change is seen as one of the major threats to the integrity of creation and to justice. It is affecting nature and communities in different parts of the world. In the WCC's understanding, human induced climate change is being precipitated primarily by the current development pattern. The prevailing economic strategy promotes endless growth and production of goods and high consumption life-styles in the richer industrialized countries and among wealthy elites throughout the world. Such patterns of development and consumption, through the increase of CO_2 emissions, deforestation, extractive industries and other means are leading to the depletion of critical natural resources, with life-threatening consequences for impoverished nations, and future generations. At the Conference of Parties (COP) of the United Nations Framework Convention on Climate Change (UNFCCC) in December 2007, in Bali, Indonesia (COP 13), the statement of the WCC "This far and no further: Act fast and act now," called for a change of paradigm "in the prevailing economic strategy of promoting endless growth and production of goods and a seemingly insatiable level of consumption … [which] are leading to the depletion of critical natural resources and to extremely dangerous implications with climate change and development (WCC 2007b, 1).

From this perspective, while stating the different dimensions of the climate change crisis (ecological, social, economic, cultural and political), delegates at COP in 2010 called for a holistic approach and stressed that:

> in the churches' perspective, justice must be the basic criterion of applied ethics in all decisions concerning the measures to cope with climate change. Although climate change is a global issue affecting all peoples and nations, those who are and will increasingly be affected by negative climate change consequences are the vulnerable communities who have contributed the least to global emissions. These include women and children, indigenous peoples, poorest communities, people with disabilities, and inhabitants of coastal low-lying areas. Vulnerable communities and states are also much more dependent on natural resources for their subsistence and do not have the means to mitigate emissions and to adapt to climate change. Their survival is at risk, and justice requires that the nations that are most responsible historically for the adverse ecological conditions should take the greatest responsibility towards the adaptation of these vulnerable communities and nations (WCC and LWF 2010, 1–2).

This ethical understanding of climate change is not only coming from religious organizations. The basis for this interpretation comes from the UNFCCC itself (UNFCCC 1992). Article 3 of the Convention presents the principles that should guide the international community when responding to climate change challenges. Principle 1 refers to critical points: the concern for present and future generations; equity as a basis for climate measures; the "common but differentiated responsibilities and respective capabilities" the leading role developed countries should have in combating climate change. Principle 2 pays special attention to the needs and circumstances of developing countries and vulnerable communities that deserve particular consideration. The "precautionary principle" is addressed in Principle 3, stressing the importance of anticipating, preventing or minimizing the causes and mitigating effects. Interestingly, the principle already responds to climate sceptics, affirming that "the lack of full scientific certainty—which is never attainable in any circumstance—should not be used as a reason for postponing such measures." The right to development, specifically sustainable development, is addressed in Principle 4, while links between climate change and economics are highlighted in Principle 5, recalling their intricate interconnection. These principles have ethical implications which have been picked up by states that are party to the Convention, NGOs, as well as academia.

25.3 Stewardship, Sustainability and Integrity: A Brief History

This understanding of Earth stewardship as the care for creation and climate justice is not something new in Christian understanding. In the 1970s the WCC started talking about "sustainable communities", responding to the threats of economic differences and ecological destruction. At a WCC world conference on science and technology for human development, in Bucharest in 1974, it was affirmed that a society will be sustainable only as long as the rate of use of non-renewable resources does not outrun the increase in resources made available, unless the emissions of pollutants are well below the capacity of the ecosystem to absorb them, and unless the need for food is at any time below the global capacity to supply it. The conference aimed at searching for a positive, pro-active response to the debate on science and technology and the future of humanity, which was sparked by the 1972 Club of Rome report, *Limits to Growth*, and emphasized the idea of a "sustainable and just society" (Robra 2002).

In the 1980s the ecological component of sustainable societies led to the affirmation of the "integrity of creation", which I mentioned above. The WCC Assembly in Vancouver in 1983 asked that "the links as well as the tensions between the goals of justice, peace, and the well-being of creation should be explored from biblical, socio-economic, and political perspectives." After the Assembly, various attempts were made to explore the challenges posed by the ecological crisis. In 1987, for instance, a meeting of theologians and scientists (including James Lovelock, who formulated the Gaia hypothesis), issued a report on "Reintegrating the Creation." During the same year, another consultation linked the integrity of creation with the theme of liberation applied to all of life, not only to humanity (Birch et al. 1990).

In 1988 a major consultation in Granvollen, Norway, enabled new and widespread attention to the threats posed to the global environment. The report of this consultation concluded: "The drive to have 'mastery' over creation has resulted in the senseless exploitation of natural resources, the alienation of the land from people and the destruction of indigenous cultures… Creation came into being by the will and love of the Triune God, and as such it possesses an inner cohesion and goodness" (Hall 2002, 274).

Granvollen, then, ensured the space for a growing ecumenical discussion around the integrity of creation in the years leading to the Justice, Peace, and Integrity of Creation (JPIC) World Convocation, which took place in Seoul, South Korea, in 1990. At this convocation ten affirmations on Justice, Peace, and Integrity of Creation were adopted. Affirmation VII summarized what integrity of creation means:

> As Creator, God is the source and sustainer of the whole cosmos. God loves the creation… We affirm that the world, as God's handiwork, has its own inherent integrity; that land, waters, air, forests, mountains and all creatures, including humanity, are "good" in God's sight. The integrity of creation has a social aspect which we recognize as peace with justice, and an ecological aspect which we recognize in the self-renewing, sustainable character of natural ecosystems (Niles 1992, 173).

Together with this affirmation, the text continues with a critique of anthropocentrism:

> We will resist the claim that anything in creation is merely a resource for human exploitation. We will resist species extinction for human benefit; consumerism and harmful mass production; pollution of land, air and waters; all human activities which are now leading to probable rapid climate change; and policies and plans which contribute to the disintegration of creation (Niles 1992, 173).

Anthropocentrism, as we said earlier, is one of the main critiques of the Judeo-Christian traditions since the famous article by Lynn White Jr, "The historical roots of our ecologic crisis", in which he accuses it of causing the domination of human beings over all of nature, and states that "especially in its Western form, Christianity is the most anthropocentric religion the world has seen"(White 1967). Against this background the following commitment of the JPIC Convocation acquires more importance:

> We commit ourselves to be members of both the living community of creation in which we are but one species, and members of the covenant community of Christ; to be full co-workers with God, with moral responsibility to respect the rights of future generations; and to conserve and work for the integrity of creation both for its inherent value to God and in order that justice may be achieved and sustained (Niles 1992, 174).

The quotations above show that, in the context of the JPIC process, the WCC affirmation takes into account the tension we referred to at the beginning between mastery (dominion) and stewardship. By affirming the integrity of creation, the document shows the relationship between the environmental and social dimensions of the earth which need to be considered simultaneously, while reacting against a purely economic approach which stresses only human benefit. Consequently, the commitment highlights the ethical dimension, stressing the rights of future generations, the intrinsic value of nature, and the justice perspective.

This analysis and commitment can be seen as a similar and parallel path to what the Long-Term Socio-Ecological Research (LTSER) network has expressed. Beyond a merely economic approach which primarily if not exclusively focuses on the economic component of social ecology, LTSER in South America has stressed the importance of noneconomic values, especially spiritual and ethical (Rozzi et al. 2012).

25.4 An Interfaith Approach

Although, as shown by Hargrove (2015), the notion of Earth stewardship may pose problems to non-Abrahamic religions, the care for creation and climate justice approach also has been reflected in some interfaith statements signed by religious leaders from faith traditions such as Buddhism, Daoism, and Sikhism.

In 2008, the Archbishop of Uppsala and Primate of the Church of Sweden, Most Reverend Anders Wejryd, called for an Interfaith Climate Summit in Uppsala. This summit adopted "The Uppsala Interfaith Climate Manifesto 2008" (Uppsala 2008), signed by 26 religious leaders. The Manifesto summarizes the common ground of different religious traditions pinpointing the role of stewards or caretakers:

> From religious traditions, with different approaches to religious life, we come together at this time in human history to assure the world of what we have in common. We all share the responsibility of being conscious caretakers of our home, planet Earth. We have reflected on the concerns of scientists and political leaders regarding the alarming climate crisis. We share their concerns (Uppsala 2008, 9).

Furthermore, the religious leaders commit themselves:

> To inform and inspire people in our own religious and cultural contexts to take responsibility for and to implement effective measures; to challenge political and business leaders where we live and work to develop comprehensive strategies and action; to focus on the struggle against global warming and draw upon our innermost religious convictions about the meaning of life. This commitment is a deeply spiritual question concerning justice, peace and hopes for a future in love and solidarity with all human beings and the whole of creation (Uppsala 2008, 15).

Justice is also presented as one of the common principles of religious traditions:

> We want to face the climate challenge with defiant optimism to highlight the core principles of all major sacred traditions of the world: justice, solidarity and compassion (Uppsala 2008, 15).

Religious leaders, thus, recognize the role scientists and politicians play in responding to the climate crisis and assume their own responsibility in informing and inspiring believers to act effectively, in challenging politicians and businesses, based on their religious convictions.

On the following year, an Interfaith Declaration on Climate Change (2009) was endorsed by a hundred organizations and hundreds of individuals including, among others the Dalai Lama, Archbishop Desmond Tutu and Cardinal Wilfrid Napier from South Africa, and Sri Sri Ravi Shankar from Bangalore, India. The Declaration proclaims that:

> The extraordinary delicacy of Nature's balance is becoming increasingly apparent, even as human actions inflict ever larger, more dangerous and potentially irreversible changes on the indivisible web of atmosphere, earth, ocean and life that is creation. Today our faiths stand united in their call to care for the Earth, and to protect the poor and the suffering. Strong action on climate change is imperative by the principles and traditions of our faiths and the collective compassion, wisdom and leadership of humanity (Interfaith Declaration 2009, 1).

The declaration recognizes that "climate change is not merely an economic or technical problem, but rather at its core is a moral, spiritual and cultural one" and affirms that "assisting vulnerable communities and species to survive and adapt to climate change fulfils our calling to wisdom, mercy, and the highest of human moral and ethical values" (Interfaith Declaration 2009, 1).

These interfaith statements show that there is a growing consensus among religious leaders on the intimate relationship between their respective faith traditions and the care for the Earth and that this is seen both as an ethical and spiritual issue. At the same time in various parts of the world faith communities are becoming more and more aware of their ecological responsibility and have responded through different initiatives like "green" churches, synagogues, mosques, and temples, and special times to pray for and reflect on the Earth. However, we cannot say that this awareness and action has been mainstreamed in religious communities. Although some of these initiatives have 20 years or more, these concerns are shared only by a very small minority among all faith communities.

25.5 Ethical and Spiritual Values for Earth Stewardship

"Human moral and ethical values," as requested in the previous paragraph by the interfaith declaration, express the recognition that the climate and ecological crises have various components, not only environmental, social, cultural, and political, but also ethical and spiritual. Faith traditions can play a relevant role in implementing a global Earth stewardship. As the Ecological Society of America (ESA)'s Earth Stewardship project recognizes, on one hand earth stewardship is rooted in religious thought and on the other faith communities are key stakeholders to implement socio-ecological transformation (Chapin et al. 2011). Indeed, as we tried to show in the previous sections, the inclusion of the environmental variable in biblical hermeneutics, theology, and ethics has produced significant changes in religious self-understanding over the last decades. These changes have influenced one of the key components of religious traditions keen to Earth Stewardship : the establishment of a common meaning. The latter includes ethical and spiritual values which, through positive messaging, have shaped cultures and societies and have encouraged communities to act.

This is the second aspect of the contribution faith communities can make to ES: their engagement in transforming lifestyles in a long-term perspective. While ES pursues an integration of ecological sciences and environmental ethics (Rozzi et al. 2012), it is relevant to state that ethics, for Christianity and for other religions, is intimately related to spirituality, which can be understood as "that attitude which puts life at the centre and defends and promotes life against all the mechanisms of death, desiccation or stagnation. The opposite of spirit, in this sense, is not the body but death, and everything associated with the system of death"(Boff 1995). Larry Rasmussen refines his understanding of the spiritual dimension as "not moral-spiritual energy in the abstract, or as the worthy and seductive subject of itself, but as directed to an earth-honouring faith." This energy calls for a conversion to earth that is not so much about "environment" or "environmentalism" as it is "the holy mystery of creation, made for and by all earth's creatures together" (Rasmussen 1996). David Hallman proposes the following as spiritual values for earth community: gratitude, humility, sufficiency, justice, love, peace, faith, and hope (Hallman 2012). From a Latin American perspective, I would also add: solidarity, resilience, and joy. These spiritual values can be seen as essentials for Earth stewardship. In the process of addressing threats to the Earth, communities need to develop countercultural values, opposing mainstream attitudes such as greed, utility, oppression, and destruction. In this sense, for instance, ecofeminism, which tries to explore the interconnections between ecology and feminism, between the domination of women and the domination of nature, has proposed the concept of healing of the Earth as most needed in today's world (Ruether 1996).

Assuming and putting into practice these values and attitudes is not an easy task. In religious terms it requires a conversion, which implies not only change of lifestyles but also deep transformation of structures and paradigms. Changing our everyday life as individuals, as families, and as communities, should go hand by hand with changes in the norms of local, national, and international legal frameworks. One's personal reduction of CO_2 emissions is of symbolic value only without a binding and ambitious legal framework that effectively incentivizes everyone else to reduce their CO_2 emissions. Thus, there is an urgent need to combine actions at the grassroots level with advocacy at the international one (Kerber 2013).

Religions have played an ambiguous role in relationship to the care for the Earth and to climate justice. They have served as inspiration for both the destruction of the Earth and for its protection, as has also been the case of other institutions and sectors of society. They continue to play a key role in many societies in the world today. Strengthening their commitment to care for the Earth and climate justice together with others in society is a fundamental factor for an effective Earth stewardship.

References

Birch C, Eakin W, MacDaniel JB (1990) Liberating life. Orbis, New York

Boff L (1995) Ecology and liberation: a new paradigm. Orbis, New York

Boff L (1997) Cry of the earth, cry of the poor. Orbis, New York

Cobb J, Griffin D (1976) Process theology: an introductory exposition. Westminster Press, Philadelphia

Conradie E (2012) Creation and salvation, 2 volumes. LIT, Berlin

Gebara I (1999) Longing for running water: ecofeminism and liberation. Fortress, Minneapolis

Hall DJ (2002) Creation. In: Lossky N, Miguez-Bonino J, Pobee J et al (eds) Dictionary of the ecumenical movement. WCC, Geneva

Hallman DG (2012) Spiritual values for earth community. WCC, Geneva

Hargrove E (2015) Stewardship versus citizenship. In: Rozzi R, Pickett STA, Callicott JB et al (eds) Earth stewardship: linking ecology and ethics in theory and practice. Springer, Dordrecht, pp 315–323

Interfaith Climate Declaration (2009) http://www.interfaithdeclaration.org. Accessed 22 May 2014

IPCC (2014) Intergovernmental panel on climate change. http://www.ipcc.ch/report/ar5/. Accessed 22 May 2014

Kerber G (2012) Leonardo Boff: towards the liberation of the whole creation. In: Conradie E (ed) Creation and salvation, vol 2. LIT, Berlin, pp 295–299

Kerber G (2013) International advocacy for climate justice. In: Globus Veldman R, Szasz A, Haluza-DeLay R (eds) How the world's religions are responding to climate change: social scientific investigations. Routledge, New York

McFague S (1993) The body of God. Fortress Press, Philadelphia

Moltmann J (1985) God in creation. SCM, London

Niles DP (1992) Between the flood and the rainbow. WCC, Geneva

Rasmussen L (1996) Earth community, earth ethics. WCC, Geneva

Robra M (2002) Sustainability. In: Lossky N, Miguez Bonino J, Pobee J (eds) Dictionary of the ecumenical Movement. WCC, Geneva, pp 1084–1185

Rozzi R, Armesto J, Gutierrez J et al (2012) Integrating ecology and environmental ethics: earth stewardship in the southern end of the Americas. Bioscience 62(3):226–236

Ruether RR (1993) Gaia and God. SCM, London

Ruether RR (ed) (1996) Women healing Earth. Orbis, New York

UNFCCC (1992) United Nations Framework Convention on Climate Change. http://unfccc.int/resource/docs/convkp/conveng.pdf. Accessed 22 May 2014

Uppsala Interfaith Climate Manifesto (2008). http://www.svenskakyrkan.se/default.aspx?id=664984. Accessed 22 May 2014

Vischer L (ed) (2007) Spirituality, creation and the ecology of the Eucharist. John Knox Series 18, Le Grand-Saconnex

White L Jr (1967) The historical roots of our ecologic crisis. Science 155(3767):1203–1207

World Council of Churches (2007a) Minutes of the meeting of the executive committee of the World Council of Churches. WCC, Geneva

World Council of Churches (2007b) Statement to the COP 13 to the UNFCCC in Bali, Indonesia, 14 Dec 2007. https://www.oikoumene.org/en/resources/documents/wcc-programmes/justice-diakonia-and-responsibility-for-creation/climate-change-water/statement-to-cop13-un-climate-conference-bali. Accessed 22 May 2014

World Council of Churches (2008) Be stewards of God's creation. Minute on global warming and climate change, by the Central Committee of the World Council of Churches. WCC, Geneva. http://www.oikoumene.org/en/resources/documents/central-committee/2008/public-issues/minute-on-global-warming-and-climate-change. Accessed 9 June 2014

World Council of Churches (2014) What is the World Council of Churches. http://www.oikoumene.org/en/about-us. Accessed 22 May 2014

World Council of Churches and Lutheran World Federation (2010) Why are the churches at the UN conference on climate change in Cancun. http://www.oikoumene.org/en/resources/documents/wcc-programmes/justice-diakonia-and-responsibility-for-creation/climate-change-water/why-are-the-churches-at-the-un-conference-on-climate-change-in-cancun. Accessed 22 May 2014

Chapter 26
World Religions, Ethics, and the Earth Charter for a Sustainable Future

Mary Evelyn Tucker

Abstract The environmental challenges we are facing require more than science, policy, economics, and law. There is a need for broader ethical responses. The search for a sustainable future requires an integration of the Earth Charter and the world religions. This integration of the principles of the Earth Charter with the virtues for human-Earth flourishing of the world's religions provides a unique synergy for rethinking a sustainable future. Such a synergy can contribute to the broadened understanding of sustainability as including economic, ecological, social, and spiritual well-being. This broadened understanding may be a basis for long-term policies, programs, and practices for a planetary future that is not only ethically sustainable, but also sustaining for human energies.

Keywords Earth Charter • Ethics • Forum on religion and ecology • Sustainability • World religions

26.1 Introduction: The Need for Ethical Approaches

The complex nature of our global environment crisis is increasingly evident as the globe is warming, as weather patterns are becoming more severe, as species are going extinct, as nonrenewable resources such as oil are being used up, as forests and fisheries are being depleted, and as water is becoming polluted or scarce. The large-scale problem of climate change is now more visible to a larger public, but the shift to a sustainable low carbon future has yet to be realized.

As the developing world attempts to raise its standard of living with rapid industrialization, there is an inevitable impact on the environment and natural resources. The result is that severe pollution of water, air, and soil is becoming more widespread in places such as India and China. Similarly, the high level of consumption of energy and resources by the developed world, especially the United States, raises serious ethical issues of equity and justice. The tension between reducing emissions for

M.E. Tucker (✉)
Yale School of Forestry and Environmental Studies, Yale University, New Haven, CT, USA
e-mail: maryevelyn.tucker@yale.edu

R. Rozzi et al. (eds.), *Earth Stewardship*, Ecology and Ethics 2,
DOI 10.1007/978-3-319-12133-8_26

environmental protection and increasing them for economic development is a source of increasing conflict and lack of trust between the developed and developing world.

Since the United Nations Conference on Environment and Development in Rio in 1992 we have witnessed a series of major UN conferences, along with the Kyoto Protocol, to try to redirect the course of development and emissions to be more equitable and sustainable. Due in part to the lack of leadership from the U.S. we have not made sufficient progress. Moreover, the worldwide increase in military spending, especially by the United States, means that less money is available for the pressing issues in the Millennium Development Goals regarding poverty and the environment (www.un.org/millenniumgoals).

Thus, the human community is still struggling to create a sustainable, low carbon future. It is becoming clear that a broader set of participants is needed to realize this goal, especially participants who are sensitive to integrating poverty alleviation with environmental protection and climate adaptation. Many religious communities have been involved in efforts to mitigate poverty, hunger, and disease, but now they are recognizing this cannot be done adequately without attention to the environment and the climate, which is deteriorating rapidly. Sufficiency of food, shelter, and health for humans will depend on a thriving atmosphere and biosphere to support life for the Earth community.

The litany of environmental, climate, and development problems is well known, but what is becoming ever more self-evident is that they cannot be solved by science, technology, law, politics, or economics alone. That is because we are more aware that environment, climate, and development issues are, in large measure, social issues. Thus "fixing" the environment or climate through technology or finance or legislation is necessary but not sufficient. Ethical approaches are also needed.

We are being pressed to see the linkage between environment and people, between healthy ecosystems and healthy social systems, between climate protection and poverty alleviation. We need truly interdisciplinary approaches and systemic thinking that includes more stakeholders. Our challenge is to create not simply low carbon societies but whole communities, where humans are not recklessly dominating and exploiting nature for material gain, but rather recognizing their profound dependence on the larger community of life. In this spirit, unrestrained economic growth that underlies the climate crisis needs to be examined. A broader ethical perspective is required so as to integrate ecology and economy, climate mitigation, adaptation, and development needs.

In short, new indicators of "progress" need to be developed. The world's religions and the Earth Charter can play a role in this redefinition with an ethical articulation of a path toward not just a low carbon society, but toward a flourishing Earth community.

26.2 New Indicators of Progress

Neo-classical economic thinking has equated economic growth with progress, despite any harm to the environment. While this thinking drives our industrial processes, economists are shifting, however gradually, to a realization that the environment and climate can no longer be seen as an externality to be ignored. Methods of cost accounting that disregard the environment are no longer viable. Instead the bottom line needs to include, not only financial profit, but also environmental health and social well-being. New measurements are being developed for this triple bottom line, such as measuring "ecosystems services." In 1997, ecological economists have estimated that it would cost some $33 trillion dollars to replicate nature's services (Costanza et al. 1997). In 2014 their estimate increased to $145 trillion (Costanza et al. 2014).

In addition, the UN Global Reporting Initiative has been formed for measuring the environmental and social impact of corporations (http://globalreporting.org). The Equator Principles have been created by a Dutch banker, Herman Mulder, for guiding banks and financial institutions in their investments (www.equator-principles.com). The "ecological footprint" provides a similar opportunity for individuals or institutions to calculate their environmental impact in a variety of areas, including use of carbon. This method was first developed by William Rees and Mathias Wackernagel and is now part of a broad international network (www.footprintnetwork.org). Religious communities have entered this arena through socially and environmentally responsible investments. Jewish and Christian groups have collaborated in forming the Interfaith Center on Corporate Responsibility based in New York (www.iccr.org).

In addition to the triple bottom line of economic profit, environmental health, and social well-being, some people are suggesting that spiritual well being is also an important component of human flourishing. They maintain that the full range of human happiness includes a sense not only of physical health, but also spiritual well being and happiness. In fact, in Bhutan the Gross National Happiness Indicator has replaced the Gross National Product Index. Developed with Buddhist principles and supported by the King, this notion has gained a wider audience than Bhutan. It is based on a conviction that there is more to social cohesion and individual fulfillment than economic competition and profit making. The Happiness Indicator takes into account other factors. For example, personal spiritual cultivation is encouraged along with community building; the quality of life is seen as more important than the quantity of material possessions; and non-material values, such as cultural and ecological integrity, are highly prized (www.grossinternationalhappiness.com). Because of this, along with many other projects named above, "progress" is being redefined as more than economic growth.

26.3 World Religions and the Earth Charter

In discussing the topic of creating a sustainable future with a low carbon society we may need a broader basis for analysis than simply economic indicators of growth. As defined by the Bruntland Commission report, *Our Common Future* (1987), it is development that meets present needs while not compromising the needs of future generations. This ethics of intergenerational equity is a necessary criterion, but may not be fully sufficient. That is because while it emphasizes balancing environmental and economic growth, it does not always take into account the full range and interaction of human-Earth flourishing. Such a broad context may be enhanced by the contribution of the world's religions, both in theory and in practice regarding poverty alleviation, environmental protection and climate mitigation. We may be able to draw on shared values as well as diversified practices of the religions. This can be done in relation to the Earth Charter, a major international document drafted in response to the needs for an integrated ethical framework for sustainable development (www.earthcharter.org).

In terms of general principles and values that the world religions offer to sustainability discussions, they can be described as broadening the category of sustainability to include past, present, and future concerns. In short, large-scale and long-term perspectives will be needed to envision sustainable ecosystems that have developed over billions of years, sustainable living for humans at present, and a sustainable future for all life. These correspond to the central concerns of the Earth Charter and the growing commitments of the world's religions to ecology, justice, and peace. They correspond to six key "values for human-Earth flourishing" shared by the world religions as they are being challenged to envision a viable future for the Earth community:

Reverence,
Respect,
Restraint,
Redistribution,
Responsibility, and
Renewal.

These values for human-Earth flourishing were first identified as the result of a 3-year conference series at Harvard on World Religions and Ecology from 1996 to 1998 (Tucker and Grim 2001).

26.4 Harvard Conferences on World Religions and Ecology

The conferences were intended to explore elements of the world's religions that highlight human-Earth relations in scripture, in ritual, and in ethics. A major goal of the series was to begin a process of retrieving, revaluating, and reconstructing the ecological dimensions of the world's religions so as to contribute to a sustaining and flourishing future for the Earth community. Over 800 international scholars and theologians of world religions participated in the conferences that included the western religions (Judaism, Christianity, and Islam), the Asian religions (Jainism, Hinduism, Buddhism, Confucianism, Daoism, and Shinto), and Indigenous religions. Ten edited volumes were published by the Harvard Center for the Study of World Religions as a result of these conferences.

26.5 Forum on Religion and Ecology at Yale

A Forum on Religion and Ecology was formed at the culminating conference at the United Nations and the American Museum of Natural History in 1998. It is now based at Yale. Moreover, a major international website was created to assist research, education, and outreach in this area. The website provides introductions to the world religions and their ecological dimensions along with annotated bibliographies of the books and articles in English on this topic. It also identifies over a hundred engaged projects of religious grassroots environmentalism. It contains a lengthy bibliography on religion and poverty issues. It includes educational materials such as syllabi, videos, CDs and DVDs (http://fore.research.yale.edu).

26.6 Field and Force of Religion and Ecology

One of the outcomes of the conference series at Harvard and the ongoing Forum work is the emerging alliance of religion and ecology both within academia and beyond. Over the dozen years since the conference series began a new field of study has emerged in colleges and secondary schools. Moreover, a new force has arisen within the religions from leaders and laity alike. Both the field and the force are contributing to a broadened perspective for a future that is not only sustainable, but also flourishing (Grim and Tucker 2014).

Within academia, religious studies departments are offering classes in this area; divinity schools and seminaries, focused on training Christian ministers, are including courses (http://www.greenseminaries.org/); and high school teachers have developed creative curriculum as well (www.rsiss.net.rsissfore.html). There are graduate programs being offered at Drew University and the University of Florida, as well as a joint Master's degree program in religion and ecology at Yale. Many environmental studies programs are encouraging the participation of religious studies and the humanities in what have been predominantly science and policy oriented programs. Moreover, a two-volume encyclopedia on religion and nature has been published and two academic journals have been launched. A Forum on Religion and Ecology was established in Canada; a Society for the Study of Religion, Nature, and Culture has been formed in the United States; and a Forum for the Study of Religion and the Environment has been created in Europe.

Within the religions, statements on the environment, on climate change and on eco-justice have been released by the major world religions and indigenous traditions. Leaders such as the Ecumenical Patriarch, the Pope, and the Dalai Lama have spoken out regarding the urgency of these issues. The Patriarch, Bartholomew, has presided over six international symposia focused on water issues and more recently climate change (www.rsesymposia.org). Rowan Williams, the head of the Anglican Church in England has written sermons on this topic (www.archbishpofcanterbury.org)

and the US Presiding Bishop for the Episcopal Church, Katherine Jefferts Schori, has testified before Congress on the risks of climate change.

Ministers and lay people are organizing projects such as fighting mountain top removal, educating children in ecology, conserving energy in the Interfaith Power and Light project (www.theregenerationproject.org). Many of these activities are depicted in the film, *Renewal* that features eight case studies of religious environmentalism across the United States (www.renewalproject.net). Catholic nuns around the world have been especially active in projects on sustainable agriculture and ecological literacy (www.sistersofearth.org; www.genesisfarm.org; McFarland 2007). In addition, in the U.S. the National Religious Partnership for the Environment has been working for 15 years with Jewish and Christian groups in the United States (www.nrpe.org), while the Alliance for the Conservation of Nature in England has established numerous ecological projects around the world (www.arcworld.org).

As this field and force has expanded there is a growing recognition from many quarters of the importance of the participation of religions in environmental programs and concerns, such as climate change. For some years, for example, scientists have been asking for religious communities to play a more active role in environmental issues. They recognize the large number of people around the world who are involved with religions. There are one billion Muslims, Christians, Hindus, and Confucians, respectively. They observe that moral authority has played an important role in many transformations of values and behavior, such as the abolition of slavery in nineteenth century England and in civil rights by Martin Luther King and other religious leaders in the United States and South Africa in the twentieth century.

26.7 Support of Scientists and Policy Makers

Moreover, scientists such as E.O. Wilson have called for an alliance between religion and science in a shared concern for the future of the environment. This was articulated in *A Warning to Humanity* in 1992 and more recently in Wilson's book, *The Creation: An Appeal to Save Life on Earth* (Wilson 2006). Similarly, biologists Paul Ehrlich and Donald Kennedy (2005) have proposed a Millennium Assessment of Human Behavior. In addition, policy think tanks, such as Worldwatch Institute in Washington DC, have encouraged the role of religions. One of their principal researchers, Gary Gardner, has published a chapter on this topic in the State of the World report of 2003 and his book *Inspiring Progress: Religious Contributions to Sustainable Development* (Gardner 2006). Moreover, the policy expert and former Dean of the School of Forestry and Environmental Studies at Yale, James Gustav Speth has also called for the participation of the world's religions in his book, *Bridge at the Edge of the World* (Speth 2008).

While religions have their problematic dimensions, including intolerance, dogmatism, and fundamentalism, they also have served as well springs of wisdom, as sources of moral inspiration, and as containers of transforming ritual practices.

Thus their tendency to be both conservators of continuity and agents of change are both apt descriptions. Religions have always played this role of conserving and transforming, balancing the dynamic tension of continuity and change for cultures over long spans of time. Indeed, human cultures are profoundly shaped by this dialectic and civilizations endure by navigating the delicate balance between tradition and modernity. Moving too deeply into traditional ways leads to fossilization and fundamentalism, while going too far into modernity can lead to superficial and inadequate responses to change.

26.8 Contributions of Religious Values and the Earth Charter Toward a Sustainable Future

We are at a moment of immense significance for the future of life on the planet and thus the world's religions may be of assistance as they move into their ecological phase. The common set of values for human-Earth flourishing identified from the Harvard conference series on World Religions and Ecology can be seen as compatible with the ethical principles of the Earth Charter. Recognizing the complementary nature of these two ethical frameworks may be helpful for linking religion, ethics, sustainability, and a low carbon future.

26.9 An Overview of the Earth Charter

The Earth Charter is both a document and a movement. It draws on scientific knowledge, legal principles, sustainability practices, ecological economics, the precautionary principle, and equity issues. In its decade long drafting process, it involved thousands of individuals and groups from around the planet and is the most inclusive civil society document ever negotiated. As a people's treaty it is a soft law document that is complemented by hard law of international covenants and laws. It has been endorsed by such international agencies as United Nations Educational Scientific and Cultural Organization (UNESCO) and the International Union for the Conservation of Nature (IUCN), the largest body of conservationists in the world. It is also endorsed by thousands of individuals and groups as well as by a number of countries and cities. The implications for the application of the Charter is seen in the *Earth Charter in Action*, a book of inspiring stories from around the world – ranging from youth to civil servants and government officials (www.earthcharterinaction.org).

The Charter was drafted by an international committee chaired by Steven Rockefeller from 1996 to 2000. A distinguished group of international figures served as Earth Charter Commissioners for the drafting process and now an Earth Charter International Council guides the activities of the Charter. There is a Secretariat and a website based in Costa Rica at the University for Peace.

The Charter offers a comprehensive framework for revisioning sustainability as balancing the needs for economic development with environmental protection. It presents an integrated set of principles to guide our emerging planetary civilization that is multinational, multicultural, and multi-religious. It provides a platform for universal commitment to the flourishing of bio-social planetary life systems along with differentiated responsibilities. There is no other global ethics that can point us toward a framework for a low carbon society.

The key components of the Earth Charter are: (1) cosmological context, (2) ecological integrity, (3) social equity, (4) economic justice, (5) democracy, (6) non-violence and peace. These six components of a sustainable future have their counterparts in the values for human-Earth flourishing that are shared among the world's religions as identified in the Harvard conference series: reverence, respect, restraint, redistribution, responsibility, and renewal. A planetary future that is "flourishing," not simply "sustainable," will be enhanced by the six components identified by the Earth Charter along with these six values of the world religions. Such a framework that integrates values for flourishing of the world's religions with the central component of global ethics in the Earth Charter may be an important context for expanding sustainability principles and practices.

26.9.1 Cosmological Context and Reverence

All cultures have been grounded in the stories they tell regarding the nature of the universe, the evolution of the Earth and of life, and the destiny of humans in this context. These cosmological stories provide accounts of the creation and evolution of life and the purpose of humans. As humans are currently trying to navigate their way between scientific accounts of evolution and the multiple religious stories of creation, the Charter articulates a broad, simple and inclusive sensibility that Earth is our home, our dwelling place.

This enlarged perspective of home may be a critical foundation for articulating a future that is both sustaining and flourishing. The Charter recognizes that we are part of a large family of life, including not only other humans but also other species. The interdependent quality of the Earth community is celebrated along with the fact that the conditions for life have been evolving for billions of years. "Humanity is part of a vast evolving universe. Earth, our home, is alive with a unique community of life. The forces of nature make existence a demanding and uncertain adventure, but Earth has provided the conditions essential to life's evolution (Earth Charter)."

Thus to speak of the broadest context for the flourishing of bio-social systems we need to be reminded of the cosmological, evolutionary story of life's emergence. The religious response to this is one of *reverence,* a quality shared by many scientists who are deeply inspired by their study of nature from cells to galaxies, enhanced now by powerful microscopes and telescopes. The intricacy and complexity of life is valued from both a spiritual and a scientific perspective. Awe and wonder become expressed through the shared experience of reverence.

26.9.2 Ecological Integrity and Respect for the Diversity of Life

The broad context for a sustaining and flourishing future from the Earth Charter is preserving ecological health and integrity. Without such a basis for healthy ecosystems there can be no long-term basis for the continuity of human life. It is expressed succinctly in the Preamble as follows: "The resilience of the community of life and the well-being of humanity depend upon preserving a healthy biosphere with all its ecological systems, a rich variety of plants and animals, fertile soils, pure waters, and clear air (Earth Charter)."

The response of the religious communities to this call for biological protection is the principle of ***respect*** for the rich diversity of life and the ecosystems that support life. Without such respect environmental exploitation will continue and we may irreversibly damage the ability of ecosystems to renew themselves. This is further spelled out in the Charter as protecting and restoring Earth's ecosystems, preventing harm through the precautionary principle, adopting effective patterns of production, consumption and reproduction, and advancing the study of ecological sustainability.

26.9.3 Social and Economic Justice, Restraint and Redistribution

The next section of the Charter highlights social and economic justice that are also key concerns of the world's religions. The religious virtues of ***restraint*** in use of resources, as well as ***redistribution*** of wealth through charitable means, complement the Charter's principles. All of the world's religions encourage moderation in personal behavior as well as in the accumulation or use of material goods. In addition, the world's religions express a strong concern for the suffering of the poor and for inequality between the wealthy and those in need. Charitable giving is valued as a fundamental religious act.

The Charter calls for eradicating poverty, equitable development, gender fairness, and non-discrimination regarding minorities and indigenous people. Thus justice is seen as balance of ecological, economic, and social factors. The term that many of the religions are using to describe this is "eco-justice" where biological and human health is seen as indispensable to one another. Indeed, preserving ecological integrity and protecting social and economic justice will require an integrated understanding of human-Earth relations.

In addition to restraint and redistribution, a broadened sense of ecological virtue is required. Women who do so much unpaid work to sustain their families, especially in developing countries, need to be valued and respected. The same applies for minorities and for indigenous peoples who have preserved valuable environmental knowledge in many parts of the world. While the religions still have a ways to go in recognizing the dignity and value of women and the communities of indigenous peoples, some progress is being made in this regard.

26.9.4 Democracy, Non-violence, and Peace, Responsibility and Renewal

Finally, the Earth Charter recognizes that democracy, non-violence, and peace are necessary ingredients for a sustaining and flourishing future. From the perspective of the religious communities, democracy requires a fundamental sense of ***responsibility*** to future generations of the community of life – human and more than human.

Non-violence and peace encourage the ***renewal*** of inner and outer peace, something that the religious communities have tried to foster for millennia. Spiritual practices such as prayer and contemplation, yoga and *tai ch'i,* and rituals and rites of passages have been developed to foster peace and non-violence for individuals and communities. Of course, it should be noted that non-violence has not always been practiced, but it is one of the reasons why Mahatma Gandhi, Leo Tolstoy, and Martin Luther King are so widely admired.

The principles in the section of the Earth Charter are: strengthening democratic institutions, promoting sustainability education, respecting animals, and promoting a culture of non-violence and peace.

26.10 Conclusion

This integration of the principles of the Earth Charter with the virtues for human-Earth flourishing of the world's religions provides a unique synergy for rethinking a sustainable future. Such a synergy can contribute to the broadened understanding of sustainability as including economic, ecological, social, and spiritual well-being. This broadened understanding may be a basis for long-term policies, programs, and practices for a planetary future that is not only ethically sustainable, but also sustaining for human energies. For at present we face a crisis of hope that we can make a transition to a viable low carbon future for the Earth community. The capacity of the world's religions to provide moral direction and inspiration for a flourishing community of life is significant. The potential of the Earth Charter to create an ethical framework for sustainable plans and practices is considerable. Together they may provide a comprehensive grounding for creating a common and shared future, and fostering an Earth stewardship initiative.

References

Costanza R, D'arge R, De Groot R et al (1997) The value of the world's ecosystem services and natural capital. Nature 387:253–260

Costanza R, de Groot R, Sutton P et al (2014) Changes in global value of ecosystem services. Glob Environ Chang 26:152–158

Ehrlich PR, Kennedy D (2005) Millennium assessment of human behavior. Science 309:562–563

Gardner G (2006) Inspiring progress: religions' contributions to sustainable development. W.W. Norton/Worldwatch Institute, New York
Grim J, Tucker ME (2014) Ecology and religion. Island Press, Washington, DC
McFarland S (2007) Green sisters: a spiritual ecology. Harvard University Press, Cambridge, MA
Speth JG (2008) Bridge at the edge of the world. Yale University Press, New Haven
Tucker ME, Grim J (eds) (2001) Religion and ecology: can the climate change? Daedalus 130 (Spec Iss, 4):1–306
Wilson EO (2006) The creation: an appeal to save life on Earth. W. W. Norton, New York
World Commission on Environment and Development [WCE] (1987) Our common future. Oxford University Press, Oxford

Chapter 27
Dorothy Stang: Monkeys Cry and the Poor Die, Earth Stewardship as Liberation Ecology

Roy H. May Jr.

Abstract Latin American liberation philosophical and theological traditions locate environmental ethics in political economy and the history of conflicts, too often violent, over the use and abuse of nature and people. It is in this historical context that Earth Stewardship should be understood in this region. Sectors of the Latin American Church, such as US naturalized Brazilian Sister Dorothy Stang in the Amazon, long have defended social justice and in recent years have integrated concern for the natural environment into their social justice agendas, often at great personal cost. Methodologically this theoretical reflection is done as a "second step" following the "first step" of active engagement on behalf of socio-ecological justice, and incorporates local realities and cultures, or "interculturality", into the formulation of liberation environmental ethics. This results in an amplified concept of moral community, understood within the framework of alterity theory, and corresponds to situational realties and struggles for socio-ecological transformation producing, what might be called, "liberation ecology" or even "liberation stewardship."

Keywords Alterity • Conflict • Liberation theology • Moral community • Religious workers

Dorothy Mae Stang, a Roman Catholic sister, labored for 30 years in the Brazilian Amazon defending the rights of poor peasants and the integrity of the rain forest. Early on she promoted reforestation: "Yes, we're reforesting with cedar, mahogany, and other noble trees from the forest. We hope to replant some 30,000 to 40,000 this year. In relation to the thousands cut down it is few, but we must begin," she wrote in a letter to her sister (Murphy 2007, p. 99). From her base in Anapu, Pará, she and others mounted a strong campaign against lumber companies that, flouting environmental laws, leveled the forests and stole land from the peasants. To defend the land rights of the poor and to conserve the forests, in 1995 she embraced the Sustainable Development Project (*Projeto de Desenvolvimento Sustentável* or PDS) and led the effort to establish the first PDS in the area. A provision of Brazilian land reform law, the PDS is a reforestation and forest protection program based on the sustainable

R.H. May Jr. (✉)
Departamento Ecuménico de Investigaciones (DEI), San José, Costa Rica
e-mail: royhmay@gmail.com

R. Rozzi et al. (eds.), *Earth Stewardship*, Ecology and Ethics 2,
DOI 10.1007/978-3-319-12133-8_27

use of forest resources by peasant farmers within legally demarcated areas. These areas or "reserves" are off-limits to the large lumber and cattle interests. When the social movements in Anapu adopted the PDS as their preferred land reform model, the PDS and Sister Dorothy became formidable obstacles to the unhealthy economic and political interests of the lumbermen (Lima 2013). Sister Dorothy reported many instances of illegal logging and testified against the loggers in court (Murphy 2007, p. 116). In retaliation the loggers threatened to destroy bridges and pollute waterways (Murphy 2007, p. 120). The conflict became violent and continued an established pattern of murder and violence that marked social tensions.[1] On the morning of February 12, 2005, while Sister Dorothy walked to a meeting with peasant farmers, near Boa Esperança, Pará, two gunmen hired by lumbermen, intercepted her and shot her six times. A little 73 year old nun who anguished over the pain of monkeys and fallen trees, and who defended peasant farmers and their rights, had to be eliminated (Murphy 2007, pp. 107–140; Lima 2013).

In Latin America, from the time of the Conquest, the natural environment has been the source of violent conflict whenever powerful economic and political interests have imposed their exploitative projects designed for their own enrichment. This history continues even today. A recent Report on Human Rights by the Inter-American Commission for Human Rights (IACHR 2011) points out that "increasing number of incidents have been reported involving threats and acts of violence against and murders of environmentalists owing to their work" (IACHR 2011, p. 133). "Many" defenders of the environment "have suffered attacks, aggression and harassment" (IACHR 2011, p. 133) and suffer violence, including murder and abduction (IACHR 2011, p. 138). Environmentalists experience these with greater frequency "mainly where there are serious tensions between the sectors that support certain industrial activities, like the extractive industries, which have enormous economic interests at stake, and those sectors that resist the implementation of projects in order to avoid the forced relocation of the communities that will be inevitable if the projects are established or to prevent the harmful effects of the contamination that the industries will produce in the waters, air, soil and subsoil" (IACHR 2011, p. 133).

Emblematic of this history is the assassination of Dorothy Stang for defending the rights of the poor and the forests that sustain their livelihoods (Murphy 2007). More recently, also symptomatic is the expulsion and denationalization of the priest Father Andrés Tamayo in Honduras following the military coup of 2008 because of his mobilizing the poor against lumber companies that ruthlessly destroy the forests, thereby reducing options for satisfying vital necessities of the poor communities (El Heraldo 2009).

---

[1] For background on rural violence in Brazil, especially the Amazon Region, see, for example: Branford and Glock (1985); for the contemporary situation and years previous, consult the annual reports, *Conflitos no campo Brasil*, published by the Pastoral Land Commission of Brazil and available in PDF format: www.cptnacional.orgbr On the persecution and murder of church workers for their defense of the land rights of the poor and indigenous people, see May 1991.

In the same vein is the impactful violent conflict that occurred between the Gnöbe-Buglé people and the national police in Chiriquí, Panamá, during February and March of 2012. The conflict was motivated by concessions that the Panamanian government granted to private companies for mining and hydroelectric development in territory belonging to the Gnöbe. The conflict caused several deaths and injuries, but resulted in the cancelation of the mining and hydroelectric concessions (Prieto 2012). Similar socio-environmental conflicts have occurred throughout the 1980s, 1990s, and 2000s in Mexico, Guatemala, Chile and other Latin American countries (Homer-Dixon 1994; Villarreal 2014; CIEL 2010).

In Chile and Argentina, Patagonia and Tierra del Fuego incarnate this history of environmental violence. From the mid-nineteenth century to the beginning of the twentieth, the owners of huge sheep ranching enterprises and gold miners systematically exterminated the Selk'ham or Ona people, forcing them from their traditional lands and even paying bounty for each Indian murdered. The genocide was virtually complete (Borrero 2001; Chapman 2007). Nature also suffered violence. Since the end of the 19th century, the sheep severely overgrazed Patagonian steppes and following their introduction to Tierra del Fuego in the mid twentieth century, beavers and rabbits have distorted ecosystems. The more recent introduction of American mink negatively affects bird populations (Jiménez et al. 2014), and salmon farming is affecting fishing communities and contaminating the pristine marine waters of the Chilean archipelago (Rozzi et al. 2012). Moreover, the Patagonia icefields, freshwater lakes, and streams, have become lucrative assets since the projection of multiple hydroelectric dams by the Chilean government (Infanti de la Mora 2008; Segura-Ortiz 2010; SICOM 2010). The Catholic bishop of Aysen, Luis Infanti de la Mora (2008, p. 48), fears that these "megaprojects...will produce grave environmental damage and irreversible social problems..." Patagonia and Tierra del Fuego again are on the verge of being centers of environmental conflict.

To the north of Patagonia, in south central Chile, various Mapuche leaders are imprisoned for their defense of their people and their land. For them, the destruction of the natural environment is a profound offense against the Mapuche people themselves. As Nils Raín, one of those jailed, explains:

> To be Mapuche means to be part of an ancestral force of nature that does not want to perish, that wants of continue living and is a spiritual force that is in the forests, in the mountains, in the rivers, in the sea. So, to be Mapuche means being a constitutive part of nature and to speak for her. It's not that the land is ours. We are the land (Raín 2011, http://periodismohumano.com/?s=nilsa+ra%C3%ADn).

This telluric "spirituality" is rooted profoundly in geography and ecology and reminds us that this region is ecologically fragile and of grand natural beauty. Even today Chilean Patagonia is one of the world's regions least affected by human intervention in historic times. Its forests, wetlands, rainwater and streams, ice fields, indigenous cultures, biodiversity, and high endemism represent one of the world's last remaining wilderness areas (Rozzi et al. 2012). Finally, it is a region of unparalleled natural beauty (De Roy 2005, pp. 136–155).

What does this history of environmental conflict teach for constructing a Latin American Earth stewardship? I put the question this way because I assume, following Ortega, that stewardship "derives from a social operation and therefore responds to determinations of place and procedure" (Ortega 2011, p. 270). In what follows, I propose several answers by suggesting elements or inputs for understanding stewardship within the framework of the Latin American liberation tradition, specifically the theology of liberation.[2] This theological movement emerged forcefully in the 1970s and, although in recent years has lost much of its original influence, has been and continues to be an influential part of Latin American political and ecclesial culture.

27.1 Environmental Thought as Act Two

I begin with methodology. What are the sources of environmental philosophical/theological thought? That is, what are the originating sources for philosophical or theological discussion of the environment? In liberation literature, the sources are concrete historical realities. Thus for the Brazilian theologian Leonardo Boff, the starting point is Earth itself (1997, p. 113). "The Earth is ill," Boff says, "and threatened" (1997, p. 1). It is "the living superorganism" (1997, p. 15), the poorest of the poor, and, therefore, we are to be in solidarity with it (1997, p. 113). Ricardo Rozzi (2012, pp. 345, 346), Chilean ecologist and philosopher, proposes that Latin American environmental ethics is "embedded in the ancient worldviews of Amerindian people" as well as those of other peoples subordinated by the socioeconomic system. Environmental philosophy should give much consideration, Rozzi argues, to the "diversity of forms of ecological knowledge and practices rooted in Amerindian, colonial, and post-colonial languages and cultural habits, which in turn are embedded in ancestral native habitats and contemporary anthropogenic habitats". No genuinely Latin American environmental philosophy can be conceived without incorporating Amerindian culture, Rozzi insists. As a second source he mentions the importance of dialogue with environmental thought from other parts of the world. A third source Rozzi proposes, and in my judgment the key for environmental ethics and the idea of Earth Stewardship, "is represented by recent social movements that are catalyzing the incorporation of environmental values into regional policies, cultural expressions, and citizen organizations" (Rozzi 2012, p. 346).

[2] In my judgment, environmental thought from Latin America should be framed within the tradition of critical thought that has produced liberation philosophy and theology. It refers to "all types of contemporary Latin American thought that analyze its own Latin American reality, is aware of the position human beings have in this reality, and adopts transformative means directed toward eliminating all forms of domination, oppression, subordination or alienation" (Sánchez-Rubio 1999, p. 125). See also Ortega 2011, note 14, for a synthesis of the origin and basis of the philosophy of liberation in Latin America. Liberation theology is the theological current of the Latin American liberationist tradition. It is understood as theology for social movements in which Christians are active participants.

This third source is key because from Latin America, in the liberationist tradition—certainly in the theology of liberation—we emphasize the primacy of praxis and action as the originating and generative source of theoretical reflection. As Ortega (2011, p. 273) explains, "The intellectual practice [of liberation] comes out of the feeling of indignation that springs from becoming aware of social, institutional, and cultural violence that is generated with the structural perversion proper to colonialism, underdevelopment and the consequential and concomitant state of cultural dependence." Social movements have incorporated environmental themes precisely because they constantly confront environmental situations that cannot be separated from social situations; situations that political economy or mode of production create and so produce "indignation."

From their struggles for social justice, activists have found it impossible to separate social struggle from environmental struggle. As they confront environmental situations, these are incorporated into social struggles. They confront situations—social as well as environmental—that produce indignation. They see, as Dorothy Stang did, that environmental struggle is "naturally allied with liberation theology and the struggles for justice in Latin America" (Murphy 2007, p. 96). So, "in the past few years a 'liberation ecology' movement, with the church as its spiritual heart, has been taking shape from Chile to Mexico" (Snell 2007; see also Infanti de la Mora 2008; Vicariato 2011). Although the church often has been on the margins of social justice, in Latin America a Christian prophetic voice has long been present.

As early as Antonio Montesinos and Bartolomé de las Casas in the sixteenth century, sectors of Latin American Christianity have vehemently protested social injustice and struggled against it. Priests, nuns, even Protestant pastors, have been involved in such struggles. Since the 1960s, this tradition has acquired a relevant role in Latin America and worldwide through liberation theology (see Dussel 1981). Dorothy Stang and Andrés Tamayo are emblematic of this social justice tradition and represent committed expressions of bottom-up stewardship. In this community-based tradition, ethics and political action are inseparable because the origin of the crisis, whether it is social or environmental, is in political economy or power relations (Ortega 2011, pp. 270–271). In this sense, the original source for understanding the meaning of stewardship must be conflict and the struggles of those who defend the Earth.

In addition to the struggles of militants such as Dorothy Stang and Andrés Tamayo and social-environmental movements, the ancestral ethos or the deep wisdom of original peoples also constitutes a point of departure for ethical environmental reflection. These too are social practices, praxis with nature that has molded social life. This ancestral praxis or deep wisdom is characterized by reciprocal exchange between nature and human society that assumes adaptive forms of relating to the environment. This is a deep source for environmental ethics.

Contemporary environmental struggles and ancestral practices (deep wisdom) are generative sources of environmental ethics because, as Ortega (2011, p. 271) insists for Latin American philosophy generally, such ethics is "thinking through a situational, critical, and consciously participatory history."

27.2 Sequence of Reflection for Liberation Ecology

This is better understood when the sequence of reflection is considered. Specifically, where does thinking about stewardship properly begin? It begins with liberative actions in defense of the environment and social justice. Following the Peruvian theologian Gustavo Gutiérrez (1973, p. 6), we can say that environmental ethics is "critical reflection on praxis:" the defense of the environment and ancestral environmental practices. This "reinforces the importance of human action as the point of departure for all reflection…geared to the transformation of the world" (Gutiérrez 1973, p. 9). As such, the "first step" for building a theory of environmental ethics or even Earth Stewardship, is commitment and struggle. This is Act One. Environmental ethics as theory "is the second step" as Gutiérrez (1973, p. 11) argues for theology. It is Act Two. Thinking follows action. Following Gutiérrez (1973, p. 11), the struggle for a just and healthy environment "does not flow as a conclusion from theological [philosophical] premises. Theology [philosophy] does not produce pastoral [environmental] activity; it reflects upon it." Environmental philosophy or theology does not produce action in defense of the environment; it reflects upon the engagement of those involved in the defense of the environment. We think from experience (Ortega 2011). Thus ethical-theological/philosophical truth will be verified in liberative actions, not by *a priori* premises or abstract truths. "Correct knowledge is contingent on right doing. Or rather, the knowledge is disclosed in the doing," José Míguez-Bonino (1975, p. 90) reminds us. So, then, ethics is an "interpretation of a determined liberative praxis" (Costadoat 2005, p. 63), such as that of Dorothy Stang, Andrés Tamayo, Bishop Infanti, or the Mapuche people.

An interpretation of the liberative praxis of these leaders and the social movements they represent, leads to understanding their active defense of the poor and the forests as "stewardship:" caring for the environment and struggling for just and healthy socio-ecological relationships. Although they do not use the term, through them, we see stewardship as active opposition to neoliberal economic policies that oppress the poor and destroy nature. As these examples illustrate, such grassroots struggles have occurred frequently and widely in Latin America during the last several decades.

So in this sense, with concrete environmental realities and social struggles as Act One, liberation ecology and Earth Stewardship are oriented toward the elimination of all forms of domination, oppression, subordination, and alienation, and as Act Two, will emerge through critical reflection at the intersection of: (1) the Earth itself and the political economy that administers it (mode of production and models of development); (2) the ancestral ethos of original peoples and biocultural landscapes; and (3) dialogue with environmental philosophies from other parts of the world reinterpreted from Latin America in order to mold them according to the different context. This is the methodology of liberation ecology.

27.3 Contextual Ethics and Material Problem

This focus suggests a type of contextual or historical ethics that takes its clues from concrete realities rather than from preconceived and universalist moral notions. Historical reality carries within itself is own normativity (May 2012). As the Brazilian ecofeminist and liberation theologian Ivone Gebara explains, "[Contextual ethics] means that it is a requirement of the historical moment in which we live, and that it unfolds from local contexts, although it connects to and opens toward a global perspective" (Gebara 1998, p. 99). Such ethics is material, not ideal. The theology of liberation assumes this type of material contextual non-universal ethics, breaking with idealist and deontological categories (May 2012; Silva-Gotay 1981, pp. 273–313).

As contextual, liberation environmental ethics corresponds to concrete situations and struggles, frequently local, without forgetting that they are part of much larger struggles. Such environmental ethics is formed from within subaltern groups that struggle for their own well-being as well as that of the environment, groups such as the poor peasants with whom Dorothy Stang worked in Brazil or Andrés Tamayo in Honduras, or the Mapuches jailed in Chile for defending their traditional territory, or even the defense of the integrity of water that Bishop Infanti leads in central and southern Chile. At the same time, a contextual focus necessitates ethics that correspond to specific ecosystemic and bioregional realities, whether they are Amazonian rain forest, sub-Antarctic temperate forests, Patagonian pampas, or high Altiplano deserts. This suggests that ethical generalizations should be avoided in order to respond to specific situations; nevertheless the contextual focus "connects to and opens toward a global perspective," as Gebara (1998, p. 99) reminds us, even while respecting and affirming the particular.

This contextual approach proposes that liberation ecology understands environmental problems as material problems that correspond to socio-historical conditions. Environmental problems are not simply problems of attitudes and personal commitments, because ideas and attitudes are rooted in material realities. Environmental ethics is political ethics because, at bottom, it is a question of political economy or mode of production. We see the political character in its purpose to transform unjust socio-historical reality and, therefore, its material character in the sense that it locates environmental problems in the configuration or structural organization of communities and in structures of power (political economy), or the material conditions of human life.

Behind national material realities is the continuing North–South contradiction and the role of neocolonialism/recolonialization, stimulated by the globalization of raw materials and markets, this in function of the enrichment of capital, whether it be foreign or national. The development of the world's wealthy continues to be predicated on the exploitation of the poor. This international structure is replicated at the national level and configures national political economies to the benefit of national elites (Robinson 2004). In every sense, this reality urges that political economy is the hermeneutic for planting environmental questions, and therefore, Earth Stewardship.

27.4 Unity of Social Questions and Environmental Questions

Certainly a basic concern of environmental ethics in the context of the theology of liberation is to maintain the unity of social questions and environmental questions. The Uruguayan environmentalist Eduardo Gudynas underlines the importance of recognizing that "human systems are in a continuous and strict interrelation with environmental systems. Neither of the two can be thought of or understood isolated from the other" (1995, p. 141; cp. Gudynas 2001). Dorothy Stang understood social struggle and environmental struggle as a single struggle. As she said in reference to the poor farmers who lived in the forest, "They have a sacrosanct right to aspire to a better life on land where they can live and work with dignity while respecting the environment" (Murphy 2007, p. 124).

Sister Dorothy lifted up the magnificence of the forest as she simultaneously demanded justice for the poor. In testimony before the state senate, she lamented the destruction of biodiversity and criticized the anti-environmental practices of both the poor farmers and the ranchers and lumber companies. But she understood environmental destruction as a consequence of social injustice.

> So this magnificent Amazon forest suffered year after year as they cut down another area of the forest and burned it off. And they kept burning off more and more. Then the big landowners began to arrive with cattle and ranches and all this forest land was further degraded. [T]he big landowners paid lots of poor farmers to invade our [PDS] reserves. We kept crying out that our reserve was being invaded (Murphy 2007, pp. 115, 116).

Then she asked the senators, "Have you ever heard a monkey sobbing in pain as his trees are being burned?" (Murphy 2007, p. 116)

From the temperate forests of southern Chile, the Mapuche leader Nilsa Raín also unifies these dimensions: "we are fighting to reconstruct a way of life harmonious with nature." Are these social justice activists? Or are they environmental activists? Both dimensions are integrally intertwined in their thought and struggles. For Dorothy Stang, Nilsa Raín, Andrés Tamayo, and many other activists, the two cannot be separated.

27.5 Dimensions of Earth Stewardship

The foregoing suggests that what is in question is the nature—human relationship. At bottom it is a question of relations and recognition. Clearly this about community and alterity.

27.5.1 Moral Community and Alterity

For Christianity,[3] "neighborliness" is the fountain of ethics: Who is my neighbor? Or, Who merits moral consideration? These are questions of moral community and alterity. Finally, then, the environmental question is that of alterity: not only how we humans relate ourselves to other humans, but also to other living beings, their habitats, landscapes and natural elements (González-Álvarez 1991).[4] For Christians in the liberation tradition—although there is much to be developed here--, this is neighborliness. Earth Stewardship in the theology of liberation will be framed by alterity. The other nonhuman, and the "face-to-face" relation, as Dussel (1988) would say, become the deep question for Earth Stewardship and the basis of liberation ecology. The parameters of moral community are widened toward an inclusive conviviality, in which not only humans receive moral consideration, but nonhumans also.

27.5.2 Socio-ecological Justice

This suggests that justice is the foundation of Earth Stewardship as liberation ecology because justice deals with community and is fundamentally determined by power relations. This also has been insisted upon by Latin American environmentalism, "to speak of socio-ecological justice, as two dimensions of justice" (Ramos-Regidor 1986, p. 109). Colombian philosophers Augusto Ángel and Felipe Ángel (2002, p. 19) affirm:

> The principal difficulty and the principal task for building an environmental society, is the establishment of a just society. Nature comes to its end when it is undermined by social injustice. It isn't possible to separate the unjust distribution of wealth from the enjoyment of earthly satisfactions without producing an impact on the environment. The excessive consumption of minority sectors on the world level and the poverty that borders on starvation of the majorities, is a social fabric through which the substance of the world evaporates.

Power, as embodied in capitalist management of science and technology, turns the Earth into a subaltern. However the same power that reduces the Earth to natural resources to be managed (wisely or otherwise), also reduces whole sectors of humanity to a resource to be managed for the benefit of powerful economic interests. Indeed this is the history of Latin America. The colonial power relationships

---

[3] In truth there is no such thing as "Christianity." Rather, there are many Christianities as expressed through time and space. They all affirm, in different ways, certain common themes, but finally vast differences exist among them as to what such themes mean and how they should be understood and lived out. My own Christian background is that of progressive Protestantism shaped by liberation theology.

[4] There is significant and growing Christian literature regarding animals as part of Christian moral community. Perhaps most notably is Linzey 1995; see also Bauckham 2011.

and worldviews that were established in the sixteenth century continue to inform the contemporary structure of power and worldviews. Colonialism was based on, and continues to prosper from, the subordination of people and nature. So just as political action seeks to empower powerless humans, Earth Stewardship—liberation ecology—also seeks to empower the non-human inhabitants of the Earth. Only such empowerment will make justice possible.

We see this dimension of socio-ecological justice in the life of Dorothy Stang and in the defense of the poor and the forests by Andrés Tamayo; it is the same dimension expressed from his jail cell by the Mapuche Nilsa Raín. Indeed, Bishop Infante makes a similar claim in relation to Patagonian water. Community, in this sense, refers to more than caring for one another. It implies commitment and struggle in favor of the subaltern and respect for the other or the distinct (Ortega 2011, p. 296). It demands that moral consideration be extended to all who live together, although doing so is contextual. This will be relevant Earth Stewardship.

27.5.3 Interculturality

Finally, environmental ethics proposed by the theology of liberation will be intercultural because, as Ortega affirms, liberation "impels us to take into account the local practices and languages that express social relations, for there we find subjugated knowledge and memories" (2011, p. 298). As I have indicated, the theology of liberation breaks with universalist-deontological ethics in favor of a contextual ethics that emerges from historical conditions. Following this liberating intuition, in recent years, pushed by the insurgence of vindicatory social movements of original peoples and Afro-descendants among others, together with the feminist insistence on diversity, an "intercultural" ethics has developed (Fornet-Betancourt 1994).

"Interculturality" as a hermeneutic for outlining environmental ethics, or thinking about Earth Stewardship, refers to other rationalities as legitimate and worthy of consideration. Additionally, a situational environmental ethics "is not only respectful of the Other but is actively committed to the liberation of the Other" and this "must be done *from* and *with* the subaltern" (Ortega 2011, p. 296; emphasis in the original). Interculturality takes with utter seriousness the "diversity of forms of ecological knowledge and practices rooted in Amerindian, colonial, and post-colonial languages and cultural habits, which in turn are embedded in ancestral native habitats and contemporary anthropogenic habitats," as Rozzi (2012, p. 346) urges.

Thus interculturality takes up the founding intuition of liberation theology that the "place" from which reflection is done is fundamentally important. In this way the "deep wisdom" of original peoples is incorporated into environmental thought. Above all, liberating environmental ethics defends the right of original peoples to practice their own ancestral ethos.

27.6 Conclusion

For Earth Stewardship as liberation ecology from Latin America, a hermeneutical key will be community or convivial life, a political term because it requires just power relations, at the same time corresponding to the ancestral ethos of original peoples that emphasize the interrelationship or "connectiveness" of all existence. This will be the moral community where the diversity of biocultures live together, where monkeys do not cry and the poor do not die. With Dorothy Stang, this Earth steward prays: "All I ask of God is His grace to help me keep on this journey, fighting for the people to have a more egalitarian life at all times and that we learn to respect God's Creation" (Murphy 2007, p. 131).

References

Ángel A, Ángel F (2002) La ética de la Tierra. Ética y medio ambiente. In: Leff E (ed) Ética, vida, sustentabilidad. Programa de las Naciones Unidas para el Medio Ambiente. Programa de Formación Ambiental para América Latina y el Caribe, Mexico

Bauckham R (2011) Living with other creatures. Green exegesis and theology. Baylor University Press, Waco

Boff L (1997) Cry of the Earth, cry of the poor (trans: Berryman P). Orbis Books, Maryknoll

Borrero LA (2001) Los Selk'ham (Ona). Evolución cultural en Tierra del Fuego. Editorial Galerna, Buenos Aires

Branford S, Glock O (1985) The last frontier, fighting over land in the Amazon. Zed Books, London

Chapman AM (2007) Los Selk'ham: la vida de los Ona en Tierra del Fuego. Emecé Editores, Buenos Aires

CIEL Center for International Law (2010) Defensoras y defensores ambientales en peligro. La situación en México y Centro América en el ámbito de la industria minera. www.ciel.org/Publications/IACHR_oct10.pdf

Costadoat J (2005) La hermenéutica en las teologías contextuales de la liberación. Teología y Vida XLVI

De Roy T (2005) The Andes: as the condor flies. Firefly Books, Richmond Hill

Dussel E (1981) A history of the church in Latin America: colonialism to liberation (trans: Neely A). William B. Eerdmans Publishing, Grand Rapids

Dussel E (1988) Ethics and community (trans: Barr R). Orbis Books, Maryknoll

El Heraldo (2009) Andrés Tamayo se marcha después de 26 años en el país. 17/11 http://archivo.elheraldo.hn/Ediciones/2009/11/18/Noticias/Andres-Tamayo-se-marcha-despues-de-26-anos

Fornet-Betancourt R (1994) Hacia una filosofía intercultural latinoamericana. DEI, San José

Gebara I (1998) Intuiciones ecofeministas. Ensayo para repensar el conocimiento y la religión. Double Clic, Montevideo

González-Álvarez LJ (1991) Ética ecológica para América Latina. Editorial El Buho, Bogotá

Gudynas E (1995) Ecology from the viewpoint of the poor. Concilium 261(5):139–149

Gudynas E (2001) Los delicados equilibrios de la conservación en América Latina. In: Primack R, Rozzi R, Feinsinger P et al (eds) Fundamentos de conservación biológica. Perspectivas latinoamericanas. Fondo de Cultura Económica, México, pp 663–664

Gutiérrez G (1973) A theology of liberation, history, politics and salvation (trans and ed: Inda C, Eagleson J). Orbis Books, Maryknoll

Homer-Dixon TF (1994) Environmental scarcities and violent conflict: evidence from cases. Int Sec 19(1):5–40, http://www.homerdixon.com/projects/evidence/evid1.htm
IACHR Inter-American Commission on Human Rights (2011) Second report on the situation of human rights defenders in the Americas. http://www.oas.org/en/iachr/reports/thematic.asp
Infanti de la Mora L (2008) Danos hoy el agua de cada día. Pastoral Letter. Vicariato Apostólico de Aysén, Coyhaique
Jiménez JE, Crego RD, Soto GE, Román I, Rozzi R, Vergara PM (2014) Potential impact of the alien American mink (*Neovison vison*) on Magellanic woodpeckers (*Campephilus magellanicus*) in Navarino Island, southern Chile. Biol Invasion 16:961–966
Lima CS (2013) Irma Dorothy Mae Stang e o Projeto de Desenvolvimento Sustentável. Memoria e resistência x o poder do latifundio. Paper presented at the II Encontro de Pesquisas e Práticas em Educaçao do Campo da Paraíba. Joao, Pessoa, 5–7 June. http://www.educacaodocampopb.xpg.com.br/IIEPPECPB2013/GT%20-%205/5.pdf
Linzey A (1995) Animal theology. University of Illinois Press, Urbana
May RH (1991) The poor of the land. A Christian case for land reform. Orbis Books, Maryknoll
May RH (2012) Ética sin principios. Otra ética posible. DEI, San José
Míguez-Bonino J (1975) Doing theology in a revolutionary situation. Fortress Books, Philadelphia
Murphy R (2007) Martyr of the Amazon. The life of Dorothy Stang. Orbis Books, Maryknoll
Ortega F (2011) Historical interpretation and ethics. Notes on the role of alterity. In: Cisneros AH, Premawardhana S (eds) Sharing values. A hermeneutics for global ethics. Globalethics.net, Geneva, pp 269–308
Prieto-Valladares JA (2012) Indignación del pueblo Gnöbe-Buglé en sus luchas por la vida y el agua (Panamá), 2012. La Voz 56:10–13
Raín N (2011) Hay más de 500 mapuches encarcelados. Periodismo humano. http://periodismohumano.com/?s=nilsa+ra%C3%ADn
Ramos-Regidor J (1986) Justicia social y justicia ecológica. Paper presented at the First Encuentro de Cultura, Ética y Religión frente al Desafió Ecológico. Primer Congreso Latino Americano de Ecología, Montevideo, 13–16 Dec
Robison WI (2004) A theory of global capitalism. Production, class, and state in a transnational world. The Johns Hopkins University Press, Baltimore/London
Rozzi R (2012) South American environmental philosophy: ancestral Amerindian roots and emergent academic branches. Environ Ethics 34(4):343–366
Rozzi R, Armesto J, Gutiérrez JR et al (2012) Integrating ecology and environmental ethics: earth stewardship in the southern end of the Americas. BioScience 62(3):226–236
Sánchez-Rubio D (1999) Filosofía, derecho y liberación en América Latina. Editorial Desclée de Brouwer, Bilbao
Segura-Ortiz P (2010) ¿RSE en la Patagonia o el precio de las voluntades? El Ciudadano http://www.elciudadano.cl/2010/03/25/20147/rse-en-la-patagonia-o-el-precio-de-las-voluntades-2/
SICOM (2010) Endesa se ha apropiado del agua de la Patagonia chilena. Interview with Luis Infanti de la Mora. http://www.youtube.com/watch?v=B9uhqwT3leI
Silva-Gotay S (1981) Hacia una ética cristiana de liberación: historización de los valores y politización de la ética. In: Silva-Gotay S, El pensamiento cristiano revolucionario en América Latina y el Caribe. Implicaciones de la teología de la liberación para la sociología de la religión. Ediciones Sígueme, Salamanca, pp 273–313
Snell MB (2007) Bulldozers and blasphemy. Sierra Magazine (Sept–Oct). http://www.sierraclub.org/sierra/200709/bulldozers.asp
Vicariato Apostólico de Aysén (2011) Aysén: huellas de una iglesia profética en la Patagonia. Documentos. Vicariato Apostólico de Aysén, Coyhaique, Chile
Villarreal C (2014) Socio-environmental and community forest conflicts in Latin America. http://www.fao.org/docrep/005/ac696e/AC696E05.htm. Accessed Jan 2014

Chapter 28
Chico Mendes and José Lutzenberger: Ecosystem Management at Multiple Scales of Government

Fernando José Rodrigues da Rocha and Fábio Valenti Possamai

Abstract Chico Mendes and José Lutzenberger together orchestrate an incredibly enthusiastic polyphonic call of encouragement for the socio-environmentalist cause. They set examples of how much "the man of the forest" and "the man of the world" can accomplish in the local and the national levels and how their idealism can set examples throughout the world. It is argued that a culture based on principles of environmental ethics should be encouraged in all educational levels throughout the globe. And this has a series of implications such as: the abandonment of the current anthropocentric, egoistic, materialistic, consumerist way of living; the abandonment of absolute national sovereignty by governments in favor of a Global Environmental Governance; the unrelentingly combat of corruption at all levels of government in many countries, including Brazil. We are subject to the interests of powerful international enterprises and the surreptitious lawful techno-bureaucracy they impose on us. And the current tendency is that their influence will grow bigger and bigger as technological novelties are put into the market. To bring it to a stop and reverse it is not an easy task, but there are examples in history that changes can take place when people have firm belief in their cause. Chico Mendes and José Lutzenberger set examples that, regardless of one's geography, or socioeconomic and political conditions, each and everyone's contribution is precious. It is the collective effort from peasants to presidents that will provide the deep, necessary changes which will make the Earth Stewardship dream come true.

Keywords Brazil • Cooperatives • Earth stewardship • Environmental philosophy • Recycling

---

F.J.R. da Rocha (✉)
Department of Philosophy, Universidade Federal do Rio Grande do Sul, Porto Alegre, Brazil
e-mail: fjrrpoa3@gmail.com

F.V. Possamai
Department of Philosophy and Religion Studies, University of North Texas, Denton, TX, USA
e-mail: fbipossamai@gmail.com

R. Rozzi et al. (eds.), *Earth Stewardship*, Ecology and Ethics 2,
DOI 10.1007/978-3-319-12133-8_28

28.1 Introduction

This chapter will present and discuss two Brazilian environmental activists who acquired international notoriety for different reasons: Chico Mendes for being acclaimed as the first environmental-cause martyr; José Lutzenberger for being the first environmental leader to occupy a place in the Brazilian Federal Government as Minister of the Environment. After the presentation of their individual trajectories, some considerations are given about their legacy, explicit or implicit, regarding northern/southern-hemisphere environmental interactions. By doing so, we hope to better integrate environmental ethics in academic programs and help create a holistic worldview that integrates the participation of different regions, disciplines, and cultures, in order to reorient global society toward a more sustainable socio-ecological trajectory.

28.2 Chico Mendes

Chico Mendes (Francisco Alves Mendes Filho) was a Brazilian rubber tapper,[1] trade union leader, and one of the most important environmental-movement leaders in South American history (Vaughn 2003). He was born in the town of Xapuri (in the Brazilian state of Acre[2]) in 1944, and was murdered by ranchers in 1988 (Rodrigues 2007). He fought to preserve the Amazon rainforest and advocated for the human rights of Brazilian peasants and indigenous people. His goal was to sustain communities of rubber tappers and indigenous people who knew how to live in the forest without wrecking it. His idea was "to live **with** the forest," not exploiting or destroying it. He proposed the establishment of 'extractive reserves' in the Amazon, which would harvest renewable resources on a sustainable basis. Chico Mendes led the Rubber Tappers Union in resisting the pressures of wealthy farmers who were extra-legally grazing their cattle on government land. Sometimes called "The Gandhi of the Amazon," his struggle caught the attention of international environmentalists who saw his resistance movement as a fight to save the rainforest.

Chico Mendes, at a very young age, had an excellent example from his father, Francisco Alves, who came to the remote Brazilian Amazon forest near the border with Bolivia and Peru in 1926, leaving the poverty of his home in the state of Ceará, Brazil. His wife, lrace Lopes Filho, was a member of a family who for generations

[1] In Portuguese: *seringueiro*.

[2] Until the beginning of the twentieth century Acre belonged to Bolivia. However, since the beginning of the nineteenth century, a large part of its population was of Brazilians who exploited rubber tree groves and who, in practice, achieved the creation of an independent territory. In 1899, Bolivians tried to gain control of the area, but Brazilians revolted and there were border confrontations, generating the episode which became known as the Acre Revolution. On November 17, 1903, with the signing over and sale in the Treaty of Petrópolis, Brazil received final possession of the region.

had made their meager living collecting the white, milky latex from the rubber trees (*Hevea* spp) in the Amazon forest. When their son, "Chico", was 9 years old, illiterate and poor like his father, he started working as a rubber tapper too. At the time, landowners did not allow their workers to build or attend schools, so his education was entirely informal. A political refugee, Euclides Fernandez Tavora, taught him how to read and write, using old magazines and a shortwave radio. Chico and his family lived in the *seringal* Cachoeira region, and in the 1970s, he became a leader in a non-violent resistance movement to defend their homes from cattle ranchers, who demanded that they leave. Along the western border, shared with Peru and Bolivia, the Brazilian government began its National Integration Program, which was intended to promote the colonization of the region with cattle ranchers and force the native people, mostly Amerindians, to relocate. Over the next 15 years, the ancient forests were set on fire to make way for farms and ranches, thus resulting in massive land erosion and loss of jobs.

Traditionally rubber tappers and their families were at the mercy of a system of debt bondage, but during the 1960s and 1970s this system faced collapse in Xapuri. Ranchers from southern Brazil began buying up rubber estates and clearing vast areas of the forest for cattle grazing. Many tappers and their families were forcibly, often brutally, evicted. Ruthless exploitation of the rainforest became the dominant policy and practice; and resistance to that exploitation was the focus of Mendes's life. From his endeavors emerged the concept of "extractive reserves," which are legally protected forest areas held in trust for people who live and work on the land in a sustainable manner. Mendes and his movement were recognized as a force not only for social justice, but also against environmental destruction. The rubber tappers were able to propose a socially equitable and environmentally sustainable development policy for the region, based on securing and improving their way of life, rather than on official investments in ranching and colonization projects that would have led both them and the forest to disaster (Gross 1989, p. 2).

Mendes gathered his fellow workers together to protest the relocation schemes, organizing blockades against bulldozers. This was the starting point for the trade union he founded in the state of Acre in 1975. Later, in 1985, he created the National Council of Rubber Tappers, which represented an expansion of the union. In cooperation with Brazilian anthropologist Mary Helena Allegretti, he organized the first national meeting of rubber tappers in Brasilia, the capital of the country. In order to secure rubber tree preserves, Mendes sought aid from environmental groups in the United States. His idea was to provide the local people with a source of income by practicing sustainable agriculture. The international environmental community recognized his work and the United Nations Environment Program awarded him the Global 500 prize in 1987. He also received the Ted Turner Better World Society Environment Award. Nevertheless, the Brazilian government and media continued to ignore him. His major success, however, was a winning effort to stop ranchers from cutting down a forest that the rubber tappers wanted to keep as a protected reserve area. Due to a coalition he built uniting the rubber tapers and leaders of the indigenous Yanomami tribe, his leadership and power became a threat to local ranchers.

He described the organization of the *Conselho Nacional de Seringueiros* (CNS or National Council of Rubber Tappers) as the first step of an organized resistance to environmental destruction in the Amazon. Chico Mendes and his companions were defending the thesis that the rainforest is worth more standing up than lying down; that extracting the richness from the forest without destroying it, is what sustainable development is all about. Extractive Reserves (*RESEX – Reservas Extrativistas*) became, after Chico Mendes, part of the Brazilian national system of protected areas. The National System for Conservation Units (*SNUC – Sistema Nacional de Unidades de Conservação*) defines them as a "Protected Area for Sustainable Use by Traditional Populations." The government decree 98.897 of January 30, 1990, signed by President Fernando Collor de Mello and Environment Minister José Lutzenberger is the legal basis for extractive reserves. It defines RESEX as "territorial space destined to self-sustainable exploration and conservation of renewable natural resources," and establishes that the executive power shall create those reservations in territorial spaces considered of ecological and social importance. The government cedes the rights of use of land and sea, and the population receives a collective title for land use. This impedes the sale of the lands that are federal properties. The responsibility for enforcement of the law is also federal. The concession guarantees access to land and sea for the future generations of the local community members.

In the case of Amazonia's rubber tappers, planners and policy-makers had a crucial challenge to face: the creation of economic and non-economic incentives which would allow non-destructive uses to become viable, and open long-term options for maintaining rainforest populations. Chico Mendes wanted to create a management system, built upon existing solidarity and cooperation, in order to encourage a shared, collective sense of responsibility and to preserve the forest as a common property resource. In order to halt or even reverse the present tendency towards depopulation of the *seringais*, extractive reserves should provide the means for rubber tappers and their families to sustain their livelihoods. Economic sustainability would provide tappers with an incentive to retain and diversify their traditional sources of income, and at the same time would encourage them to preserve the forest. Discussions and initial efforts to improve the income-generating capacity of extractive reserves have focused on two areas: the improvement of product quality, and support services for existing rubber and Brazil-nut industries. In addition, however, it will be necessary to diversify the range of productive activities to include sustainable forest management, agroforestry, and use of other non-timber forest products, such as medicinal plants.

While forging partnerships with other green groups, Chico Mendes insisted that people should not be regarded as separate from nature but, instead, be considered integral components of the natural landscape. He was a Brazilian pioneer in what is now called environmental and social justice, promoting the rights of communities to help shape their destinies from the ground up. Despite threats, Mendes refused to leave his home state of Acre, in the westernmost part of the Brazilian Amazon forest, for safer terrain. And so his life was cut short by a single shotgun blast. On the evening of Thursday, December 22, 1988, Mendes was assassinated in his

Xapuri home by Darly Alves da Silva, a rancher. The shooting took place exactly one week after Mendes' 44th birthday, when he had predicted that he would not live until Christmas. Mendes was the 19th rural activist to be murdered that year in Brazil. Many felt that although the trial was proceeding against the actual killers, the involvement of the Ranchers' Union, the Rural Democratic Union, and the Brazilian Federal Police was ignored. In December 1990, Darly Alves da Silva, his son Darly Alves da Silva Jr., and their ranch hand, Jerdeir Pereia, were sentenced to 19 years in prison for their part in Mendes' assassination. Chico Mendes was well aware of the threat to his own life; perhaps he foresaw his death. In a letter written shortly before his assassination, he wrote: "If a messenger from heaven came down and guaranteed me that my death would help to strengthen our struggle it would even be worth it. But experience teaches us the opposite ... I want to live" (Mendes 1989, p. 6).

In recent years, some of his compatriots have risen to prominence. As the new millennium began, the daughter of a rubber tapper from Acre, Marina Silva, became the Brazilian Minister of the Environment. A forest engineer and former political advisor of Mendes, Jorge Viana, was elected Acre's governor. And although Brazil's first working class President, Luis Inácio Lula da Silva, was criticized early in his administration by environmental groups for allowing deforestation rates to increase, his presidency clearly signaled a great transition. Lula once stood trial in military court alongside Mendes for their respective unions activities. Mendes' philosophy has been adapted by rural Amazonian communities of small farmers and settlers, including those who were lured by offers of free land during the effort to construct new roads in the Amazon region by the military dictatorship in the 1960s. Some of these farmers, seeing the limits of the old methods of cutting, burning, planting, and moving on, have embraced new forms of agriculture that can be sustained on fragile Amazonian soils. As Chico Mendes stated:

> Our struggle will continue until all our areas are guaranteed, until indigenous people have their land guaranteed. The forest is our mother, our source of life and in order to save it, we will do everything we can until the end.... At first I thought I was fighting to save rubber trees, and then I thought I was fighting to save the Amazon rainforest. Now I realize I am fighting for our humanity (Revkin 2004, p. 201).

Perhaps the most significant element of Mendes' legacy is the enhanced power and voice acquired by the organizations related to him and the rubber tappers' cause: the National Council of Rubber Tappers and the Amazon Work Group. As a result a new generation of environmental leaders and activists came onto the scene. Furthermore, the political conditions for potential change have never been better, for state and federal policies which promote and support sustainability have been framed. The poverty, degradation, and destruction of the Amazon forest and its peoples are amongst the greatest of current socio-environmental challenges. It was only after the death of Chico Mendes that the 970,750-ha Chico Mendes Extractive Reserve, accommodating 3,000 families, was created.

One of Chico Mendes' earliest allies from the "other Brazil" – the developed, industrialized south – was the late José Lutzenberger, an agronomist who became the country's leading ecologist. He was appointed Minister of the Environment shortly after Mendes' death. He saw the Amazon as a smaller mirror of the global environment. Lutzenberger affirmed that "It's a complicated system that can take a lot of abuse, but you get to a point where suddenly things fall apart. It's like pushing a long ruler toward the edge of a table. Nothing happens, nothing happens, nothing happens—and then suddenly the ruler falls to the floor" (Revkin 2004, p. 102).

28.3 José Lutzenberger

José Antônio Lutzenberger was born in Porto Alegre, the state capital of Rio Grande do Sul, in southern Brazil, on December 17th, 1926. His father was a distinguished German architect, painter, and professor, who was invited to work for a German engineering company installed in Brazil. His mother was the granddaughter of a German immigrant who became prosperous in the cattle and farming business.

Best known by his family name, or simply as Lutz, he attended upper class private catholic high schools, one run by Jesuits, the other by Marists. He learned German at home, and was also fluent in English, French, and Spanish. He graduated as an agriculture engineer at the Federal University of Rio Grande do Sul, reputed to be one of the best universities in Brazil. He continued his education at Louisiana State University, USA, where he obtained an MA in agro-chemistry and edaphology.

On his return to Brazil, he spent 4 years working for a fertilizer company in Porto Alegre. In 1957 he was invited by CIBA-GEIGY to go to Germany. He soon became a technician and executive in the area of agrotoxins. After 2 years in Germany, he was transferred to Venezuela, and later on to Morocco. His market area covered such countries as Spain, Portugal, Canary Islands, Ceuta, Colombia, Ecuador, Algeria, and Tunisia.

During his work abroad, he continued studying subjects such as mathematics, biology, history, and history of the religions. Among his favorite authors were Albert Schweitzer, Thomas Berry, and Francis Chaboussou, an agronomist at the French National Institute of Agricultural Research. Lutzenberger's reading of Rachel Carson's *Silent Spring* had a great impact on him, as she recounted the damage caused by the agrochemical industry to the environment. His range of interests was wide, but it was the Gaia Hypothesis proffered by James Lovelock and Lynn Margulis that Lutzenberger adopted as his basic theoretical framework.

When BASF (*Badische Anilin und Soda Fabrik*) invested in the pesticide business, it became harder and harder for him to reconcile his professional activities and his personal ideas about environmental protection. In 1971 he made the decision to leave BASF; and he came back to Porto Alegre.

On his return, he found fertile soil for his preservationist ideas. Henrique Luis Roessler, another German descendent, was already active in environmental causes for over a decade; and Augusto César Cunha Carneiro, also an environmental activist, was in close contact with Lutzenberger. The idea of creating a society for the defense of nature had ripened. Lutzenberger had been in touch with the Sierra Club and the National Audubon Society in the USA, as well as with environmental groups in Europe. These were the models chosen to follow and in 1971, the *Associação Gaúcha de Proteção ao Ambiente Natural* (AGAPAN), the Gaucha Association for the Protection of the Natural Environment was founded. It was the first ecological association in Brazil and South America, and was taken as a model for the creation of many others. Lutzenberger soon became the best known member of the group, thanks to his strong personality and scientifically based arguments. His persuasiveness, sincerity, and enthusiasm led him to become the most outspoken advocate of the environment in Brazil and the first Brazilian environmental activist, after Chico Mendes, to acquire international visibility. He soon built up a network of contacts that encompassed the five continents. Among his personal friends he counted Herman Daly, Amory Lovins, Ross Jackson, and Hazel Henderson. Prince Charles himself was one of those who encouraged Lutzenberger's activities.

When Lutzenberger came back to Brazil, the country was still ruled by the military regime, which fostered efforts to promote industrial growth and mega infrastructure projects, such as the Itaipú dam and the Transamazon highway. Taking advantage of the favorable international economic conditions, heavy loans were taken. The civil society was under censorship, not allowed to organize itself and take part in policy decisions.

Fernando Collor de Mello, who in March 15, 1991 invited Lutzenberger to be his Minister of the Environment, was the first democratically elected civil president of Brazil, after 30 years of military regime. Fernando Collor in his presidential campaign focused on restoring the credibility of public administration in Brazil. He wanted the nation, under his administration, to build up an image as a serious, reliable country, so that he could attract international investment. At that time, Brazil was being pilloried as a major environmental villain, due to the devastation of the Amazon rainforest. Seeking useful green credentials, he accepted Prince Charles's suggestion that he engage Lutzenberger as his Environmental Minister. The presidential invitation was accepted. Lutzenberger believed that he could achieve more in government than outside of it, although for him, that decision was a very difficult one. According to Fernandes and Valença (2004, p. 246):

> he tried to implement a radically new model of environmental policy, abandoning the technocratic approach and the mega projects, and inviting the local populations to participate. Consequently, he soon was in conflict with the reigning system. His main opponents were the ruralists' representatives in Congress and the lumber industry men.

He figured that the existing government policies, inherited from the military regime, were not protecting the Amazon forest, but, on the contrary, encouraged its degradation by miners, loggers, and other corporate interests. In addition to that, he was against the megaprojects cherished by the military regime, such as nuclear-weapons production and the construction of gigantic dams. He soon attracted strong opposition from the Supreme School of War. The National Forest Service (IBAMA, *Instituto Brasileiro do Meio Ambiente e dos Recursos Naturais*), politicians, and lobbyists also joined forces with the military against Lutzenberger.

He was very straight forward and severe in his criticism of many long-established interest groups and their practices. One of his immediate plans was to take strong corrective measures against the lumber industry. He was attributing responsibility for wrongdoings to no less powerful groups than the Order of Brazilian Lawyers, the General Procuracy of the Republic, and many NGOs. Government members tried to dissuade him of his purpose, but in vain. During the Brazilian representation at the UN, in March, 1992, he announced that Brazil's National Forest Service was a "lumberman's branch" and a "corruption center." Furthermore, he also stated that the money contributed to Brazil by international environmental NGOs was being diverted to corrupt individuals. He advised them to stop contributing to projects in Brazil.

Before that, in 1991, in a meeting with Austria's Chancellor, Franz Vranitzki, President Collor, using a typical third-world-country discourse, said that Brazil was a poor country in need of help from rich ones. Surprisingly, taking advantage of the fact that President Collor did not know German, Lutzenberger addressed the Austrian Minister in his native language and said: "…We have an unbelievably rich country … We have all possible resources. But we are a very poor country. Unbelievably poor. One cannot imagine how poor we are in decent politicians" (Bones and Hasse 2002, p. 27).

Lutzenberger clearly was not endowed with the diplomatic skills, which, if he were, would have allowed him to easily move among the diverse areas of activity that his position as a minister required. Perhaps, he never cared to. The newspapers announced that he was about to fall from power. A few months before the Earth Summit, Rio 92, which he had helped to organize, he was replaced by Professor José Goldenberg, a physicist, with long experience in governmental agencies.

The evaluation of Lutzenberger as Minister is controversial. Some say that with him a great opportunity was wasted, others that he was a historical landmark. As Environment Minister, Lutzenberger's main achievements were to persuade:

(A) President Collor to recognize and protect the land of the Yanomami Indians in the northwest Amazon Basin, creating a 36,000 square-mile sanctuary, and to expel the wildcat gold miners who were decimating the Indians and destroying the rainforest habitat;
(B) Prince Charles, a strong supporter of environmental causes, to visit the Amazon, and to host an environmental gathering aboard of the Royal Yacht;
(C) the Brazilian Government to abandon the Atomic Bomb Project, to sign the Antarctic Treaty and the Wales Convention, and to implement the Brazilian Environmental Code.

As environmental activist, Lutzenberger's main achievements were:

1. to wage a successful campaign for the propagation of the philosophical foundations of the environmental movement;
2. to install 40-million-dollars' worth of antipollution equipment in a paper pulp factory in Porto Alegre;
3. to stop abusive destruction of city green areas and the indiscriminate use of agrotoxins and transgenics in plantations;

4. to stop water pollution by industries;
5. to stop the development of nuclear energy.

As an environmental activist, on the one hand, Lutzenberger campaigned against (i) the privatization of potable water resources, (ii) the use of fossil fuels, (iii) the population explosion, (iv) the use of Gross Domestic Product (GDP) as measure of development, (v) the paradigm of continued progress, and (vi) the control of the world's farmers by transnational corporations. On the other hand, he campaigned for the creation of three state preservation parks, a state law to regulate the use of agrotoxins, the Agronomic Prescription Directory, waste selection mills, and small waste recycle enterprises.

Lutzenberger was also a far sighted entrepreneur. He created the Gaia Foundation to serve as a role model and teaching institution in the areas of sustainable development, regenerative agriculture, environmental education, and urban waste recycling. As its president, he signed an environmental assistance contract with the Amazon State to promote sustainable development through rational exploration of natural forest, fishing, and mineral resources.

His other organizations – *Vida Produtos Biológicos Ltda* (Life Biological Products Ltd.) and *Tecnologia Convivial Ltda* (Convivial Technology Ltd.) – were created to carry out research and develop technologies for recycling paper pulp solid residues, as well as the organic waste of tanneries, slaughterhouses, and packing plants. Now merged under the name of *Vida* (Life), they have over 3,000 clients and an annual gross income of about 3.5 billion US dollars (Jornal do Comércio 2012, p. 4).

In his 31 years of environmental militancy Lutzenberger delivered over 80 lectures in Brazil and 40 abroad. In recognition for his contribution to the environmental cause, he received over 40 awards from countries such as Austria, Bolivia, Brazil, Germany, Italy, Spain, and the People's Republic of China.

Most of Lutzenberger's ideas and campaigns reached the general public through newspapers and magazine articles. He did not focus on organizing his writings for publication in book form. He trusted his friends and collaborators to undertake this task. His book production consists of nine titles. *Fim do futuro? Manifesto ecológico brasileiro* (The end of the Future? Brazilian Ecological Manifesto, 1976) is probably his foundational text.

Lutzenberger was a melting pot of influences. Due to his family origins, he had a German mind set and grew up in an important German community in Brazil. Then he was influenced by the American environmental movements, while studying in Louisiana for his MA degree. Later, he contacted European green organizations during his stay in Germany. It is easy to understand that on his return to Brazil, he put into practice what he had learned in the northern hemisphere. Lutzenberger was not the only person to bring American and European environmental ideals to Brazil. Many former political refugees from the military dictatorship came back after the country's re-democratization. Fernando Gabeira, for example, acquired special prominence, as he went directly into politics and founded a Brazilian version of the European Green Party.

One can say that all UN conferences and reports had an impact in Brazil. After the 1970s, the environmental movement and its developments became more and more international. Lutzenberger, for one, had contacts and personal friends on five continents.

It is in the international political level that one may find unbalanced perspectives from the northern and southern hemispheres, regardless of the fact that such a distinction is not so pertinent anymore. To take just one example: the low carbon economy target. In the present international system, the world is divided into three economic superpowers, namely, the USA, the European Union, and China. Another group of five great economic powers is formed by Japan, India, Brazil, Russia, and South Korea. In practice, the conflict of interests is not a matter of north versus south. It is the European Union versus all the others, because the EU emphasizes the need to create a World Environmental Organization, as powerful as the World Trade Organization. South Korea and Japan sympathize with the idea, but do not embrace it. Brazil, regardless of some recent feeble efforts, still has a highly polluting energy matrix; and consequently does not have much interest in contributing to the construction of a Global Environmental Governance. On the other side of the spectrum we find the United States and China, which are very powerful in the global geopolitical scene, but are in strong opposition to the adoption of international antipollution laws.

Lutzenberger was aware that the political fight was not sufficient. It needed transformation in the education system and culture. He was an eager reader in many areas of knowledge. This gave him a lot of ammunition to make pungent criticisms of the present universities performance in the education of our youth. Lutzenberger strongly advocated the importance of environmental philosophy, claiming that it was fundamental as far as future generations are concerned. He pointed out that due to the conservative curricula that students in general are subject to, almost all finish their schooling in complete ignorance of both science and technology. It looks like, he said, that universities have only succeeded in deadening the student's critical faculties and sensibilities. In view of the compartmentalization of knowledge adopted by most universities, students obtain their degrees without even having a single class about the theories of evolution (sometimes even forbidden) and ecology.

Furthermore, he regarded the prevailing reductionism as a major problem. Especially in economics, he said, we need a completely new paradigm. "If this approach to university education continues, young people will only grow up to become agents of the grotesque, suicidal juggernaut that modern industrial civilization has become. That needs to be changed" (Lutzenberger 1996, p. 42).

Lutzenberger adopted and disseminated the idea that the "Living Planet" should be seen as a whole, that life is a continuous chain of interactions, which goes from bacteria to the most sophisticated organisms, all of which are interdependent. *Mutatis mutandis*, to build an effective approach to Earth stewardship, one must overcome the deleterious religious and philosophical theories that postulate anthropocentrism as well as the Cartesian dualism. The drive for universality, for homogeneity, even for equality should also be discarded. It is only through the acceptance of the existing differences in geographical regions, cultures, abilities, demands, needs, and expectations that a global society can pave its way towards sustainability, a perspective that concurs with the perspectives of biocultural ethics and Earth Stewardship proposed in the regions of southern Chile (Rozzi et al. 2012; Rozzi 2013).

28.4 Implications for an Earth Stewardship Initiative

Fair and long lasting solutions can only be achieved through the honest participation of all members in all the chains of governmental decisions.

The principle of responsibility proposed by Hans Jonas fits in well with the Earth Stewardship approach: everyone is responsible for maintaining our planet's health for our own and for future generations. This sense of responsibility has been fostered by the direct participation of diverse community members in program decisions. We need to recover the spirit of the first Earth Day 1970, in which 20 million Americans took part. Electronic communications resources, such as the internet and social media, have been offering new and very effective ways for individuals and communities to participate in processes of decision-making.

Nevertheless, for an Earth Stewardship initiative, such as the one proposed by the Ecological Society of America (Chapin et al. 2011), to succeed many important and urgent steps should be taken. A culture based on principles of environmental ethics should be encouraged in all educational levels throughout the globe. This implies the abandonment of the current anthropocentric, egoistic, materialistic, consumerist way of living. Governments should give away part of their sovereignty in favor of a Global Environmental Governance, for environmental problems do not respect frontiers, they are everyone's. And as far as many countries are concerned, including Brazil, corruption at all levels of government should be unrelentingly combated.

No one can say that facing the interests of powerful international enterprises and the surreptitious lawful techno bureaucracy they impose on us is an easy task, but there are examples in history that changes can take place when people have firm belief in their cause. Chico Mendes and José Lutzenberger set examples that, regardless of one's geography, or socioeconomic and political conditions, each and everyone's contribution is precious. It is the collective effort from peasants to presidents that will provide the deep, necessary changes which will make the Earth Stewardship dream come true.

References

Bones E, Hasse G (2002) Pioneiros da Ecologia. Já, Porto Alegre

Chapin FS III, Pickett STA, Power ME et al (2011) Earth stewardship: a strategy for social-ecological transformation to reverse planetary degradation. J Environ Stud Sci 1:44–53

Fernandes E, Valença MM (2004) Brasil urbano. Mauad, Rio de Janeiro

Gross T (1989) in Mendes, Chico. Fight for the forest: Chico Mendes in his own words. Latin American Bureau (Research and Action) Ltd, London

Jornal do Comércio (2012) Issue of May 14, 2012. Porto Alegre, Brazil

Lutzenberger J (1976) Fim do futuro? Manifesto ecológico brasileiro. Movimento, Porto Alegre

Lutzenberger J (1996) Science, technology, economics, ethics and environment. In: Callicott JB, da Rocha FJR (eds) Earth summit ethics. SUNY Press, New York

Mendes C (1989) Fight for the forest: Chico Mendes in his own words. Latin American Bureau (Research and Action) Ltd., London

Revkin A (2004) The burning season: the murder of Chico Mendes and the fight for the Amazon Rain Forest. Island Press, Washington, DC

Rodrigues G (2007) Walking the forest with Chico Mendes: struggle for justice in the Amazon. University of Texas Press, Austin

Rozzi R (2013) Biocultural ethics: from biocultural homogenization toward biocultural conservation. In: Rozzi R, Pickett STA, Palmer C et al (eds) Linking ecology and ethics for a changing world: values, philosophy, and action. Springer, Dordrecht/Heidelberg/New York/London, pp 9–32

Rozzi R, Armesto JJ, Gutiérrez J et al (2012) Integrating ecology and environmental ethics: Earth stewardship in the southern end of the Americas. BioScience 62(3):226–236

Vaughn J (2003) Environmental activism: a reference handbook. ABC-CLIO, Santa Barbara

Chapter 29
On Frank Golley's International and Interdisciplinary Insights for a Twenty-First Century Earth Stewardship Based on Environmental Ethics

Alan P. Covich

Abstract Current interest in developing a worldview to enhance Earth stewardship recognizes the importance of a multicultural perspective based on environmental ethics and a global understanding of the value of biodiversity and ecosystem processes. Frank Golley was a champion in developing and implementing ecosystem concepts based on "nature-centered thinking". His environmental ethics-based principles emphasized "connectedness" among people and their environment that included the value of cultural differences in responding to natural and human-driven disturbances. Golley concluded that long-term, large-scale international studies are essential in evaluating the vulnerability of species and their habitats as well as ecosystem processes. Some ecological disturbances are easy to observe (e.g., floods, fires, hurricanes) while others (e.g., gradual loss of species, slow spread of invasive species) take longer to study but most require long-term research before their full impacts are known. Moreover, cumulative effects and indirect effects of complex interactions require multi-disciplinary research to sort out the causes and effects of changing ecosystem structure and function, often at a global scale. Research in tropical ecosystems demonstrated the need for organizing site-based research for extensive comparative studies. Golley's international leadership accelerated progress in enhancing the understanding of how these long-term changes in tropical rainforests and agroecosystems can affect local populations and global connections with other ecosystems. Today ongoing studies of both wet and dry tropical forests are providing important data related to ecosystem services needed by local communities. Golley's linking of ecosystem ecology, landscape ecology, and environmental ethics helped to define pragmatic sustainable "response systems" as a way to resolve complex tradeoffs among conflicting perspectives and his approach can continue to help develop ideas for Earth stewardship.

Keywords Ecosystem services • Environmental ethics • Long-term ecosystem research • Stewardship

---

A.P. Covich (✉)
Odum School of Ecology, Institute of Ecology, University of Georgia, Athens, GA, USA
e-mail: alanc@uga.edu

R. Rozzi et al. (eds.), *Earth Stewardship*, Ecology and Ethics 2,
DOI 10.1007/978-3-319-12133-8_29

29.1 Introduction

The challenges of sustaining the planet require additional ways for professional ecologists and others to contribute to the emerging concepts of Earth Stewardship based on interdisciplinary and international collaboration (Chapin et al. 2015 in this volume [Chap. 12]; Rozzi et al. 2012). For more than a century ecologists, conservation biologists, and many others have debated how to respond to the losses of biodiversity and habitats as well as the planet's carrying capacity (Hutchinson 1948; Di Pasquo 2013; Kingsland 2015 in this volume [Chap. 2]). Along with his international colleagues, Frank Golley brought people together to focus on the importance of species and their values to societies around the world (Fig. 29.1). Golley's global perspective on environmental issues that threatened the planet's sustainability was based on first-hand experiences in many parts of the world and the results of his active involvement in large-scale, tropical research studies. Golley and his network of international colleagues developed pragmatic views that emphasized the total value of ecosystems. These new ideas and methods combined perspectives from anthropology, economics, ecology, and environmental ethics that continue to remain highly relevant to the ongoing discussions today regarding the roles of professional ecologists and conservationists. There is an increasing need to communicate more effectively to a wider audience regarding the value of natural ecosystems and the provisions of goods and

Fig. 29.1 Frank Golley in 1991 on campus at the University of Georgia's Institute of Ecology in Athens (Photo by Media Services, University of Georgia)

services that depend on sustaining biodiversity. Although progress over the last few decades is clear, some of the same concerns raised previously still remain as challenges, especially for more effective inclusion of people and the appreciation of their ecosystem connections (Golley 1978, 1991; Likens 1998).

29.2 The growing Importance of People in the Global Ecosystem

Golley's contributions were important in redirecting professional ecologists to extend their perspectives to view humans as part of the planetary ecosystem (Golley 1993, 1994, 1998). Golley's close working relationships with colleagues such as Gene Odum and H.T. Odum (Craige 2002), his participation in international research projects, and his leadership in professional societies, provided a basis for developing his philosophy. His views on the role of professional ecologists and their needs for developing their own ethical codes grew from a deeply focused appreciation of cultural and biological diversity and his leadership roles in several international professional societies (Golley 1978, 1983, 1984).

Frank Golley's approaches to ecological research included recognizing the importance of intrinsic values in resolving environmental conflicts through use of ecosystem and landscape-level concepts, especially the value of "connectedness" (Golley 1998). Based on his research on the development of the ecosystem concept, Golley concluded that the acceptance of this idea helped initiate "a dialogue about how humans value nature." This concept "provided a basis for moving beyond strictly scientific questions to deeper questions of how humans should live with each other and the environment. In that sense, the ecosystem concept continues to grow and develop as it serves a larger purpose" (Golley 1993).

The ecosystem concept contributed to interdisciplinary collaborations among participants in the UNESCO's Man and the Biosphere Program and other large projects where environmental values were considered, in some cases for the first time (Odum 1977; Golley and Hadley 1981; Golley 1993). The multiple ways in which different individuals debated, and many eventually agreed on, the importance of this ecosystem concept helped to re-define the primary role that the ecosystem concept played in their research and teaching (Mitman 1992; Wright et al. 1995; McDonnell and Pickett 1993). The ecosystem approach remains highly useful in conservation biology and much of current ecological science (Likens and Lindenmayer 2012; Perring et al. 2013).

By the 1960s and into the 1970s ecologists began new debates about the value of ecological concepts that seemed inadequate for resolving global environmental conflicts (MacIntosh 1987; Hagen 1989; Di Pasquo 2013). Ecologists were moving beyond only considering direct effects, to a more thorough consideration of the impacts of unexpected indirect effects. They no longer used the "balance of nature" metaphor that originated many decades earlier (Egerton 1973; Hagen 1992; Pickett 2013). Golley emphasized that it is essential to have a broad view that includes consideration of both past and present human impacts. He noted that human activity had already affected every ecosystem in some way and that pristine "untouched" ecosystems were increasingly rare or no longer existent. Because human impacts occurred over many centuries, comparisons with non-human ecosystems often benefited from considering paleoecological data (e.g., Bush et al. 2000; McLauchlan et al. 2013). Students were also introduced to systems analysis where the boundaries were effectively defined and sometimes open to additional, previously undescribed inputs (e.g., energy subsidies from fossil fuels). More studies considered applied problems such as the global cycling of radioactive isotopes, especially strontium (Odum 1951; Limburg 2004). The values of different people toward large-scale environmental impacts became a topic of growing international concern (Mooney et al. 2013; Patten 2014).

Atmospheric testing of nuclear weapons, pollution of waterways, and unsafe applications of pesticides changed how ecologists viewed their own environmental ethics and values (Covich 2015). This awareness and the contributions of their students grew throughout the decades as the environmental impacts of wars became a focus of ecological research. The national and international professional societies attempted to meet these new needs and once again Golley contributed important ideas and energy to making these transitions successful.

29.3 Development of International Ecosystem Research Programs

Golley realized the importance of human-dominated ecosystems and the complexity of different ethical relationships in many parts of the world. His international experience underscored the value of direct field observations and the importance of getting investigators and policy makers into natural settings where they could see first-hand the beauty and complexity along a continuum of human-dominated ecosystems. He emphasized the connections among cities and their surrounding forests that related to watershed management as well as the management and protection of biodiversity for its own intrinsic value.

The early phases of ecological research on populations, communities, and ecosystems often relied on methods and metaphors from physiology (Mitman 1992; Hagen 2008). Focusing on birds and mammals, Golley's first research was on the physiological ecology of individual animals. He then used this experience to begin the transition to whole ecosystem research using organism-based concepts of metabolism to examine effects of added nutrients on growth, energetic pathways, and waste production. This physiologically based view of ecosystems was shared by several other ecologists and emphasized the significance of highly variable rates of dynamic processes such as productivity, organic matter decomposition, and energy flow. These rates were often accelerated or diminished by how people and other biotic forces, as well as abiotic factors, interacted spatially and temporally.

This whole ecosystem view initially also used an early metaphor when biotic communities were studied as "super organisms: undergoing predictable development over time (Kingsland 1995; Golley 1993)". Although the metaphor of a "super organism" faded as an influence (Burgess 1981; MacIntosh 1987), it intrigued community and ecosystem ecologists in their early thinking. G.E. Hutchinson (1940) noted that "If ... the community is an organism, it should be possible to study the metabolism of that organism." He conducted studies of the "intermediate metabolism" of phosphorus and nitrogen cycles in lakes, using radioactive phosphorus-32 as a tracer (Hutchinson 1941). This shift to biogeochemical cycling became a major emphasis in ecosystem science that relied more on isotopic tracers. While serving as the first director of the University of Georgia's Savannah River Ecology Laboratory from 1962 to 1967, Golley worked with a group of ecologists who developed new concepts using radioactive tracers to study nutrient cycling and energy flows (Odum and Golley 1963). These new techniques were needed to conduct basic research as well as to study any releases of radioactive materials from nuclear power plants and from atmospheric testing. This research helped to set the stage for large-scale ecosystem studies in the International Biological Program (Golley 1993; Coleman 2010). These studies compared rates of productivity and measured micro-concentrations (parts per billion) of essential nutrients and contaminants (Kwa 1993; Golley 2001; Creager 2013).

Golley actively contributed to the concept of biospheric metabolism during the International Geophysical Year in 1957–1958 that led to the International Geosphere-Biosphere Program with an increased international research network in the 1980s and 1990s (Kwa 2005; Steffen et al. 2004; Mooney et al. 2013). This interdisciplinary Earth science program documented daily and seasonal patterns of global metabolism, as concentrations of atmospheric oxygen and carbon dioxide changed from the equator to the poles through the year in response to changes in solar energy and temperatures (Melillo et al. 1993; Mooney 1996). These discoveries were pushed forward by high-resolution remote sensing, improved dissolved gas detectors, and faster computer modeling, all combining to result in a new perspective about life on "spaceship Earth", especially once people viewed images of the Earth from the Moon. This advanced technology still required international programs to conduct additional local "ground-truth" field studies at multiple scales to test the observations and the new model predictions. This new technology enhanced studies of local disturbances that created gaps in biotic distributions across the mosaic of habitats (Shugart 1998; Turner and Chapin 2005; Turner 2010). In that sense the new research based on remote sensing continued to benefit from models based on local patchiness, a concept developed by Monica Turner, one of Golley's former doctoral students.

The responses of many different complex, adaptive ecosystems to disturbance have stimulated new integrative concepts that relate directly to managing the global ecosystem (Levin 1998, 1999). Today several national and international programs (Gosz et al. 2010; Waide and Thomas 2013) are building collaboration and helping to organize and provide online data bases. For example, the International Long-Term

Research Program (ILTER) that began in 1993 is documenting ecosystem changes over daily, annual, and decadal time scales, and from micro to macro spatial scales, across 37 national networks (Maass and Equihua 2015 this volume [Chap. 14]). Another example is the National Climate Assessment that periodically summarizes changes in the climate impacts on a wide range of ecosystems (Melillo et al. 2014).

Planning by the UN Environment Program's Intergovernmental Platform on Biodiversity and Ecosystem Services (IPBES) can help create some consensus on methods and uses of economic valuations, especially as related to climate change (Larigauderie and Mooney 2010; Granjou et al. 2013). Agreements among the 118 national representatives will be useful and access to the online reports in six languages can enhance communication. The role of the International Union for Conservation of Nature (IUCN) and other non-governmental agencies will be essential to provide reliable information on biodiversity. Other recognized needs include more transparency and inclusiveness in the IPBES discussions (Hotes and Opgenoorth 2014).

29.4 Environmental Ethics: Practice What You Teach

During most of the twentieth century, the widespread intellectual fragmentation in universities prevented most biologists from dealing comprehensively with environmental ethics. The definitions of intrinsic, instrumental, and systemic values used by philosophers such as Holmes Rolston (1988, 1994), Baird Callicott (1984), and Bryan Norton (1986) stimulated discussions of environmental values among social and natural scientists (De Laplante 2004; Taylor 2005; Reiners and Lockwood 2010).

Today, the integration of social and ecological disciplines has grown, but creative tension often centers on the appropriate use of economic valuation methods for comparing complex, adaptive systems and inter-generational responsibilities (Daily et al. 2000; Farley and Costanza 2010; Dendoncker et al. 2014). There remains an urgent need for innovative studies of environmental ethics and intrinsic values of species and ecosystems that relate to Earth stewardship (see Fu et al. 2011; Aguirre Sala 2015 this volume [Chap. 15]).

Early in his career, Golley developed his philosophy of inclusive values that had long-lasting impacts on many students as he engaged them in discussions of "nature-centered thinking." For example, he explored environmental ethics in his seminars at the Mediterranean Institute of Agronomy in Zaragoza, Spain, where he emphasized the importance of humans in the water cycle. In other seminars around the world, he connected ethical concerns with loss of biodiversity and cultural diversity in tropical forests. Golley also contributed much to teaching ecosystem concepts as chair of the Education Committee of the Ecological Society of America during 1962–1963. Golley, inspired by a suggestion from Peter Raven, was instrumental in fostering the University of Georgia's requirement that all freshmen enroll in an environmental literacy course. Golley concluded that the key to success in linking ecological ideas with a wider community depended on ecological education and called for a greater effort to enhance environmental ethics and environmental literacy:

> Thinking ecologically means synthesizing the many fields of human knowledge into a coherent world view. Ultimately, the scientific ecologist includes in his or her purview ethics, values, and politics. As a consequence, there will never be an overall consensus on the form and objectives of ecological science (Keller and Golley 2000).

Even while Golley served as the Institute of Ecology's second director, he taught several courses and built the program, even designing the program's building. Later in his career, while serving as undergraduate coordinator for the ecology major, Golley recevied the Institute's Outstanding Advisor Award in 1997. He was dedicated to developing environmental ethics in the curriculum throughout his 43 year career in the Institute of Ecology. Golley's twenty-question quiz began by asking "which way was north?" There is a compass on the floor of the outside atrium of the ecology building pointing north to remind them. Students enjoyed knowing that he grew much of his food on his farm and even made his own shoes to show the importance of "connectedness." Golley (1998) emphasized that "environmental values should derive from what we know and what we can do. Practically, they should not conflict with ecological and environmental knowledge. Our ultimate objective is to build a way of thinking and acting which is scientifically consistent and satisfies human needs without destroying the environment." He defined cultural values as "a collective sense of the social whole." His course notes were transformed into a widely read book on ecological concepts, with implications for environmentalism and ethics (Golley 1998).

Golley was one of the founding members of the University of Georgia's Environmental Ethics Certificate Program in 1983. He helped organize the Fourth Annual Conference in Philosophy on campus in Athens in 1971, and was among the first ecologists at the University to focus on environmental ethics. As William Blackstone (1974), Golley's colleague in the Philosophy Department, noted: "we must broaden our evaluational perspective to include the entire range of values which are essential not only to the welfare of man but also to the welfare of other living things and to the environment that sustains all life." Golley helped organize additional international conferences on ethics by bringing together many active leaders in this newly emerging field (Ferré and Hartel 1994; Dallmeyer and Ike 1998). These discussions were energized by collaboration with many colleagues, including Dorinda Dallmeyer, who currently directs of the Environmental Ethics Certificate Program. Golley's perspective on the emerging ideas of deep ecology and their relationships with ecology as a science suggested that "there is a close parallel between the two sets of concepts and one supports the other" (Golley 1987a). His view about "connectedness" developed in part from his interest in deep ecology. Golley recognized human-centeredness of intrinsic value was insufficient and that non-human centeredness of intrinsic value meant *everything* has value, *independent* of human valuing. Humans *are* part of the ecosystem, and have ethical responsibilities to consider all other values.

29.5 Organizing Professional Societies to Consider Environmental Ethics

Since the 1970s as awareness of environmental problems accelerated professional societies developed codes of ethics to deal with socio-environmental challenges. At his ESA Annual Meeting Presidential Address, Golley criticized the fragmentation of ecological science (Golley 1978). He recognized that ecologists "view the world as a collection of interlocking systems." His concern was how ecologists could better develop new ideas: "It is no surprise that United States ecologists have been preoccupied with competition theory and have tended to pay less attention to social interaction, mutualism and symbolism…ecologists have an opportunity to aid in the revitalization of our society in the largest sense." Differences in views among professional ecologists are still strongly debated and create an ongoing need for development of new ways for dealing with environmental ethics as well as professional ethics.

There is increased interest in looking back as well as forward to see how much ecologists have learned during the last century about environmental values as the Ecological Society of America (ESA) celebrates its centennial in 2015, 2 years after the British Ecological Society's centenary (Callicott 2015). The incubation of many ideas within professional organizations such as the ESA, has resulted in diverse groups forming new organizations such as The Nature Conservancy, the Society of Conservation Biology, and others (Burgess 1981; Tjossem 1994). These organizations have considerable overlap in memberships and are creating new ways of communicating with the public and a new generation of highly motivated students regarding environmental ethics.

In the 1970s Golley also served as the president of the International Society for Tropical Ecology (ISTE) and as the vice president on planning for the Organization of Tropical Studies. His experience in many tropical countries provided important opportunities to communicate the value of biodiversity and the importance of large-scale, long-term studies. This experience also likely contributed to his appreciation of cultural diversity and the different ways ecological values entered into decision making in the developing world where traditional ecological knowledge often remained important and reflected very different value systems from those of industrialized nations.

During his presidency of ISTE, Golley organized a major symposium in India on energy flow and primary productivity (Golley and Golley 1972). Later he helped organize additional international workshop on tropical ecology in Costa Rica (Farnsworth and Golley 1974) and Venezuela (Golley and Medina 1975), resulting in recommendations to the National Science Foundation for future research to evaluate the loss of primary rainforests and their biodiversity. All these activities and the growing network of tropical researchers contributed to the emerging perspective of a "fragile ecosystem" concept as part of UNESCO's Man and the Biosphere Program. This research was based on comparing regional differences in an ecosystem's geologic age, soils, and slopes as well as the scales of past and recent human-driven disturbances (Gómez-Pompa et al. 1974; Golley et al. 1975; Golley and Hadley 1981). Although controversial from its initial conception, the idea was widely accepted that some lowland tropical ecosystems were relatively "fragile" and had limited capacities to respond to large-scale disturbances. Although small-scale disturbances appeared to be important in maintaining species diversity by opening up patches for dispersal by regional species, large-scale disturbances often led to establishing pastures and plantations that increased the immediate economic value of the land. These persistent landscape-level conversions had long-term costs derived from the cumulative losses of extensive forest cover and watershed protection as well as declines in biodiversity.

Golley explained his views on valuation of ecosystems in his presidential address at the International Association of Ecology (INTECOL) Fifth Congress of Ecology in 1990 at Yokohama, Japan, where he emphasized that "to solve global issues we need new ways to value nature, environmental goods and services and social-environmental processes which do not rest entirely on monetary currency, and can handle non-market values, while not assuming an endless process of economic growth to solve environmental problems." He set out an optimistic perspective that emphasized the roles of various international associations that were emerging to link ecology with human activities such as landscape ecology, restoration ecology, agro-ecology, and ecological economics. His views reinforced a general perspective within the International Council for Science and other organizations that strived to integrate the natural and social sciences.

Although Golley worked to help unify ecology by integrating other natural and social sciences, he acknowledged the inevitable creation of sub-disciplines: "these boundaries will be fuzzy, and like those in nature, will shift with time" (Golley 1987b). Golley's interest in spatial dynamics resulted in his serving the new International Association of Landscape Ecology (IALE) as the first editor-in-chief of the journal *Landscape Ecology* from 1987 to 1997 (Barrett et al. 2014). Earlier, Golley and Monica Turner, a former doctoral student, organized the first annual US-IALE meeting in 1986 that led to forming the US chapter.

29.6 Disturbances, Ethics and Ecosystem Processes

The current focus on the creation of novel ecosystems (Hobbs et al. 2013; Perring et al. 2013) builds on Golley's concepts related to ethical considerations regarding how people change their environment, for example, by introducing non-native species for the sake of novelty and curiosity without considering all the environmental impacts (Simberloff 2014). The resulting patterns of species distributions are influenced by legacies of disturbances that combine to influence how different species disperse and recolonize following major disturbances. This resiliency is a major component of ecosystems (Pickett et al. 1994; Cuddington and Beisner 2005; Pickett 2013). The complexity of disturbances and species interactions over multiple spatio-temporal scales creates a mosaic of patchy habitats and alters ecosystem processes across the landscape. The fragmentation and loss of suitable habitats are increasing rates of species extinction in many areas (Lovejoy 2002; Raven et al. 2011; Pimm et al. 2014) and threatening the capacity of ecosystems to sustain their essential processes (Covich et al. 2004; Naeem et al. 2009; Loreau 2010). The Millennium Ecosystem Assessment (2005) provided a global view of ecosystem services that ecologists, economists, and many others can evaluate by considering alternative future in the context of environmental ethics that Golley championed (Carpenter et al. 2009).

29.7 Learning from Long-Term Research on Tropical Ecosystems

Frank Golley contributed to developing concepts linking ecosystems, landscapes, and historic legacies to how people manage whole ecosystems. Golley (1993) suggested that these "response systems" are dynamic ecosystems whose "state at any particular time is contingent upon its history and the environment…it has a reciprocal relation to its environment and is not merely responding to it." This view is distinct from the earlier static perspective that dominated previous uses of the concept. Golley emphasized the importance of long-term information about an ecosystem's history and specific antecedent effects for understanding the past as part of the basis for considering management options. Golley was effective in communicating this view within a wide international community as well as at the US National Science Foundation during his 2-year appointment as director of the growing ecosystem program.

Golley's contributions also reflect the importance of environmental ethics from an international, multi-cultural perspective and the need for long-term research leading to better understanding of ecosystems and the sustained production of goods and services based on biodiversity. As a result of his international research and teaching, Golley was involved in studies of different management approaches to tropical forest and agricultural ecosystems. His work with the World Bank, UNESCO, and the National Science Foundation provided many opportunities to make the point that healthy ecosystems are of great value to people, and that these values are based on sustaining the biosphere's diversity.

A series of studies in tropical locations yielded new ideas about how ecosystems function following different types of disturbance. These studies illustrate how Golley viewed the importance of different rates of ecological and cultural changes in Neotropical ecosystems in Puerto Rico, Mexico, and Venezuela (Boxes 29.1, 29.2, and 29.3). He recognized the need for a diverse set of site-based, long-term studies to generate comparable information on how the "response systems" varied following disturbances. Golley compared results from different sites and developed his ideas on the importance of connections and hierarchical ordering among many direct and indirect interrelationships (Wu 2014). Golley's focus on "connectedness" provides some lessons for resolving similar challenges today. These studies also illustrated how people became involved in one way or another as participants in the research teams or as residents in the study sites. Their learning from each other and their sharing information and ideas, especially with diverse students, were critical given that there was little previous information on which to build in these early studies. Many of these programs that began decades ago are still working on new

Box 29.1. Luquillo, Puerto Rico

One of the first large-scale tropical ecosystem research projects was located in the Luquillo Mountains of Puerto Rico, where Golley collaborated with a group of ecologists organized by H.T. Odum in the early 1960s. The project followed an earlier study funded by the Rockefeller Foundation with a focus on large-scale measures of metabolism of mangrove forest and the rain forest (Golley et al. 1962; Odum and Jordan 1970; Lugo 2004). With a goal of understanding the functioning of rainforest ecosystems, a large clear-plastic cylinder was built to enclose a stand of trees. Oxygen production and consumption were measured to track day and night metabolism and transpiration (Jordan 2001; Lugo 2004). Solar energy, rainfall, temperature, and soil nutrients were measured to determine variations over time in the total metabolism of the soil and the forest trees.

The project was part of a response to international concern about the safe development of nuclear power and its peaceful applications (Creager 2013). A main objective of the El Verde research was to measure the effects of a brief release of gamma ray radiation from a cesium-137 source on a small area of the Luquillo Experimental Forest (now also known as the El Yunque National Forest), part of the US National Forests. The US Atomic Energy Commission supported this study of an "acute exposure" (that might occur from an accidental release of radiation from a power plant) to compare effects with a slow, "chronic release" of cesium-137 in a temperate forest on Long Island, New York at the Brookhaven National Laboratory (Woodwell 1962), and related studies at Oak Ridge, Tennessee (Creager 2013). Beginning in 1960, research in Puerto Rico included collections of leaf litter and fauna before and after the radiation release, to quantify how the biota (encompassing a high diversity of trees) might differ in daily and monthly leaf-fall rates. These data on primary productivity allowed for a wide range of comparisons with other tropical forests (Jordan 1983; Lugo and Heartsill-Scalley 2014).

Additional collections of leaf litter, gas exchange measurements, and studies on the diversity the biota continue as part the US National Science Foundation's Long-Term Ecological Research Program. The early baseline data remain useful several decades later (Harris et al. 2012; Lugo and Heartsill-Scalley 2014). Current studies focus on the long-term effects of variable rainfall and the effects of changes in the frequency of hurricanes and droughts (Brokaw et al. 2012;

(continued)

Box 29.1. (continued)
Lugo et al. 2012; Willig et al. 2012). Shifts in the pattern and amounts of monthly and inter-annual leaf fall, as well as rainfall and temperature data document climatic variability (González et al. 2013). The lesson here is that long-term data provide important information that emerges from analyses over time when answering one set of questions that can be useful in generating results to address new questions. The value of the data increases over time as the environmental conditions change in ways that were not considered initially.

Box 29.2. Chamela, Mexico
Another example provides a comparison between tropical and temperate ecosystems with a link to Golley through his colleague, Carl Jordan, and one of Jordan's former graduate students, Manuel Maass from Mexico (Jordan 2001). When Maass entered the University of Georgia's doctoral program with an interest in tropical ecology, he was surprised that his graduate committee (Carl Jordan, Frank Golley and Wayne Swank) advised him to learn forest hydrology by working at the Coweeta Hydrologic Laboratory in North Carolina, a US Forest Service Experimental Forest. Later, Maass helped to develop the Chamela Long-Term Ecological Research Program, the Chemela-Cuixmala Biosphere Reserve. A series of long term studies on productivity, biogeochemistry and hydrologic variability resulted in decades of ecological research (Vose and Maass 1999; Maass et al. 2002) that included a comprehensive study of this tropical dry forest in Jalisco, Mexico (Maass et al. 2005).

A major focus is on inter-annual variations in rainfall that strongly influence the differences between the wet and dry seasons in the forest's primary productivity. The impacts of Pacific hurricanes and occurrence of El Niño and La Niña events have major impacts on droughts and floods that affect rain-fed agriculture and local water supplies. This long-term research includes collaboration with the populations around the watershed, and a strong focus on ecosystem services that are provided to the surrounding communities such as erosion and flood control, and pollination services for agricultural production. This site is part of the International Long-Term Research Program (ILTER) now chaired by Manuel Maass. The ILTER network includes 37 national networks and nearly 600 academic groups working at sites across five continents (Maass and Equihua 2015 this volume [Chap. 14]). A lesson from the development of this type of site-based research is that students and former students are essential to conduct multi-decadal, multi-generational research. A deep understanding regarding the need to document and share data, as well the reliability of the team's efforts, takes time and leadership to develop.

Box 29.3. San Carlos de Rio, Venezuela

The San Carlos Project started in 1974 as a 10-year collaboration among Frank Golley, Carl Jordan, Hans Klinge from Germany, and Ernesto Medina from Venezuela (Golley et al. 1971; Jordan 2001). They and their graduate students, such as Chris Uhl and Florencia Montagnini, were among some of the first to study nutrient cycling in the Amazon (Uhl and Jordan 1984). These former students are working today with their students on documenting ecosystem services in the tropics (Montagnini et al. 2013) to advance sustainability and an ethical and ecological consciousness (Uhl et al. 1990; Uhl 2003).

A major contribution of the project was documentation that rapid nutrient cycling occurred rapidly in this lowland rainforest. This discovery was a major explanation for the luxurious green forest that impressed early travelers to the Amazon. The use of isotopic tracers documented that nutrients were held in the living forest biomass, and then released to temporary storage in the soils and back to living roots once the dead organic matter was rapidly broken down by fungi and bacteria. The rapid cycling was due to the symbiosis of tree roots with mycorrhizal fungi that increased nutrient uptake by roots. The concentrations of nutrients were extremely limited, and different from the higher nutrient concentrations observed in some other tropical regions where soils were relatively deep and geologically young. The weathered soils of the lowland Amazon were only productive when those species of plants and their associated biota were present because they had evolved to use the nutrients efficiently (Jordan 1987). Large-scale clearing of the forest for agricultural crops could more often create pastures. Therefore, the rate of regrowth of the rain forest did follow the resilient pathways that characterized smaller patches of disturbances such as wind damage or localized fires.

Indigenous farmers recognized that the soil nutrients were depleted within a few years after patches of the rainforest were cut and burned for small agricultural plots. To continue to grow their crops, they moved to other areas of older forests that had accumulated nutrients over time, and rotated their farming. Until populations grew large, this rotation apparently allowed sufficient time for the forest to recover in many regions. In some areas of the tropics, the selective cutting of trees left certain species in place if they were to be used for house construction or food. This "filtering" or traditional "management" practices in selecting certain forest species left a legacy of useful plants in the modern ecosystem (Gómez-Pompa et al. 1974). This example of an international team effort provided insights about differences in nutrient cycling on ecosystem productivity. Current interest is focused on how more prolonged droughts may increase the extent and frequency of fires that will further alter primary production that affects the carbon balance of the rainforests and the atmosphere (Gatti et al. 2014).

challenges because many students were included in these early studies and some are now leading the research.

29.8 The Future Needs for Integrating Multiple Value Systems

Golley's research contributions helped to develop a wider appreciation of ecological and cultural values. Over several decades other professional ecologists, including many of his former students, changed their personal value systems and modified their self-defined academic, non-governmental organizational, or governmental agency roles. Like Golley, they also actively supported a need to test ecological ideas by getting involved in applied problems, and in generating long-term data to test emerging concepts. These ecologists continue to support the emerging recognition that long-term, site-based research in many different locations can help conservationists and environmentalists meet the growing challenges involved in Earth stewardship.

The Earth Stewardship initiative recently begun by the ESA (Power and Chapin 2009; Chapin et al. 2011, 2015 in this volume [Chap. 12]), has deep roots that grew over several decades (MacIntosh 1987; Callicott 2015). Greater integration of environmental education to consider diverse, culturally important, non-market values in decision making will need to expand learning of environmental ethics (Rozzi 1999, 2013; Palmer 2012). These efforts increasingly are effective at the international level with more online multi-lingual and open-access journals, although the distribution and access in the developing world is still not equitable. This imbalance affects the long-term capacity to sustain biodiversity that is often both disproportionately high and under increasing risks in many tropical countries and other developing regions. The need to help create new partnerships, is an ongoing challenge for professional societies, non-governmental organizations, and universities. The number of people interested in resolving current environmental issues has increased but this growth still requires collaboration regarding the pluralistic view of ecology in the decades ahead.

29.9 Conclusion

Frank Golley's many contributions linked ecosystem science, landscape ecology, and environmental ethics. His insights continue to provide valuable perspectives on the need for long-term, multi-cultural appreciation of the integrity of natural and human-dominated ecosystems. Major challenges remain. Creating wider understanding of the current knowledge derived from international long-term research is important. Increased environmental literacy and implementation of new insights continue to be essential needs that Golley and his colleagues worked to extend internationally. Many of the students who contributed to these early studies are mentoring their own students and the cycle of life-long learning continues. This learning cycle needs to expand if stewardship is to succeed.

References

Aguirre Sala J (2015) Hermeneutics and field environmental philosophy: integrating ecological sciences and ethics into earth stewardship. In: Rozzi R, Chapin FS III, Callicott JB et al (eds) Earth stewardship: linking ecology and ethics in theory and practice. Springer, Dordrecht, pp 235–247

Barrett GW, Barrett TL, Wu JG (eds) (2014) History of landscape ecology in the United States. Springer, New York

Blackstone WT (ed) (1974) Philosophy and environmental crisis. University of Georgia Press, Athens

Brokaw N, Crowl T, Lugo A et al (eds) (2012) A Caribbean forest tapestry - the multi-dimensional nature of disturbance and response. Oxford University Press, New York

Burgess RL (1981) United States. In: Kormody EJ, McCormick JF (eds) Handbook of contemporary developments in world ecology. Greenwood Press, Westport, pp 67–101

Bush MB, Miller MC, De Oliveira PE et al (2000) Two histories of environmental change and human disturbance in eastern lowland Amazonia. The Holocene 10:543–553

Callicott JB (1984) Non-anthropocentric value theory and environmental philosophy. Am Philos Q 21:299–309

Callicott JB (2015) The centennial return of stewardship to the Ecological Society of America. In: Rozzi R, Chapin FS III, Callicott JB et al (eds) Earth stewardship: linking ecology and ethics in theory and practice. Springer, Dordrecht, pp 161–171

Carpenter SR, Mooney HA, Agard J et al (2009) Science for managing ecosystem services: beyond the Millennium Ecosystem Assessment. Proc Natl Acad Sci 106:1305–1312

Chapin FS III, Pickett STA, Power ME et al (2011) Earth stewardship: a strategy for social-ecological transformation to reverse planetary degradation. J Environ Stud Sci 1:44–53

Chapin FS III, Pickett STA, Power ME et al (2015) Earth stewardship: an initiative by the Ecological Society of America to Foster Engagement to Sustain Planet Earth. In: Rozzi R, Chapin FS III, Callicott JB et al (eds) Earth stewardship: linking ecology and ethics in theory and practice. Springer, Dordrecht, pp 173–194

Coleman DC (2010) Big ecology: the emergence of ecosystem science. University of California Press, Berkeley

Covich AP (2015) Projects that never happened: ecological insights from Darien, Panama. Bull Ecol Soc Am (In press)

Covich AP, Austen MC, Barlocher F et al (2004) The role of biodiversity in the functioning of freshwater and marine benthic ecosystems. BioScience 54:767–775

Craige BJ (2002) Eugene Odum: ecosystem ecologist and environmentalist. University of Georgia Press, Athens

Creager ANH (2013) Life atomic- a history of radioisotopes in science and medicine. University of Chicago Press, Chicago

Cuddington K, Beisner BE (2005) Kuhnian paradigms lost: embracing the pluralism of ecological theory. In: Cuddington K, Beisner BE (eds) Ecological paradigms lost: routes of theory change, vol 2. Academic, New York, pp 419–428

Daily GC, Söderqvist T, Aniyar S et al (2000) The value of nature and the nature of value. Science 289:395–396

Dallmeyer DG, Ike AF (eds) (1998) Environmental ethics and the global marketplace. University of Georgia Press, Athens
De Laplante K (2004) Toward a more expansive conception of ecological science. Biology and Philosophy 19:263–281
Dendoncker N, Keune H, Jacobs S et al (2014) Inclusive ecosystem services valuation. In: Jacobs S, Dendoncker N, Keune H (eds) Ecosystem services: global issues, local practices. Elsevier, Amsterdam, pp 3–12
Di Pasquo F (2013) Una historia de la problemática ambiental y de sus efectos sobre la ecología disciplinar. Scientiae Studia 11:557–581
Egerton FN (1973) Changing concepts in the balance of nature. Q Rev Biol 48:322–350
Farley J, Costanza R (2010) Payments for ecosystem services: from local to global. Ecol Econ 69:2060–2068
Farnsworth EG, Golley FB (eds) (1974) Fragile ecosystems: evaluation of research and application in the Neotropics. Springer, New York
Ferré F, Hartel P (1994) Ethics and environmental policy: theory meets practice. University of Georgia Press, Athens
Fu BJ, Su C-H, Wei Y-P et al (2011) Double counting in ecosystem services valuation: causes and countermeasures. Ecol Res 26:1–14
Gatti LV, Gloor M, Miller JB, Doughty CE et al (2014) Drought sensitivity of Amazonian carbon balance revealed by atmospheric measurements. Nature 506:76–80
Golley FB (1978) The Humpty-Dumpty problem. Bull Ecol Soc Am 59(3):128–130
Golley FB (ed) (1983) Tropical rain forest ecosystems: structure and function, Ecosystems of the world 14A. Elsevier, Amsterdam
Golley FB (1984) Historical origins of the ecosystem concept in biology. In: Moran E (ed) The ecosystem concept in anthropology. Westview Press, Boulder, pp 33–49
Golley F (1987a) Deep ecology from the perspective of ecological science. Environ Ethics 9:45–55
Golley FB (1987b) Introducing landscape ecology. Landsc Ecol 1:1–3
Golley FB (1991) Reasoning form ecological knowledge to problem solving. In: Sontag MS, Wright SD, Young GL, Grace M (eds) Human ecology: strategies for the future. Society for Human Ecology, Fort Collins, pp 52–60
Golley FB (1993) A history of the ecosystem concept in ecology: more than the sum of the parts. Yale University Press, New Haven
Golley FB (1994) Rebuilding a humane and ethical decision system for investing in natural capital. In: Janson AM, Hammer M, Folke C, Constanza R (eds) Investing in natural capital: the ecological economics approach to sustainability. Island Press, Washington, DC, pp 169–178
Golley FB (1998) A primer for environmental literacy. Yale University Press, New Haven
Golley FB (2001) Establishing the network. In: Barrett GW, Barrett TL (eds) Holistic science: the evolution of the Georgia Institute of Ecology (1940–2000). Taylor and Francis, New York, pp 38–67
Golley PM, Golley FB (eds) (1972) Tropical ecology with an emphasis on organic production. Institute of Ecology, Athens
Golley F, Hadley M (1981) The tropical forest, a rich but fragile resources. UNESCO Courier April; 13–16
Golley FB, Medina E (eds) (1975) Tropical ecological systems: trends in terrestrial and aquatic research. Springer, New York
Golley FB, Odum HT, Wilson RF (1962) The structure and metabolism of a Puerto Rican red mangrove forest in May. Ecology 43:9–19
Golley FB, Olein MD, Hoy DR (1971) Cognized environments of San Carlos valley settles. Rev Geogr 74:33–50
Golley FB, McGinnis JT, Clements RG et al (1975) Mineral cycling in a tropical moist forest. University of Georgia Press, Athens
Gómez-Pompa A, Anaya AL, Golley F et al (1974) Recovery of tropical ecosystems. In: Fragile ecosystems. Springer, Berlin/Heidelberg, pp 113–138

González G, Willig MR, Waide RB (2013) Ecological gradient analyses in a tropical landscape: multiple perspective and emerging themes. Ecol Bull 54:13–20

Gosz RJ, Waide R, Magnuson J (2010) Twenty-eight years of the US-LTER program: experience, results, and research questions. In: Muller F, Schubert H, Klotz S (eds) Long term ecological research: between theory and application. Springer Science and Business Media, Dordrecht/Heidelberg/New York/London, pp 54–75

Granjou C, Mauz I, Louvel S et al (2013) Assessing nature? The genesis of the Intergovernmental Platform on Biodiversity and Ecosystem Services (IPBES). Sci Technol Soc 18:9

Hagen JB (1989) Research perspectives and the anomalous status of modern ecology. Biol Philos 4:433–455

Hagen JB (1992) An entangled bank: the origins of ecosystem ecology. Rutgers University Press, New Brunswick

Hagen JB (2008) Teaching ecology during the environmental age, 1965–1980. Environ Hist 13:704–723

Harris NL, Lugo AE, Brown S et al (2012) Luquillo experimental forest: research history and opportunities. USDA EFR-1, Washington, DC

Hobbs RJ, Higgs ES, Hall C (eds) (2013) Novel ecosystems: intervening in the new ecological world. Wiley, New York

Hotes S, Opgenoorth L (2014) Trust and control at the science-policy interface in IPBES. BioScience 64:277–278

Hutchinson GE (1940) Bio-ecology. Ecology 21:267–268

Hutchinson GE (1941) Limnological studies of Connecticut. IV. Mechanisms of intermediate metabolism in stratified lakes. Ecol Monogr 11:21–60

Hutchinson GE (1948) Circular causal systems in ecology. Ann N Y Acad Sci 50:221–246

Jordan CF (1983) Productivity of tropical rain forest ecosystems and the implications for their use as future wood and energy sources. In: Golley FB (ed) Tropical rain forest ecosystems: structure and function. Elsevier, Amsterdam, pp 117–136

Jordan CF (1987) Amazonian rain forests: ecosystem disturbance and recovery. Springer, New York

Jordan CF (2001) Tropical ecology with the institute of ecology. In: Barrett GW, Barrett TL (eds) Holistic science: the evolution of the Georgia institute of ecology (1940–2000). Taylor and Francis, New York, pp 261–276

Keller DR, Golley FB (2000) The philosophy of ecology: from science to synthesis. University of Georgia Press, Athens

Kingsland SE (1995) Modeling nature: episodes in the history of population ecology, 2nd edn. University of Chicago Press, Chicago

Kingsland SE (2015) Ecological science and practice: dialogues across cultures and disciplines. In: Rozzi R, Chapin FS III, Callicott JB et al (eds) Earth stewardship: linking ecology and ethics in theory and practice. Springer, Dordrecht, pp 17–26

Kwa C (1993) Radiation ecology, systems ecology, and management of the environment. In: Science and nature: essays in the history of environmental science. British Society for the History of Science, Oxford, pp 213–249

Kwa C (2005) Local ecologies and global science: discourses and strategies of the International Geosphere-Biosphere Programme. Soc Stud Sci 35:923–950

Larigauderie A, Mooney H (2010) The intergovernmental science-policy platform on biodiversity and ecosystem services: moving a step closer to an IPCC-like mechanism for biodiversity. Curr Opin Environ Sustain 1(2):9–14

Levin SA (1998) Ecosystems and the biosphere as complex adaptive systems. Ecosystems 1:431–436

Levin SA (1999) Fragile dominion. Perseus Publishing, Reading

Likens GE (1998) Limitations to intellectual progress in ecosystem science. In: 7th Cary conference: successes, limitations and frontiers in ecosystem science. Springer, New York, pp 247–271

Likens GE, Lindenmayer DB (2012) Integrating approaches leads to more effective conservation of biodiversity. Biodivers Conserv 21:3323–3341

Limburg KE (2004) The biogeochemistry of strontium: a review of H.T. Odum's contributions. Ecol Model 1778:31–33

Loreau M (2010) From populations to ecosystem. The theoretical foundations for a new ecological synthesis. Princeton University Press, Princeton
Lovejoy TE (2002) Biodiversity: threats and challenges. In: O'Riordan T, Stoll-Kleemann S (eds) Biodiversity, sustainability, and human communities: protecting beyond the protected. Cambridge University Press, Cambridge, pp 33–45
Lugo AE (2004) HT Odum and the Luquillo Experimental Forest. Ecol Model 178:65–74
Lugo AE, Heartsill-Scalley T (2014) Research in the Luquillo experimental forest has advanced understanding of tropical forests and resolved management issues. In: Hayes DC et al (eds) USDA Forest Service experimental forests and ranges. Springer, New York, pp 435–461
Lugo AE, Waide RB, Willig MR et al (2012) Ecological paradigms for the tropics: old questions and continuing challenges. In: Brokaw N, Crowl TA, Lugo AE et al (eds) A Caribbean forest tapestry- The multi-dimensional nature of disturbance and response. Oxford University Press, Oxford, pp 3–41
Maass M, Equihua M (2015) Earth stewardship, socioecosystems, the need for a transdisciplinary approach and the role of the International Long Term Ecological Research network (ILTER). In: Rozzi R, Chapin FS III, Callicott JB et al (eds) Earth stewardship: linking ecology and ethics in theory and practice. Springer, Dordrecht, pp 217–233
Maass JM, Jaramillo V, Martínez-Yrízar A, García-Oliva F, Pérez-Jiménez A, Sarukhán J (2002) Aspectos funcionales del ecosistema de selva baja caducifolia en Chamela, Jalisco. In: Noguera FA, Vega-Rivera JH, García-Aldrete AN et al (eds) Historia Natural de Chamela. UNAM, Mexico, pp 525–542
Maass JM, Balvanera P, Castillo A et al (2005) Ecosystem services of tropical dry forests: insights from longterm ecological and social research on the Pacific Coast of Mexico. Ecol Soc 10(1):1–23
MacIntosh RP (1987) Pluralism in ecology. Annu Rev Ecol Syst 18:321–341
McDonnell MJ, Pickett STA (1993) Humans as components of ecosystems: the ecology of subtle human effects and populated areas. Springer, New York
McLauchlan KK, Williams JJ, Engstrom DR (2013) Nutrient cycling in the palaeorecord: fluxes from terrestrial to aquatic ecosystems. The Holocene 23:1635–1643
Melillo JM, McGuire AD, Kicklighter DW, Moore B III, Vörösmarty CJ, Schloss AL (1993) Global change and terrestrial net primary productivity. Nature 363:234–240
Melillo JM, Richmond TC, Yohe GW (eds) (2014) Climate change impacts in the United States: the third national climate assessment. U.S. Global Change Research Program. Washington, DC. doi:10.7930/J0Z31WJ2
Millennium Ecosystem Assessment (2005) Ecosystems and human well-being: synthesis. Island Press, Washington, DC
Mitman G (1992) The state of nature: ecology, community, and American social thought, 1900–1950. University of Chicago Press, Chicago
Montagnini F, Ibrahim M, Restrepo EM (2013) Silvopastoral systems and climate change mitigation in Latin America. Bois et Forets des Tropiques 316:3–16
Mooney HA (1996) Ecosystem physiology: overview and synthesis. In: Walker BH, Steffen WL (eds) Global change and terrestrial ecosystems. Cambridge University Press, Cambridge
Mooney HA, Duraiappah A, Larigauderie A (2013) Evolution of natural and social science interactions in global change research programs. Proc Natl Acad Sci U S A 110:3665–3672
Naeem S, Bunker DE, Hector A, Loreau M, Perrings C (eds) (2009) Biodiversity, ecosystem functioning, and human wellbeing: an ecological and economic perspective. Oxford University Press, Oxford
Norton BG (ed) (1986) The preservation of species. Princeton University Press, Princeton
Odum HT (1951) The stability of the world strontium cycle. Science 114:407–411
Odum EP (1977) The emergence of ecology as a new integrative discipline. Science 195:1289–1293
Odum EP, Golley FB (1963) Radioactive tracers as an aid to the measurement of energy flow at the population level in nature. In: Schultz V, Klement AW Jr (eds) Radioecology. Chapman and Hall, London, pp 403–410

Odum HT, Jordan CF (1970) Metabolism and evaportranspiration of the lower forest in a giant cylinder. In: Odum HT, Pigeon RF (eds) A tropical rainforest: a study of irradiation and ecology at El Verde, Puerto Rico. US Atomic Energy Commission TID-24270, Springfield, pp I65–I189

Palmer MA (2012) Socio-environmental sustainability and actionable science. BioScience 62:5–6

Patten BC (2014) Systems ecology and environmentalism: getting the science right. Part I: Facets for a more holistic nature book of ecology. Ecol Model 293:4–21

Perring MP, Standish RJ, Hobbs RJ (2013) Incorporating novelty and novel ecosystems into restoration planning and practice in the 21st century. Ecol Process 2:18

Pickett STA (2013) The flux of nature: changing worldviews and inclusive concepts. In: Rozzi R, Pickett STA, Palmer C et al (eds) Linking ecology and ethics for a changing world: values, philosophy, and action. Springer, Dordrecht/Heidelberg/New York/London, pp 265–280

Pickett STA, Kolasa J, Jones CG (1994) Ecological understanding. Academic, San Diego

Pimm SL, Jenkins CN, Abell R et al (2014) The biodiversity of species and their rates of extinction, distribution, and protection. Science 344:1246752

Power ME, Chapin FS III (2009) Planetary stewardship. Front Ecol Environ 7:399

Raven PH, Chase JM, Pires JC (2011) Introduction to special issue on biodiversity. Am J Bot 98:333–335

Reiners WA, Lockwood JA (2010) Philosophical foundations for the practice of ecology. Cambridge University Press, Cambridge

Rolston H III (1988) Environmental ethics: duties and values in the natural world. Temple University Press, Philadelphia

Rolston H III (1994) Conserving natural value. Columbia University Press, New York

Rozzi R (1999) The reciprocal links between evolutionary – ecological sciences and environmental ethics. BioScience 49:911–921

Rozzi R (2013) Biocultural ethics: from biocultural homogenization toward biocultural conservation. In: Rozzi R, Pickett STA, Palmer C et al (eds) Linking ecology and ethics for a changing world: values, philosophy, and action. Springer, Dordrecht/Heidelberg/New York/London, pp 9–32

Rozzi R, Armesto JJ, Gutiérrez J et al (2012) Integrating ecology and environmental ethics: earth stewardship in the southern end of the Americas. BioScience 62:226–236

Shugart HH (1998) Terrestrial ecosystems in changing environments. Cambridge University Press, Cambridge

Simberloff D (2014) Introduced species, homogenizing biotas and cultures. In: Rozzi R, Pickett STA, Palmer C et al (eds) Linking ecology and ethics for a changing world: values, philosophy, and action. Springer, Dordrecht/Heidelberg/New York/London, pp 33–48

Steffen W, Sanderson RA, Tyson PD et al (2004) Global change and the earth system: a planet under pressure. Springer, New York

Taylor PJ (2005) Unruly complexity: ecology, interpretation, engagement. University of Chicago Press, Chicago

Tjossem SF (1994) Preservation of nature and academic respectability: tension in the Ecological Society of America, 1915–1979. Ph.D. Dissertation, Cornell University, Ithaca

Turner MG (2010) Disturbance and landscape dynamics in a changing world. Ecology 91:2833–2849

Turner MG, Chapin FS III (2005) Causes and consequences of spatial heterogeneity in ecosystem function. In: Lovett GM, Jones CG, Turner MG et al (eds) Ecosystem function in heterogeneous landscapes. Springer, New York, pp 9–30

Uhl C (2003) Developing ecological consciousness: paths to a sustainable future. Rowman and Littlefield, Oxford

Uhl C, Jordan CF (1984) Succession and nutrient dynamics following forest cutting and burning in Amazonia. Ecology 65:1476–1490

Uhl C, Buschbacher ND, Clark R et al (1990) Studies of ecosystem response to natural and anthropogenic disturbances provide guidelines for designing sustainable land-use systems in

Amazonia. In: Anderson AB (ed) Alternatives to deforestation: steps towards sustainable use of the Amazon rain forest. Columbia University Press, New York, pp 24–42

Vose JM, Maass JM (1999) A comparative analysis of hydrologic responses of tropical deciduous and temperate deciduous watershed ecosystems to climatic change. In: Aguirre-Bravo CF, Rodriguez C (eds) North American science symposium: toward a unified framework for inventorying and monitoring forest ecosystem resources, Proceedings RMRS-P-12. U.S. Department of Agriculture, Forest Service, Rocky Mountain Research Station, Fort Collins, pp 292–298

Waide RB, Thomas MO (2013) Long-term ecological research network. In: Earth system monitoring. Springer, New York, pp 233–268

Willig MR, Bloch CP, Covich AP et al (2012) Long-term research in the Luquillo mountains: synthesis and foundations for the future. In: Brokaw N, Crowl TA, Lugo AE et al (eds) A Caribbean forest tapestry – the multi-dimensional nature of disturbance and response. Oxford University Press, Oxford, pp 361–441

Woodwell GM (1962) Effects of ionizing radiation on terrestrial ecosystems. Science 138:572–577

Wright SD, Borden R, Bubolz M et al (1995) Human ecology: progress through integrative perspectives. Society for Human Ecology, Fort Collins

Wu J (2014) Hierarchy theory: an overview. In: Rozzi R, Pickett STA, Palmer C et al (eds) Linking ecology and ethics for a changing world: values, philosophy, and action. Springer, Dordrecht/Heidelberg/New York/London, pp 281–302

Index

R. Rozzi et al. (eds.), *Earth Stewardship*, Ecology and Ethics 2,
DOI 10.1007/978-3-319-12133-8